Nonlinear Networks: Theory and Analysis

OTHER IEEE PRESS BOOKS

Clearing the Air: The Impact of the Clean Air Act on Technology
Edited by John C. Redmond, John C. Cook, and A. A. J. Hoffman
Active Inductorless Filters
Edited by Sanjit K. Mitra
A Practical Guide to Minicomputer Applications
Edited by Fred F. Coury
Power Semiconductor Applications, Volume I: General Considerations
Edited by John D. Harnden, Jr. and Forest B. Golden
Power Semiconductor Applications, Volume II: Equipment and Systems
Edited by John D. Harnden, Jr. and Forest B. Golden
Semiconductor Memories
Edited by David A. Hodges
Minicomputers: Hardware, Software, and Applications
Edited by James D. Schoeffler and Ronald H. Temple
Digital Signal Processing
Edited by Lawrence R. Rabiner and Charles M. Rader
Laser Theory
Edited by Frank S. Barnes
Integrated Optics
Edited by Dietrich Marcuse
Literature in Digital Signal Processing: Terminology and Permuted Title Index
Edited by Howard D. Helms and Lawrence R. Rabiner
Laser Devices and Applications
Edited by Ivan P. Kaminow and A. E. Siegman
Computer-Aided Filter Design
Edited by George Szentirmai
Key Papers in the Development of Information Theory
Edited by David Slepian
Technology and Social Institutions
Edited by Kan Chen
Key Papers in the Development of Coding Theory
Edited by E. R. Berlekamp
Automatic Test Equipment: Hardware, Software, and Management
Edited by F. Liguori
Stability of Large Electric Power Systems
Edited by R. T. Byerly and E. W. Kimbark
Computer Communications
Edited by P. E. Green, Jr. and R. W. Lucky

Nonlinear Networks: Theory and Analysis

Edited by
Alan N. Willson, Jr.

Associate Professor of Engineering and Applied Science
University of California, Los Angeles

A volume in the IEEE PRESS Selected Reprint Series,
prepared under the sponsorship of the
IEEE Circuits and Systems Society.

IEEE PRESS

The Institute of Electrical and Electronics Engineers, Inc. New York

International Standard Book Numbers: Clothbound: 0-87942-045-6
Paperbound: 0-87942-046-4

Library of Congress Catalog Card Number 74-19558

PRINTED IN THE UNITED STATES OF AMERICA

Contents

Contents

Introduction

This book provides under one cover a collection of papers in which a theory of the equations of nonlinear networks is developed. The theory is mainly concerned with the relationship between 1) the properties of the nonlinear circuit equations and 2) the properties of the device models that characterize the elements ot the network, and the nature of their interconnection. By the phrase *properties of the equations* we include, for example, the structure of the equations, the issue of whether or not the equations have a solution, the issue of whether or not the solution is unique, the continuity and boundedness of the solution, and, in the case of differential equations, certain aspects of the solution's behavior that are related to the network's stability. These properties of a network's equations are intimately related to the behavior of the network as a whole. If, for example, a network is to behave as a bistable memory element, the equations that describe the circuit's dc behavior certainly must possess more than one solution.

By the phrase *properties of the device models* we mean properties of the mathematical relations that characterize the elements of the circuit. In the case of a nonlinear resistor, for example, such properties include the continuity of the resistor's v versus i curve, the monotonicity or nonmonotonicity of the curve, the fact that the curve saturates (or does not saturate) for large values of voltage or current, etc. Also considered are such physical properties as the device's *passivity.*

The theory presented here is thus concerned with the development of a means for achieving a sound understanding of certain facets of a circuit's behavior, as they result from the nature of the circuit's elements and its topological structure. The method of accomplishing this is through an analytical study of the circuit's equations. The theory is also concerned with a study of some of the attributes of the network equations that are important when it is desired to solve the equations with the aid of a digital computer.

The book is organized, roughly, in the same manner as the organization of the first paper, "Some Aspects of the Theory of Nonlinear Networks," which surveys the material in the remaining papers. Thus, although there is a degree of overlap, the papers basically occur in three categories. The first eight papers that follow the survey paper are concerned, primarily, with networks of *two-terminal elements*, that is, networks of nonlinear resistors, inductors, capacitors, and independent sources. The next nine papers are concerned with networks which contain *transistors* as well as the previously mentioned two-terminal elements. Finally, the remaining five papers contain results that are pertinent to the study of nonlinear networks containing *multiterminal elements* of a rather general type.

To help illustrate the nature of some of the material contained herein we shall now briefly discuss a few examples of nonlinear circuits with reference to some of the theorems of the survey paper.

We begin our discussion with the network of Fig. 1. It represents a simple example of the kind of circuit that is accommodated by Theorems 1, 2, and 3. The circuit is constructed from nonlinear resistors and independent sources. Each resistor is characterized by a strictly monotone increasing function which maps the real line onto itself (that is, which maps one of the axes in the v–i plane onto the other axis).

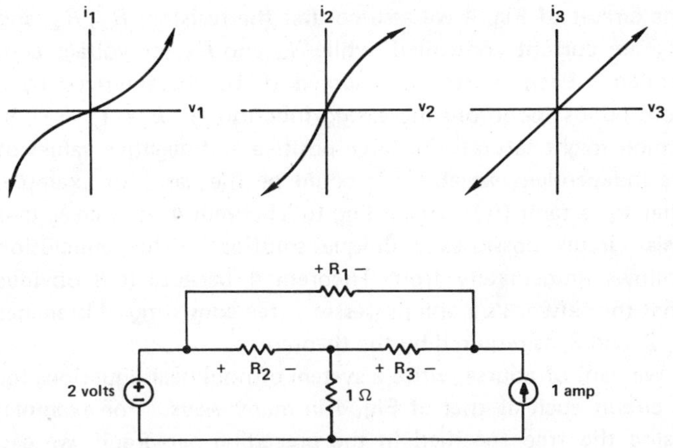

Fig. 1. Example of a nonlinear network which possesses a unique solution.

Any such circuit is known to possess a *unique solution*, by which we mean a unique set of branch voltages and branch currents that simultaneously satisfy the topological constraints imposed by Kirchhoff's voltage and current laws, and the "Ohm's-law" constraints that are imposed by the nature of the elements that comprise each of the branches. As a very special case of this fact, there follows the well-accepted result that any circuit constructed from positive-valued linear resistors and independent sources possesses a unique solution.

If, in a resistive two-terminal-element circuit, one of the resistors has a v versus i characteristic that is not strictly monotone increasing, then it is quite possible that the circuit could have several solutions. For any given value of $R > 0$, for example, the circuit of Fig. 2 will have either one, two, or three solutions, depending upon the value E of the voltage source.

If the nonlinear v versus i characteristics of the resistors saturate, then it is possible that a circuit's equations could have *no* solution. A trivial example of this situation is provided by the circuit of Fig. 3, which has no solution when the value I of the independent current source is more negative than I_0.

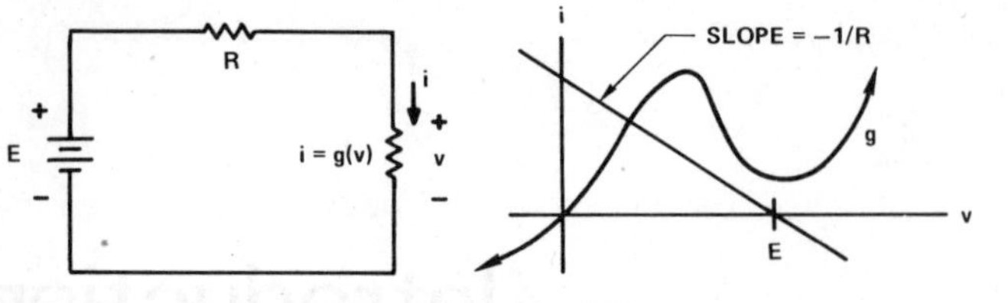

Fig. 2. Example of a nonlinear network which might possess more than one solution.

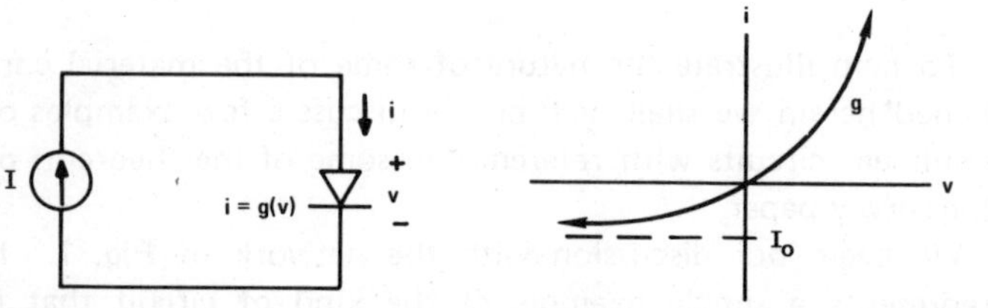

Fig. 3. Example of a nonlinear network which might possess no solution.

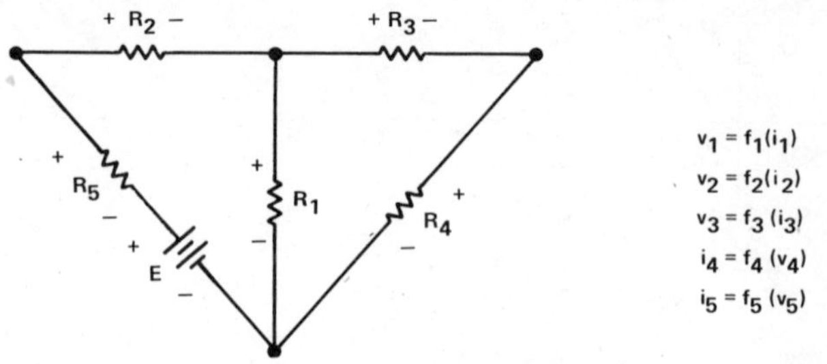

Fig. 4. Example of a nonlinear network which possesses a unique solution.

In contrast to the behavior illustrated by the circuit of Fig. 3, it happens that many circuits do possess a unique solution for all values of their independent sources even though their resistors' v versus i characteristics saturate. For example, in the circuit of Fig. 4 we assume that the resistors R_1, R_2, and R_3 are current controlled, while R_4 and R_5 are voltage controlled. Each resistor is assumed to be characterized by a continuous monotone-increasing function f_k, $k = 1, \cdots, 5$, which might saturate for large positive and negative values of its independent variable. It could be the case, for example, that $v_1 = \tanh(i_1)$. According to Theorem 4, it is clear that this circuit possesses a unique solution. This conclusion follows immediately from Theorem 4 because it is obvious that the network's graph possesses a tree consisting of branches 1, 2, and 3, as required by the theorem.

We can, of course, write a system of nonlinear equations for a circuit such as that of Fig. 4 in many ways. For example, using the tree specified in the preceding paragraph, we can write the fundamental cut-set equations

$$\begin{aligned} i_1 + f_4(v_4) + f_5(v_5) &= 0 \\ i_2 + f_5(v_5) &= 0 \\ i_3 - f_4(v_4) &= 0. \end{aligned}$$

Then, using the Ohm's-law relations specified in Fig. 4, we obtain

$$\begin{aligned} v_1 &= f_1[-f_4(v_4) - f_5(v_5)] \\ v_2 &= f_2[-f_5(v_5)] \\ v_3 &= f_3[f_4(v_4)]. \end{aligned}$$

Finally, we can express the link voltages v_4, v_5 in terms of the tree-branch voltages

$$\begin{aligned} v_4 &= v_1 - v_3 \\ v_5 &= v_1 + v_2 - E \end{aligned}$$

which yields the following set of three nonlinear equations in the three variables v_1, v_2, v_3:

$$\begin{aligned} v_1 &= f_1[-f_4(v_1 - v_3) - f_5(v_1 + v_2 - E)] \\ v_2 &= f_2[-f_5(v_1 + v_2 - E)] \\ v_3 &= f_3[f_4(v_1 - v_3)]. \end{aligned} \quad (1)$$

These equations can be solved for v_1, v_2, v_3 by using, for example, an iterative technique (e.g., Newton's method). Finally, the values of all of the remaining branch variables can then be found by simply substituting the values of v_1, v_2, and v_3 into the preceding relations. It is interesting to note that as a consequence of Theorem 4 we are assured that the equations (1) possess a unique solution no matter which continuous monotone-increasing functions f_k appear in these equations, and no matter what value is assumed by the independent voltage source E. If one were simply given a system of nonlinear equations in the variables v_1, v_2, v_3, it would in general be a nontrivial task to establish the existence of a unique solution to the equations.

It is easy to give examples of circuits for which the (sufficient) topological criterion of Theorem 4 is inadequate for determining the existence of a unique solution. Consider, for example, the circuit of Fig. 5. Since this circuit has only *one* current-controlled resistor, and since any tree of its graph possesses *two* branches, the existence of a unique solution for this circuit cannot be inferred by a direct appeal to Theorem 4. Nevertheless, the circuit *does* possess a unique solution for any given value I of the independent current source, as can be determined, for example, from an analysis of the circuit equations

$$\begin{pmatrix} f_1(v_1) \\ f_2(v_2) \\ f_3(i_3) \end{pmatrix} + \begin{bmatrix} 0 & 0 & 1 \\ 0 & 0 & 1 \\ -1 & -1 & 0 \end{bmatrix} \begin{pmatrix} v_1 \\ v_2 \\ i_3 \end{pmatrix} = \begin{pmatrix} I \\ I \\ 0 \end{pmatrix}$$

according to the results of Theorem 5. Alternatively, one can deduce the existence of a unique solution directly from the topological structure of this simple circuit by applying Theorem 6.

For certain networks of nonlinear resistors in which some of the resistors' characteristics are nonmonotonic (the characteristic of a tunnel diode, for example), the issue of determining whether or not *at least one* solution exists can be resolved by techniques that are similar to those previously mentioned. Theorems 7 and 8 are addressed to such issues.

In the circuit shown in Fig. 6, L_2, C_3, and R_4 are positive-valued linear elements. L_1 and R_5 are nonlinear elements characterized by the relations

$$\begin{aligned} i_1 &= \tanh(\varphi_1) \\ v_5 &= \exp(3i_5). \end{aligned}$$

This circuit is an example of a nonlinear *RLC* network. For such a circuit it is often desired to write so-called normal-form

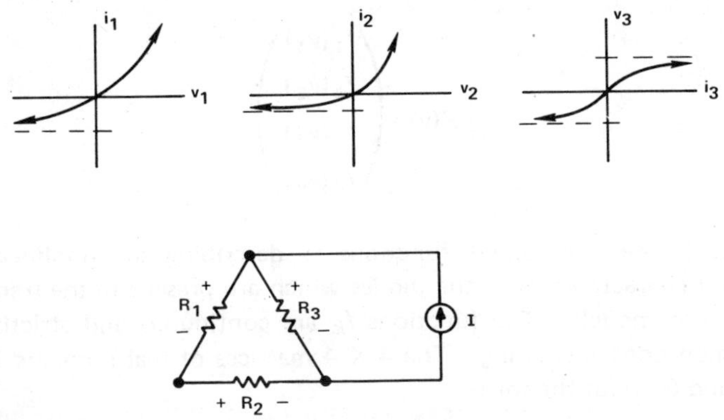

Fig. 5. Example of a nonlinear network which possesses a unique solution.

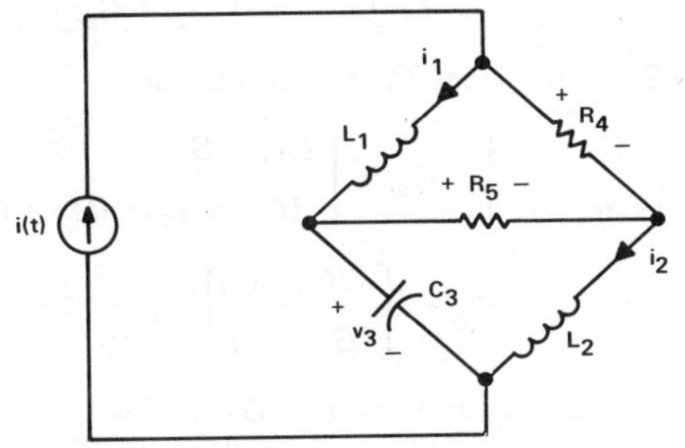

Fig. 6. Example of a nonlinear RLC network.

differential equations (state equations) to describe the circuit's behavior. Theorems 9, 10, and 11 are concerned with determining when and how this can be done, and related issues. It happens that normal-form equations can easily be written for this circuit—we have

$$\frac{d\varphi_1}{dt} = [i(t) - \tanh(\varphi_1)]\ R_4 - \exp\{3[\tanh(\varphi_1) + i_2 - i(t)]\}$$

$$\frac{di_2}{dt} = \frac{1}{L_2} v_3 - \frac{1}{L_2} \exp\{3[\tanh(\varphi_1) + i_2 - i(t)]\}$$

$$\frac{dv_3}{dt} = \frac{1}{C_3} i(t) - \frac{1}{C_3} i_2 .$$

The structure of this circuit is such that these normal-form differential equations can be written down by inspection. For other circuits whose structure is not quite so simple, it is useful to have a systematic procedure to rely on. We find, in fact, that the matter of central importance in such a procedure is again the issue of solving nonlinear algebraic equations. For example, in the case of the circuit shown in Fig. 7(a), normal-form differential equations can be written whenever it is possible to solve for the variables v_L and i_C in the dc circuit shown in Fig. 7(b). This, in turn, is accomplished by solving the nonlinear equation

$$v_s(t) = [f(v) + i_L]\ R + v + v_C \tag{2}$$

for the variable v, as a function of v_s, i_L and v_C. Once v is found then $i_C = f(v)$ and $v_L = v + v_C$ are obtained immediately. Equation (2) may be rewritten as

$$Rf(v) + v = v_s - Ri_L - v_C$$

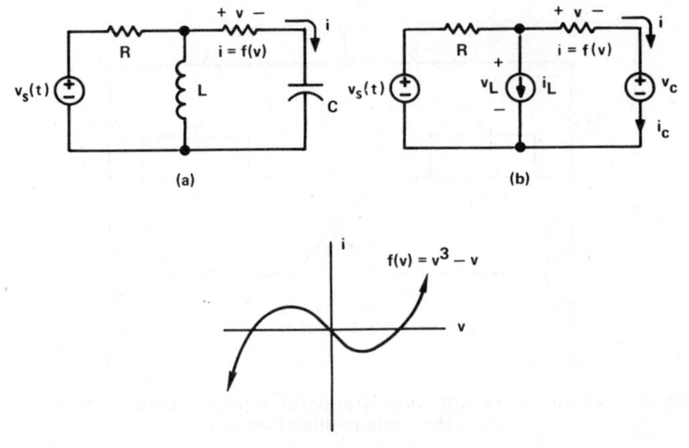

Fig. 7. Example of a nonlinear RLC network and the associated resistive network whose solution is required in the analysis.

which, assuming $f(v)$ is described by the curve shown in Fig. 7(c), has a unique solution v for any given values of v_s, i_L, and v_C if and only if $0 \leqq R \leqq 1$. To view this in a slightly different manner, the mapping $Rf(v) + v$ then, and only then, possesses an inverse (mapping the whole real line onto itself). We denote this inverse by g; hence $v = g(v_s - Ri_L - v_C)$. The resulting normal-form differential equations are

$$\frac{di_L}{dt} = \frac{1}{L} [v_C + g(v_s - Ri_L - v_C)]$$

$$\frac{dv_C}{dt} = \frac{1}{C} f[g(v_s - Ri_L - v_C)] .$$

No matter how one obtains normal-form differential equations for this circuit, the aforementioned restriction on the range of permissible values for the resistor R will be encountered and must be dealt with.

In addition to it being a crucial factor in the formulation of the normal-form differential equations for nonlinear networks, one is usually required to deal with the theory of the nonlinear algebraic equations of *resistive* networks in at least two other respects in order to solve these equations. First, in order to obtain the initial conditions for the differential equations, it is necessary, in general, to solve such nonlinear algebraic equations. Second, nonlinear algebraic equations of this type must be solved at each time step when using the popular *implicit* numerical integration algorithms.

When circuits containing transistors are considered, even when all of the circuit's nonlinearities are characterized by strictly monotone increasing functions, the issue of determining whether or not a unique solution exists cannot, in general, be resolved by a topological analysis of the circuit, as can be done in the case of, say, networks of nonlinear resistors. The actual *values* of the parameters involved in the element characterizations must also be taken into consideration. A rather extensive theory concerning the properties of the equations of transistor networks has been formulated recently, and the major results are contained in Theorem 12 through Theorem 26. An example of the application of one such result is the following.

The circuit of Fig. 8 is well known to be capable of performing as a bistable multivibrator when appropriate values

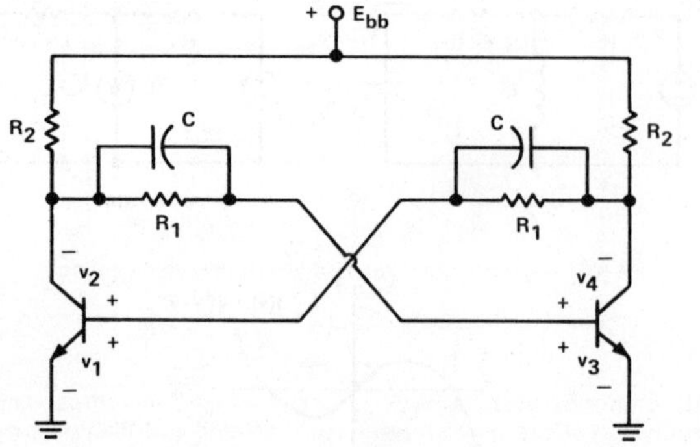

Fig. 8. Example of a nonlinear transistor network which can possess more than one equilibrium point.

are chosen for the components. Although the occurrence of the phenomenon of bistability is of course critically dependent upon the fact that the transistors are *nonlinear* devices, the usual analyses of this circuit, which can be found in the literature in many places, have a tendency to obscure this fact. These analyses involve the characterization of each transistor by one of several rather crude linear models. They then proceed to switch between various pairs of these models to suit the convenience of the analysis. (The nonlinear character of the circuit is of course accounted for by the switching process.) The lack of rigor in such analyses is justified by the fact that useful results are quite easily obtained. The serious circuit analyst, however, is usually left with a somewhat queasy feeling concerning the validity of the whole procedure. A more serious criticism, however, is that it is not clear from this type of analysis whether or not the results would be altered by the use of a more precise transistor model (for example, the transistors' reverse current gain α_r is usually ignored completely). Furthermore, the rather ad hoc nature of the analysis does not provide one with a clearly defined procedure to use on a large variety of circuits.

The circuit of Fig. 8, with the transistors being modeled by the simple large-signal dc transistor model of Fig. 9, is an example of a circuit that can easily be analyzed by the use of the results contained in this book. The analysis is quite simple and quite general. It could easily be applied to any transistor circuit. We begin with the observation, mentioned earlier, that the dc circuit that is obtained by omitting the capacitors from the circuit of Fig. 8 is required to have more than one solution if there is to be any *possibility* for the given circuit to have two *stable* equilibrium points (that is, to be a *bistable* circuit).[1] The circuit's dc equations can easily be written in the form

$$TF(v) + Gv = c \tag{3}$$

where v is a vector whose components are the four base-emitter and base-collector voltages shown in Fig. 8, and $F(v)$ is the nonlinear mapping defined by

[1]When the parameter values of the circuit elements are chosen such that this circuit is bistable, it then happens that the dc equations have *three* solutions, only two of which correspond to *stable* equilibrium points.

$$F(v) = \begin{pmatrix} f_1(v_1) \\ f_2(v_2) \\ f_3(v_3) \\ f_4(v_4) \end{pmatrix}$$

with the component functions f_k describing the nonlinear v–i characteristics of the diodes which are present in the transistor models. The functions f_k are continuous and strictly monotone increasing. The 4×4 matrices of real numbers T and G are of the form

$$T = \left[\begin{array}{c|c} T_1 & 0 \\ \hline 0 & T_1 \end{array}\right], \quad G = \left[\begin{array}{c|c} G_a & G_b \\ \hline G_b & G_a \end{array}\right]$$

where the 2×2 matrices T_1, G_a, and G_b are

$$T_1 = \begin{bmatrix} 1 & -\alpha_r \\ -\alpha_f & 1 \end{bmatrix}, \quad G_a = \begin{bmatrix} 2G_1 + G_2 & -(G_1 + G_2) \\ -(G_1 + G_2) & (G_1 + G_2) \end{bmatrix},$$

$$G_b = \begin{bmatrix} -2G_1 & G_1 \\ G_1 & 0 \end{bmatrix}$$

(where, of course, $G_1 = 1/R_1$ and $G_2 = 1/R_2$), and the vector c is

$$c = \begin{pmatrix} 1 \\ -1 \\ 1 \\ -1 \end{pmatrix} G_2 E_{bb}.$$

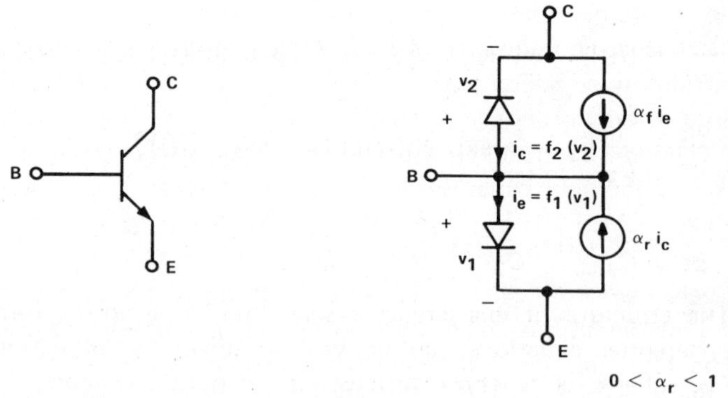

Fig. 9. Simple large-signal dc transistor model.

In order for (3) to have more than one solution it is necessary, according to the result of Theorem 14, for the matrix $T^{-1}G$ to have at least one *negative* principal minor. We can write

$$T^{-1}G = \left[\begin{array}{c|c} T_1^{-1}G_a & T_1^{-1}G_b \\ \hline T_1^{-1}G_b & T_1^{-1}G_a \end{array}\right]$$

and, by exploiting the special structure of this matrix, we can determine with a relatively small amount of algebra when a principal minor can be negative. (The special structure of the matrix

$T^{-1}G$, which results from the circuit's symmetrical topology and from the assumption that a pair of identical transistors are used, is in no way a crucial requirement of the analysis. It simply allows us to achieve a substantial savings in the amount of algebra that is necessary.) It is, in fact, rather easy to show that each principal minor will be nonnegative for all values of α_r, α_f in the interval (0, 1) and all $G_1 \geqslant 0, G_2 \geqslant 0$, except, possibly, for the one 2×2 principal minor obtained from $T^{-1}G$ by deleting the first and third rows and columns.[2] For this principal minor to be negative, it is necessary and sufficient that

$$\det \begin{bmatrix} (1-\alpha_f)(G_1+G_2) & \alpha_f G_1 \\ \alpha_f G_1 & (1-\alpha_f)(G_1+G_2) \end{bmatrix} = (1-\alpha_f)^2\,(G_1+G_2)^2 - \alpha_f^2 G_1^2 < 0$$

which is, of course, equivalent to the inequality

$$\frac{\alpha_f}{1-\alpha_f} > 1 + \frac{R_1}{R_2}. \tag{4}$$

Thus it is necessary that the parameter values be chosen such that the inequality (4) holds, in order for the circuit of Fig. 8 to have more than one equilibrium point. This requirement is, in fact, identical to the criterion for bistability, $R_1 < (\beta - 1)R_2$, which is usually obtained for this circuit by the type of analysis described earlier.

For certain special classes of circuits it is possible to determine from our theory that bistability is impossible, based only on a consideration of a circuit's topological structure. For example, according to Theorem 20, bistability is impossible in *any* one-transistor circuit. Another example is provided by the circuit of Fig. 10, which is discussed in more detail in the survey paper. Without writing any equations, without even specifying the numerical values for any of the parameters, it is possible to conclude that the dc equations of *this* circuit cannot possess more than one solution.

Theorem 27 through Theorem 30 of the survey paper pertain to nonlinear equations of a very general sort:

$$f(x) = y \tag{5}$$

where here f is taken to be simply a mapping of points x in n-dimensional Euclidean space into points y in the same space. The fact that these theorems presume no strong *structural* hypotheses concerning the nonlinear mapping f allows them to be of quite general applicability. As an example of the use of one of these theorems we shall now show how we could conclude that the equations (1) possess a unique solution without

[2]To show that the 4×4 principal minor is always nonnegative, it is helpful to use the identity

$$\det \begin{bmatrix} A & B \\ B & A \end{bmatrix} = \det (A+B) \cdot \det (A-B)$$

where A and B are square matrices. Because of the special structure of $T^{-1}G$, it is only necessary to check two of the four 3×3 principal minors. These are quite easily evaluated and shown to be nonnegative. It is clear that all of the 1×1 principal minors are nonnegative; and, once again due to the special structure of $T^{-1}G$, it is only necessary to check four of the 2×2 principal minors. Three of these are easily seen to be nonnegative.

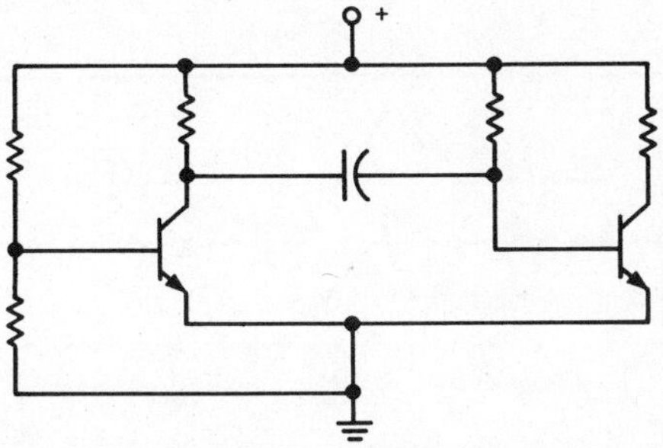

Fig. 10. Example of a nonlinear transistor network whose topology is such that it can possess no more than one equilibrium point.

relying upon our knowledge of the topological structure of the circuit of Fig. 4, from which these equations were derived. We shall use Theorem 30. The equations are first rewritten in the form

$$\begin{aligned} v_1 - f_1[-f_4(v_1 - v_3) - f_5(v_1 + v_2 - E)] &= 0 \\ v_2 - f_2[-f_5(v_1 + v_2 - E)] &= 0 \\ v_3 - f_3[f_4(v_1 - v_3)] &= 0. \end{aligned}$$

Next, we view the left-hand side of these equations as comprising the mapping $f\colon E^3 \to E^3$ of (5). The Jacobian matrix of this mapping is easily found to be

$$\left[\frac{\partial f}{\partial x}\right] = \begin{bmatrix} 1 + d_1 d_4 + d_1 d_5 & d_1 d_5 & -d_1 d_4 \\ d_2 d_5 & 1 + d_2 d_5 & 0 \\ -d_3 d_4 & 0 & 1 + d_3 d_4 \end{bmatrix}$$

where

$$d_1 = \left.\frac{df_1(\xi)}{d\xi}\right|_{\xi = -f_4(v_1 - v_3) - f_5(v_1 + v_2 - E)}, \cdots,$$

$$d_5 = \left.\frac{df_5(\xi)}{d\xi}\right|_{\xi = v_1 + v_2 - E.}$$

Since each of the nonlinear functions $f_1, \cdots, f_5$ is assumed to be monotone increasing, it is clear that each of the d_k, $k = 1, \cdots, 5$, is nonnegative. The required sequence of principal minors of $[\partial f/\partial x]$ is easily found to be

$$\begin{aligned} \det J_1 &= 1 + d_1 d_4 + d_1 d_5 \\ \det J_2 &= 1 + d_1 d_4 + d_1 d_5 + d_2 d_5 + d_1 d_2 d_4 d_5 \\ \det J_3 &= 1 + d_1 d_4 + d_1 d_5 + d_2 d_5 + d_1 d_2 d_4 d_5 + d_3 d_4 \\ &\quad + d_1 d_3 d_4 d_5 + d_2 d_3 d_4 d_5 \end{aligned}$$

from which it is clear that the hypotheses of Theorem 30 are satisfied with a choice of $\epsilon = 1$. Hence we conclude that (1) has a unique solution.

Much of the work contained in the papers included in this book has not yet appeared in textbook form. It is hoped that a collection such as this will be useful to those students, and others, who are interested in learning about these results, and who might also want to do research in this area. The present state of the development of our knowledge will be readily seen to be incomplete. There is indeed much room for extensions and generalizations of these results, as well as for additional fundamental contributions.

Some Aspects of the Theory of Nonlinear Networks

ALAN N. WILLSON, JR.

Invited Paper

***Abstract*—In the development of network theory over the years, the primary focus of attention has been in the area of linear systems. Several reasons for this emphasis can easily be cited, but perhaps the foremost reason is that it has long been thought that, except in certain very special cases, little progress toward a rigorous definitive theory could be expected once the hypothesis of linearity is discarded. The recent success in the use of numerical methods for computing solutions of the equations for specific nonlinear networks (the importance of which is not to be minimized) has, furthermore, resulted in a certain complacency on the part of many engineers who occasionally need to solve network problems. One senses their outlook as being, basically, that whenever a particular nonlinear problem arises, one need only then run, data in hand, to the computer. Somewhat ironically, however, the development of computer-aided network analysis techniques has also been a prime impetus for many of the recent theoretical investigations in the field of nonlinear networks, and although much remains to be done, a rather comprehensive body of knowledge in this area has begun to take form.**

A number of related recent contributions to the theory of nonlinear networks are reviewed here. As distinct from the computational aspects of the network analysis problem, we discuss work whose primary purpose is to yield an understanding of the nature of the equations that describe the behavior of nonlinear networks, and to identify and relate certain properties of the network elements, and the manner of their interconnection, to properties of the equations and their solutions. In addition, we do frequently touch on the problem of computation since, as has already been implied, it is indeed one of the purposes of the work discussed here to provide more of a theoretical foundation on which to base the numerical analyses.

This invited paper is one of a series planned on topics of general interest—The Editor.

Manuscript received March 2, 1973; revised April 27, 1973.

The author was with Bell Laboratories, Murray Hill, N. J. He is now with the Electrical Sciences and Engineering Department, University of California at Los Angeles, Los Angeles, Calif. 90024.

I. Introduction

THIS PAPER presents a tutorial review of certain aspects of the theory of nonlinear networks. More specifically, we provide here an integration of a number of contributions that have been directed toward the development of insight into, and understanding of, the nature of these networks and their mathematical analyses. With a few notable exceptions, all of these contributions have appeared in the technical literature within the last ten years.

Our treatment is intended primarily for the nonspecialist in the field of network theory. Thus a minimum of back-

Reprinted from *Proceedings of the IEEE*, vol. 61, pp. 1092–1113, August 1973.

ground knowledge is assumed, and the development is rather self-contained. Numerous references to the literature are given, to serve as a guide for the reader interested in obtaining a deeper level of understanding.

By the word "network" we refer only to electrical networks. One should not, however, lose sight of the fact that the theory often applies as well to other types of systems. Among the many analogous systems that are familiar to engineers one finds, for example, hydraulic networks, mechanical systems, and networks describing traffic flow. It is perhaps fair to say, however, that electrical networks are outstanding for the extent to which they require consideration of nonlinear phenomena. Not only is it the case that so-called second-order effects must often be considered in an otherwise linear structure, but also nonlinearities are frequently exploited (indeed, introduced!) in order to permit the design of networks possessing certain types of behavior.

Another characteristic that tends to distinguish electrical networks from the network models of other types of systems is the frequent occurrence in electrical networks of "coupling elements" (by which we mean elements that provide coupling between the electrical variables associated with two different branches of the network—a coupling *in addition to* that which is merely a result of the topological constraints imposed on those variables by Kirchhoff's laws). Whereas it is relatively rare to find elements that are analogous to, say, controlled sources in the structure of certain other types of systems, one must *usually* deal with such active and nonreciprocal elements in the analysis of electrical networks. The use of a controlled source is essential, for example, in even the most elementary description of a transistor.

Although the analysis of distributed nonlinear electrical systems such as networks containing transmission lines has received some attention in the recent literature, we restrict our consideration here to *lumped* electrical networks. The behavior of lumped networks can be described by systems of algebraic and ordinary differential equations, while the study of more difficult nonlinear partial differential equations or functional-differential equations is required in the treatment of distributed systems.

The analysis of nonlinear phenomena is by no means a new branch of science or mathematics. Many studies of nonlinear systems are now classics. Excellent treatments of the famous work of such notable contributors as Poincaré, Lyapunov, and Van der Pol, to name but a few, are to be found in many books. The works mentioned in [1]–[3] are several that might profitably be consulted. It is not our purpose here to review the application to electrical networks of these classical analytical techniques, although the applications are numerous. Rather, we are concentrating on more recent work which for the most part has not yet appeared in book form.

The digital computer has, of course, had a revolutionary effect on the nature of network analysis. Insofar as quantitative studies are concerned, the computer has greatly extended the engineer's ability to determine very precise and extensive information concerning a network's behavior. Engineers are also finding the computer to be quite a useful aid in the design of networks. Much work has been done recently in both of these areas. In computer-aided analysis many recent contributions have been concerned with techniques for extending the computer's capabilities to handle efficiently the numerically very difficult systems of differential equations describing nonlinear networks. In computer-aided design many contributions have been concerned with the development and application of techniques for using the computer to optimize the choice of parameter values in nonlinear networks. As important as these contributions are, it is not the purpose of this paper to review those aspects of the study of nonlinear networks (although our discussion will touch on these areas at several points). Several recent issues of IEEE publications have been devoted to the subject of computer-aided design and analysis, and the reader is referred particularly to the review papers [4]–[6] in those publications.

One common characteristic of many of the classical studies in nonlinear analysis and the computer-aided network analyses is that one usually starts in either case with a mathematical characterization of the system, a set of equations, and then proceeds (either analytically or numerically) to relate in a quantitative manner certain attributes of the system's behavior to the values of the system's parameters. In contrast, we shall be concerned here with what might be considered certain more fundamental aspects of the study of nonlinear networks. We shall discuss work whose primary purpose is to yield an understanding of the nature of the equations that describe the behavior of nonlinear networks, and to identify and relate certain properties of the network elements, and the manner of their interconnection, to properties of the equations and their solutions.

II. Networks of Two-Terminal Elements

One manner of classifying network elements is according to the number of terminals they possess. According to this classification the simplest type of element is the two-terminal element. Such an element may be characterized simply by a relationship between its two fundamental electrical variables: the voltage v across its terminals, and the current i that flows through the element from one terminal to the other. For a resistor or for an independent voltage or current source the relationship is algebraic,[1] while derivatives are required for the characterization of an inductor or a capacitor.

The simplest two-terminal resistors are those in which the relationship between their voltage and current is expressed by specifying the value of one of these variables as a single-valued function of the other variable. This is, by far, the most common situation. A tunnel diode, for example, is a nonlinear resistor that is usually characterized in this manner. Thus the tunnel diode's current and voltage are related by the equation $i=f(v)$, where the function f is defined by a curve in the v–i plane such as that shown in Fig. 1. An element characterized by this type of relation is said to be *voltage controlled*. An analogous definition applies to *current-controlled* elements. In general, it is not necessary that the element's characteristic curve pass through the origin in the v–i plane. Usually the domain of a function f which characterizes a voltage-controlled or current-controlled resistor is the entire real line. There are occasions, however, where it is convenient to restrict the domain to a proper subset of the real line—this

[1] Throughout this work, for the sake of conciseness, we somewhat abuse the strict mathematical definition of the word "algebraic." Our use of this word is intended simply to distinguish those relations between several variables which involve the use of (perhaps *transcendental*) nonlinear mappings, but which do not involve time derivatives or integrals of the variables.

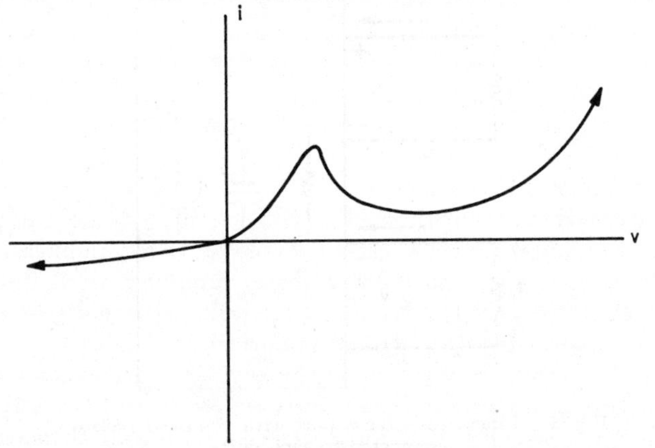

Fig. 1. Typical v–i curve characterizing a tunnel diode.

sometimes happens, for example, in the characterization of a Zener diode.

If neither of a resistor's variables can be expressed as a single-valued function of the other variable, then a more general method of characterization is required. An element having a quite pathological v–i curve can, for example, be constructed by the simple series connection of two tunnel diodes. This observation is made in [7], where Chua and Rohrer consider ***unicursal elements***. Unicursal resistors are two-terminal elements that may be characterized parametrically by a pair of equations $v=f_1(\rho)$, $i=f_2(\rho)$, where f_1 and f_2 are continuous functions of bounded variation, defined for all values of the parameter ρ lying within some (perhaps infinite) interval of the real line.

An independent voltage source is a two-terminal element whose voltage has a prescribed value while the current flowing through the source may assume any value. An independent current source has an analogous definition. Some writers have found it convenient to view such elements as two-terminal resistors with a v–i characteristic curve that is a straight line parallel to one of the axes in the v–i plane. While we shall remain aware of such an interpretation, we shall not formally adopt this point of view here.

It is convenient to characterize nonlinear two-terminal inductors and capacitors in a manner similar to the preceding by introducing the additional variables ϕ, denoting magnetic flux linkages, and q, denoting electric charge. Thus an inductor is characterized by an algebraic relation between the variables ϕ and i which might be thought of as a curve in the ϕ–i plane, with the additional relationship $v=d\phi/dt$. Similarly, a capacitor is characterized by an algebraic relation between the variables q and v, with the additional relationship $i=dq/dt$.

In theory, one further relationship of a similar type is possible among the variables v, i, ϕ, and q. Namely, one could consider two-terminal elements characterized by an algebraic relationship between the variables q and ϕ, with the additional relationships $i=dq/dt$ and $v=d\phi/dt$ also assumed to hold. Chua [8] has recently studied the properties of such two-terminal elements, which he refers to as ***memristors***. We do not consider memristors here since such a physical device has not yet been discovered.[2]

[2] Chua has, however, specified active-circuit realizations of such devices.

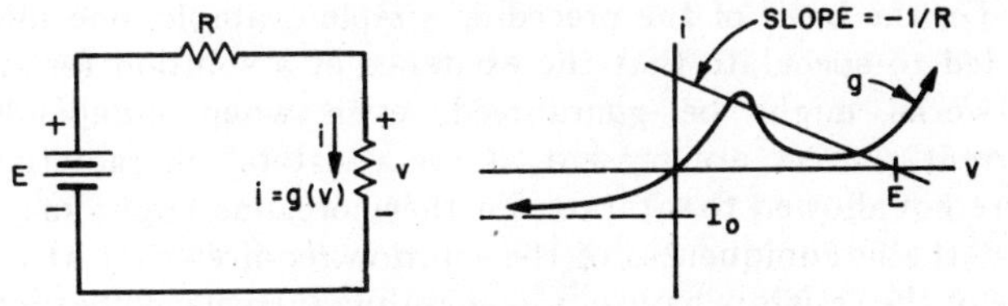

Fig. 2. Simple network illustrating Theorem 1.

In some of the work considered here, the two-terminal elements have been allowed to be time varying. In these cases the same types of characterizations described previously still apply, with the additional provision that the algebraic relation (between, say, q and v in the case of a capacitor) is allowed to vary with time.

A. Resistive Networks

We now consider networks in which only two-terminal resistors and independent sources are present. Perhaps the most fundamental questions that might be asked about such networks are those concerning whether or not there exist values for the branch voltages and currents which satisfy the constraints imposed by the resistor characteristics while also satisfying Kirchhoff's voltage and current laws. Furthermore, it is natural to also inquire whether such values of the branch variables are unique. In 1947, Duffin [9] proved several basic theorems for such networks. Duffin restricted his attention to resistors characterized by continuous functions that express the resistor's current as a function of its voltage. He proved[3] the following.

Theorem 1: In a network consisting of independent voltage sources and voltage-controlled resistors, there exists at least one set of branch voltages and currents satisfying the resistor constraints and Kirchhoff's voltage and current laws, provided that each resistor's v–i characteristic function g is continuous and satisfies

$$\int_0^x g(v)dv \to +\infty \quad \text{as} \quad x \to \pm\infty.$$

The simple network shown in Fig. 2, in which $R>0$, is an example of a network to which the preceding theorem applies. Note that in this example the question of the existence of a solution for the network[4] (indeed, the determination of a solution) can be resolved graphically by simply observing whether or not the "load line" $i=(E-v)/R$ intersects the nonlinear resistor's characteristic curve. It is clear, by this example, that the uniqueness of a solution cannot also be guaranteed by the hypotheses of Theorem 1. Here, for a suitable choice of values for E and R, as many as three solutions are possible. It is also clear, from considering this example, that the theorem would not be true if independent current sources were allowed to be present in the network—if the voltage source were replaced by an independent current source of I amperes, where $I<I_0$, then no solution would exist.

[3] Here, and in the following, we freely paraphrase the work contained in the references. This is necessary in order to maintain a sense of continuity in our presentation, since we shall quote the work of many writers where a variety of notations and points of view have been adopted.

[4] Here, and in the following, we use a phrase such as "solution for the network" to refer to a solution of a set of equations from which the values of all of the network's branch variables can directly be determined.

On the basis of the preceding simple example, one might be led to speculate that the existence of a solution for such networks might be guaranteed, even when independent current sources are present, if the resistors' characteristics were not allowed to saturate. Furthermore, one might suspect that the nonuniqueness of the solution requires that at least one of the resistors have a nonmonotone-increasing characteristic. One's intuition would be correct in this case, for Duffin has also proved[5] [9] the following.

Theorem 2: In a network consisting of independent voltage and current sources and voltage-controlled resistors, there exists *at least one* set of branch voltages and currents satisfying the resistor constraints and Kirchhoff's voltage and current laws, provided that each resistor's v–i characteristic function g is continuous and satisfies $g(v)\to+\infty\ (-\infty)$ as $v\to+\infty\ (-\infty)$.

Theorem 3: In a network consisting of independent voltage and current sources and voltage-controlled resistors, there exists *at most one* set of branch voltages and currents satisfying the resistor constraints and Kirchhoff's voltage and current laws, provided that each resistor's v–i characteristic function g is strictly monotone increasing.[6]

It follows, of course, as a corollary of Theorems 2 and 3 that a network of nonlinear resistors, each of which is characterized by a continuous strictly monotone-increasing function that maps the real line onto itself (each resistor could thus be viewed as being either voltage or current controlled) and independent voltage and current sources, has a unique solution. In particular, this proves the existence and uniqueness of a solution for any *linear* network of positive resistors and independent sources.

References [10]–[12] contain results that are related to the aforementioned theorems of Duffin. In particular, related proofs that are different from Duffin's, and results pertaining to the computation of solutions, are given. Furthermore, Minty's work [11], [12] contains certain generalizations concerning the kinds of resistor characteristics that can be accommodated.

It is clear that many resistive networks containing strictly monotone-increasing voltage- or current-controlled elements whose characteristics saturate, also have unique solutions. This occurs when the network's topology is such that the noninvertibility of the v–i characteristic functions of certain elements is of no consequence. Furthermore, since saturating v–i characteristics are often used in models of common two-terminal elements (semiconductor diodes, for example, are often characterized by an equation of the form $i=I_0(e^{Kv}-1)$ where I_0 and K denote positive constants), it is important to be able to identify these networks.

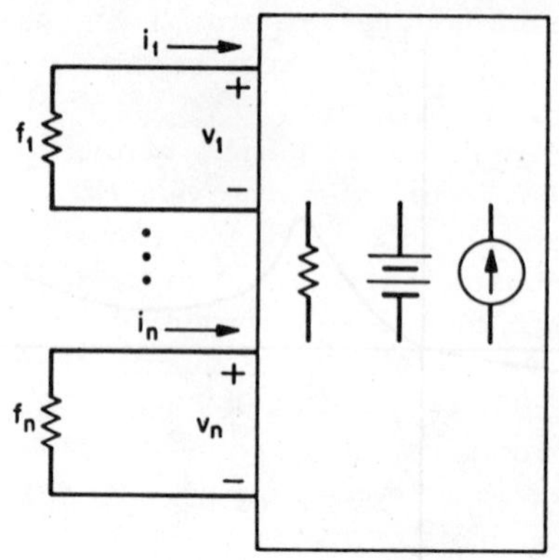

Fig. 3. Linear resistive n-port with nonlinear resistors connected to each port.

An early attempt at dealing with such networks is contained in a paper by Desoer and Katzenelson [13]. They proved the following.

Theorem 4: A sufficient condition for the existence of a unique solution for a network consisting of time-varying[7] voltage-controlled and current-controlled resistors characterized by continuous monotone-increasing functions and independent voltage and current sources, is that the resistor network formed by short-circuiting all voltage sources and open-circuiting all current sources has a tree (a *forest,* if the network consists of more than one piece) such that all tree branches correspond to current-controlled elements and all links correspond to voltage-controlled elements.

It should be noted that the characteristics of the nonlinear resistors considered in Theorem 4 need not be *strictly* monotone increasing, and of course they may saturate. If it happens that some of the resistors are characterized by strictly monotone-increasing functions that map the real line *onto* itself, then these resistors can be viewed as being either current controlled or voltage controlled, and hence several choices are possible in the search for a tree as required by the theorem.

A totally different type of approach to this problem is considered by Sandberg and Willson in [14], [15], where necessary and sufficient conditions for the existence of a unique solution of the network equations are determined. They view the network as having the (perfectly general) structure shown in Fig. 3. Each of the nonlinear resistors, connected to the ports of a linear n-port containing only nonnegative resistors and independent sources, is considered to be characterized by either an equation of the form $i_k=-f_k(v_k)$ or of the form $v_k=-f_k(i_k)$, where, for $k=1,\cdots,n$, f_k is a continuous strictly monotone-increasing mapping of the real line into itself. They also show that the problem of determining a solution for the network is equivalent to the problem of solving the equation

$$AF(x)+Bx=c \tag{1}$$

where, for $k=1,\cdots,n$, the component x_k of the vector[8] $x=(x_1,\cdots,x_n)^T$ corresponds to the port variable (i.e., either current or voltage) at the kth port that is the controlling

[5] In Duffin's statement of Theorems 2 and 3, he explicitly requires that a certain "chain condition" be satisfied by the network. The essence of this condition is to require that the network consist of one piece. Such a statement is necessary in his formulation of the theorems since he also admits into consideration network branches consisting of resistors with an identically zero conductance function—that is, branches that are open circuits. Clearly, the voltages are not uniquely specified across such branches in a network consisting of several pieces. A related matter is the following: We take into consideration in Theorems 1–3 networks constructed by an arbitrary interconnection of the specified elements. We make the tacit assumption, however, that the interconnection is not such that self-contradictory constraints are imposed on the structure by the independent sources. Such a situation would result, for example, if two voltage sources of different values were connected in parallel.

[6] A function g is *monotone increasing* if $g(v_1)\leq g(v_2)$ for all real numbers $v_1<v_2$; it is *strictly monotone increasing* if $g(v_1)<g(v_2)$ for all $v_1<v_2$.

[7] It is clear that the results of Theorems 1–3 also hold when the elements are permitted to be time varying.

[8] Here, and throughout the paper, the superscript T is used to denote the *transpose* of a vector or a matrix.

variable of the kth nonlinear resistor. The nonlinear mapping F characterizes the nonlinear resistors and is defined, for all n-vectors x, by $F(x) = (f_1(x_1), \cdots, f_n(x_n))^T$. The symbols A and B in (1) denote $n \times n$ matrices of real numbers, and c denotes a real n-vector; these provide the characterization of the linear portion of the network, the resistive n-port of Fig. 3. Such a characterization is known [16] to exist for an arbitrary n-port constructed of nonnegative resistors and independent sources. Moreover, due to the nonnegativity (i.e., the *passivity*) of the n-port's linear resistors, it follows that the pair of matrices (A, B) possesses the following special property: *Whenever the vectors x, y are such that $Ax = By$, then $x^T y \geq 0$.* We call pairs of matrices (A, B) that possess this property *passive pairs*.

In the following theorem, which is adequate for the type of problem addressed here, but which is in fact a corollary of a considerably more general result contained in [15], the notation[9] $\mathcal{B}(F) \cap \mathcal{N}(B) = \{\theta\}$ simply means that there exist no real n-vectors $x \neq \theta$ for which 1) $Bx = \theta$ (i.e., x is in the null space of B) and 2) $\lim_{\rho \to \infty} \|F(\rho x)\| < \infty$ (see [15] for further details). That is, the nature of the matrix B sets specific limitations on the manner in which the nonlinear functions f_k are permitted to saturate.

Theorem 5: Let F be a nonlinear mapping as described above and let (A, B) be a passive pair of real $n \times n$ matrices. Then there exists a unique solution of (1) for each real n-vector c if and only if $\mathcal{B}(F) \cap \mathcal{N}(B) = \{\theta\}$. If $\mathcal{B}(F) \cap \mathcal{N}(B) \neq \{\theta\}$, then there exists some real n-vector c such that (1) has no solution.

As explained before, the condition $\mathcal{B}(F) \cap \mathcal{N}(B) = \{\theta\}$ of Theorem 5 relates the nature of the nonlinear resistor v–i characteristic functions f_k (that is, whether or not they saturate for large positive or negative values of the controlling variable) to the pertinent aspects of the topological structure of the network (the nature of the sign pattern of the components of the real n-vectors in the null space of B is the crucial issue here, and this is determined solely by the network's topology). Although it is actually not difficult to determine whether or not the condition $\mathcal{B}(F) \cap \mathcal{N}(B) = \{\theta\}$ holds for a given network, the following result, contained in a recent paper [17] by Desoer and Wu, makes the topological aspects of the problem more transparent.

Theorem 6: A network of voltage-controlled and current-controlled resistors, with each resistor characterized by a continuous strictly monotone-increasing function, has a unique solution when independent sources of arbitrary value are connected to the network (using only a pliers-type entry[10] for the voltage sources and a soldering-iron-type entry[10] for the current sources) if and only if the network possesses the following two properties.

1) Every loop of current-controlled resistors either contains at least one type-U element (i.e., an element whose v–i characteristic is unbounded for both large positive and large negative values of the controlling variable), or else contains at least two type-H elements (a type-H element is one whose v–i characteristic is bounded, i.e., saturates, for either large positive or large negative values of the controlling variable, but is unbounded for large values of the controlling variable having the opposite sign) that are not similarly directed.[11]

2) Every cut set of voltage-controlled resistors either contains at least one type-U element, or else contains at least two type-H elements that are not similarly directed.

Desoer and Wu also present related results in [17] concerning the existence of a solution when the resistors' v–i characteristics are (not necessarily *strictly*) monotone increasing, and results concerning the continuity of the network's solution as a function of the values of the sources. Furthermore, they present algorithms for checking the preceding conditions 1) and 2).

We now direct our attention to networks containing resistors whose characteristics are not necessarily monotone increasing. The simple example of Fig. 2 shows that we can expect such networks perhaps to have more than one solution. Theorem 1 states conditions which assure the existence of at least one solution provided the network contains only voltage-controlled resistors and independent *voltage* sources. A result that applies to the more general case has been proved by Sandberg and Willson [15]. They formulated the problem in the same manner as explained in the above discussion concerning Theorem 5, the only difference being that now the continuous functions f_k are only required to be strictly monotone increasing for all values of the independent variable with magnitude greater than some (arbitrary) positive number. We say that such functions are *eventually strictly monotone increasing*. The tunnel diode v–i characteristic of Fig. 1 is an example of such a function. They proved the following theorem, which is also only a corollary of a much more general result contained in [15].

Theorem 7: Let F be a nonlinear mapping as described previously with the functions f_k eventually strictly monotone-increasing, and let (A, B) be a passive pair of real $n \times n$ matrices. Then there exists at least one solution of (1), for each real n-vector c, if $\mathcal{B}(F) \cap \mathcal{N}(B) = \{\theta\}$.

Desoer and Wu [17] also applied the techniques used in the proof of Theorem 7 to obtain the following result.

Theorem 8: A sufficient condition for the existence of at least one solution for a network consisting of voltage-controlled and current-controlled resistors characterized by continuous eventually strictly monotone-increasing functions and independent voltage and current sources, is that the resistor network formed by short-circuiting all voltage sources and open-circuiting all current sources possesses the properties 1) and 2) of Theorem 6.

Having established the existence (and perhaps uniqueness) of a solution for a given network, one is likely then to want to compute a (the) solution. Since, as explained in the Introduction, we do not intend to deal extensively with the computational aspects of these problems here, we restrict our treatment of this issue to the following few remarks. The problem consists of solving certain nonlinear equations in several variables; the recent book by Ortega and Rheinboldt [18] contains much information on this subject. In the literature

[9] Here, and in the remainder of the paper, we use the symbol θ to denote the n-vector, each of whose components is zero. The Euclidean norm of an n-vector x is denoted by $\|x\|$. That is, $\|x\| = (x_1^2 + \cdots + x_n^2)^{1/2}$.

[10] The concise phrases ***pliers-type entry*** and ***soldering-iron-type entry***, apparently due to Guillemin, are reasonably standard in the circuit theory literature. Their meaning should be intuitively clear. The reader is referred to a discussion [70, p. 92 ff.], where it is shown that one need only, in general, consider pliers-type entries for voltage sources and soldering-iron-type entries for current sources.

[11] To say that the type-H current-controlled elements are not similarly directed is simply to specify that the series combination of the two elements, oriented as they occur in the loop, would have a composite characteristic function of type U. An analogous definition applies to type-H voltage-controlled elements.

on nonlinear networks, Minty's paper [12] presents a computation scheme, and Katzenelson [19] has described an iterative technique for dealing with networks whose elements are described by continuous, monotone, piecewise-linear functions. In [20] Katzenelson and Seitelman apply a contraction-mapping solution algorithm due to Sandberg to certain nonlinear network equations, and algorithms concerned with Newton-type methods are discussed by Willson in [21]. Finally, Rohrer discusses the use of the "successive secant" method and the method of false position in [22]. Results contained in the references mentioned in later sections of this paper, dealing with computation methods for the equations of more general classes of networks, can also be applied to these problems.

B. RLC Networks

We now consider networks consisting of two-terminal resistors, inductors, capacitors, and independent voltage and current sources. For such networks a fundamental problem is to formulate a system of differential (and perhaps algebraic) equations which describes the behavior of the network. Related problems concern the determination, from the nature of the differential equations, of certain of the network's properties, such as the determination of whether a solution exists and is unique for all time $t \geq 0$, whether the network admits solutions with "finite escape time," what kind of asymptotic (in time) behavior can be expected of the solution, and whether or not the network is stable. Finally, one often desires to actually simulate the network's behavior by numerically solving the differential equations with the aid of a digital computer.

Many results, addressed to these related questions, are to be found in the literature on ordinary differential equations. These results typically consider systems of differential equations in the so-called "normal form," that is, equations having the form $dx/dt = f(x, t)$, where x is an n-vector whose components are the system's dynamic variables, and where f denotes a nonlinear mapping of an open set in the $(n+1)$-dimensional Euclidean space E^{n+1} into E^n. For this reason much attention has been given to the matter of determining when it is possible to describe the behavior of a network by a system of differential equations in the normal form.[12] (It is not difficult to give simple examples of pathological networks for which normal-form description is impossible—see [7].)

Brayton and Moser [24] consider nonlinear *RLC* networks as having the form shown in Fig. 4; that is, the network is viewed as consisting of an n-port whose elements comprise only nonlinear (i.e., *not necessarily* linear) resistors and independent sources, with the nonlinear inductors and capacitors connected across each port.

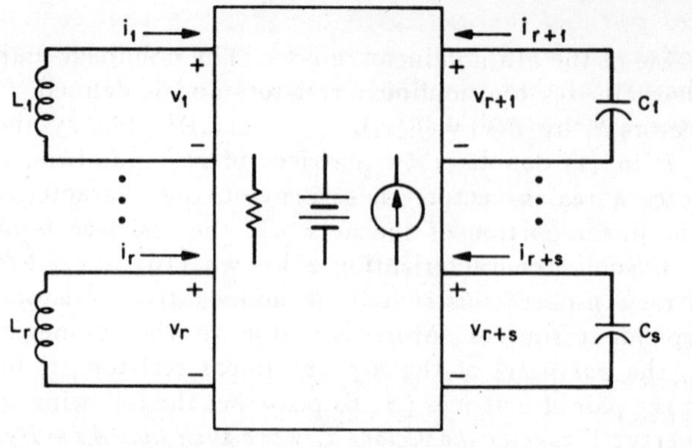

Fig. 4. Nonlinear *RLC* network.

They consider only networks containing voltage-controlled nonlinear capacitors and current-controlled nonlinear inductors with characteristic curves described by differentiable functions having derivatives that are everywhere nonnegative. Letting $i = (i_1, \cdots, i_r)^T$ and $v = (v_{r+1}, \cdots, v_{r+s})^T$, where the port variables that are the components of these vectors are identified in Fig. 4, the mappings $L(i)$ and $C(v)$ are defined by $L(i) = (L_1(i_1), \cdots, L_r(i_r))^T$ and $C(v) = (C_{r+1}(v_{r+1}), \cdots, C_{r+s}(v_{r+s}))^T$, where $L_k(i_k)$, for $k = 1, \cdots, r$, and $C_k(v_k)$, for $k = r+1, \cdots, r+s$ denote the values of the derivatives of the respective inductor and capacitor ϕ–i and q–v characteristic functions at the points i_k and v_k.

Brayton and Moser assume that the nonlinear n-port of Fig. 4 can be characterized by a nonlinear mapping of $(r+s)$-dimensional Euclidean space into itself, which expresses the values of the port variables $v_1, \cdots, v_r, i_{r+1}, \cdots, i_{r+s}$, in terms of the values of the remaining (independent) port variables, the components of the vectors i and v. To determine whether or not such a characterization is possible for a particular network is, of course, an issue that is resolved by the kinds of considerations discussed in Section II-A. A class of networks for which it can be determined from topological considerations alone that the corresponding n-port of Fig. 4 possesses such a characterization, is the class of networks which Brayton and Moser discuss in detail in [24] and which they have called *complete networks*. These are networks of two-terminal voltage-controlled and current-controlled elements for which the network's graph possesses a tree containing all of the capacitive branches and none of the inductive branches, each resistive tree branch corresponds to a current-controlled resistor, each resistive link corresponds to a voltage-controlled resistor, and finally, for which the locations of the resistive branches are such that there exist no loops in which resistive branches appear *both* as tree branches and as links.

In any case, whenever the nonlinear n-port in Fig. 4 can be characterized in terms of the independent port variables that are the components of the vectors i and v, then the network's behavior is described by a system of differential equations of the form

$$L_k(i_k)\frac{di_k}{dt} = f_k(i, v, t), \qquad k = 1, \cdots, r$$

$$C_k(v_k)\frac{dv_k}{dt} = f_k(i, v, t), \qquad k = r+1, \cdots, r+s \qquad (2)$$

which immediately can be put into normal form.[13]

A principal result of [24] concerns a property of the functions on the right-hand side of (2) that derives from the fact that the elements contained in the n-port of Fig. 4 are *reciprocal elements*. Brayton and Moser show that there exists a scalar-valued function $P(i, v)$, which they call the

[12] It should be mentioned here, however, that recent research concerning the computational aspects of the network simulation problem have indicated a trend away from this type of mathematical formulation. See, for example, [23].

[13] The normal-form equations hold everywhere except at such points as might exist where for some k either $L_k = 0$ or $C_k = 0$.

mixed potential function, with the property that each of the functions f_k, $k=1, \cdots, r+s$, can be written as a partial derivative of that function. More precisely, they show that (2) takes the special form

$$L_k(i_k)\frac{di_k}{dt} = \frac{\partial P}{\partial i_k}, \qquad k = 1, \cdots, r$$

$$C_k(v_k)\frac{dv_k}{dt} = -\frac{\partial P}{\partial v_k}, \qquad k = r+1, \cdots, r+s.$$

Moreover, they show how the mixed potential function can be constructed from the given network. The process involves the construction of the *current potential functions* and the *voltage potential functions* for certain elements of the network,[14] and the expression of the sum of these functions in terms of the variables i and v. In general, the construction of the mixed potential function $P(i, v)$ requires the solution of implicit equations. A simple procedure for its construction, however, can be given for certain classes of networks. The reader is referred to [24] for details, where, in particular, it is shown that for complete networks the mixed potential function takes the form

$$P(i, v) = -A(i) + B(v) + i^T Dv \tag{3}$$

where A and B are scalar-valued functions and D denotes a constant matrix.

The principal application of the concept of the mixed potential function, as the reader who is familiar with Lyapunov's "second method" has perhaps anticipated, concerns its use in determining stability criteria for nonlinear networks. Brayton and Moser explore such applications in [24]. The reader is also referred to [27] and [28] for related results. The following theorem concerning the stability of nonlinear networks is one of several that appear in [24]. It applies to networks that have a mixed potential function of the special form (3) (e.g., complete networks), with the further restriction that the function $A(i)$ in (3) is assumed to be of the form $\frac{1}{2}i^T Ai + i^T a$, with A a constant symmetric matrix and a a constant vector. (The mixed potential function will have this special structure if the network is a complete network and if the first r equations of (2) are linear.)

Theorem 9: If A is a positive definite matrix, if $B(v) + \|Dv\| \to \infty$ as $\|v\| \to \infty$, and if there exists $\delta > 0$ such that $\|L^{1/2}(i)A^{-1}DC^{-1/2}(v)\| \leqq 1-\delta$ for all i, v, then all solutions of (2) approach the set of equilibrium points as $t \to \infty$.

Brayton and Moser give examples of the application of their stability results to particular networks and, by way of these examples, show that the results are, in a certain sense, sharp. A collection of other results concerning the stability of nonlinear networks characterized by normal-form differential equations is to be found in [29, ch. 8].

Several other notable contributions concerned with the writing of normal-form equations for nonlinear *RLC* networks are contained in the papers of Chua and Rohrer [7], Stern [30], Desoer and Katzenelson [13], and Desoer and Wu [17].

The approach taken by Chua and Rohrer in [7] starts with the selection of a *normal tree* for the graph of the network at issue. The concept of a normal tree is due to Bryant [31]; it is a tree chosen so as to include the maximum number of capacitive branches and the minimum number of inductive branches. Bryant shows that the choiçe of a normal tree leads to a fundamental circuit matrix and a fundamental cut-set matrix [32] having a particularly convenient structure. (See the discussion in Section IV-B for details.) By analyzing the structure of these matrices and the corresponding implications concerning the form of certain nonlinear relationships, Chua and Rohrer prove the following theorem.

Theorem 10: A network of differentiable, unicursal, two-terminal *RLC* elements has a normal-form representation when independent current sources and voltage sources are connected to the network in an arbitrary manner (using only a pliers-type entry for the voltage sources and a soldering-iron-type entry for the current sources) provided there exists a normal tree such that:

1) each tree inductance is current controlled, with a nonzero derivative everywhere;
2) each link capacitance is voltage controlled, with a nonzero derivative everywhere;
3) the nonlinear mapping F is one-to-one.

Furthermore, a network of differentiable, unicursal, two-terminal *RLC* elements has a normal-form representation in which the dynamic variables consist of capacitor voltages and inductor currents only if the network possesses a normal tree for which the preceding conditions 1)–3) hold.

The nonlinear mapping F referred to in condition 3) of Theorem 10 is an algebraic relation expressing a certain linear combination of tree-capacitance branch voltages and link-inductance branch currents (the appropriate linear combination being determined by the network's topology) as a function of the independent variables appearing in the characterization of the unicursal resistors. (The mapping F is a specialization, for unicursal elements, of the mapping defined by the R equations derived in the more detailed discussion in Section IV-B.) Condition 3) simply guarantees that the inverse of this mapping exists. In principle, condition 3) corresponds to the assumption, in the preceding development of Brayton and Moser's normal-form equations, that the nonlinear n-port of Fig. 4 can be characterized in terms of a certain set of independent port-variables.

For many *RLC* networks the conditions of Theorem 10 are easily verified by inspection, except for condition 3). In general, the verification of that condition requires the use of techniques such as those discussed in Section II-A. In certain special cases this is also a trivial matter. For example, as Chua and Rohrer point out [7, theorem 2], it is sufficient, for a given network, that a certain set of easily identified resistors should have v–i characteristic functions that are strictly monotone-increasing mappings of the real line onto itself. This follows from the results contained in Theorems 2, 3, and 4.

An extension of Theorem 10 is also given in [7], which considers the normal-form description of certain networks containing controlled sources. We shall defer a more thorough consideration of such networks until Section IV.

Stern [30] also approaches the problem of determining normal-form network equations by the selection of a normal tree. He takes time-varying elements into consideration, and his treatment of the problem is at a sufficient level of generality to accommodate networks containing multiterminal elements. His results will therefore be discussed in more detail in Section IV.

The paper [13] by Desoer and Katzenelson considers nonlinear *RLC* networks in which each element is assumed to be characterized by a single nonlinear function that may also be

[14] The concepts of *current potential function* and *voltage potential function* appear in the earlier work of W. Millar [25] and C. Cherry [26], where they are given the names *content* and *cocontent*, respectively.

time varying (e.g., the capacitors are either voltage controlled with $q=f(v, t)$ or else charge controlled with $v=f(q, t)$). They derive sufficient conditions for the existence of a normal-form characterization of the network by an approach that involves considerations related to the concept of a normal tree. The conditions are topological in nature and concern the structure of three one-element-kind (i.e., either R, L, or C) subnetworks which are "extracted" from the given network. In particular they require the existence of certain trees associated with these subnetworks for which (as in Theorem 4) the elements associated with certain "crucial" branches are characterized by strictly monotone-increasing characteristic functions with specific kinds of controlling variables. The reader is referred to [13, theorem IV] for a statement of the theorem. We state here a theorem of Desoer and Wu [17] which, for networks of elements with *increasing* characteristics, represents a generalization of the Desoer and Katzenelson theorem.

Theorem 11: Let $\mathfrak{N}$ denote a network of monotone-increasing, two-terminal, RLC elements characterized by time-varying characteristic functions that are continuously differentiable in both the controlling variable and time. Let $\mathfrak{N}_L$, $\mathfrak{N}_C$, $\mathfrak{N}_R$ denote three subnetworks derived from $\mathfrak{N}$ as follows: for the inductive subnetwork $\mathfrak{N}_L$, replace by short circuits all elements of $\mathfrak{N}$ except the inductors; for the capacitive subnetwork $\mathfrak{N}_C$, replace by open circuits all elements of $\mathfrak{N}$ except the capacitors; for the resistive subnetwork $\mathfrak{N}_R$, replace all capacitors of $\mathfrak{N}$ by short circuits and replace all inductors of $\mathfrak{N}$ by open circuits.

Suppose the following four conditions are satisfied by $\mathfrak{N}_L$ (respectively, $\mathfrak{N}_C$, $\mathfrak{N}_R$).

1) Every loop[15] of current-controlled inductors (respectively, charge-controlled capacitors, current-controlled resistors) contains at least one element that is strictly monotone increasing.

2) Every cut set[16] of flux-controlled inductors (respectively, voltage-controlled capacitors, voltage-controlled resistors) contains at least one element that is strictly monotone increasing.

3) Every loop of current-controlled inductors (respectively, charge-controlled capacitors, current-controlled resistors) either contains at least one type-U element,[17] or else contains at least two type-H elements that are not similarly directed.

4) Every cut set of flux-controlled inductors (respectively, voltage-controlled capacitors, voltage-controlled resistors) either contains at least one type-U element, or else contains at least two type-H elements that are not similarly directed.

The network then has a normal-form representation when independent current sources and voltage sources are connected to the network in an arbitrary manner (using only a pliers-type entry for the voltage sources and a soldering iron-type entry for the current sources).

Desoer and Wu also prove that the normal-form equations have a unique solution on some interval $[t_0, t_1)$, for any initial conditions, provided that the independent sources are characterized by regulated functions of time.

In [69] Sandberg considers the analysis of periodically time-varying electrical networks containing independent voltage and current sources, linear (not necessarily lumped) time-invariant elements, and time-varying nonlinear resistors. He shows that when certain natural conditions are satisfied and the linear time-invariant portion of the network is passive, then the network's equations possess a unique solution, an approximation to which can be computed by certain well-known truncation techniques (which involve the approximation of periodic signals by a finite number of their Fourier components). Furthermore, upper bounds on the truncation error are also determined, and it is shown that the approximate solution approaches the exact solution as the extent of the truncation is reduced to zero.

[15] A self-loop (which might exist in $\mathfrak{N}_L$ due to the short-circuiting of elements in $\mathfrak{N}$) is regarded as a loop.

[16] A cut set may contain only a single branch (due to the open-circuiting of certain elements in $\mathfrak{N}$).

[17] See Theorem 6 for the definition of type-U elements, type-H elements, and the concept of "similarly directed" type-H elements.

III. Transistor Networks

According to the classification of networks suggested in Section II (based upon the number of terminals the network's elements possess), the next class of networks that it is logical to consider is the class of networks constructed from elements having at most three terminals. Such elements as vacuum tubes and various kinds of transistors are examples of commonly encountered three-terminal elements. Of these, the bipolar transistor (which we subsequently refer to simply as *the transistor*) is perhaps the three-terminal element which is presently, in many senses, the most important. For this reason we devote our attention in this section of the paper to a consideration of networks containing only transistors and the previously mentioned two-terminal elements. We note, however, that (as will be explained in detail in Section IV-A) many of the techniques and results discussed here are also applicable to networks containing other three-terminal, and indeed n-terminal, elements.

A. The DC Transistor Model

The characterization of a three-terminal network element is not nearly so simple as is the case for a two-terminal element—where all that is involved, basically, is the specification, in some convenient form, of a curve in the v–i plane. Even the dc characterization of a transistor has been accomplished in many quite different ways, and rather complicated models can result. (This is particularly true if one wishes to account for certain effects that are due to the *distributed* nature of the physical device.)

In Fig. 5, a commonly used large-signal dc transistor model is displayed. It is easily verified that the voltage and current variables defined in that figure obey the following relationships:

$$\begin{pmatrix} i_1 \\ i_2 \end{pmatrix} = \begin{bmatrix} 1 & -\alpha_r \\ -\alpha_f & 1 \end{bmatrix} \begin{pmatrix} f_1(v_1) \\ f_2(v_2) \end{pmatrix}$$

$$\begin{pmatrix} v_1 \\ v_2 \end{pmatrix} = \begin{pmatrix} \bar{v}_1 \\ \bar{v}_2 \end{pmatrix} - \begin{bmatrix} r_e + r_b & r_b \\ r_b & r_c + r_b \end{bmatrix} \begin{pmatrix} i_1 \\ i_2 \end{pmatrix}.$$

Each of the parameters α_f and α_r may assume any value in the open interval (0, 1). The parameters r_b, r_c, and r_e, which account for lead resistances, are sometimes omitted by device modelers (their presence is sometimes accounted for by including appropriate additional resistors in the network to which the transistor model is connected). To accommodate these various points of view, we therefore specify only that the values of the parameters r_b, r_c, and r_e be nonnegative. Thus any or all of them may be zero.

Depending upon whether the transistor being modeled is a p-n-p or an n-p-n, the graph of each of the functions f_1 and

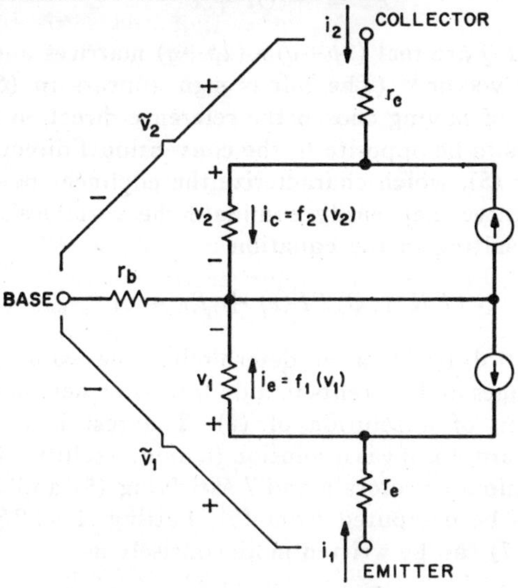

Fig. 5. Large-signal dc transistor model.

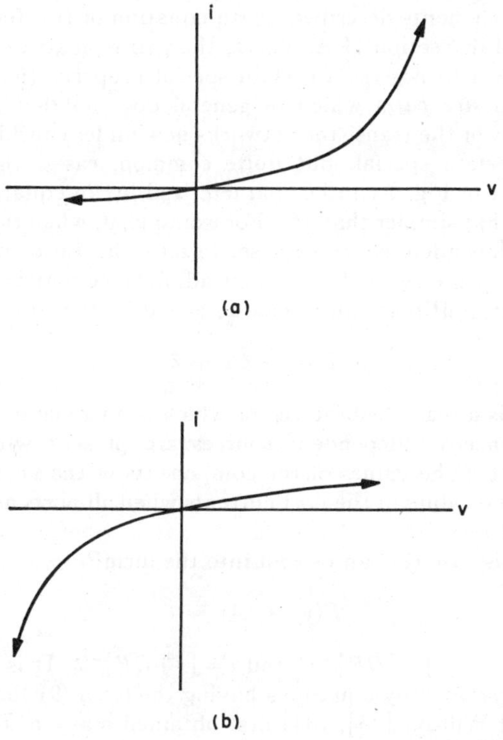

Fig. 6. General shape of the functions f_1 and f_2: (a) p-n-p transistor; (b) n-p-n transistor.

f_2 has one of the general shapes shown in Fig. 6 (at least, for values of $|v|$ that are "not too large"). Often these functions are described by an equation of the form

$$f_k(v) = m_k(e^{n_k v} - 1), \qquad k = 1, 2 \tag{4}$$

where m_k and n_k are appropriately chosen constants, both being positive for a p-n-p transistor and both negative for an n-p-n. On the other hand, for example, a piecewise-linear representation is sometimes specified for f_1 and f_2.

The nature of the functions f_1 and f_2 for large values of $|v|$ depends upon the assumptions the modeler is willing to make and the effects he is interested in considering. For large negative (in the p-n-p case) values of v, for example, the graph of f_k approaches—according to (4)—the horizontal asymptote $i = -m_k$. Thus if the modeler chooses to use (4) to describe f_k for all values of v, the range of f_k will not be the entire real line. If, on the other hand, the effect of ohmic surface leakage across the p-n junction is included in the model, the graph of the function f_k will asymptotically approach a straight line having a small, but positive, slope. The range of such a function is, obviously, the whole real line. One might also wish to include the effect of avalanche breakdown in the reverse-biased region. If this is done, the graph of f_k will have a shape reminiscent of that of a Zener diode in the $v < 0$ part of its domain.

In the forward-biased region there are also effects, particularly apparent for large values of v, which the modeler may or may not wish to recognize. For example, there is the so-called high-level injection phenomenon which tends to decrease the value of the forward current and which, using (4), is usually accounted for by a decrease in the magnitude of n_k for large values of v. In addition, there is the effect of the ohmic resistance of the crystal, which tends to reduce the value of forward current for large values of v.

From the point of view of the device modeler, the question of whether or not to include some of the effects just mentioned is often a minor issue—for many networks the behavior will be essentially the same whether or not, say, surface leakage is accounted for in the transistor model. From the point of view of the network analyst, however, the situation is somewhat different. For example, the matter of whether or not the functions f_k map the real line *onto* the real line can, in some cases, make the difference between whether or not there exists a solution of the network's equations. Similarly, other results that have been reported recently also seem to depend upon the graphs of the functions f_k having certain special properties.

It seems safe to say that no matter which "special effects" are included (or omitted) in the description of the transistor, the functions f_k may at least be considered to be continuous strictly monotone-increasing mappings of the real line into itself. For the purpose of formulating the equations for transistor networks, this is the only hypothesis that we shall make. When additional hypotheses regarding the nature of these functions are needed (to obtain certain results concerning properties of these equations), those hypotheses will be mentioned explicitly. In each case it will be clear that the additional hypotheses are, in some appropriate sense, rather weak.

Similar remarks can be made for the diodes that might also be present in transistor networks. Thus we assume that each diode is described by an equation of the type $i = f(v)$, where, at this point, we assume only that the function f is a continuous strictly monotone-increasing mapping of the real line into itself.

B. DC Equations for Transistor Networks

We consider here dc networks constructed from transistors, voltage-controlled nonlinear resistors with strictly monotone-increasing characteristic functions (which include the most common nonlinear two-terminal resistor, the semiconductor diode), linear (positive) resistors, and independent sources. We first consider writing a set of equations whose solution determines the voltage and the current in every branch of the network.

With no loss in generality we consider the network to have

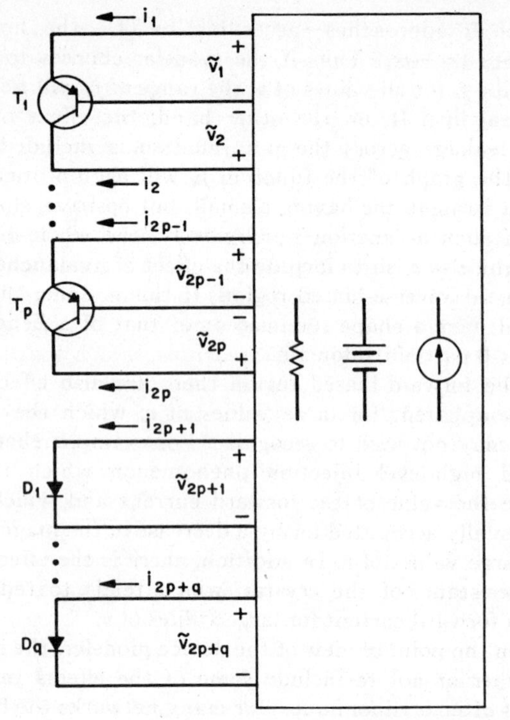

Fig. 7. Canonical form of a dc transistor network.

the canonical structure shown in Fig. 7; that is, we consider the network to be a multiport containing resistors and independent sources, with p transistors and q diodes connected to the ports.

According to our characterization of the transistor by the model of Fig. 5, the nonlinear portion of the network can be characterized in terms of the vectors of port variables $\tilde{v}=(\tilde{v}_1, \cdots, \tilde{v}_{2p+q})^T$, $i=(i_1, \cdots, i_{2p+q})^T$ by the equations

$$i = TF(v)$$
$$v = \tilde{v} - Ri \tag{5}$$

where T denotes a block-diagonal matrix with p 2×2 diagonal blocks of the form

$$\begin{bmatrix} 1 & -\alpha_r^{(k)} \\ -\alpha_f^{(k)} & 1 \end{bmatrix}, \qquad \text{for } k = 1, \cdots, p$$

and one diagonal block which is the $q\times q$ identity matrix, while R denotes a block-diagonal matrix with p 2×2 diagonal blocks of the form

$$\begin{bmatrix} r_e^{(k)} + r_b^{(k)} & r_b^{(k)} \\ r_b^{(k)} & r_c^{(k)} + r_b^{(k)} \end{bmatrix}, \qquad \text{for } k = 1, \cdots, p$$

and one $q\times q$ diagonal block which is a matrix whose elements are all zeros. The nonlinear mapping F has the form $F(v) = (f_1(v_1), \cdots, f_{2p+q}(v_{2p+q}))^T$, where each of the functions f_k is a continuous strictly monotone-increasing mapping of the real line into itself.

As mentioned in Section II-A, it is proved in [16] that any linear multiport of the type shown in Fig. 7 can be characterized in terms of its port-variable vectors $\tilde{v}$ and i by an equation of the form

$$P\tilde{v} = -Qi + c \tag{6}$$

where P and Q are real $(2p+q)\times(2p+q)$ matrices and c is a real $(2p+q)$-vector.[18] (The minus sign appears in (6) as a consequence of having chosen the reference direction for the port currents to be opposite to the conventional direction.)

By using (5), which characterize the nonlinear portion of the network, we may easily eliminate the variables $\tilde{v}$ and i from (6), resulting in the equation

$$(PR + Q)TF(v) + Pv = c. \tag{7}$$

The central problem in determining the values of all branch voltages and currents in a dc transistor network is the determination of a solution of (7). The rest is relatively straightforward, for if v is a solution (a unique solution) of (7), then the (unique) vectors $\tilde{v}$ and i satisfying (5) and (6) can immediately be computed from (5). Letting $A=(PR+Q)T$ and $B=P$, (7) can be written more concisely as

$$AF(v) + Bv = c. \tag{8}$$

Certain results pertaining to nonlinear networks of *two-terminal* elements described by an equation of this form were discussed in Section II-A. There, the pair of matrices (A, B) was known to possess a certain special property (namely, it was a *passive pair*) which in general does not hold for the equations of the transistor networks now under consideration.

In certain special, but quite common, cases, the linear multiport in Fig. 7 can be characterized by an equation that is somewhat simpler than (6). For example, if, when the values of all independent sources are set to zero, the linear multiport of Fig. 7 possesses a short-circuit admittance matrix G, then the linear multiport can be characterized by the equation

$$i = -G\tilde{v} + \tilde{c}$$

where $\tilde{c}$ is a real constant vector which is, in general, nonzero since nonzero independent sources are present within the multiport. (The values of the components of the vector $\tilde{c}$ are simply the values of the port currents when all ports are short-circuited.)

In this case (7) can be cast into the form[19]

$$F(v) + Av = \tilde{c} \tag{9}$$

where $A=T^{-1}[I+GR]^{-1}G$ and $c=[I+GR]^{-1}\tilde{c}$. It is for networks described by equations having the form (9) that Sandberg and Willson [33], [34] first obtained many of the transistor network results that will be discussed here.

C. P_0 Matrices and $\mathcal{W}_0$ Pairs

Two fundamental concepts in our subsequent discussion of results concerning the transistor network equations (8) and (9) are the concept of a P_0 matrix and the related concept of a pair of matrices of class $\mathcal{W}_0$. The class of matrices called P_0 was defined by Fiedler and Pták [35]. They proved that the following properties of a real square matrix A are equivalent.

1) All principal minors of A are nonnegative.

2) For each vector $x\neq\theta$, there exists an index k such that $x_k\neq 0$ and $x_k(Ax)_k\geqq 0$.

[18] We also note that many other linear multiports (e.g., many multiports containing controlled sources) also possess such a characterization.

[19] It is clear that the matrix T is nonsingular, and it is proved in [34] that the matrix $[I+GR]$ must also be nonsingular.

3) For each vector $x \neq \theta$, there exists a diagonal matrix $D_x \geqq 0$[20] such that[21] $\langle x, D_x x \rangle > 0$ and $\langle Ax, D_x x \rangle \geqq 0$.

4) Every real eigenvalue of A, as well as of each principal submatrix of A, is nonnegative.

Sandberg and Willson proved [33], [34] that another property can be added to this list of equivalent properties.

5) $\det (D+A) \neq 0$ for every diagonal matrix $D > 0$.

The class of all matrices possessing one (and hence all) of the above properties is called P_0. The class of P_0 matrices should not be confused with the class of positive semidefinite matrices. Although any positive semidefinite matrix is also a P_0 matrix, the class of P_0 matrices is a larger class. This derives from the fact that no requirements concerning symmetry are specified in the definition of the class of P_0 matrices.

We now define a related class of *pairs of matrices*. This class was first studied by Willson [16], [36]. (See also [37] for related results.) The pairs of matrices in this class possess properties that are in a sense analogous to the properties of a single P_0 matrix. For each pair of real $n \times n$ matrices (A, B), the set of matrices $\mathcal{C}(A, B)$ consists of all $n \times n$ matrices that can be constructed by juxtaposing columns taken from either A or B while maintaining the original relative ordering of the columns. Thus, denoting by M_k the kth column of a matrix M, we have $M \in \mathcal{C}(A, B)$ if and only if for each $k = 1, \cdots, n$, either $M_k = A_k$ or $M_k = B_k$. Obviously, $\mathcal{C}(A, B)$ contains at most 2^n matrices (there will be less than 2^n matrices in $\mathcal{C}(A, B)$ if, for one or more values of k, $A_k = B_k$). The pair of real $n \times n$ matrices (M, N) is said to be a *complementary pair taken from* $\mathcal{C}(A, B)$ if and only if both M and N are members of $\mathcal{C}(A, B)$ and for each $k = 1, \cdots, n$, either $M_k = A_k$ and $N_k = B_k$, or else $M_k = B_k$ and $N_k = A_k$. It is obvious that (A, B) is a complementary pair taken from $\mathcal{C}(A, B)$. It is also clear that $\mathcal{C}(A, B) = \mathcal{C}(B, A)$ and, moreover, that if (M, N) is any complementary pair taken from $\mathcal{C}(A, B)$, then $\mathcal{C}(M, N) = \mathcal{C}(A, B)$. Furthermore, for each $M \in \mathcal{C}(A, B)$, there exists $N \in \mathcal{C}(A, B)$ such that (M, N) is a complementary pair. The following properties of a pair of real square matrices (A, B) are equivalent.

1) $\det (AD+B) \neq 0$ for every diagonal matrix $D > 0$.

2) There exists a matrix $M \in \mathcal{C}(A, B)$ such that $\det M \neq 0$ and such that $\det M \cdot \det N \geqq 0$ for all $N \in \mathcal{C}(A, B)$.

3) For each vector $x \neq \theta$, there exists an index k such that either $(A^T x)_k \neq 0$ or $(B^T x)_k \neq 0$, and such that $(A^T x)_k (B^T x)_k \geqq 0$.

4) For each vector $x \neq 0$ there exists a diagonal matrix $D_x \geqq 0$ such that either $\langle A^T x, D_x A^T x \rangle > 0$ or $\langle B^T x, D_x B^T x \rangle > 0$ (that is, such that $\langle A^T x, D_x A^T x \rangle + \langle B^T x, D_x B^T x \rangle > 0$), and such that $\langle A^T x, D_x B^T x \rangle \geqq 0$.

5) For each complementary pair of matrices (M, N) taken from $\mathcal{C}(A, B)$, each real value of λ that satisfies $\det (M - \lambda N) = 0$ is nonnegative.

6) There exists a complementary pair of matrices (M, N) taken from $\mathcal{C}(A, B)$ such that $M^{-1} N \in P_0$.

7) There exists a matrix $M \in \mathcal{C}(A, B)$ such that $\det M \neq 0$; and, for any complementary pair of matrices (M, N) taken from $\mathcal{C}(A, B)$ with $\det M \neq 0$, $M^{-1} N \in P_0$.

The class of all pairs of matrices possessing one (and hence all) of the above properties is called $\mathcal{W}_0$.

[20] Here, and in the following, the phrase *diagonal matrix* $D \geqq 0$ (*diagonal matrix* $D > 0$) is used to describe a square matrix with only nonnegative elements (positive elements) on the main diagonal, and with all other elements having the value zero.

[21] Here, and in the following, the notation $\langle x, y \rangle$ is used to denote the *scalar product* of two real n-vectors x and y; that is, $\langle x, y \rangle = x^T y = x_1 y_1 + \cdots + x_n y_n$.

D. Existence and Uniqueness of Solutions

For each positive integer n, we denote by $\mathfrak{F}^n$ that collection of mappings of the n-dimensional Euclidean space E^n *onto* itself defined by $F \in \mathfrak{F}^n$ if and only if there exist, for $k = 1, \cdots, n$, strictly monotone-increasing functions f_k mapping the real line onto itself such that for each $x = (x_1, \cdots, x_n)^T \in E^n$, $F(x) = (f_1(x_1), \cdots, f_n(x_n))^T$.

The following theorem, proved by Sandberg and Willson [33], provides a necessary and sufficient condition for the existence of a unique solution of (9) for all mappings F defined by strictly monotone-increasing functions f_k, characterizing the transistor and diode nonlinearities, which map the real line *onto* itself, and for all real constant vectors c.

Theorem 12: If A is an $n \times n$ matrix of real numbers, then there exists a unique solution of (9) for each $F \in \mathfrak{F}^n$ and for each $c \in E^n$ if and only if $A \in P_0$.

The following theorem, generalizing Theorem 12, is proved in [16].

Theorem 13: If A and B are $n \times n$ matrices of real numbers, then there exists a unique solution of (8) for each $F \in \mathfrak{F}^n$ and for each $c \in E^n$ if and only if $(A, B) \in \mathcal{W}_0$.

The following theorem, a generalization of an earlier result (a special case of [33, corollary 1]), is proved in [16]. It shows that the condition $(A, B) \in \mathcal{W}_0$ guarantees the *uniqueness* of a solution (if a solution exists) even in cases where the nonlinear functions f_k do not necessarily map *onto* the entire real line (as is the case, for example, for the semiconductor diode function (4)).

Theorem 14: If $F(x) = (f_1(x_1), \cdots, f_n(x_n))^T$, where each of the functions f_k is strictly monotone increasing, and if $(A, B) \in \mathcal{W}_0$, then there exists at most one solution of (8) for each $c \in E^n$.

Whenever $(A, B) \notin \mathcal{W}_0$, it follows from the proof of the "only if" part of Theorem 13 that there exists a mapping $F \in \mathfrak{F}^n$ and a vector $c \in E^n$ such that (8) has more than one solution. On the other hand, however, if the mapping F is fixed, then even if $(A, B) \notin \mathcal{W}_0$, it can still happen that there exists a unique solution of (8) for all $c \in E^n$. The next theorem to be considered, which is proved in [16] and which generalizes an earlier result of Sandberg [38] pertaining to equations of the type (9), shows that when the *fixed* mapping F possesses another special property (rather than assuming that $F \in \mathfrak{F}^n$), then the nonuniqueness of solutions of (8) follows, for some $c \in E^n$, whenever $(A, B) \notin \mathcal{W}_0$. Moreover, it happens that under these hypotheses there exists some $c \in E^n$ such that (8) has two solutions with the norm of their difference being equal to any previously specified positive value. The special property that is assumed to hold is the following: *for each of the functions f_k defining the mapping F, it is possible to draw a straight line having any given positive slope, and any given length, between some pair of points on the graph of f_k.* We denote by $\mathcal{E}^n$ that collection of mappings F possessing this property. Note, in particular, that mappings F constructed from functions f_k of the type (4) possess this property; hence the following theorem emphasizes the fundamental nature of the condition $(A, B) \in \mathcal{W}_0$ concerning the uniqueness of solutions for the equations of transistor networks.

Theorem 15: Let $F \in \mathcal{E}^n$, let $(A, B) \notin \mathcal{W}_0$ be a pair of real $n \times n$ matrices, and let δ be a positive constant. Then, for some $c \in E^n$ there exist solutions $\hat{v}, \bar{v}$ of (8) satisfying $\|\hat{v} - \bar{v}\| = \delta$.

As explained in our discussion concerning networks containing only two-terminal resistive elements, it can happen that the network equations have no solution in certain situations in which the functions characterizing some of the nonlinear resistors do not map *onto* the entire real line. Sandberg and Willson prove in [34], as a special case of a more general result, that for networks described by an equation of the form (9), with $A \in P_0$ and each of the functions f_k that define the mapping F, continuous and strictly monotone-increasing (but not necessarily mapping the real line *onto* the entire real line), a sufficient condition for the existence of a unique solution of the equation for each right-hand-side vector c is that the matrix A is nonsingular. Using the aforementioned property 6), which characterizes $\mathcal{W}_0$ pairs, it follows that an obvious corresponding sufficient condition for the existence of a unique solution for networks described by an equation of the form (8) is that both of the matrices A, B should be nonsingular and that $(A, B) \in \mathcal{W}_0$.

In [39] Sandberg proves a theorem providing necessary and sufficient conditions for the existence of a unique solution for transistor networks that are described by (9), with $A = T^{-1}G$, and with all nonlinear functions f_k having the semiconductor-diode-type characterization (4).[22] The theorem also uses the fact that the transistor is a passive device, that is, that physically realizable transistors do not supply energy to any network to which they are connected. This fact is expressed mathematically by the inequality $\langle v, TF(v)\rangle \geqq 0$ for all real n-vectors v.[23] The theorem is related to Theorem 5, and the reader is referred to the discussion immediately preceding that theorem for more details concerning the notation.

Theorem 16: Let $F(v) = (f_1(v_1), \cdots, f_n(v_n))^T$, where each of the functions f_k is of the type (4), and let $\langle v, TF(v)\rangle \geqq 0$ hold for all $v \in E^n$. Then there exists a unique solution of (9) for each n-vector c if and only if $T^{-1}G \in P_0$ and $\mathcal{B}(F) \cap \mathfrak{N}(G) = \{\theta\}$. If $T^{-1}G \notin P_0$, then there are at least two solutions of (9) for some n-vector c, and if $T^{-1}G \in P_0$ but $\mathcal{B}(F) \cap \mathfrak{N}(G) \neq \{\theta\}$, then for some n-vector c there is no solution of (9).

E. Results Concerning Continuity, Boundedness, and Computation

For transistor networks, as well as for many other systems whose behavior is described by an equation having the form (8), the vector c may be regarded as the system's input, and the vector x may be regarded as the system's response, or output. Those properties that one might expect well-behaved systems to possess are likely to include continuity and boundedness; thus one might expect 1) "small" changes to result, in the value of a system's output, when "small" changes are made to the value of the system's input, and 2) a bounded sequence of input vectors to yield a bounded sequence of outputs. The following two theorems, taken from [16], show that such properties are indeed possessed by well-behaved systems described by an equation having the form (8).

Theorem 17: For each $F \in \mathfrak{F}^n$ and each pair of real $n \times n$ matrices $(A, B) \in \mathcal{W}_0$, the solution v of (8) is a continuous function of the vector c.

Theorem 18: If (A, B) is a pair of real $n \times n$ matrices, then $(A, B) \in \mathcal{W}_0$ if and only if for each $F \in \mathfrak{F}^n$ and each unbounded sequence of points $x^1, x^2, \cdots$ in E^n, the corresponding sequence $c^1, c^2, \cdots$ $(c^k = AF(x^k) + Bx^k,\ k = 1, 2, \cdots)$ is unbounded.

Many methods can be applied for computing solutions of (8). Some sources of information concerning the topic of computation have been mentioned at the end of Section II-A. In addition, certain results pertaining to useful computation schemes for special classes of equations having the form (8) are to be found in [33] and [37].

[22] It is remarked in [39], however, that certain weaker assumptions would suffice here.

[23] The reader is also referred to [40], where Gopinath and Mitra derive conditions for passivity in terms of the parameter values of the transistor model.

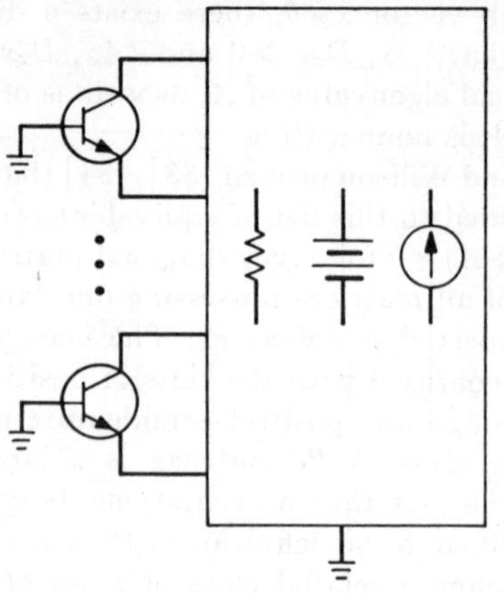

Fig. 8. Common-base transistor network.

The following theorem, taken from [16], shows that with the slight additional requirement that the characteristic functions for all nonlinear resistors be *continuously differentiable* strictly monotone-increasing mappings of the real line onto itself (denoted by $F \in \mathfrak{F}^n \cap C^1$), there exists a computation scheme that will find the solution of (8) whenever its existence and uniqueness can be proved by Theorem 13. The theorem is an extension of an earlier result concerning equations of the form (9), due to Gersho [41].

Theorem 19: Let M be an arbitrary positive definite symmetric matrix, and let $Q: E^n \rightarrow E^1$ be defined by

$$Q(x) = [AF(x) + Bx - c]^T M [AF(x) + Bx - c]$$

where $F \in \mathfrak{F}^n \cap C^1$, $(A, B) \in \mathcal{W}_0$, and $c \in E^n$. For each $x \in E^n$ and each $\gamma \geqq 0$ let

$$g(x, \gamma) = \begin{cases} \dfrac{Q(x) - Q[x - \gamma \nabla Q(x)]}{\gamma \|\nabla Q(x)\|^2}, & \gamma > 0 \\ 1, & \gamma = 0 \end{cases}$$

where $\nabla Q(x)$ denotes the gradient of Q at the point x. Then, if δ is any real number satisfying $0 < \delta \leqq \frac{1}{2}$, and if x^0 is an arbitrary point in E^n, the sequence $\{x^k: k = 0, 1, 2, \cdots\}$ converges to the solution of (8), where (for $k = 0, 1, 2, \cdots$) the x^k satisfy

$$x^{k+1} = x^k - \gamma^k \nabla Q(x^k)$$

each γ^k being any real number that satisfies $\delta \leqq g(x^k, \gamma^k) \leqq 1 - \delta$ if $g(x^k, 1) < \delta$, or $\gamma^k = 1$ if $g(x^k, 1) \geqq \delta$.

F. "Common-Base" Transistor Networks

To illustrate the manner in which some of the preceding theory can be used to obtain an understanding of how the behavior of certain networks is dictated by their topological structures alone, we now consider a result proved in [33] and [34] concerning "common-base" transistor networks.

Consider first the class of transistor networks that may be viewed as having the special structure shown in Fig. 8, that is, networks having some node (which we call "ground") to which the base terminal of each transistor is connected. If it happens that the linear multiport in Fig. 8 can be characterized in terms of a short-circuit admittance matrix G, then it

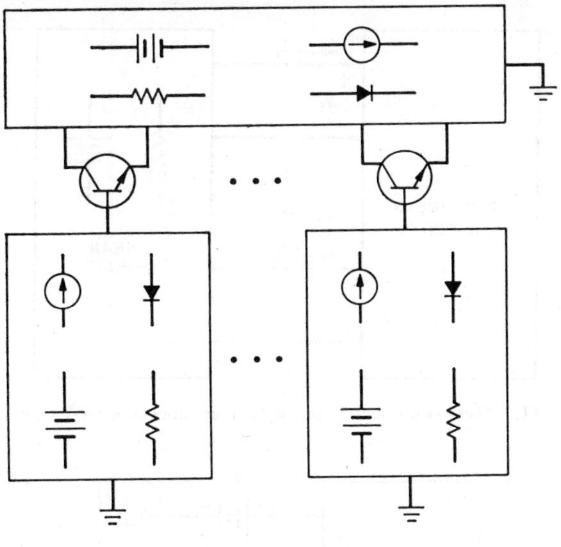

Fig. 9. A special class of transistor networks.

is well known that, as a result of the special common-ground port structure, the (symmetric) matrix G will have the form

$$G = \begin{bmatrix} G_{11} & G_{12} & G_{13} \cdots \\ G_{21} & G_{22} & G_{23} \cdots \\ & \cdots & \end{bmatrix} = \begin{bmatrix} (g_{11}+g_{12}+g_{13}+\cdots) & -g_{12} & -g_{13} \cdots \\ -g_{12} & (g_{12}+g_{22}+g_{23}+\cdots) & -g_{23} \cdots \\ & \cdots & \end{bmatrix}.$$

In particular, the matrix G will be *weakly column-sum dominant*, which means simply that the elements G_{ij} of each column will be such that $G_{jj} \geqq \sum_{i\neq j} |G_{ij}|$. It is also clear, from its special structure, that the matrix T of (5) is *strongly column-sum dominant*, i.e., that for each column of T, $T_{jj} > \sum_{i\neq j} |T_{ij}|$.

Let us temporarily assume that lead resistors are not present in the transistor model and hence the equation for the network has the form (9) with $A = T^{-1}G$. It is proved in [33] that whenever T is strongly column-sum dominant and G is weakly column-sum dominant, then $T^{-1}G \in P_0$. Thus it follows from the preceding theory that the equation has at most *one* solution.

This result is extended in [34] to apply to a more general class of networks—namely, to the class of networks that can be viewed as having the structure shown in Fig. 9. Here diodes are also allowed within the multiports, and two-terminal subnetworks may be present in series with the base lead of each transistor. Furthermore, the transistor model may include lead resistors. For such networks, it is shown in [34] that the aforementioned uniqueness property still holds.

An example of the application of the above result concerns the behavior of the emitter-coupled multivibrator shown in Fig. 10(a). A circuit having this structure is known to be capable of monostable or astable operation. It follows from the preceding result, by simple inspection of the network's topology, that bistable operation is impossible. Clearly, the circuit's equilibria are specified by the solutions of the equivalent dc network of Fig. 10(b). This network, however, having the structure specified in Fig. 9, can have *at most* one solution.

Another example of the application of the preceding theory concerns networks containing only *one* transistor. *All* such networks certainly have the form specified in Fig. 9.

The following theorem summarizes the preceding results.

Theorem 20: One cannot synthesize a bistable network which consists of resistors, inductors, capacitors, diodes, independent voltage and current sources, and either one transistor (modeled as in Fig. 5) or an arbitrary number of transistors with the network having the structure of Fig. 9 when all capacitors are open-circuited and all inductors are short-circuited.

It is important to recognize that the elements admitted into consideration in Theorem 20 do not include ideal transformers. The theorem would not then be true, as has been noted recently by Baranyi and Newcomb [42], who have applied the preceding theory to the study of one-transistor networks containing ideal transformers.

G. Existence of "At Least One" Solution

An example presented in Section II-A shows that when an independent current source is present in a simple network of voltage-controlled two-terminal nonlinear resistors, the network may have no solution. Furthermore, it is also shown there that this situation does not arise in a similar circuit having only an independent voltage source. We now discuss a result, proved by Sandberg and Willson in [43], which shows that the existence of *at least one* solution is indeed a general consequence of dc networks having only *voltage-controlled* nonlinear elements and independent *voltage* sources. More precisely, the result concerns dc networks containing nonlinear elements that are members of a rather general class of voltage-controlled multiterminal elements possessing certain quite reasonable and, from a physical point of view, quite unrestrictive properties; in particular, it is shown in [43] that transistors and diodes possess the required properties.

When only independent *voltage* sources are present, the general transistor network of Fig. 7 can be thought of as a special case of the nonlinear network of Fig. 11.[24] For this network we assume that the nonlinear multiport is characterized by a mapping $\mathfrak{N}$; that is, the vectors of port variables $v=(v_1, \cdots, v_n)^T$, $i=(i_1, \cdots, i_n)^T$ are assumed to obey the relation

$$i = \mathfrak{N}(v) \tag{10}$$

where the nonlinear mapping $\mathfrak{N}$ possesses the following properties.

1) $\mathfrak{N}$ is a continuous mapping from E^n into E^n.

2) For each vector $c \in E^n$ there exists a real number $R>0$ such that $\langle v, \mathfrak{N}(v+c)\rangle \geqq 0$ for all $v \in E^n$ satisfying $\|v\| \geqq R$.

3) If $\tilde{N}$ and $\hat{N}$ denote disjoint subsets of the set of indices $N=\{1, \cdots, n\}$, with $\tilde{N} \cup \hat{N} = N$, then it follows from $v_k = v_k'$ for all $k \in \hat{N}$, and[25] $\mathfrak{N}_k(v) = \mathfrak{N}_k(v')$ for all $k \in \tilde{N}$, that $v_k = v_k'$ for all $k \in \tilde{N}$.

4) If $\tilde{N}$ and $\hat{N}$ denote disjoint subsets of the set of indices $N=\{1, \cdots, n\}$, with $\tilde{N} \cup \hat{N} = N$, and if, for each $k \in \hat{N}$, the

[24] To allow this point of view, we consider that the transistor model does not account for lead resistances. We assume here that such resistances are accounted for by the presence of appropriate resistances in the linear part of the network.

[25] We denote by $\mathfrak{N}_k$ the kth component of the mapping $\mathfrak{N}$.

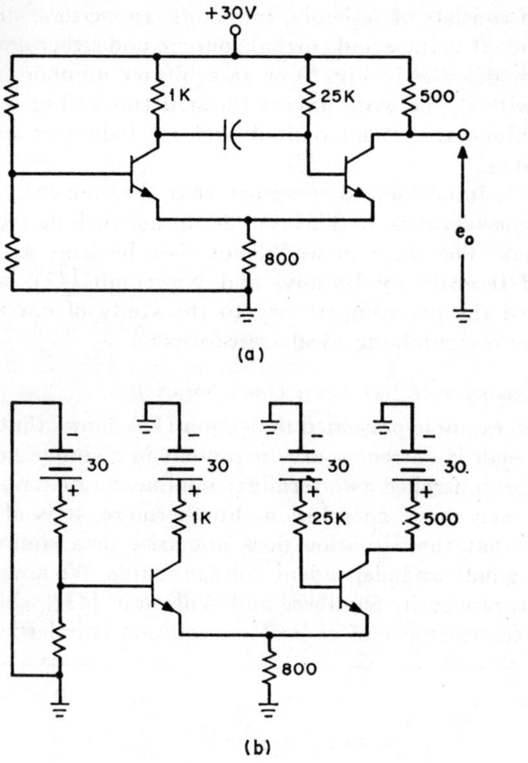

Fig. 10. Example circuit: (a) emitter-coupled multivibrator; (b) equivalent dc network.

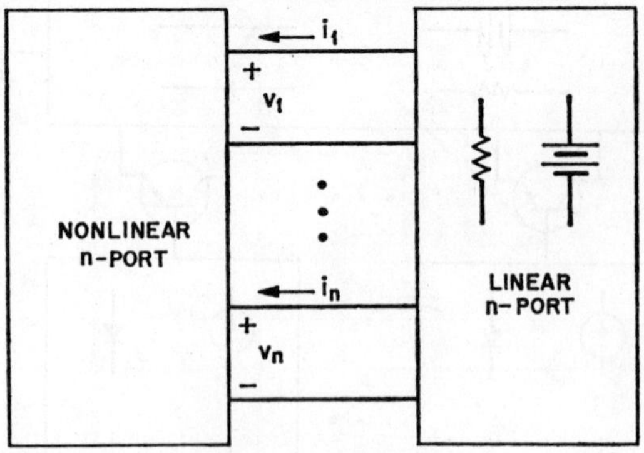

Fig. 11. Network constructed by interconnecting two *n*-ports.

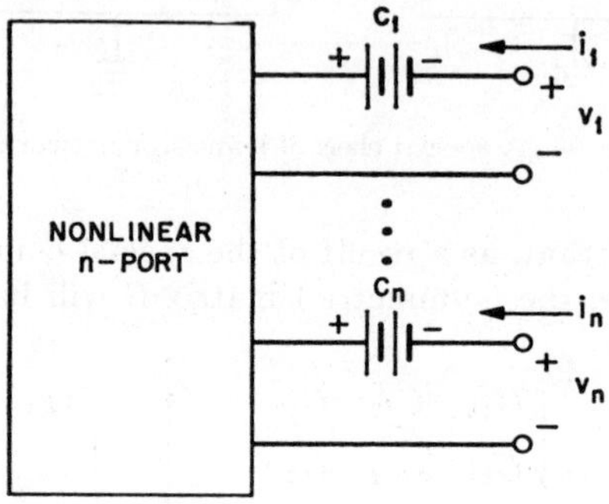

Fig. 12. Physical interpretation of property 2) for nonlinear *n*-ports.

real numbers c_k are given, then there exists $v \in E^n$ such that $v_k = c_k$ for each $k \in \hat{N}$ and $\mathfrak{N}_k(v) = 0$ for each $k \in \bar{N}$.

The physical interpretation of property 1), continuity, is clear. Property 2) specifies that if voltage sources of arbitrary fixed values c_k are connected in series with each port to form a new *n*-port (as shown in Fig. 12), then that *n*-port is "eventually passive." That is, for all port voltage vectors v with $\|v\|$ sufficiently large, the direction of net power flow is *into* the *n*-port.

Properties 3) and 4) specify, in particular, that the mapping $\mathfrak{N}$ is one-to-one and that $\mathfrak{N}$ contains the origin θ in its range. Moreover, they specify that the *m*-ports formed by attaching voltage sources of arbitrary values to an arbitrary collection of $n-m$ ports, and then considering only the remaining m ports as being externally accessible, have those same properties. Thus for all such *m*-ports, no two different excitations (sets of port voltages) produce the same response (set of port currents), and there always exists some excitation for which the response is zero.

It is reasonable to expect that properties 1)–4) are possessed by many realistic models of nonlinear devices. Moreover, it is shown in [43] that they are possessed by transistors and diodes.

The requirement that the independent sources contained within the linear multiport of Fig. 11 be only of the *voltage* type is shown in [43] to imply that the *n*-vector c in the characterization (6) of a linear multiport is contained in the range of the matrix operator P for that characterization.[26] It then follows that the linear *n*-port possesses a characterization of the form

$$P(v - c') = -Qi \tag{11}$$

where c' is a constant real *n*-vector and P and Q are real $n \times n$ matrices. The minus sign appears in (11) due to the unconventional (relative to the linear *n*-port) reference direction of the port-current vector.

Using (10) and (11) and making the substitution $w = v - c'$, we easily obtain the equation

$$Q\mathfrak{N}(w + c') + Pw = \theta. \tag{12}$$

We assume that each of the resistors contained in the linear *n*-port of Fig. 11 has a nonnegative resistance. As mentioned in Section II-A, it then follows that the pair of matrices (P, Q) is a *passive pair* (that is, that $Px = Qy$ implies $x^T y \geq 0$).

The following theorem, concerning (12), thus guarantees the existence of at least one solution for nonlinear networks having the structure of Fig. 11. (In particular, it guarantees the existence of at least one solution for all transistor–resistor–voltage source networks.)

Theorem 21: If (P, Q) is a passive pair of real $n \times n$ matrices and if the nonlinear mapping $\mathfrak{N}$ possesses the aforementioned properties 1)–4), then for each real *n*-vector c', there exists *at least one* *n*-vector w such that (12) is satisfied.

H. Dynamic Equations of Transistor Networks and Their Numerical Solution

In Fig. 13 our dc transistor model has been modified to take into account the presence of certain nonlinear junction capacitances. (We assume that the transistor lead resistances are taken into account by their inclusion in the circuit to which the transistor model is connected.) Such a dynamic

[26] This is not true, in general, if independent *current* sources are permitted in the linear multiport; see [43] for a counterexample. Of course, it remains true, however, if the *n*-port's topology is such that the independent current sources could be eliminated by an appropriate application of Thévenin's theorem.

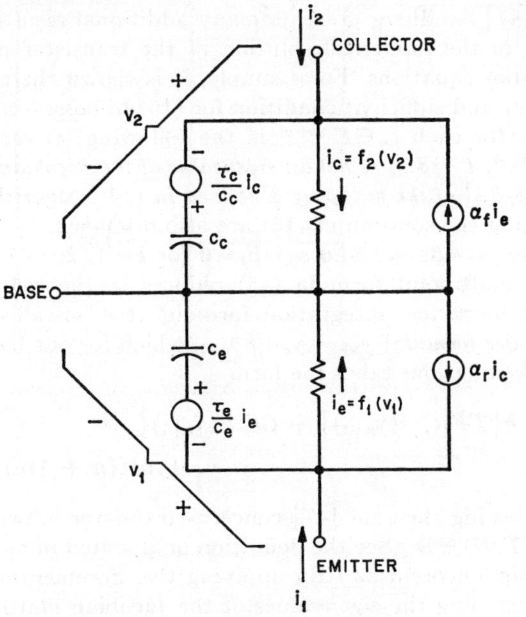

Fig. 13. Dynamic transistor model.

transistor model has been described by Gummel [44] and Koehler [45], and Sandberg, in his basic work concerning the equations of dynamic transistor networks, from which most of our discussion of this topic is taken, has shown [46], [47], using this transistor model, that the dynamic behavior of networks of the type shown in Fig. 7[27] can be described by an equation of the form

$$\frac{du}{dt} + TF[C^{-1}(u)] + GC^{-1}(u) = b(t) \qquad (13)$$

provided that the linear multiport of Fig. 7 can be characterized in terms of an admittance matrix G. The matrices T and G and the mapping F in (13) are defined as in Section III-B, and each component C_k of the mapping C is of the form $C_k(v) = c_k v_k + \tau_k f_k(v_k)$, where the function f_k is the nonlinear diode characteristic for one of the transistor junctions (which is assumed here to satisfy $f_k(0) = 0$ and to be a continuously differentiable strictly monotone-increasing mapping of the real line into itself) and the values of the positive constants c_k and τ_k are specified by the corresponding values of either c_e and τ_e or c_c and τ_c for that junction. The symbol $b(t)$ denotes a vector that may be time varying (due to the presence of time-varying independent sources in the network). It is assumed that $b(t)$ is bounded and continuous on $0 \leqq t < \infty$. These assumptions are sufficient to guarantee that for any initial condition u_0, (13) has a unique solution $u(t)$ which exists for all $t \geqq 0$.

It is a well-known property of asymptotically stable linear systems that the asymptotic behavior of the system's response is dependent only upon the asymptotic nature of the system's forcing function. This property by no means holds for nonlinear systems in general. A certain class of transistor networks whose behavior is described by the nonlinear differential equation (13), however, does possess this property. We direct our attention to those networks for which the matrices T and G are such that for *some* diagonal matrix $D > 0$, both DT and DG are strongly column-sum dominant (see Section III-F). We denote by $\mathfrak{D}$ the class of all pairs of matrices which possess this property.[28] The following result, concerning such networks, is proved in [46].

Theorem 22: If $(T, G) \in \mathfrak{D}$, and $\hat{u}(t)$, $\tilde{u}(t)$ are solutions of (13) corresponding, respectively, to the forcing functions $\hat{b}(t)$, $\tilde{b}(t)$, and if $\|\hat{b}(t) - \tilde{b}(t)\| \to 0$ as $t \to \infty$, then $\|\hat{u}(t) - \tilde{u}(t)\| \to 0$ as $t \to \infty$.

An interesting corollary of Theorem 22 is the following: *whenever $(T, G) \in \mathfrak{D}$ and the forcing function $b(t)$ approaches some constant vector b_∞ as $t \to \infty$, then there exists a constant vector u_∞, independent of the initial condition u_0, such that $u(t) \to u_\infty$, as $t \to \infty$; in particular, if $b(t) \to \theta$, then $u(t) \to \theta$.*

The following theorem, which provides an upper bound for the rate of decay of transients in transistor networks whose independent sources have constant values for $t \geqq 0$, is also proved in [46].

Theorem 23: If $(T, G) \in \mathfrak{D}$, and $b(t) = b_\infty$ for $t \geqq 0$, then

$$\sum_{k=1}^{2p+q} d_k \,|\, u_k(t) - u_{\infty k} \,|$$

$$\leqq \exp(-Kt) \sum_{k=1}^{2p+q} d_k \,|\, u_k(0) - u_{\infty k} \,|, \qquad t \geqq 0$$

for every set of positive constants $d_1, d_2, \cdots, d_{2p+q}$ such that

$$0 < K \triangleq \min_j \min \left\{ \frac{1}{\tau_j}(1 - \bar{d}_j d_j^{-1} \alpha_j), \; \frac{1}{C_j}\left(G_{jj} - \sum_{i \neq j} d_i d_j^{-1} \,|\, G_{ij} \,| \right) \right\}$$

in which $-\alpha_j$ is the nonzero off-diagonal term in the jth column of T, for $j = 1, \cdots, 2p$, and $\alpha_j = 0$ for $j = 2p+1, \cdots, 2p+q$; also, for $j = 1, \cdots, 2p$, $\bar{d}_j = d_{j+1}$ for j odd, and $\bar{d}_j = d_{j-1}$ for j even. Moreover, $(T, G) \in \mathfrak{D}$ implies that there exist positive constants d_j, $j = 1, \cdots, 2p+q$, such that $K > 0$.

A similar result, providing a *lower* bound for the rate of decay of transients in transistor networks, is also given in [46].

Gopinath and Mitra [40] also consider the dynamic equations of transistor networks. They exploit the "passivity" of the transistor and show, using the Lyapunov theory, that when each element of a transistor network is passive, the *unforced* system is globally asymptotically stable and the *forced* system is bounded input–bounded output stable.

The numerical solution of (13), which is, of course, only a special case of the normal-form equation

$$\frac{du}{dt} = \phi(u, t)$$

namely, the case in which

$$\phi(u, t) = b(t) - TF[C^{-1}(u)] - GC^{-1}(u) \qquad (14)$$

is often accomplished by use of one of the general linear multi-

[27] The dynamic semiconductor diode model has an analogous form to that of the transistor. Also, a straightforward extension of the results to be discussed here which permits a similar treatment for certain networks containing additional nonlinear capacitors and inductors is described in [46].

[28] It happens that $(T, G) \in \mathfrak{D}$ for many transistor networks described by (9); see [46] for examples. It is easy to show that the class $\mathfrak{D}$ is contained in the class $\mathcal{W}_0$.

point formulas

$$y_{n+1} = \sum_{k=0}^{r} \alpha_k y_{n-k} + h \sum_{k=-1}^{r} \beta_k \tilde{y}_{n-k} \qquad (15)$$

in which

$$\tilde{y}_{n-k} = \phi(y_{n-k}, (n-k)h).$$

Numerical integration formulas of the preceding type, with $\beta_{-1} > 0$, are of considerable use in applications involving the typically "stiff systems" of differential equations encountered in the analysis of nonlinear transistor networks [48]. With $\beta_{-1} > 0$, y_{n+1} is defined *implicitly* through

$$y_{n+1} - h\beta_{-1}\phi(y_{n+1}, (n+1)h) = \sum_{k=0}^{r} \alpha_k y_{n-k} + h \sum_{k=0}^{r} \beta_k \tilde{y}_{n-k}$$

in which the right-hand side depends on y_{n-k} only for $k \in \{0, 1, 2, \cdots, r\}$. In the application of this formula to (13), we have

$$y_{n+1} + h\beta_{-1}\{TF[C^{-1}(y_{n+1})] + GC^{-1}(y_{n+1})\} = z_n \qquad (16)$$

in which

$$z_n = \sum_{k=0}^{r} \alpha_k y_{n-k} + h \sum_{k=0}^{r} \beta_k \tilde{y}_{n-k} + h\beta_{-1}b((n+1)h).$$

Clearly, it makes sense to use an implicit integration method requiring the solution of (16) at each step only if there is reason to suspect that a solution of (16) exists. By letting $x_{n+1} = C^{-1}(y_{n+1})$, it is easily shown that (16) is equivalent to

$$[\tau + h\beta_{-1}T]F(x_{n+1}) + [c + h\beta_{-1}G]x_{n+1} = z_n. \qquad (17)$$

Recall that the matrices τ and c are both diagonal, with positive diagonal elements; thus the matrix $[\tau + h\beta_{-1}T]^{-1}[c + h\beta_{-1}G]$ is nonsingular and is a P_0 matrix for all *sufficiently small* $h > 0$. Consequently, (17) possesses a unique solution for all sufficiently small $h > 0$ and all $z_n \in E^{(2p+q)}$. (We need not assume here that the component functions f_k of the mapping F necessarily map onto the entire real line—see the paragraph immediately following Theorem 15.) Furthermore, it is shown in [43] that, with the additional assumption that the transistor and diode models are passive, there exists *at least one* solution of (17) for all $h > 0$ and all $z_n \in E^{(2p+q)}$. Finally, the following theorem, proved by Sandberg in [38], shows that (17) possesses a *unique* solution for all $h > 0$ and all $z_n \in E^{(2p+q)}$ for a certain class of transistor networks. For the matrix T associated with any given transistor network, we define $\mathfrak{J}(T)$ to be the set of all matrices derived from T by replacing each of the parameters $\alpha_f^{(k)}$, $\alpha_r^{(k)}$ by any real numbers $\delta_f^{(k)}$, $\delta_r^{(k)}$ lying within the intervals $0 < \delta_f^{(k)} \leqq \alpha_f^{(k)}$, $0 < \delta_r^{(k)} \leqq \alpha_r^{(k)}$.

Theorem 24: If $M^{-1}G \in P_0$ for all $M \in \mathfrak{J}(T)$, then[29] $(T + D_1)^{-1}(G + D_2) \in P_0$ for all diagonal matrices $D_1 \geqq 0$, $D_2 \geqq 0$.

As explained in [38], Theorem 15 and Theorem 24 imply that whenever the dc equation (9) possesses at most one solution for each $c \in E^{(2p+q)}$ and for each set of transistor α's which are "not larger than" the given α's, then (17) possesses a unique solution for all $h > 0$ and all $z_n \in E^{(2p+q)}$.

[29] See [47, theorem 5] for an interesting set of conditions that are equivalent to the condition $M^{-1}G \in P_0$ for all $M \in \mathfrak{J}(T)$.

In [47] Sandberg presents many additional results which pertain to the numerical solution of the transistor network differential equations. For example, it is shown there that a necessary and sufficient condition for (16) to possess a unique solution for each $z_n \in E^{(2p+q)}$ is the following: *at each point* $u \in E^{(2p+q)}$, $1/(h\beta_{-1})$ *is not an eigenvalue of the Jacobian matrix* $J_\phi = [\partial\phi/\partial u]$ *of the mapping* ϕ *defined in* (14). Algorithms for computing the solution of (16) are also discussed.

When $\alpha_0 = \beta_{-1} = 1$ and $\alpha_k = \beta_k = 0$ for $k = 1, 2, \cdots, r$, the general multipoint formula (15) reduces to the well-known implicit numerical integration formula (the so-called *backward Euler formula*) $y_{n+1} = y_n + h\tilde{y}_{n+1}$, which for our transistor network equations takes the form

$$y_{n+1} + h\{TF[C^{-1}(y_{n+1})] + GC^{-1}(y_{n+1})\} = y_n + hb((n+1)h). \qquad (18)$$

The following theorem [47] concerns transistor networks for which $(T, G) \in \mathfrak{D}$. (See the definition of $\mathfrak{D}$ stated immediately preceding Theorem 22.) By applying the aforementioned result concerning the eigenvalues of the Jacobian matrix J_ϕ, it can be shown that $(T, G) \in \mathfrak{D}$ guarantees the existence of a unique solution y_{n+1} of (18) for each right-hand side vector.

Theorem 25: Let the diagonal matrix $D > 0$ be such that DT and DG are both strongly column-sum dominant matrices. For $j = 1, 2, \cdots, (2p+q)$, let f_j be a continuous (not necessarily strictly) monotone-increasing mapping of the real line into itself with $f_j(0) = 0$. Let the mappings F and C be defined as in the discussion pertaining to (13) at the beginning of Section III-H. Let h be a positive constant, and suppose that the sequences $\{y_n\}$, $\{w_n\}$ satisfy

$$y_{n+1} + h\{TF[C^{-1}(y_{n+1})] + GC^{-1}(y_{n+1})\} = y_n + w_n$$

for all $n \geqq 0$. Then there exists $\delta > 0$, depending only on the c_j, the τ_j, T, G, and D, such that[30]

$$\|Dy_n\|_1 \leqq (1 + \delta h)^{-n}\|Dy_0\|_1 + \sum_{k=1}^{n} (1 + \delta h)^{-k}\|Dw_{(n-k)}\|_1$$

for all $n \geqq 1$, and

$$\|D(y_n - \hat{y}_n)\|_1 \leqq (1 + \delta h)^{-n}\|D(y_0 - \hat{y}_0)\|_1 + \epsilon \sum_{k=0}^{n} (1 + \delta h)^{-k}$$

for all $n \geqq 1$, in which $\{\hat{y}_n\}$ is any sequence in $E^{(2p+q)}$ with the property that $\|D(\hat{y}_n - y_n^*)\|_1 \leqq \epsilon$ for all $n \geqq 1$, with ϵ a positive constant, and the sequence $\{y_n^*\}$ such that

$$y_{n+1}^* + h\{TF[C^{-1}(y_{n+1}^*)] + GC^{-1}(y_{n+1}^*)\} = \hat{y}_n + w_n$$

for all $n \geqq 0$.

The preceding theorem shows that whenever $(T, G) \in \mathfrak{D}$, then for all $h > 0$ the sequence $y_1, y_2, \cdots$ of (18) is bounded whenever the sequence $b(h), b(2h), \cdots$ is bounded, and $y_n \to \theta$ whenever $b((n+1)h) \to \theta$. Furthermore, the theorem shows that at each step in the computation of an approximate solution of (18), the accumulated round-off error is bounded, and an estimate of the bound is given which shows that the

[30] The norm $\|v\|_1$ of a $(2p+q)$-vector v is defined by

$$\|v\|_1 = \sum_{k=1}^{2p+q} |v_k|.$$

effect of the initial round-off error decays exponentially, and that the effect of the local round-off error does not build up indefinitely.

The following related result, also proved in [47], provides a conceptually interesting uniform bound on the norm of the difference between corresponding points along the solutions of (13) and (18).

Theorem 26: Let the diagonal matrix $D>0$ be such that DT and DG are both strongly column-sum dominant matrices. Let b denote a real continuously differentiable $(2p+q)$-vector valued function of t, for $0 \leq t < \infty$, with b and db/dt bounded on $[0, \infty)$. For $j=1, 2, \cdots, (2p+q)$, let f_j be a continuously differentiable function, mapping the real line into itself, with $(d/dx)f_j(x)>0$ for all x, and with $f_j(0)=0$. Let the mappings F and C be defined as in the discussion pertaining to (13) at the beginning of Section III-H. Let u denote the solution of (13), and define the sequence $\{u_n\}$ by $u_n=u(nh)$ for all $n \geq 0$, where h is an arbitrary positive constant. Let the sequence $\{y_n\}$ be the solution of (18). Then there exist positive constants δ and ρ, both independent of h, such that

$$\| D(u_n - y_n) \|_1 \leq (1 + \delta h)^{-n} \| D(u_0 - y_0) \|_1 + \rho h,$$

for all $n \geq 1$.

In [49] Desoer and Haneda apply a concept known as *the measure of a matrix* (a real-valued mapping which in some respects is analogous to a matrix norm) to obtain generalizations of several of the preceding results.

IV. Networks Containing Multiterminal Elements

Having considered networks with only two-terminal elements in Section II, and transistor networks (a special class of networks having multiterminal elements) in Section III, we now direct our attention to the most general case: multiterminal-element networks.

A. Networks Containing Controlled Sources

Perhaps the most fundamental type of multiterminal element is the (linear) controlled source, that is, the voltage or current source whose value is not a constant or an independent function of time, but rather a constant multiple of the value of the voltage or the current associated with some other branch in the network.[31] Transistor networks (using the transistor model of Fig. 5), for example, can be viewed simply as networks of controlled sources and nonlinear two-terminal elements. Indeed, it is *usually* the case that a physical multiterminal element is modeled by an interconnection of controlled sources and nonlinear *two-terminal* elements.

By taking this point of view, it then follows (from the *linearity* of the controlled sources) that the network can be thought of as simply a linear multiport (which is, in general, active and nonreciprocal) with a nonlinear *two-terminal* element connected at each port. In the dc case, for example, the problem of determining the network's branch voltages and currents can then be reduced to the study of an equation having the form $AF(x)+Bx=c$, where $F(x)=(f_1(x_1), \cdots, f_n(x_n))^T$—an equation about which a great deal is known (as discussed in Sections II and III)—whenever it is the case that the linear multiport can be characterized in terms of its port variables, the components of the vectors x and y, by an equation of the form

$$Ay + Bx = c \tag{19}$$

and provided that each of the nonlinear two-terminal resistors can be characterized by a function f_k of the port variable x_k.

The problem of characterizing a linear multiport in the general case in which controlled sources are present is nontrivial, and a characterization of the form (19) is not always possible.[32] The study of linear networks having controlled sources is generally more difficult than is the case when controlled sources are not present. One reason for this is that many issues that could formerly be resolved by topological means alone can no longer be handled so simply when controlled sources are present. In such networks the actual element values have to be considered and algebraic problems also arise. The reader is referred to Purslow's paper [51] for an extensive discussion of these matters.

In spite of the aforementioned potential difficulties, it is nevertheless likely to be the case for all but the most pathological circuits that the kind of approach previously outlined can be carried out, and thus many of the results discussed in the previous sections can profitably be applied. One class of multiterminal-element networks for which it is particularly obvious that extensions of the results of Section III are possible, is the class of transistor networks containing, for example, multiple-emitter transistors.

B. Normal-Form Differential Equations for Networks of Interconnected "Like-Kind" Elements

We now direct our attention to another approach to the study of nonlinear multiterminal-element networks—an approach that has been considered by several authors. The fundamental idea is to view the network as an interconnection of three "like-kind" nonlinear multiterminal elements, with each branch of the network being associated with one of these elements. Thus we have: 1) a multiterminal resistor, characterized by a nonlinear algebraic relation among the resistive branch voltages and currents; 2) a multiterminal inductor, characterized by a nonlinear algebraic relation among the inductive branch flux linkages and currents; and 3) a multiterminal capacitor, characterized by a nonlinear algebraic relation among the capacitive branch voltages and charges.

Stern [29], [30] allows the multiterminal elements to be time varying. The "Ohm's law" relations for the three elements are thus, in the general case:

$$f_r(i_r, v_r, t) = \theta$$

$$f_l(i_l, \phi_l, t) = \theta, \qquad v_l = \frac{d}{dt}\phi_l$$

$$f_c(v_c, q_c, t) = \theta, \qquad i_c = \frac{d}{dt}q_c$$

where i_r, v_r are n_r-vectors (there are n_r branches for the resistive element), and f_r is a mapping of the $(2n_r+1)$-dimensional

[31] We call controlled sources multiterminal elements since the values of branch variables in at least two branches of the network are involved in their characterization. It would perhaps be more precise to call a *pair* of elements consisting of the controlled source and the two-terminal element contained in the controlling branch (a resistor, for example) the multiterminal element.

[32] The *nullator* [50], a one-port whose only admissible combination of port voltage and current variables is the pair (0, 0), is an example of such a multiport.

Euclidean space $E^{(2n_r+1)}$ into the m_r-dimensional Euclidean space E^{m_r}. The most common situation[33] is, of course, that in which $m_r = n_r$, that is, the case in which there are n_r equations in the $2n_r+1$ (including time) variables. The relations for the inductive and capacitive elements are similarly defined.

To accommodate independent sources in the network, we consider each branch of each multiterminal element to be connected in series with an independent voltage source, and we consider each of these series connections to be connected in parallel with an independent current source. Of course, it will usually be the case that the values of most of these sources are zero. (Alternatively—and this is the approach actually adopted by Stern—we could simply consider the independent sources to be a part of the *resistive* multiterminal element.)

The $(n_r+n_l+n_c)$-vectors v, i, $e(t)$, and $j(t)$ have components which correspond, respectively, to the branch voltages and branch currents of the multiterminal elements and the values of the corresponding independent voltage and current sources. The Kirchhoff voltage and current laws $B(v-e(t))=\theta$, $Q(i-j(t))=\theta$ are linear relations among the branch variables; they express the fact that the elements are interconnected and specify the nature of the interconnection. Concerning the linear graph of the network, in which each edge corresponds to one of the branches of a multiterminal element (including its associated independent sources), an edge is called a capacitive edge if it corresponds to a branch of the capacitive element, etc.

Bryant's concept of a *normal tree* for the network's linear graph [31] was mentioned in Section II-B. A normal tree is simply a tree[34] of the graph, chosen so as to include the maximum number of capacitive branches and the minimum number of inductive branches. Having selected such a tree, the components of the vectors v and i are ordered as follows: $v=(v_S^T, v_R^T, v_L^T, v_C^T, v_G^T, v_\Gamma^T)^T$, $i=(i_S^T, i_R^T, i_L^T, i_C^T, i_G^T, i_\Gamma^T)^T$, where the components of the vectors v_S, i_S correspond to branch variables associated with edges of the graph that are capacitive links, and the components of the vectors v_R, i_R correspond to branch variables associated with edges that are resistive links. Similarly, the components of the vectors having subscripts L, C, G, Γ, respectively, correspond to branch variables associated with edges that are inductive links, capacitive tree branches, resistive tree branches, and inductive tree branches. Due to the fact that the normal tree contains a *maximum* number of capacitive branches and a *minimum* number of inductive branches, and due to the above ordering of the components of the vectors v and i, it easily follows that the matrices B and Q in the Kirchhoff's law relations have the form

$$B = \left[\, I \mid F \,\right] = \begin{bmatrix} I_{SS} & 0 & 0 & F_{SC} & 0 & 0 \\ 0 & I_{RR} & 0 & F_{RC} & F_{RG} & 0 \\ 0 & 0 & I_{LL} & F_{LC} & F_{LG} & F_{L\Gamma} \end{bmatrix}$$

and $Q=[-F^T \mid I]$. For many networks, certain of the submatrices appearing in the preceding description of the matrices B and Q will not be present. If, for example, the network's graph has no loops containing only capacitive branches, then the normal tree can be chosen such that there are no capacitive links. Hence in this case the submatrix F_{SC} will not appear. A similar remark applies to the submatrix $F_{L\Gamma}$ for networks whose graph has no inductive cut sets. A reasonable choice of dynamic variables in the general case are the components of the vectors q and ϕ, defined by

$$q = q_C - F_{SC}{}^T q_S$$
$$\phi = \phi_L + F_{L\Gamma}\phi_\Gamma. \tag{20}$$

The total number of dynamic variables thus equals the number of independent capacitive branch voltages plus the number of independent inductive branch currents.

We recognize, of course, that

$$i_r = \begin{pmatrix} i_R \\ i_G \end{pmatrix}, \quad v_r = \begin{pmatrix} v_R \\ v_G \end{pmatrix}, \quad i_l = \begin{pmatrix} i_L \\ i_\Gamma \end{pmatrix},$$

etc. Thus it is trivial to use the preceding Kirchhoff's and Ohm's law relations, along with (20), to obtain the equations:

$$f_r\left(\begin{pmatrix} i_R \\ i_G \end{pmatrix}, \begin{pmatrix} v_R \\ v_G \end{pmatrix}, t\right) = \theta$$

$$v_R + F_{RC}v_C + F_{RG}v_G = e_R(t) + F_{RC}e_C(t) + F_{RG}e_G(t)$$

$$i_G - F_{RG}{}^T i_R - F_{LG}{}^T i_L = j_G(t) - F_{RG}{}^T j_R(t) - F_{LG}{}^T j_L(t) \quad (R \text{ equations})$$

$$f_l\left(\begin{pmatrix} i_L \\ i_\Gamma \end{pmatrix}, \begin{pmatrix} \phi_L \\ \phi_\Gamma \end{pmatrix}, t\right) = \theta$$

$$\phi = \phi_L + F_{L\Gamma}\phi_\Gamma$$

$$i_\Gamma - F_{L\Gamma}{}^T i_L = j_\Gamma(t) - F_{L\Gamma}{}^T j_L(t) \quad (L \text{ equations})$$

$$f_c\left(\begin{pmatrix} v_S \\ v_C \end{pmatrix}, \begin{pmatrix} q_S \\ q_C \end{pmatrix}, t\right) = \theta$$

$$q = q_C - F_{SC}{}^T q_S$$

$$v_S + F_{SC}v_C = e_S(t) + F_{SC}e_C(t) \quad (C \text{ equations})$$

$$\frac{d}{dt}q = F_{RC}{}^T i_R + F_{LC}{}^T i_L + j_C(t) - F_{SC}{}^T j_S(t) - F_{RC}{}^T j_R(t) - F_{LC}{}^T j_L(t)$$

$$\frac{d}{dt}\phi = -F_{LC}v_C - F_{LG}v_G + e_L(t) + F_{LC}e_C(t) + F_{LG}e_G(t) + F_{L\Gamma}e_\Gamma(t). \tag{21}$$

Clearly, if it is then possible to solve the C equations for v_C as a function of q and t, and to solve the L equations for i_L as a function of ϕ and t, and finally to solve the R equations for i_R and v_G as functions of v_C, i_L, and t (and hence, using the solutions of the L and C equations, as functions of q, ϕ, and t), these solutions, when substituted into the right-hand side of (21), yield the normal-form dynamic equations that describe the network's behavior. Thus the only potentially nontrivial aspect of the problem of obtaining the network's normal-form equations is the solving of each of the sets of algebraic R, L, and C equations.

It is not unreasonable to expect that for many networks it will be possible to solve the R, L, and C equations since, for

[33] There are many conceivable exceptions; the nullator and norator [50], for example, are resistive 1-ports having $m_r=2$ and $m_r=0$, respectively.

[34] A *forest*, if the graph has more than one piece.

example, the L equations in the most common situation, i.e., when $m_l = n_l$, comprise (treating the time variable t as a parameter) $2n_l$ equations in $2n_l + n_\phi$ unknowns (where n_ϕ denotes the number of components in the vector ϕ). Since the three sets of R, L, and C equations have a similar form, it suffices in our subsequent discussion to restrict our attention to one of these sets of equations. We choose to discuss the R equations.

C. Solution of R Equations

Depending upon the nature of the particular network at issue, the problem of solving the R equations can range in difficulty from being an entirely trivial problem to being one of considerable difficulty. In any specific analysis problem, additional information concerning the nature of the mapping f_R and the nature of the topological relationships, as expressed by the specific structure of the matrices F_{RC}, F_{RG}, etc., will be known. The difficulty of solving the R equations is determined primarily by these specifics. Given only the general form of the R equations, it is not even necessarily the case that a solution exists or is unique, let alone that we can *solve* the equations. In order to make definitive statements concerning the solution of the R equations, it is helpful, therefore, to consider the implications of certain typical additional assumptions.

In [30] Stern considers several specific situations, each involving certain assumptions concerning the nature of the mapping f_r and the network's topology. He derives conditions for each of these situations which ensure that a unique solution of the R equations exists, and proves that certain computational schemes can be used to solve the equations. In each situation Stern reduces the problem of solving the R equations to the problem of solving a certain equation which in general has the form $f(x, y, t) = \theta$ for the n-vector x as a function of the n-vector y, and t. In one situation, for example, this equation can in fact be written as $f(x, t) = y$, which Stern shows to have a unique solution for all y, t provided the mapping f is continuous in t and has continuous bounded partial derivatives in x, and provided the Jacobian matrix $J_f = [\partial f/\partial x]$ of the mapping f is uniformly positive definite in x. (A square matrix $M(x, t)$ is defined to be *uniformly positive definite* in x if it is symmetric and if, for each t, there exists $\mu > 0$ such that $u^T M u \geqq \mu u^T u$ for all u and all x.) Moreover, Stern shows that in this situation the solution can be obtained by a convergent iteration scheme (convergent for all sufficiently small $h > 0$) of the form $x_{k+1} = x_k - h[f(x_k, t) - y]$. In [52] Ohtsuki and Watanabe show that it is not necessary to require, in the definition of a uniformly positive definite matrix J_f, that J_f be *symmetric*, in order to prove that $f(x, t) = y$ has a unique solution.

In another situation Stern shows that it suffices to solve an equation of the form $x = f(y + Ax, t)$, where A is an $n \times n$ matrix. He proves that a unique solution of this equation exists for all y, t provided the mapping $f(u, t)$ is continuous in t and has continuous bounded partial derivatives in u, and provided the matrix $[I - (\partial f/\partial u)A]$ is uniformly Hadamard in u. (A square matrix $M(u, t)$ is defined to be *uniformly Hadamard* in u, if, for each t, there exists $\mu > 0$ such that $M_{ii} - \sum_{j \neq i} |M_{ij}| \geqq \mu$ for all i, and for all u.) In this situation Stern also shows that the solution can be computed by use of the convergent iteration scheme $x_{k+1} = x_k - \Delta[f(y + Ax_k, t) - x_k]$, where the diagonal matrix $\Delta > 0$ is chosen such that the right-hand side is a "contraction mapping."

In [53] Varaiya and Liu also consider the problem of solving the R equations. They assume that (a choice of a normal tree is possible such that) the mapping f_r provides no coupling between "link variables" and "tree-branch variables," and furthermore they assume that the mapping f_r is defined, in terms of two continuously differentiable mappings f and g, in the form

$$i_R = f(v_R, t)$$
$$v_G = g(i_G, t).$$

Then the problem of solving the R equations is equivalent to the problem of solving, for each t, a set of equations having the form

$$\begin{aligned} x + F_{RG} g(y) &= u \\ - F_{RG}{}^T f(x) + y &= w \end{aligned} \qquad (22)$$

for the vectors x, y in terms of the vectors u, w. Varaiya and Liu prove that there exists a unique solution of (22) (and hence of the R equations) provided the Jacobian matrices $J_f = [\partial f/\partial x]$ and $J_g = [\partial g/\partial y]$ are both positive semidefinite for all x, y, and provided either 1) one of the matrices J_f, J_g is symmetric and positive definite for all values of (respectively) x, y, or 2) one of the matrices J_f, J_g is diagonal for all values of (respectively) x, y. They also show that the solution can be obtained as the limit of the solution of a globally asymptotically stable differential equation. Certain results pertaining to the stability of the resulting normal-form equations (21) are also presented in [53].

D. Solution of Equations $f(x) = y$

As explained in Section IV-C, the central problem in obtaining the normal-form differential equations for networks of interconnected "like-kind" elements is often simply the problem of solving certain nonlinear equations which we write in the general form

$$f(x) = y \qquad (23)$$

where f is a nonlinear mapping of the n-dimensional Euclidean space E^n into itself.[35] A fundamental result which provides useful conditions for determining the existence of a unique solution of (23) is the following.

Theorem 27: The mapping f is a homeomorphism[36] of E^n onto E^n if and only if 1) f is a local homeomorphism,[35] and 2) $\|f(x)\| \to \infty$ as $\|x\| \to \infty$.

A useful corollary of Theorem 27 is the following.

Corollary: The continuously differentiable mapping f is a diffeomorphism[37] of E^n onto E^n if and only if 1) det $[\partial f/\partial x] \neq 0$ at each point in E^n, and 2) $\|f(x)\| \to \infty$ as $\|x\| \to \infty$.

The preceding corollary appears in a paper [54] by Palais. Holzmann and Liu [55] were apparently the first to recognize the relevance of Theorem 27 and its corollary to the type of network problem discussed in this section. In [56] Wu and Desoer present proofs of Theorem 27 and the corollary which use only elementary mathematical concepts. The theorem is

[35] For convenience of notation, we no longer explicitly display the independent variable t.

[36] A mapping f is defined to be a *homeomorphism of E^n onto E^n* if f is a continuous mapping of E^n onto E^n and if there exists a continuous inverse mapping f^{-1} of E^n onto E^n. A mapping f from E^n into E^n is defined to be a *local homeomorphism* if, for each $x \in E^n$, there exist open neighborhoods $U(x)$, $V(f(x))$ in E^n such that the restriction of f to U is a continuous mapping of U onto V with a continuous inverse mapping of V onto U.

[37] A mapping f is defined to be a *diffeomorphism of E^n onto E^n* if f is a continuously differentiable mapping of E^n onto E^n with a continuously differentiable inverse mapping of E^n onto E^n.

also to be found in [18, section 5.3], where an elementary proof is also given. Chua and Lam present a result in [57] which states that except for the case in which $n=2$ (where, in general, the following is known to be false), the preceding corollary remains valid upon replacing the term "diffeomorphism" by the term "homeomorphism," and upon replacing condition 1) by the condition: *det* $[\partial f/\partial x] \geq 0$ *at each point in* E^n, *with equality holding only on a set of isolated points.*

Sandberg has shown (see [47, appendix]) that if the mapping f of the preceding corollary is *twice* continuously differentiable, then conditions 1) and 2) of the corollary guarantee the existence of steepest descent as well as Newton-type algorithms which converge to the solution of the equation $f(x)=\theta$.

The following *global implicit function theorem* is due to Kuh and Hajj [58]. It can be proved by a rather straightforward application of the preceding corollary.

Theorem 28: Let f be a continuously differentiable mapping of $E^m \times E^n$ into E^m. Then there exists a unique continuously differentiable mapping g of $E^m \times E^n$ into E^m with the property that $g(y, u)=x$ for all $x, y \in E^m$, $u \in E^n$ satisfying $f(x, u)=y$, provided that 1) det $[\partial f/\partial x] \neq 0$ at each point in $E^m \times E^n$, and 2) for each $u \in E^n$, $\|f(x, u)\| \to \infty$ as $\|x\| \to \infty$.

It is often the case, in the use of Theorem 27 or its corollary to ascertain the existence of a unique solution of (23) for a specific class of mappings f arising from network problems, that the information concerning the specific structure of f is rather easily used to determine that condition 1) of, say, the corollary holds. The more difficult part of the problem is usually that of showing that the "growth condition" 2) holds. This is one reason why the following results, which provide *sufficient* conditions for determining that a unique solution of (23) exists for certain classes of mappings f, are of considerable interest.

In [59] Vehovec presents a theorem stating that a sufficient condition for a continuously differentiable mapping f, having a Jacobian matrix $[\partial f/\partial x]$ whose elements are bounded on E^n, to be a diffeomorphism of E^n onto E^n, is that there exists a real number $\epsilon > 0$ such that $|\det [\partial f/\partial x]| \geq \epsilon$ for all $x \in E^n$. The following theorem, proved by Sandberg [47], is a somewhat stronger result.

Theorem 29: If f is a continuously differentiable mapping of E^n into E^n with a Jacobian matrix $[\partial f/\partial x]$ whose elements are bounded on E^n, and if there exist real numbers $a>0$ and $b \geq 0$ such that[38] $|\det [\partial f/\partial x]| \geq 1/(a+b\|x\|)$ for all $x \in E^n$, then f is a homeomorphism of E^n onto E^n.

It is easy to construct counterexamples which show that without the hypothesis that the elements of the Jacobian matrix are bounded, Theorem 29 would be false. The following theorem, proved in [60] by Fujisawa and Kuh, does not require that the elements of the mapping's Jacobian matrix be bounded. It is required here, however, that the ratios of n different determinants be examined.

Theorem 30: Let f be a continuously differentiable mapping of E^n into E^n. For $k=1, \cdots, n$, let J_k denote the matrix consisting of the elements in the first k rows and first k columns[39] of the Jacobian matrix $[\partial f/\partial x]$. Suppose there exists a positive constant ϵ such that

$$|\det J_1| \geq \epsilon, \quad \left|\frac{\det J_2}{\det J_1}\right| \geq \epsilon, \cdots, \left|\frac{\det J_n}{\det J_{n-1}}\right| \geq \epsilon,$$

for all $x \in E^n$. Then there exists a unique solution of (23) for each $y \in E^n$.

Fujisawa and Kuh show [60] that the condition referred to in Section IV-C concerning the work of Stern and the work of Ohtsuki and Watanabe—that the Jacobian matrix $[\partial f/\partial x]$ be uniformly positive definite—is but a special case of the requirement that the matrices J_k satisfy the "ratio conditions" of Theorem 30. They also show that the results of Varaiya and Liu, discussed in Section IV-C, concerning the existence of a unique solution of (22), can be extended somewhat by the use of Theorem 30. In particular, since (22) has a unique solution for all u, w provided $x+F_{RG}g[w+F_{RG}{}^T f(x)]=u$ (respectively, $y-F_{RG}{}^T f[u-F_{RG}g(y)]=w$) has a unique solution for all u, w, and since, as shown in [60], a sufficient condition for a mapping of the form $z+\phi(z)$ to satisfy the ratio conditions of Theorem 30 is that the Jacobian matrix $J_\phi=[\partial\phi/\partial z]$ is a P_0 matrix (see Section III-C) for all z, it follows that (22) has a unique solution provided J_f (respectively, J_g) is a nonnegative diagonal matrix and provided $F_{RG}J_gF_{RG}{}^T \in P_0$ for all y (respectively, $F_{RG}{}^T J_f F_{RG} \in P_0$ for all x). This generalizes one of the aforementioned results of Varaiya and Liu in the sense that a sufficient condition for $F_{RG}J_gF_{RG}{}^T \in P_0$ or $F_{RG}{}^T J_g F_{RG} \in P_0$ is that, respectively, J_g or J_f is a positive semidefinite matrix.

Another facet of the problem of solving equations having the form (23) arising in the study of nonlinear networks is the solution of such equations when the mapping f of E^n into E^n is continuous and piecewise linear. As mentioned in Section II-A, Katzenelson [19] has proposed a solution algorithm for the equations of networks of two-terminal, continuous, strictly monotone-increasing, piecewise-linear resistors. Ohtsuki and Yoshida [61] and Fujisawa and Kuh [62] have considered the case of resistive multiterminal-element networks. A mapping f of E^n into E^n is defined by Fujisawa and Kuh to be piecewise linear if it is possible to partition E^n into a finite number of convex regions whose boundaries are hyperplanes, and if in each region the Jacobian matrix $[\partial f/\partial x]$ is constant and nonzero. They show that a continuous piecewise-linear mapping f of E^n into E^n is a homeomorphism of E^n onto E^n if and only if, for any unit vector $\alpha \in E^n$ and any $x \in E^n$, there exists exactly one nonzero vector β such that $f(x+\nu\beta)=f(x)+\nu\alpha$ for all sufficiently small positive ν. As an extension of Theorem 30 for continuous piecewise-linear mappings, they show that a sufficient condition for f to be a homeomorphism of E^n onto E^n is that the leading principal minors of all Jacobian matrices are nonzero and are all of the same sign. The problem of computing a solution of (23) for piecewise-linear mappings f is also discussed in [61] and [62]. In particular, Katzenelson's iterative computation scheme is considered, and particular attention is given in [62] to the issue of determining how the computation should proceed when an iterate falls on the boundary of one of the regions.

In [63] Chua considers two methods of computing the solutions of equations for networks containing two-terminal piecewise-linear resistors and controlled sources. One of his methods can accommodate networks having multiple solutions. This method is capable, in theory, of computing all solutions for the network, and appears to be capable of doing

[38] In the statement of this theorem appearing in [47], the absolute value symbol does not appear. It is clear, however, that the same proof given there remains valid in this slightly more general formulation.

[39] Since any mapping of E^n onto itself defined by a simple relabeling of the coordinates is a homeomorphism of E^n onto E^n (in fact, a diffeomorphism), it follows as a corollary of Theorem 30 that *any* nested sequence of n submatrices of the Jacobian matrix $[\partial f/\partial x]$ can in fact be used as the sequence $J_1, \cdots, J_n$ in Theorem 30. This observation is the crucial concept in the generalization of Theorem 30 that is also given in [60].

the computation rather efficiently in the case of networks having only a few nonlinear elements, even though each element is characterized by a piecewise-linear curve having many segments.

Other techniques of solving certain equations having the form (23) have been mentioned elsewhere in this paper, and the reader is also referred to [18] for a collection of useful computation techniques.

V. Concluding Remarks

In an attempt to provide a reasonably comprehensive survey of a number of related recent contributions to the field of nonlinear network theory, it has been necessary to restrict the scope of our coverage to a somewhat narrow domain. This has resulted in the exclusion of many notable contributions to the general field which, in the writer's view, are slightly tangential to the development of the central concepts presented here. Examples of work that might have been discussed in a more exhaustive treatment are the studies of the effects of parasitic elements on the equations of nonlinear networks [64], [65], and the development of techniques for the synthesis of nonlinear element characteristics [66].

Although a rather complete theory of the equations of nonlinear networks is beginning to take form, the attentive reader will surely have noticed many gaps in the theory and points of departure for further extensions. It is the sincere hope of this writer that readers will thereby be encouraged to join in the task of advancing the theory. One aspect of the theory that is in particular need of development concerns the equations of networks which have more than one solution. Although scattered results concerning rather special situations are to be found in the literature, very little has been accomplished in the way of developing generally applicable methods for determining the number of solutions possessed by the equations of a given network.

Finally, we wish to mention that some recent work [67], [68] has indicated that certain concepts taken from the field of differential geometry might profitably be applied to the study of nonlinear networks. It is expected that future work in this direction will be of considerable value.

VI. Acknowledgment

The author wishes to thank his colleagues L. O. Chua, E. S. Kuh, and I. W. Sandberg, who were kind enough to read, and prepare helpful comments on, an earlier version of this paper.

References

[1] W. J. Cunningham, *Introduction to Nonlinear Analysis.* New York: McGraw-Hill, 1958.
[2] S. Lefschetz, *Differential Equations: Geometric Theory*, 2nd ed. New York: Wiley.
[3] N. Minorsky, *Nonlinear Oscillations.* Princeton, N. J.: Van Nostrand, 1962.
[4] W. J. McCalla and D. O. Pederson, "Elements of computer-aided circuit analysis," *IEEE Trans. Circuit Theory*, vol. CT-18, pp. 14–26, Jan. 1971.
[5] G. C. Temes and D. A. Calahan, "Computer-aided network optimization: The state-of-the-art," *Proc. IEEE*, vol. 55, pp. 1832–1863, Nov. 1967.
[6] S. W. Director, "Survey of circuit-oriented optimization techniques," *IEEE Trans. Circuit Theory*, vol. CT-18, pp. 3–10, Jan. 1971.
[7] L. O. Chua and R. A. Rohrer, "On the dynamic equations of a class of nonlinear RLC networks," *IEEE Trans. Circuit Theory*, vol. CT-12, pp. 475–489, Dec. 1965.
[8] L. O. Chua, "Memristor—The missing circuit element," *IEEE Trans. Circuit Theory*, vol. CT-18, pp. 507–519, Sept. 1971.
[9] R. J. Duffin, "Nonlinear networks. IIa," *Bull. Amer. Math. Soc.*, vol. 53, pp. 963–971, Oct. 1947.
[10] G. Birkhoff and J. B. Diaz, "Non-linear network problems," *Quart. Appl. Math.*, vol. 13, pp. 431–443, Jan. 1956.
[11] G. J. Minty, "Monotone networks," *Proc. Royal Soc.* (London), Ser. A, vol. 257, pp. 194–212, Sept. 1960.
[12] ——, "Solving steady-state nonlinear networks of 'monotone' elements," *IRE Trans. Circuit Theory*, vol. CT-8, pp. 99–104, June 1961.
[13] C. A. Desoer and J. Katzenelson, "Nonlinear RLC networks," *Bell Syst. Tech. J.*, vol. 44, pp. 161–198, Jan. 1965.
[14] I. W. Sandberg, "Necessary and sufficient conditions for the global invertibility of certain nonlinear operators that arise in the analysis of networks," *IEEE Trans. Circuit Theory*, vol. CT-18, pp. 260–263, Mar. 1971.
[15] I. W. Sandberg and A. N. Willson, Jr., "Existence and uniqueness of solutions for the equations of nonlinear dc networks," *SIAM J. Appl. Math.*, vol. 22, pp. 173–186, Mar. 1972.
[16] A. N. Willson, Jr., "New theorems on the equations of nonlinear dc transistor networks," *Bell Syst. Tech. J.*, vol. 49, pp. 1713–1738, Oct. 1970.
[17] C. A. Desoer and F. F. Wu, "Nonlinear monotone networks," *SIAM J. Appl. Math.*, to be published.
[18] J. M. Ortega and W. C. Rheinboldt, *Iterative Solution of Nonlinear Equations in Several Variables.* New York: Academic Press, 1970.
[19] J. Katzenelson, "An algorithm for solving nonlinear resistor networks," *Bell Syst. Tech. J.*, vol. 44, pp. 1605–1620, Oct. 1965.
[20] J. Katzenelson and L. H. Seitelman, "An iterative method for solution of networks of nonlinear monotone resistors," *IEEE Trans. Circuit Theory*, vol. CT-13, pp. 317–323, Sept. 1966.
[21] A. N. Willson, Jr., "On the solutions of equations for nonlinear resistive networks," *Bell Syst. Tech. J.*, vol. 47, pp. 1755–1773, Oct. 1968.
[22] R. A. Rohrer, "Successive secants in the solution of nonlinear network equations," in *Mathematical Aspects of Electrical Network Analysis* (SIAM–AMS Proc., vol. 3). Providence, R.I.: Amer. Math. Soc., 1971.
[23] G. D. Hachtel, R. K. Brayton, and F. G. Gustavson, "The sparse tableau approach to network analysis and design," *IEEE Trans. Circuit Theory*, vol. CT-18, pp. 101–113, Jan. 1971.
[24] R. K. Brayton and J. K. Moser, "A theory of nonlinear networks" (Pts. I and II), *Quart. Appl. Math.*, vol. 22, pp. 1–33, 81–104, Apr. and July 1964.
[25] W. Millar, "Some general theorems for non-linear systems possessing resistance," *Phil. Mag.*, vol. 42, pp. 1150–1160, Oct. 1951.
[26] C. Cherry, "Some general theorems for non-linear systems possessing reactance," *Phil. Mag.*, vol. 42, pp. 1161–1177, Oct. 1951.
[27] J. Moser, "On nonoscillating networks," *Quart. Appl. Math.*, vol. 25, pp. 1–9, Apr. 1967.
[28] R. K. Brayton, "Necessary and sufficient conditions for bounded global stability of certain nonlinear systems," *Quart. Appl. Math.*, vol. 29, pp. 237–244, July 1971.
[29] T. E. Stern, *Theory of Nonlinear Networks and Systems.* Reading, Mass.: Addison-Wesley, 1965.
[30] ——, "On the equations of nonlinear networks," *IEEE Trans. Circuit Theory*, vol. CT-13, pp. 74–81, Mar. 1966.
[31] P. R. Bryant, "The order of complexity of electrical networks," *Proc. Inst. Elec. Eng.* (London), Monograph 335E, vol. 106C, pp. 174–188, June 1959.
[32] S. Seshu and M. B. Reed, *Linear Graphs and Electrical Networks.* Reading, Mass.: Addison-Wesley, 1961.
[33] I. W. Sandberg and A. N. Willson, Jr., "Some theorems on properties of dc equations of nonlinear networks," *Bell Syst. Tech. J.*, vol. 48, pp. 1–34, Jan. 1969.
[34] ——, "Some network-theoretic properties of nonlinear dc transistor networks," *Bell Syst. Tech. J.*, vol. 48, pp. 1293–1311, May–June 1969.
[35] M. Fiedler and V. Pták, "Some generalizations of positive definiteness and monotonicity," *Numer. Math.*, vol. 9, pp. 163–172, Dec. 1966.
[36] A. N. Willson, Jr., "A useful generalization of the P_0 matrix concept," *Numer. Math.*, vol. 17, pp. 62–70, Mar. 1971.
[37] ——, "A note on pairs of matrices and matrices of monotone kind," to be published in *SIAM J. Numer. Anal.*, vol. 10, Sept. 1973.
[38] I. W. Sandberg, "Theorems on the analysis of nonlinear transistor networks," *Bell Syst. Tech. J.*, vol. 49, pp. 95–114, Jan. 1970.
[39] ——, "Conditions for the existence of a global inverse of semiconductor-device nonlinear-network operators," *IEEE Trans. Circuit Theory*, vol. CT-19, pp. 34–36, Jan. 1972.
[40] B. Gopinath and D. Mitra, "When are transistors passive?" *Bell Syst. Tech. J.*, vol. 50, pp. 2835–2847, Oct. 1971.
[41] A. Gersho, "Solving nonlinear network equations using optimization techniques," *Bell Syst. Tech. J.*, vol. 48, pp. 3135–3138, Nov. 1969.
[42] A. Baranyi and R. Newcomb, "Multiple dc solution one-transistor circuits," *IEEE Trans. Circuit Theory*, vol. CT-19, pp. 626–628, Nov. 1972.
[43] I. W. Sandberg and A. N. Willson, Jr., "Existence of solutions for

the equations of transistor–resistor–voltage source networks," *IEEE Trans. Circuit Theory*, vol. CT-18, pp. 619–625, Nov. 1971.
[44] H. K. Gummel, "A charge-control transistor model for network analysis programs," *Proc. IEEE* (Lett.), vol. 56, p. 751, Apr. 1968.
[45] D. Koehler, "The charge-control concept in the form of equivalent circuits, representing a link between the classic large signal diode and transistor models," *Bell Syst. Tech. J.*, vol. 46, pp. 523–576, Mar. 1967.
[46] I. W. Sandberg, "Some theorems on the dynamic response of nonlinear transistor networks," *Bell Syst. Tech. J.*, vol. 48, pp. 35–54, Jan. 1969.
[47] ——, "Theorems on the computation of the transient response of nonlinear networks containing transistors and diodes," *Bell Syst. Tech. J.*, vol. 49, pp. 1739–1776, Oct. 1970.
[48] I. W. Sandberg and H. Shichman, "Numerical integration of systems of stiff nonlinear differential equations," *Bell Syst. Tech. J.*, vol. 47, pp. 511–527, Apr. 1968.
[49] C. A. Desoer and H. Haneda, "The measure of a matrix as a tool to analyze computer algorithms for circuit analysis," *IEEE Trans. Circuit Theory*, vol. CT-19, pp. 480–486, Sept. 1972.
[50] H. J. Carlin and D. C. Youla, "Network synthesis with negative resistors," *Proc. IRE*, vol. 49, pp. 907–920, May 1961.
[51] E. J. Purslow, "Solvability and analysis of linear active networks by use of the state equations," *IEEE Trans. Circuit Theory*, vol. CT-17, pp. 469–475, Nov. 1970.
[52] T. Ohtsuki and H. Watanabe, "State-variable analysis of RLC networks containing nonlinear coupling elements," *IEEE Trans. Circuit Theory*, vol. CT-16, pp. 26–38, Feb. 1969.
[53] P. P. Varaiya and R. Liu, "Normal form and stability of a class of coupled nonlinear networks," *IEEE Trans. Circuit Theory*, vol. CT-13, pp. 413–418, Dec. 1966.
[54] R. S. Palais, "Natural operations on differential forms," *Trans. Amer. Math. Soc.*, vol. 92, pp. 125–141, July 1959.
[55] C. A. Holzmann and R. Liu, "On the dynamical equations of nonlinear networks with n-coupled elements," in *Proc. 3rd Annu. Allerton Conf. Circuit and System Theory*, pp. 536–545, 1965.
[56] F. F. Wu and C. A. Desoer, "Global inverse function theorem," *IEEE Trans. Circuit Theory*, vol. CT-19, pp. 199–201, Mar. 1972.
[57] L. O. Chua and Y.-F. Lam, "Global homeomorphism of vector-valued functions," *J. Math. Anal. Appl.*, vol. 39, pp. 600–624, Sept. 1972.
[58] E. S. Kuh and I. N. Hajj, "Nonlinear circuit theory: Resistive networks," *Proc. IEEE*, vol. 59, pp. 340–355, Mar. 1971.
[59] M. Vehovec, "Simple criterion for the global regularity of vector-valued functions," *Electron. Lett.*, vol. 5, pp. 680–681, Dec. 1969.
[60] T. Fujisawa and E. S. Kuh, "Some results on existence and uniqueness of solutions of nonlinear networks," *IEEE Trans. Circuit Theory*, vol. CT-18, pp. 501–506, Sept. 1971.
[61] T. Ohtsuki and N. Yoshida, "DC analysis of nonlinear networks based on generalized piecewise-linear characterization," *IEEE Trans. Circuit Theory*, vol. CT-18, pp. 146–152, Jan. 1971.
[62] T. Fujisawa and E. S. Kuh, "Piecewise-linear theory of nonlinear networks," *SIAM J. Appl. Math.*, vol. 22, pp. 307–328, Mar. 1972.
[63] L. O. Chua, "Efficient computer algorithms for piecewise-linear analysis of resistive nonlinear networks," *IEEE Trans. Circuit Theory*, vol. CT-18, pp. 73–85, Jan. 1971.
[64] C. A. Desoer and M. J. Shensa, "Networks with very small and very large parasitics: Natural frequencies and stability," *Proc. IEEE*, vol. 58, pp. 1933–1938, Dec. 1970.
[65] L. O. Chua and G. R. Alexander, "The effects of parasitic reactances on nonlinear networks," *IEEE Trans. Circuit Theory*, vol. CT-18, pp. 520–532, Sept. 1971.
[66] L. O. Chua, "Linear transformation converter and its application to the synthesis of nonlinear networks," *IEEE Trans. Circuit Theory*, vol. CT-17, pp. 584–594, Nov. 1970.
[67] R. K. Brayton, "Nonlinear reciprocal networks," in *Mathematical Aspects of Electrical Network Analysis* (SIAM–AMS Proc., vol. 3). Providence, R. I.: Amer. Math. Soc., 1971.
[68] C. A. Desoer and F. F. Wu, "Trajectories of nonlinear *RLC* networks: A geometric approach," *IEEE Trans. Circuit Theory*, vol. CT-19, pp. 562–571, Nov. 1972.
[69] I. W. Sandberg, "On truncation techniques in the approximate analysis of periodically time-varying nonlinear networks," *IEEE Trans. Circuit Theory*, vol. CT-11, pp. 195–201, June 1964.
[70] E. A. Guillemin, *Introductory Circuit Theory*. New York: Wiley, 1953.

NONLINEAR NETWORKS. IIa

R. J. DUFFIN

A network is a collection of conducting wires and batteries arbitrarily interconnected. Kirchhoff [1, 6][1] gave a topological-type proof that the currents in the wires are uniquely determined for wires obeying Ohm's law. (Ohm's law is a linear law stating that current and potential drop are proportional.) If the wires obey a nonlinear law, more than one distribution of current is in general possible. For some engineering application a multiplicity of states is desirable as, for example, in counting circuits and oscillators. For other applications it is essential that only one state be possible. It is seldom intuitively evident, however, whether or not a given nonlinear network will have multiple states. Hence, it appears that a qualitative mathematical treatment of nonlinear networks should be of some practical importance [5].

A large class of conductors used in engineering are such that the current through the conductor and the potential drop across the conductor are nondecreasing functions of one another. Such conductors we shall term *quasi-linear*. Examples are: selenium, copper oxide, silicon carbide (thyrite), and thermionic rectifiers [9]. The main result of this note is the proof that ***a network of quasi-linear conductors has a stable state of currents, and this state is unique.***

A stable state of currents in a network must satisfy Kirchhoff's laws, which simply are statements of the conservation of electricity and the single valuedness of the potential function. Maxwell [8] discovered two concise ways of expressing these laws: the junction equations and the mesh equations. More or less as a digression we shall show that the mesh equations may be put in the same functional form as the junction equations if and only if the network is planar.

The formulation of mechanical analogs to electric networks has received considerable attention in the literature because of the transfer of techniques suggested by the analogy. We discuss here a different type of analog which we call an ***elastic network***. An elastic network is a collection of springs connected to each other at junction points. Forces are applied to the junction points to hold the network in a stretched condition. A tennis net is an example. Electric networks are analogous to one-dimensional elastic networks. Planar or spatial

Presented to the Society, September 15, 1945; received by the editors September 8, 1945, and, in revised form, March 18, 1947.

[1] Numbers in brackets refer to the references cited at the end of the paper.

elastic networks are more general than electric networks and may give rise to nonlinear problems even for springs obeying Hooke's law.

1. Maxwell's junction equations. In order to simplify notation and to avoid ambiguity the following trivial restrictions shall be imposed on the configuration of the networks considered. A *proper network* is a set of $n+1$ junction points $(0, 1, \cdots, n)$, $n \geqq 1$, connected by conducting wires such that each wire connects exactly two distinct junction points and no more than one wire directly connects the same two junction points.

The conductivity function $g(x)$ of a wire is the experimentally determined relation between the current w through the wire and the potential drop x across it, $w = g(x)$. If a wire obeys Ohm's law, $g(x) = kx$ where k is a positive constant, the conductivity.

The conductivity function of the wire connecting junction points i and j will be designated as $g_{ij}(x)$ and is such that if w_{ij} is the current flowing from i to j and v_i and v_j are the potentials of these junctions then

$$w_{ij} = g_{ij}(v_i - v_j). \tag{1}$$

By the conservation of electricity, $w_{ij} = -w_{ji}$; so $g_{ij}(x) = -g_{ji}(-x)$. If there is no wire directly connecting i and j then $g_{ij}(x) \equiv 0$.

Suppose currents u_i flow into the junction points from an external source. If a stable state of such a network exists, no electricity may pile up at a junction point so

$$\begin{aligned} u_0 &= \quad 0 \quad + g_{01}(v_0 - v_1) + g_{02}(v_0 - v_2) + \cdots + g_{0n}(v_0 - v_n), \\ u_1 &= g_{10}(v_1 - v_0) + \quad 0 \quad + g_{12}(v_1 - v_2) + \cdots + g_{1n}(v_1 - v_n), \\ &\cdots\cdots\cdots\cdots\cdots\cdots\cdots\cdots\cdots\cdots\cdots, \\ u_n &= g_{n0}(v_n - v_0) + g_{n1}(v_n - v_1) + g_{n2}(v_n - v_2) + \cdots + \quad 0. \end{aligned} \tag{2}$$

These are Maxwell's junction equations. Since $g_{ij}(x) + g_{ji}(-x) \equiv 0$ it follows that $\sum_{i=0}^{n} u_i \equiv 0$, and so at least one of the equations is dedependent. Moreover, because the potentials v_i enter only as paired differences, one of them may be given an arbitrary value.

If equations (2) are solvable for the potentials, then relations (1) determine the distribution of current in the wires when the currents entering the junction are given.

It should be noted that the definition given of the conductivity function is broad enough to allow batteries to be included in the network. Thus, if for one value of x the relation $g_{ij}(x) = 0$ holds, then $-x$ may be regarded as the value of the potential jump of a battery inserted in the wire connecting points i and j. Thus, for an isolated network excited by batteries equations (2) become

$$0 = \sum_{j=0}^{n} g_{ij}(v_i - v_j), \qquad i = 0, 1, \cdots, n. \tag{3}$$

In a non-proper network two or more wires may directly connect the same junction points. This case may be reduced to a proper network since the conductivity function for wires connected in parallel is the sum of the separate functions. The class of functions to be considered in what follows has the property that the sum of two functions also belongs to the class; hence, there is no essential restriction in considering only proper networks.

2. Existence and uniqueness conditions. The junction equations will now be considered abstractly. However, all variables and constants shall be assumed real.

THEOREM 1. *Equations* (3) *have at least one solution in the variables* v_i *if*:

(a) $g_{ij}(x) \equiv -g_{ji}(-x)$.

(b) *The functions* $g_{ij}(x)$ *are continuous for all* x.

(c) *For each pair* (i, j) *either* $g_{ij}(x) \equiv 0$ *or* $\int_0^x g_{ij}(t)dt \to +\infty$ *as* $x \to \pm\infty$.

PROOF. We shall say that a set of functions $\{g_{ij}(x)\}$, $i, j = 0, 1, \cdots, n$, satisfies the *chain condition* if for each integer i there exists an ordered sequence of integers $i, a, b, c, \cdots, e, f$ (dependent on i) such that no function of the sequence $g_{ia}, g_{ab}, g_{bc}, \cdots, g_{ef}, g_{f0}$ vanishes indentically. First, suppose that the chain condition is satisfied and define $G_{ij}(x) = \int_0^x g_{ij}(t)dt$. Then by conditions (a) and (c) it is clear that there is an unbounded increasing function $h(x)$ such that either $G_{ij}(x) \equiv 0$ or $G_{ij}(x) \geqq h(|x|)$. It may be assumed, moreover, that $h(x)$ is independent of i and j and that $h(0)$ is negative.

Let $v_0 = 0$ and define $\psi(v_1, v_2, \cdots, v_n) = \sum_{i=0}^{n}\sum_{j=0}^{n} G_{ij}(v_i - v_j)$. Consider the value of ψ on an n-dimensional cube whose corners are at the points $(\pm l, \pm l, \cdots, \pm l)$, $l > 0$. For some value of i it follows that $v_i = \pm l$. Hence, $(v_i - v_a) + (v_a - v_b) + \cdots + (v_e - v_f) + (v_f - v_0) = (v_i - v_0) = \pm l$. All integers of the chain sequence $i, a, b, \cdots, f, 0$ may be assumed distinct; so there are at most n adjacent pairs. Thus, for some adjacent pair, say (d, e), it follows that $|v_d - v_e| \geqq l/n$ and $G_{de}(v_d - v_e) \geqq h(l/n)$. An estimate of ψ may be made from this single term, $\psi \geqq h(l/n) + (n+1)^2 h(0)$. For l sufficiently large this shows that ψ is everywhere greater on the surface of the cube than it is at the center. In the interior, ψ must have a minimum point and at this point

$$0 = 2^{-1}\partial\psi/\partial v_k = \sum_{j=0}^{n} g_{kj}(v_k - v_j), \qquad k = 1, 2, \cdots, n.$$

These are precisely equations (3) except for the equation $k=0$, which is dependent on the others.

If the chain condition is not satisfied, equations (3) are split into two sets. The first set includes the equation 0 and all equations i such that a nonvanishing chain $g_{ia}, g_{ab}, \cdots, g_{ef}, g_{f0}$ does exist. If g_{ix} appears in the first set as a nonvanishing function so also does g_{xi}, because the chain $g_{xi}, g_{ia}, g_{ab}, \cdots, g_{ef}, g_{f0}$ is nonvanishing. This shows that the equations and variables appearing in the first set have the same labels. The second set, being what is left over, also has the same labels on the equations and the variables appearing. Clearly both sets have the same form as equations (3) but are of lower dimension than n. This splitting process is repeated on the second set, and so on. Finally, there results a division into a number of mutually exclusive sets, and each set either satisfies the chain condition or vanishes identically. This completes the proof.

In the network language the chain condition means that there is a chain of wires connecting every pair of junction points. In other words, the network is one piece.

THEOREM 2. *Equations* (2) *have a solution in the variables* v_i *for any choice of the constants* u_i *such that* $\sum_{i=0}^{n} u_i = 0$ *if*:

(a) $g_{ij}(x) = -g_{ji}(-x)$.

(b) *The function* $g_{ij}(x)$ *are continuous for all* x.

(c) *For each pair* (i, j) *either* $g_{ij}(x) \equiv 0$ *or* $g_{ij}(x) \to +\infty(-\infty)$ *as* $x \to +\infty(-\infty)$.

(d) *The chain condition is satisfied.*

PROOF. Define $\psi_1 = \sum_{i=0}^{n}\sum_{j=0}^{n} G_{ij}(v_i - v_j) - 2\sum_{i=0}^{n} u_i v_i$ $(v_0 = 0)$. Either $G_{ij}(x) \equiv 0$ or $G_{ij}(x) \geqq h(|x|)$, where now $h(x)$ is an increasing function such that $h(x)/x \to +\infty$ as $x \to +\infty$. Proceeding as in the proof of Theorem 1 gives the inequality

$$\psi_1 \geqq h(l/n) + (n+1)h(0) - l\sum_{i=1}^{n} |u_i| \to +\infty \qquad \text{as } l \to +\infty.$$

This shows that ψ_1 has a minimum at a point inside a large cube; hence at this point

$$0 = 2^{-1}\partial\psi_1/\partial v_k = \sum_{j=0}^{n} g_{kj}(v_k - v_j) - u_k, \qquad k = 1, 2, \cdots, n.$$

This completes the proof.

THEOREM 3. *Equations* (2) *may not have more than one solution in the variables* v_i, $i=1, 2, \cdots, n$, *with* $v_0=0$ *if*:

(a) $g_{ij}(x) \equiv -g_{ji}(-x)$.

(b) *For each pair* (i, j) *either* $g_{ij}(x) \equiv 0$ *or* $g_{ij}(x)$ *is an increasing function for all* x.

(c) *The chain condition is satisfied.*

PROOF. Let $u_i = \sum_{j=0}^{n} g_{ij}(v_i - v_j)$ and $u_i' = \sum_{j=0}^{n} g_{ij}(v_i' - v_j')$ then

$$\sum_{i=0}^{n} (u_i - u_i')(v_i - v_i')$$

$$= \sum_{i=0}^{n}\sum_{j=0}^{n} \{g_{ij}(v_i - v_j) - g_{ij}(v_i' - v_j')\}(v_i - v_i')$$

$$= \sum_{i=0}^{n}\sum_{j=0}^{n} \{g_{ji}(v_j - v_i) - g_{ji}(v_j' - v_i')\}(v_j - v_j')$$

$$= \sum_{i=0}^{n}\sum_{j=0}^{n} \{g_{ij}(v_i - v_j) - g_{ij}(v_i' - v_j')\}(v_j' - v_j)$$

$$= 2^{-1}\sum_{i=0}^{n}\sum_{j=0}^{n} \{g_{ij}(v_i - v_j) - g_{ij}(v_i' - v_j')\}\{(v_i - v_j) - (v_i' - v_j')\}.$$

If $u_i = u_i'$, the left side vanishes. Each term of the last series on the right side is non-negative, because g_{ij} is a nondecreasing function; hence,

$$\{g_{ij}(v_i - v_j) - g_{ij}(v_i' - v_j')\}\{(v_i - v_j) - (v_i' - v_j')\} = 0.$$

Under the assumption that g_{ij} is an increasing function, these factors must vanish together, so $\{(v_i - v_j) - (v_i' - v_j')\} = 0$. For a chain sequence,

$$\{(v_i - v_a) - (v_i' - v_a')\} + \{(v_a - v_b) - (v_a' - v_b')\} + \cdots + \{(v_f - v_0) - (v_f' - v_0')\} = v_i - v_i'.$$

Because each expression in braces vanishes, $v_i = v_i'$, and the proof is completed.

3. **Maxwell's mesh equations.** In actual engineering applications the junction equations are seldom employed but rather the alternative method of the mesh equations is used. A direct proof of the equivalence of the two methods does not seem to be available in the literature (we shall not consider this question either).

We now derive the mesh equations in the special case of an isolated network which can be diagramed in a plane without crossed wires. Let the regions (meshes) formed be numbered from 0 to m (including the exterior region). Let the resistance function be defined as the inverse of the conductivity function $g(x)$, and let $r_{ij}(w)$ be the sum of the resistance functions of the wires common to region i and region j. Let w_i be a cyclic current flowing around region i. The sign of w_i is determined by a clockwise convention for the interior regions and by a counterclockwise convention for the exterior regions. The potential is a single-valued function, so the net change in potential around the region i vanishes. Thus

$$0 = \sum_{j=0}^{m} r_{ij}(w_i - w_j), \qquad i = 0, 1, \cdots, m. \tag{4}$$

These are the mesh equations for a planar network. After the fictitious currents w_i are found, the actual currents are given immediately by the expressions $w_i - w_j$. Obviously, equations (4) are of the same form in the variables w_i as equations (3) are in the variables v_i.

It can be shown that any distribution of current which satisfies the conservation of electricity could be realized by assigning suitable values to a set of n of the cyclic currents w_i. Such a set is called a *complete set* [1, 6].

To write the mesh equations for a non-planar network, some complete set of cyclic currents is selected and a similar procedure is followed [3]. However, a theorem of MacLane [7] on "graphs" implies that some wire of a non-planar network must be traversed by at least three of the cyclic currents. This proves the following theorem.

THEOREM 4. *The mesh equations for an isolated network can be expressed in the same form as the junction equations if and only if the network is planar.*

To form mesh equations for a non-isolated network, it is simply necessary to add fixed non-cyclic currents which enter at one junction, follow some path, and leave at another junction.

4. Elastic networks. A proper elastic network is a set of $n+1$ junction points $(0, 1, \cdots, n)$, $n \geqq 1$, connected by springs such that each spring connects exactly two distinct junction points and no more than one spring directly connects the same two junction points. The junction points are the end points of the springs. The force function $f(x)$ of a spring is the force required to stretch a spring so that the distance between its end points is x. Let $f_{ij}(x) = f_{ji}(x)$ be the force function of the spring connecting junction points i and j. If $\mathbf{r}_i$ is the position vector of the ith junction point, the components of force which the junction i exerts on the spring (i, j) are

$$\overset{x}{w}_{ij} = f_{ij}(|r_i - r_j|)\frac{x_i - x_j}{|r_i - r_j|},$$

$$\overset{y}{w}_{ij} = f_{ij}(|r_i - r_j|)\frac{y_i - y_j}{|r_i - r_j|},$$

$$\overset{z}{w}_{ij} = f_{ij}(|r_i - r_j|)\frac{z_i - z_j}{|r_i - r_j|}.$$

It is convenient to express these relations in the single vector equation $w_{ij}=g_{ij}(r_i-r_j)$, where $g_{ij}(r)$ is a vector function whose magnitude is $|f_{ij}(|r|)|$ and whose direction in parallel to that of r. Clearly $g_{ij}(r) = -g_{ij}(-r)$ if $r \neq 0$, and to have this relation hold for $r=0$ it must be assumed that $f_{ij}(0)=0$.

If u_i is the total force applied to the ith junction from outside, the equations of equilibrium are

$$u_i = \sum_{j=0}^{n} g_{ij}(r_i - r_j), \qquad i = 0, 1, \cdots, n. \tag{5}$$

These are analogous to Maxwell's junction equations (2) but express the conservation of force rather than the conservation of electricity.

We shall say that a spring is ***quasi-linear*** if its force function $f(x)$ is a continuous, increasing, and unbounded function and if $f(0)=0$.

THEOREM 5. *If ν of the junctions ($\nu \geqq 1$) of a one piece network of quasi-linear springs are held fixed and given forces are applied to the other junctions then there is an equilibrium configuration, and this configuration is unique.*

PROOF. We may suppose that the junctions $(0, 1, \cdots, \nu-1)$ are held fixed and that $r_0=0$. Define $G_{ij}(x)=\int_0^x f_{ij}(t)dt$ and

$$\psi = \sum_{i=0}^{n}\sum_{j=0}^{n} G_{ij}(|r_i - r_j|) - 2\sum_{i=\nu}^{n} u_i \cdot r_i.$$

(Note that $G_{ij}(x)$ is the potential energy of the spring.) Then ψ is a function of $3(n-\nu+1)$ variables. As before, on a $3(n-\nu+1)$-dimensional cube of edge $2l$, one of the variables takes the value $\pm l$; for definiteness, suppose that it is the x coordinates of the radius vector r_i. Then, considering a chain sequence $i, a, b, \cdots, f, 0$,

$$(x_i - x_a) + (x_a - x_b) + \cdots + (x_e - x_f) + (x_f - x_0) = \pm l.$$

For some adjacent pair, say (d, e), we have $|r_d - r_e| \geqq |x_d - x_e| \geqq l/n$. Then, as before, it follows that ψ is everywhere greater on a sufficiently large cube than it is at the center. Setting the partial derivatives of ψ with respect to the coordinates equal to zero at a minimum point gives exactly equations (5) from ν to n. This proves the existence of a solution.

As in the proof of Theorem 3, one obtains the identity

$$\tag{6} \begin{aligned} &\sum_{i=0}^{n} (u_i - u_i')\cdot(r_i - r_i') \\ &= 2^{-1}\sum_{i=0}^{n}\sum_{j=0}^{n} \{g_{ij}(r_i - r_j) - g_{ij}(r_i' - r_j')\}\cdot\{(r_i - r_j) - (r_i' - r_j')\}. \end{aligned}$$

The expression on the left is clearly zero because either $(u_i - u_i') = o$ or $(r_i - r_i') = o$. Each term on the right is of the form $\{g_{ij}(p) - g_{ij}(p')\}\cdot\{p - p'\}$, and $g_{ij}(p) = cp$ where c is a positive scalar, $|p|c = f_{ij}(|p|)$. Thus,

$$\tag{7} \begin{aligned} \{g_{ij}(p) - g_{ij}(p')\}\cdot\{p - p'\} &= \{cp - c'p'\}\cdot\{p - p'\} \\ &= c|p|^2 - (c + c')p\cdot p' + c'|p'|^2 \\ &\geq c|p|^2 - (c + c')|p||p'| + c'|p'|^2 \\ &= \{c|p| - c'|p'|\}\{|p| - |p'|\} \\ &= \{f_{ij}(|p|) - f_{ij}(|p'|)\}\{|p| - |p'|\} \geq 0. \end{aligned}$$

Thus, each term on the right side of equation (6) is non-negative, and because the sum is zero each term must vanish. It follows that relation (7) is actually an equality and that $\{f_{ij}(|p|) - f_{ij}(|p'|)\}\{|p| - |p'|\} = 0$. Thus, $|p| = |p'|$ if $f_{ij}(x) \neq 0$. But the vectors p and p' are parallel otherwise, $p\cdot p' < |p||p'|$; $|p| \neq 0$ and relation (7) could not be an equality; hence $p = p'$ if $f_{ij}(x) \neq 0$.

The remainder of the uniqueness proof parallels the proof of Theorem 3. Note that a similar theorem is valid for an electric network of quasi-linear conductors.

The analogy between elastic and electric networks developed here suggests a method which, though somewhat abstract, is nevertheless quite practical. The idea is simply to set up the analog of Maxwell's mesh equations for elastic networks or even for more complicated problems in statics such as trusses. Without the analogy it would appear difficult to invent such a method, as the concept of using fictitious circulating forces in the meshes, for variables, is rather fantastic.

The writer has built a mechanical model of a quasi-linear electric network (a wheatstone bridge circuit) by constraining the junctions of an elastic network to move along vertical rods. Hanging weights on the junctions gives "applied currents." It is clear that, even with springs obeying Hooke's law, a nonlinear problem arises.

The discrete boundary value problems discussed here suggest analogous considerations for continuous media. In a later note such nonlinear Dirichlet problems will be treated.

References

1. W. Ahrens, Math. Ann. vol. 49 (1897) p. 311.

2. C. E. Clark, Bull. Amer. Math. Soc. vol. 47 (1941) p. 769.

3. W. Feussner, Annalen der Physik vol. 9 (1902) p. 1304, vol. 15 (1904) p. 385.

4. L. Graetz, *Handbuch der Electrizität und des Magnetismus*, Leipsig, 1912, vol. 2, p. 55.

5. G. P. Harnwell, *Principles of electricity and electromagnetism*, McGraw-Hill, 1938, pp. 138, 167.

6. G. Kirchhoff, Pogendorff Annalen vol. 72 (1847) p. 497; *Collected works*, p. 22.

7. S. MacLane, Fund. Math. vol. 28 (1936) p. 22; Duke Math. J. vol. 3 (1937) p. 466.

8. J. C. Maxwell, *A treatise on electricity and magnetism*, 3d ed., pp. 403–408.

9. Western Electric Company, *Varistors: their characteristics and uses*, reprints from Bell Laboratories Record.

Carnegie Institute of Technology

Nonlinear *RLC* Networks

By CHARLES A. DESOER* and JACOB KATZENELSON

(Manuscript received July 16, 1964)

This article considers the question of existence and uniqueness of the response of nonlinear time-varying RLC networks driven by independent voltage and current sources. It is proved that under certain conditions the response exists, is unique, and is defined by a set of ordinary differential equations satisfying some Lipschitz conditions. These conditions are of two types: (1) the network elements must have characteristics which satisfy suitable Lipschitz conditions and (2) the network must satisfy certain topological conditions. It should be noted that elements with nonmonotonic characteristics are allowed and that the element characteristics need to be continuous but not differentiable.

I. INTRODUCTION

This article considers the questions of existence and uniqueness of the response of nonlinear time-varying *RLC* networks. It is proved that under conditions imposed on the network elements and the network topology the response exists, is unique, and is defined by a set of ordinary differential equations satisfying some Lipschitz conditions. Thus, from the conditions imposed on the network it follows that the response of a network of this class is continuous whenever the sources applied to the network are continuous functions of time. In other words, for the class of networks under consideration, jump phenomena (of the type that occur in relaxation oscillators) are excluded.[1]

One motivation for studying this problem is the construction of nonlinear network models for physical devices and processes. The behavior of these models is often investigated by simulation studies performed on digital computers. It is clear that in order to get meaningful answers the existence and uniqueness of the model's response have to be assured. The simulation study requires the setting up of an appropriate set of differential equations and their integration. As networks of the class

* On leave of absence from the Department of Electrical Engineering, University of California, Berkeley, California.

considered here do not have jump phenomena, their equations can be integrated by some standard subroutines.

This article may be viewed as an extension to the *RLC* case of the articles by R. J. Duffin[2,3,4] and G. Birkhoff and J. B. Diaz[5] which were devoted to nonlinear resistive networks. We make heavy use of topological considerations and had to extend the techniques developed for the linear case by many people[6,7,8] P. R. Bryant in particular.[9,10] For further references see Ref. 16.

In the next section, we classify the network elements and exhibit the basic assumptions which hold for the remainder of the article. Some simple nonlinear circuits are also considered. Section III presents some standard reductions of sources and the definition of determinateness. Section IV deals with one-element-kind networks; its theorems are generalizations of Duffin's work and include some of his theorems as corollaries. The main result of the article is Theorem IV in Section V, which states the conditions under which a nonlinear *RLC* network is determinate. The conditions are of two types: (*i*) every characteristic has to satisfy suitable Lipschitz conditions and (*ii*) the network has to satisfy certain topological conditions. It has to be noted that, first, elements with nonmonotonic characteristics are allowed and, second, that each characteristic has to be representable by a function which is continuous but not necessarily differentiable. Finally, in Section VI we introduce a symbolic notation which allows us to write the differential equations for the nonlinear case in a manner which resembles that of the linear case.

II. ELEMENTS AND SIMPLE CIRCUITS

2.1 *Elements*

We assume that the reader has some familiarity with network theory, so that the basic concepts need not be defined.[11,12] A network may be considered as a set of points, called *nodes*, and a set of connecting *branches*. Each branch represents a physical *two-pole*. We assume that the voltage drop across each two-pole and the current through each two-pole can be measured at any time. The sign conventions are shown in Fig. 1: if, with respect to some arbitrary reference, the potential of A is larger (smaller) than the potential of B, then v is positive (negative); if the current actually flows in the direction of the arrow (opposite to the arrow) then i is positive (negative). Thus the product vi gives the power delivered by the outside world to the two-pole under consideration.

In most of the following, the branches consist of either a single source or a single *element* such as a resistor, an inductor or a capacitor. For each of these elements we shall adopt very broad definitions which we will narrow down in stating specific results. A two-pole is called a *resistor* if it is defined, for each t, by a set of ordered pairs (v, i), where v and i are finite numbers representing all the possible values, at time t, of the voltage and the current associated with the resistor. If the set of ordered pairs is independent of t, the resistor is said to be *time-invariant.* The set of (v, i) is called the *characteristic* of the resistor; for example, the characteristic of an *ideal diode* is given by

$$\{(0,i): 0 \leqq i < \infty\} \, U \, \{(v,0): -\infty < v \leqq 0\} \, .$$

A resistor is called *current-controlled* if, for all time and all currents in the interval $(-\infty, \infty)$, the voltage $v(t)$ is a function* of the current $i(t)$ and time t (we shall write $v(t) = \mathcal{R}(i(t),t)$), and the function $\mathcal{R}(i,t)$ is a piecewise-continuous function† of t for each fixed number i. A *voltage-*

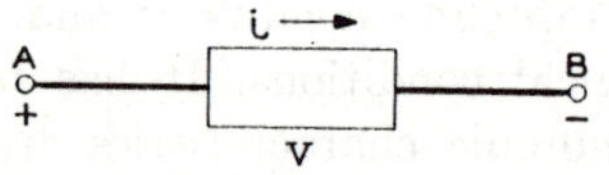

Fig. 1 — Sign conventions for two-pole.

controlled resistor is defined in the dual manner. For example, a voltage source is a current-controlled resistor and a current source is a voltage-controlled resistor. If a resistor is *current-controlled* and *time-invariant* then the characteristic can be represented by a function $v = \mathcal{R}(i)$. A resistor is called a *one-to-one resistor* if, for each t, the voltage is related to the current by a one-to-one mapping from $(-\infty, \infty)$ onto $(-\infty, \infty)$ which may depend on time.

A two-pole is called an *inductor* if it is defined, for each t, by a set of ordered pairs (φ,i) which represent the instantaneous flux and current associated with the inductor. The voltage across the inductor is given by $v = d\varphi/dt$. The *current-controlled* inductor, the *flux-controlled* inductor and the *one-to-one* inductor are defined as in the case of resistors. In the first two cases, if the elements are time-invariant, we shall write $\varphi = \mathcal{L}(i)$ and $i = \Gamma(\varphi)$, respectively.

* Unless specifically indicated, we follow modern usage: each function is single-valued; i.e., to each element of its domain it associates one and only one element of its range.

† A vector-valued function of time is said to be piecewise-continuous if it is continuous in every finite interval except at a finite number of points where it is discontinuous. At these points the function has a finite limit on the left as well as on the right.

A two-pole is called a *capacitor* if it is defined, for each t, by a set of ordered pairs (q,v) which represent the instantaneous charge and voltage associated with the capacitor. The current through the capacitor is given by $i = dq/dt$. The *charge-controlled capacitor*, the *voltage-controlled capacitor* and the *one-to-one capacitor* are defined as in the case of resistors. In the first two cases, if the elements are time invariant, we shall write $v = \mathfrak{D}(q)$ and $q = \mathfrak{C}(v)$, respectively.

Throughout the article we consider only elements whose characteristics can be represented, at all times, by a function defined on the interval $(-\infty, \infty)$. For example, Fig. 2(a), (b) and (c) represents the characteristics at time t of three time-varying resistors; we consider only resistors of the type shown in Fig. 2(a) and (b), since they are current- and

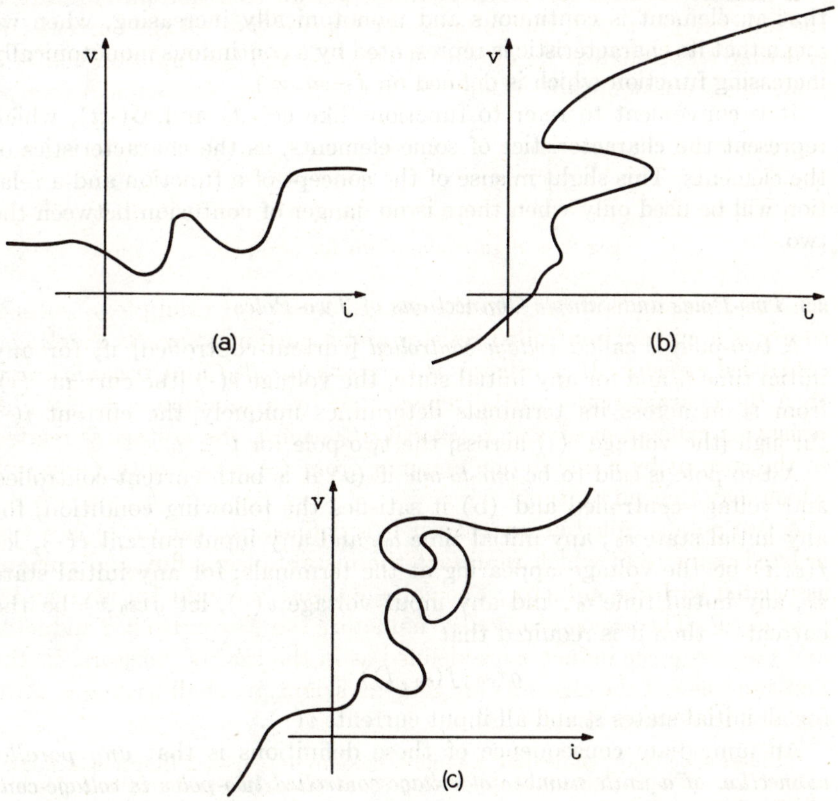

Fig. 2 — Characteristics at time t of three time-varying resistors: (a) and (b) are current- and voltage-controlled, respectively, while (c) is neither current- nor voltage-controlled.

voltage-controlled; these characteristics can be represented by

$$v(t) = \mathcal{R}(i(t),t), \quad \text{and} \quad i(t) = \mathcal{G}(v(t),t),$$

respectively. The characteristics of Fig. 2(c) cannot be represented in this way, and resistors of this type will not be considered.

Throughout the paper, whenever time-varying network elements are considered, it is assumed that the functions $\mathcal{R}(\cdot,t)$, $\mathcal{G}(\cdot,t)$, $\mathcal{L}(\cdot,t)$, $\Gamma(\cdot,t)$, $\mathfrak{D}(\cdot,t)$, $\mathcal{C}(\cdot,t)$ are piecewise-continuous functions of t for all fixed values of their first argument.

In addition to resistors, capacitors and inductors, our networks include voltage and current sources. Throughout this article we shall assume that the voltages of the voltage sources and currents of current sources are regulated functions of time.* For convenience we shall say that an element is continuous and monotonically increasing, when we mean that its characteristic is represented by a continuous monotonically increasing function which is defined on $(-\infty,\infty)$.

It is convenient to refer to functions like $\mathcal{R}(\cdot,t)$ and $\mathfrak{D}(\cdot,t)$, which represent the characteristics of some elements, as the characteristics of the elements. This slight misuse of the concept of a function and a relation will be used only when there is no danger of confusion between the two.

2.2 *Two-Poles and Simple Connections of Two-Poles*

A two-pole is called *voltage-controlled* [current-controlled] if, for any initial time t_0 and for any initial state, the voltage $v(\cdot)$ [the current $i(\cdot)$] from t_0 on across its terminals determines uniquely the current $i(\cdot)$ through [the voltage $v(\cdot)$ across] the two-pole for $t \geqq t_0$.

A two-pole is said to be *one-to-one* if (a) it is both current-controlled and voltage-controlled and (b) it satisfies the following condition: for any initial state s_0, any initial time t_0, and any input current $i(\cdot)$, let $f(s_0,i)$ be the voltage appearing at the terminals; for any initial state s_0, any initial time t_0, and any input voltage $v(\cdot)$, let $g(s_0,v)$ be the current — then it is required that

$$g(s_0, f(s_0, i)) = i$$

for all initial states s_0 and all input currents $i(\cdot)$.

An immediate consequence of these definitions is that *any parallel connection of a finite number of voltage-controlled two-poles is voltage-controlled.*

* A vector-valued function of time is said to be regulated when, for all t, it has a limit on the left as well as a limit on the right.[13] A step function and a rectangular wave are regulated functions.

Consider the case where there are only two two-poles in the parallel connection. Let them be characterized by the functions

$$i_k = F_k(v, s_k(t_0)), \qquad (k = 1,2),$$

where v is the voltage across the parallel connection, i_k is the current through the kth two-pole, $s_k(t_0)$ is the state of the kth two-pole at time t_0. The v_k and i_k are real-valued functions defined on $[t_0, \infty)$. Kirchoff's current law implies that the current i through the parallel connection is given by

$$F_1(v, s_1(t_0)) + F_2(v, s_2(t_0));$$

hence, for fixed $(s_1(t_0), s_2(t_0))$, i is a function of v. This argument obviously extends, by induction, to the case where there are a finite number of two-poles.

A dual argument would show that *any series connection of a finite number of current-controlled two-poles is current-controlled.*

A parallel connection of current-controlled two-poles is not necessarily current-controlled. Refer to Fig. 3, which shows the characteristics of two current-controlled resistors. The dashed line shows the characteristic of the parallel connection: depending on the operating point there may be three distinct values of the voltage for the same input current. Dually,

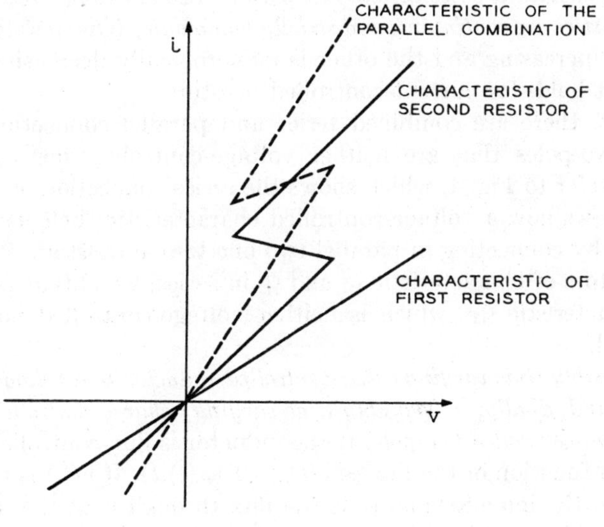

Fig. 3 — Parallel connection of two current-controlled resistors.

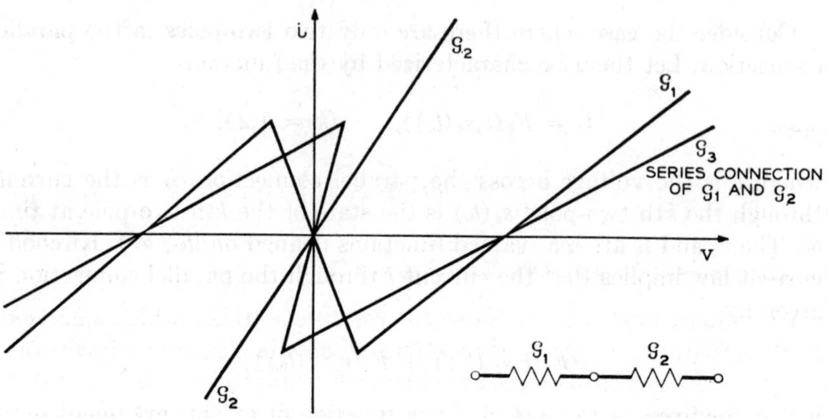

Fig. 4 — Series connection of two voltage-controlled resistors.

a series connection of voltage-controlled two-poles is not necessarily voltage-controlled.

To assume that each two-pole is one-to-one is not enough to cause both arbitrary parallel connections and arbitrary series connections to be one-to-one. Indeed, the well known characterization of continuous functions of bounded variation[14] implies that *any voltage-controlled resistor characteristic,* $i(t) = \mathcal{R}(v(t),t)$, *that is continuous and of bounded variation in* v *can be obtained by connecting in parallel two one-to-one resistors whose characteristics are continuous and strictly monotonic.* (One resistor is monotonically increasing and the other is monotonically decreasing.) A dual statement holds for current-controlled resistors.

In fact, there are combined series and parallel connections of one-to-one two-poles that are neither voltage-controlled nor current-controlled. Refer to Fig. 4, which shows the series connection of $\mathcal{G}_1$ and $\mathcal{G}_2$. Fig. 5 shows how a voltage-controlled characteristic such as $\mathcal{G}_1$ may be obtained by connecting in parallel two one-to-one resistors. Putting the two resistors of characteristic $\mathcal{G}_1$ and $\mathcal{G}_2$ in series, we obtain (see Fig. 4) the characteristic $\mathcal{G}_3$, which is neither voltage-controlled nor current-controlled.

A (possibly time-varying) flux-controlled inductor is a voltage-controlled two-pole and, dually, a (possibly time-varying) charge-controlled capacitor is a current-controlled two-pole. If the inductor is flux-controlled, the current i is a function of the flux φ: $i(t) = \Gamma(\varphi(t),t)$. If $v(\cdot)$ is the voltage applied to the inductor and φ_0 is the flux through it at the initial time t_0, then by Lenz's law

$$\varphi(t) = \int_{t_0}^{t} v(t')dt' + \varphi_0$$

hence,

$$i(t) = \Gamma\left(\int_{t_0}^{t} v(t')dt' + \varphi_0 , t\right) \quad \text{for all} \quad t.$$

2.3 *Examples of One-To-One Two-Poles*

We present here a set of sufficient conditions under which some elementary parallel or series connections of circuit elements constitute a two-pole which is either current-controlled, voltage-controlled or one-to-one. As the reader expects, quite specific assumptions will have to be

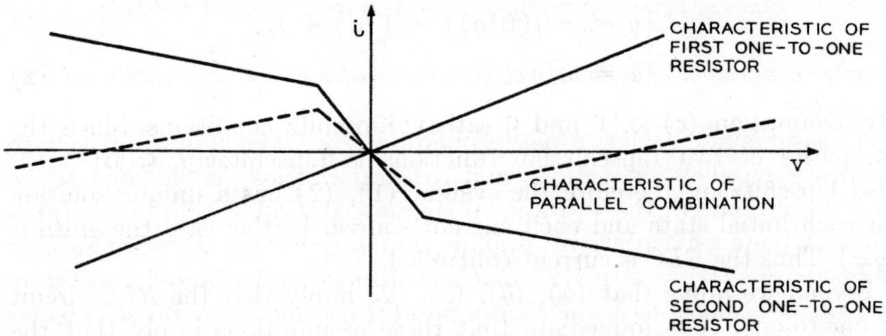

Fig. 5— Parallel connection of two one-to-one resistors.

made on the characteristics of the elements in order for the circuit to be a one-to-one two-pole.

The elements that we are going to consider are capacitors, resistors and inductors. Let us rank order these elements together with voltage sources and current sources in the following way: E,C,R,L,J. We shall say that a resistor is higher in rank than an inductor or a current source but lower in rank than a capacitor or a voltage source.

Until the end of this section, to simplify the discussion and without loss of generality, elements are assumed to be time-invariant.

Theorem: Consider the following circuits: the parallel RC, the parallel RL, the parallel LC and the parallel RLC circuit.

(A) If (a) the highest-ranked element is current-controlled (charge-controlled in the case of the capacitor),

(b) all other elements are voltage-controlled (flux-controlled in the case of the inductor), and

(*c*) *the characteristics of all elements satisfy a Lipschitz condition according to Table I,*

then the parallel circuit is current-controlled.

(*B*) *If, in addition* (*d*) *the highest ranked element is one-to-one, then each parallel circuit is one-to-one.*

Proof: We shall consider only the *RLC* circuit, since the proofs of the simpler cases follow in a similar way.

First let us prove that (*a*), (*b*), and (*c*) imply that the circuit is current-controlled. Let i_s be the source current. Then with the usual notation

$$i_s = \dot{q} + \mathcal{G}(v) + \Gamma(\varphi)$$

or, equivalently, using the fact that the capacitor is charge-controlled

$$\begin{cases} \dot{q} = -\mathcal{G}(\mathfrak{D}(q)) - \Gamma(\varphi) + i_s . & (1) \\ \dot{\varphi} = \mathfrak{D}(q) & (2) \end{cases}$$

By assumption (*c*) $\mathfrak{D}$, Γ and $\mathcal{G}$ satisfy Lipschitz conditions. Since the composite of two Lipschitzian functions is Lipschitzian, $\mathcal{G}(\mathfrak{D}(\cdot))$ is also Lipschitzian; therefore the system (1), (2) has a unique solution for each initial state and each current source. In this case the state is (q,φ). Thus the *RLC* is current-controlled.

Second we prove that (*a*), (*b*), (*c*), (*d*) imply that the *RLC* circuit is one-to-one. It is immediate that these assumptions imply that the *RLC* circuit is voltage-controlled. It remains to show that it is one-to-one.

Call $q_1(\cdot)$, $\varphi_1(\cdot)$, and $v_1(\cdot)$ the charge, flux and voltage resulting from the initial state (q_0, φ_0) at time t_0 and the input current i_s. The functions $q_1(\cdot)$ and $\varphi_1(\cdot)$ are the corresponding solutions of (1) and (2); $v_1(t) = \dot{\varphi}_1(t) = \mathfrak{D}(q_1(t))$. We have to show that, starting from the same initial state (q_0, φ_0) at time t_0, the input current resulting from the applied voltage v_1 is precisely i_s.

Let q_2, φ_2 and i_2 be the resulting charge, flux and input current. It is immediate that $v_1(t) = \dot{\varphi}_2(t) = \mathfrak{D}(q_2(t))$. Since $\varphi_2(t_0) = \varphi_0$, we have

TABLE I

Circuit	Highest-Ranked Element	Characteristics That Satisfy Lipschitz Conditions
RL	*R*	$\mathfrak{R}(i)$, $\Gamma(\varphi)$
RC	*C*	$\mathfrak{D}(q)$, $\mathcal{G}(v)$
LC	*C*	$\mathfrak{D}(q)$, $\Gamma(\varphi)$
RLC	*C*	$\mathfrak{D}(q)$, $\mathcal{G}(v)$, $\Gamma(\varphi)$

$\varphi_2 = \varphi_1$. Since the capacitor is one-to-one, q_2 is uniquely defined by the relation above in terms of v_1, hence $q_2 = q_1$. Finally, by Kirchhoff's current law

$$i_2 = \dot{q}_2 + \mathcal{G}(v_2) + \Gamma(\varphi_2)$$
$$= \dot{q}_1 + \mathcal{G}(\mathfrak{D}(q_1)) + \Gamma(\varphi_1).$$

The last expression is precisely i_s by (1). Therefore $i_2 = i_s$. This concludes the proof that the parallel *RLC* circuit is a one-to-one two-pole. The dual case is covered by the following

Theorem: Consider the following circuits: the series RL, the series RC, the series LC and the series RLC circuit.

(A) If

(a) the lowest-ranked element is voltage-controlled (flux-controlled for inductor),

(b) all other elements are current-controlled (charge-controlled for capacitors), and

(c) the characteristics of all elements satisfy Lipschitz conditions according to Table II,

then the series circuit is voltage-controlled.

(B) If in addition (d) the lowest-ranked element is one-to-one,

then each series circuit is one-to-one.

The proof is similar to that of the previous theorem and is therefore omitted.

III. REDUCTION OF THE NETWORK

Throughout the article we consider networks consisting of nonlinear time-varying resistors, capacitors, inductors (without mutual inductance) and independent sources. We shall label by $\mathfrak{N}$ the network under consideration. Usually, we consider each element and each source as constituting a branch of $\mathfrak{N}$. We denote by E,C,R,L,J the set of branches of $\mathfrak{N}$ which are voltage sources, capacitors, resistors, inductors and cur-

TABLE II

Circuit	Lowest-Ranked Element	Characteristics That Satisfy Lipschitz Conditions
RL	*L*	$\mathfrak{R}(i)$, $\Gamma(\varphi)$
RC	*R*	$\mathfrak{D}(q)$, $\mathcal{G}(v)$
LC	*L*	$\mathfrak{D}(q)$, $\Gamma(\varphi)$
RLC	*L*	$\mathfrak{D}(q)$, $\mathfrak{R}(i)$, $\Gamma(\varphi)$

rent sources, respectively. In our discussion, certain networks derived from $\mathfrak{N}$ will play an important role. In order to refer to them conveniently, let us define the following notations: Let "A" be a subset of the set of branches of $\mathfrak{N}$. Let us define[9]

$\mathfrak{N}_A$ to be the network derived from $\mathfrak{N}$ by removing all branches except the ones which are members of A,

$\mathfrak{N}_{(A)}$ to be the network derived from $\mathfrak{N}$ by replacing branches of set A by a short circuit, and

$\mathfrak{N}_{(A)^*}$ to be the network derived from $\mathfrak{N}$ by removing the branches of set A.

We shall use these notations as well as combinations of them. For example, $\mathfrak{N}_{(E)C}$ is the network derived from $\mathfrak{N}$ by first replacing branches of set E (the voltage sources) by short circuits and then removing all elements which do not belong to set C. Similarly, $\mathfrak{N}_{(E)(J)^*}$ is the network derived from $\mathfrak{N}$ by shorting all voltage sources and removing all current sources.

$S*\mathfrak{N}$ is defined to be the network derived from $\mathfrak{N}$ by separating it into the maximum number of separable subnetworks.

Throughout the article we assume that, first, for any cut set of current sources only, the source currents satisfy Kirchoff's current law, and, second, for any loop of voltage sources only, the source voltages satisfy Kirchoff's voltage law.

Without loss of generality we consider networks that are connected and nonseparable. This assumption does not exclude the possibility that $\mathfrak{N}_{(E)(J)^*}$ be both unconnected and/or separable. In the following we shall prove that without a loss of generality we can restrict the discussion to a network $\mathfrak{N}$ such that $\mathfrak{N}_{(E)(J)^*}$ is both connected and nonseparable. The proof consists of an algorithm which changes the configuration and reduces $\mathfrak{N}$ into a network $\mathfrak{N}'$ which has the following properties:

(*i*) $\mathfrak{N}'$ consists of connected subgraphs, $\mathfrak{N}_i'$, such that for each one of them, $\mathfrak{N}'_{i(E)(J)^*}$ is connected and nonseparable.

(*ii*) For all branches of $\mathfrak{N}$ and $\mathfrak{N}'$ which are not sources, any set of branch currents and voltages is a solution of $\mathfrak{N}$ if and only if it is also a solution of $\mathfrak{N}'$ (when the latter is driven by the corresponding sources).

(*iii*) Current sources of $\mathfrak{N}'$ are linearly related to the current sources of $\mathfrak{N}$. The same is true for voltage sources.

The step-by-step reduction of the network $\mathfrak{N}$ to $\mathfrak{N}'$ is done as follows.

(1) From each loop which consists of voltage sources only, remove one voltage source.

(2) In each cut set which consists of current sources only, replace one of the current sources by a short circuit.

The resulting network is connected; it has a tree which includes *all* the voltage sources as tree branches and *all* the current sources as links.

(3) Each current source J whose fundamental loop includes more that one tree branch is removed from the network and is replaced by a set of current sources identical to J, each one placed in parallel with a tree branch of the fundamental loop.

(4) All current sources that are in parallel with voltage sources are removed.

(5) Any parallel connection of current sources is replaced by one equivalent current source.

(6) Consider each fundamental cut set defined by a voltage source. For each one of them insert in every link a voltage source equal to that which is in the tree branch and, finally, short circuit the tree branch voltage source.

(7) In each link, replace any series connection of voltage sources by one equivalent voltage source.

(8) Separate the network into the maximum number of connected, nonseparable subgraphs.

The resulting network is called $\mathfrak{N}'$. Property (*ii*) follows from the fact that all the steps of the above algorithm do not change the source contribution to any of the fundamental loop equations or the cut set equations. Property (*i*) follows from the fact that all current sources are links of $\mathfrak{N}'$ and all voltage sources are in a link. Property (*iii*) follows from steps (5) and (7). Finally, observe that $S*\mathfrak{N}_{(E)(J)*}$ is identical with $\mathfrak{N}'_{(E)(J)*}$.

It is well known that the state of the network is completely determined by all the voltages, fluxes, charges, and currents in the branches of the network. In the case of linear networks it is well known that certain proper subsets of these variables may be chosen as the state. For special classes of nonlinear *RLC* networks similar subsets will be indicated in the sequel.

We call a *solution* of an *RLC* network any set of voltages and currents of resistors, charges and voltages of capacitors, fluxes and currents of inductors which satisfy the Kirchhoff's laws and the branch characteristics. A network $\mathfrak{N}$ is said to be *determinate* if for any value of the initial state $\mathbf{s}_0$, given at any initial time t_0, and for any value of the sources $\mathbf{E}(\cdot)$, $\mathbf{J}(\cdot)$, there exists one and only one solution for $t \geqq t_0$ on some nonvanishing interval $[t_0, t_\alpha)$.

In the following section we shall describe a broad class of nonlinear *RLC* networks which are determinate.

IV. ONE-ELEMENT-KIND NETWORK

The purpose of this section is to establish a set of sufficient conditions under which a nonlinear (possibly time-varying) resistor network driven by a set of independent current sources and voltage sources has, for all possible inputs, one and only one set of branch voltages and branch currents that satisfy Kirchhoff's laws. Conditions under which the solution satisfies a Lipschitz condition with respect to the sources are also given.

The analysis of nonlinear resistor networks is almost identical with that of nonlinear capacitor networks or nonlinear inductor networks. Since the nonlinear resistors are the most flexible elements, we shall develop our analysis in terms of resistor networks.

Let us start by making three preliminary remarks:

(*i*) Given a resistor network together with an arbitrary distribution of current sources, it is always legitimate to assume that there are no cut sets of current sources only. (Dually, that there are no loops of voltage sources only.)

(*ii*) Any voltage source in series with a resistor may always be absorbed into a suitably redefined branch characteristic. Refer to Fig. 6, where v_1 and v_2 are the node voltages of nodes 1 and 2 referred to the same datum. Let the current through the resistor be given by its characteristic $g(v,t)$; since $g(v,t) = g(v_1 - v_2 - e,t)$ and since $e(\cdot)$ is a known function of time, we may introduce a new branch characteristic $g_{12}(\cdot,\cdot)$ specified at each instant of time by

$$g_{12}(v_1 - v_2,t) \triangleq g(v_1 - v_2 - e(t),t).$$

In other words, the voltage source e has been absorbed into the time dependence of g_{12}. A similar reasoning applies to a current-controlled resistor in series with a voltage source.

The dual case can be taken care of in the same manner: in this case, a current source which is in parallel with either a voltage-controlled or a current-controlled resistor can be absorbed into the branch.

Thus, without loss of generality, a network of nonlinear resistors and sources can be thought of as a network of nonlinear time-varying resist-

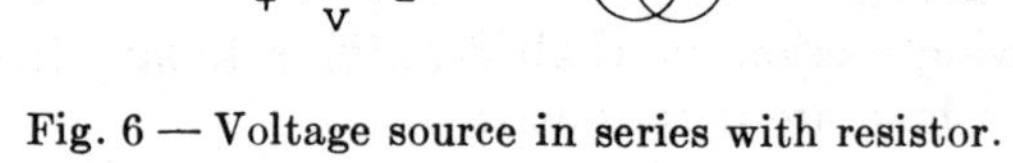

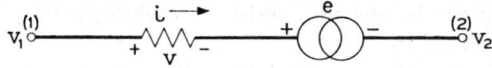

Fig. 6 — Voltage source in series with resistor.

ors with the understanding that the sources have been absorbed in the branch characteristics.

(*iii*) Thus when, as in Theorems I and II below, we consider a network of nonlinear time-varying resistors, we include the case of a network made up of time-varying resistors and of independent sources. There is no loss of generality in considering only connected networks, since it amounts to considering successively each separate part of an unconnected network.

We turn now to the statement of the main theorems.

Theorem I (Existence and Uniqueness): Consider a connected nonseparable network $\mathfrak{N}$ *of nonlinear (possibly time-varying) resistors. In case the resistor joining node* α *to node* β *is voltage-controlled, its characteristic is defined by the function* $g_{\alpha\beta}(\cdot,\cdot)$ *such that* $g_{\alpha\beta}(v_\alpha(t) - v_\beta(t),t)$ *is the current flowing through it at time t from node* α *to node* β*; here* v_α *and* v_β *are the node-to-datum voltages of nodes* α *and* β*. Similarly, if this resistor is current-controlled, its characteristic is defined by the function* $r_{\alpha\beta}(\cdot,\cdot)$ *such that* $r_{\alpha\beta}(i_{\alpha\beta}(t),t)$ *is the voltage difference between node* α *and node* β *at time t; here* $i_{\alpha\beta}$ *is the current through the resistor measured positively if it flows from* α *to* β.

If

(*a*) *there exists a tree* $\mathfrak{T}$ *such that all its tree branches are current-controlled and all its links are voltage-controlled,*

(*b*) *for all* α,β*, all t and all x in* $(-\infty,\infty)$

$$g_{\alpha\beta}(x,t) = -g_{\beta\alpha}(-x,t) \text{ if } (\alpha,\beta) \text{ is a link}$$

$$r_{\alpha\beta}(x,t) = -r_{\alpha\beta}(-x,t) \text{ if } (\alpha,\beta) \text{ is a tree branch}$$

(*c*) *for all links and all t,* $g_{\alpha\beta}(\cdot,t)$ *is a monotonically (not necessarily strictly) increasing continuous function defined on* $(-\infty,\infty)$*, and for all tree branches and all t,* $r_{\alpha\beta}(\cdot,t)$ *is a monotonically (not necessarily strictly) increasing continuous function defined on* $(-\infty,\infty)$.

Then,

for all current-sources i^s *connected between any pair of nodes and for all voltage sources* e^s *connected in series with network branches there exists one and only one set of branch voltages and branch currents that satisfy the Krichhoff laws and the branch characteristics.*

The conclusion of Theorem I can also be stated as follows: any network $\mathfrak{N}'$, formed from $\mathfrak{N}$ by inserting any set of voltage sources in series with any branch and any set of current sources between any node pair, is *determinate*.

Assumption (*b*) is a consequence of the physical meaning of the functions g and r and of the sign conventions: from a physical point of view they do not restrict the nonlinear resistors in any way. The two corollaries that follow are special cases of Theorem I. Corollary I is an extension of Theorems 2 and 3 of Duffin,[3] and is implied by his 1948 paper.[4] Such an extension has been pointed out by I.W. Sandberg.[15]

Corollary 1: Consider a connected network of nonlinear voltage-controlled (possibly time-varying) resistors.

If

(a) for all branches and all t, $g_{\alpha\beta}(\cdot,t)$ is a monotonically (not necessarily strictly) increasing, continuous function defined on $(-\infty,\infty)$, and

(b) there exists a tree $\mathfrak{J}$ such that all its branches have $g_{\alpha\beta}$'s which are, for all t, monotonically increasing one-to-one mappings of $(-\infty,\infty)$ onto $(-\infty,\infty)$,

then the conclusion of Theorem I holds.

Proof: The conclusion follows directly from Theorem I since the tree branches have $g_{\alpha\beta}$'s that are, for all t, monotonically increasing one-to-one mappings of $(-\infty,\infty)$ onto $(-\infty,\infty)$; hence the tree branches are also current-controlled resistors satisfying assumption (c) of Theorem I.

Corollary 2: Consider a connected network of nonlinear, current-controlled (possibly time-varying) resistors.

If

(a) for all branches and all t, $r_{\alpha\beta}(\cdot,t)$ is a monotonically (not necessarily strictly) increasing, continuous function defined on $(-\infty,\infty)$, and

(b) there exists a tree $\mathfrak{J}$ such that its links have $r_{\beta\alpha}$'s which are, for all t, monotonically increasing one-to-one mappings of $(-\infty,\infty)$ onto $(-\infty,\infty)$, *then* the conclusion of Theorem I holds.

We consider now the extension of the Thévenin and Norton equivalent circuits to nonlinear resistive networks. If we pick an arbitrary node pair of such a network $\mathfrak{N}$, we may regard these nodes as the terminals of a two-terminal network: we shall call the characteristic of this two-terminal network *the input characteristic of* $\mathfrak{N}$ *at these two nodes*. Dually, if we pick a branch and insert two terminals in series with it, we obtain a two-terminal network: we shall call the characteristic of this two-terminal network *the branch-input characteristic of* $\mathfrak{N}$.

Theorem II (Thévenin and Norton equivalent circuits): Consider a network $\mathfrak{N}$ *satisfying the requirements of Theorem I together with the same kind of source distribution.*

Then

(a) the input characteristic of $\mathfrak{N}$ at any node pair is that of a current-controlled resistor whose characteristic is a continuous, monotonically increasing function defined on $(-\infty, \infty)$. This characteristic may be represented by the Thévenin equivalent circuit of Fig. 7(a): a series combination of a voltage source and a monotonically increasing current-controlled resistor whose characteristic passes through the origin.

(b) The branch-input characteristic of $\mathfrak{N}$ at any branch is that of a voltage-controlled resistor whose characteristic is a continuous, monotonically increasing function defined on $(-\infty, \infty)$. This characteristic may be represented by the Norton equivalent circuit of Fig. 7(b): a parallel combination of a current source and a monotonically increasing voltage-controlled resistor whose characteristic passes through the origin.

Let us consider some special cases of Theorem II.

Corollary 3: Consider a connected network of nonlinear (possibly time-varying) resistors satisfying assumptions (*a*), (*b*) and (*c*) of Theorem I.

(a) If, in addition, the characteristics of the tree branches of $\mathfrak{T}$ are strictly increasing, then the input characteristic at any node pair is that of a strictly increasing current-controlled resistor. If the characteristics of all tree branches of $\mathfrak{T}$ are continuous, monotonically increasing, one-to-one mappings of $(-\infty, \infty)$ onto $(-\infty, \infty)$ then so is the input characteristic at any node pair.

(b) If the characteristics of the links of the tree $\mathfrak{T}$ are strictly increasing, then the branch-input characteristic is that of a strictly increasing voltage-controlled resistor. If the characteristics of all links of $\mathfrak{T}$ are continuous, monotonically increasing, one-to-one mappings of $(-\infty, \infty)$ onto $(-\infty, \infty)$, then so is any branch-input characteristic.

Proof of Theorems I and II: The proof of these two theorems is divided

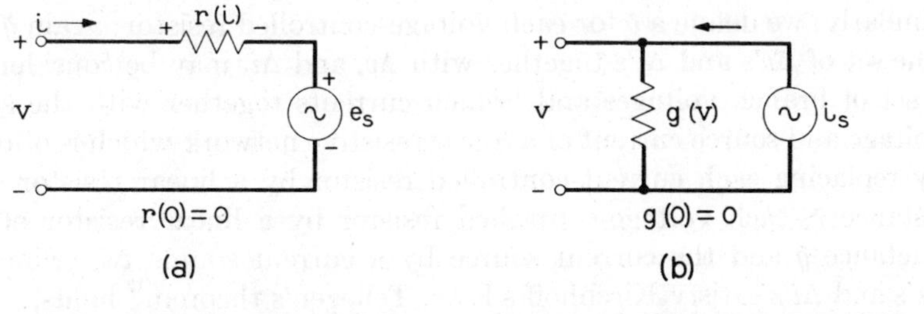

Fig. 7 — (a) Thévenin equivalent circuit: series combination of voltage source and a monotonically increasing current-controlled resistor whose characteristic passes through the origin. (b) Norton equivalent circuit: parallel combination of current source and monotonically increasing voltage-controlled resistor whose characteristic passes through the origin.

into two parts: in part one, we show that if Theorem I holds for a k-node network then Theorem II is true for a k-node network. In part two, we use this implication to prove Theorem I by induction.

The statement of the theorem allows time-varying resistors (and hence includes independent sources); however, in order to have as simple a notation as possible, we write down the proof as if all resistors were time-invariant.

Part One: We show that, for any integer $k \geqq 2$, if Theorem I holds for a k-node network then the input characteristic at any node pair is that of the Thévenin equivalent circuit specified in Theorem II (a). Let us connect the node pair under consideration to a current source i_s (see Fig. 8); this current source is viewed as an additional link, since it is a voltage-controlled resistor. By assumption, to each i_s there is one and only one set of branch currents and voltages that satisfy Kirchhoff's

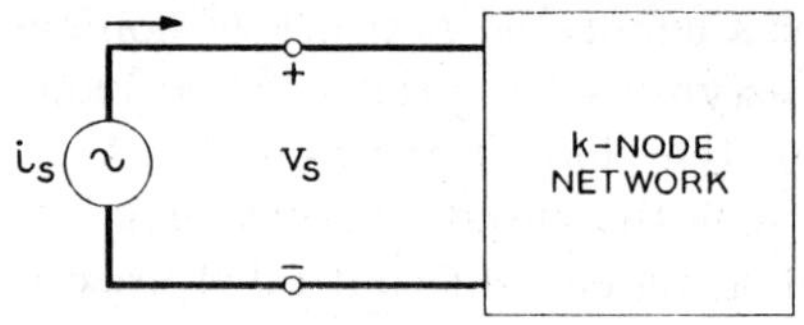

Fig. 8 — Node pair connected to current source.

laws and the branch characteristics. Consider two distinct values of i_s, namely, i_s and i_s'. Let the corresponding branch variables be $\mathbf{v},\mathbf{i}$ and $\mathbf{v}',\mathbf{i}'$. For each current-controlled branch define a number $\bar{r}$ (which depends on $\mathbf{i}$ and $\mathbf{i}'$) by the relation

$$v - v' \triangleq \Delta v = r(i) - r(i') \triangleq \bar{r} \cdot (i - i') = \bar{r} \cdot \Delta i.$$

Since all the current-controlled branches are monotonically increasing, $\bar{r} \geqq 0$. (If $\Delta i = 0$, $\bar{r}$ may be taken to be any nonnegative number.) Similarly, we define a $\bar{g}$ for each voltage-controlled resistor; again $\bar{g} \geqq 0$. The set of Δv's and Δi's together with Δv_s and Δi_s may be considered as a set of branch voltages and branch currents together with the source voltage and source current of a *linear* resistive network which is obtained by replacing each current-controlled resistor by a linear resistor of resistance $\bar{r}$, each voltage-controlled resistor by a linear resistor of conductance $\bar{g}$ and the current source by a current source Δi_s. Since the Δv's and Δi's satisfy Kirchhoff's laws, Tellegen's theorem[12] holds,

$$\Delta v_s \cdot \Delta i_s = \sum \Delta v \Delta i$$

where the sum is over all resistive branches.

Since all branches have monotonically increasing characteristics, this is a sum of nonnegative terms and $\Delta v_s \Delta i_s \geqq 0$. In other words, $\Delta i_s > 0$ implies that $\Delta v_s \geqq 0$: that is, the Thévenin equivalent circuit has a current-controlled monotonically increasing characteristic. The continuity of the characteristic follows from the following considerations: irrespective of the values of the $\tilde{r}$'s and $\tilde{g}$'s, the fact that assumption (c) of Theorem I requires them to be nonnegative implies that the current transfer ratio from the current source to any branch has a magnitude no larger than unity;[11] hence $\Delta i_s \to 0$ implies $\Delta i \to 0$ for all branches. Since the tree branches have continuous characteristics, it follows that, for them, $\Delta v \to 0$ and, by Kirchhoff's voltage law, the same holds for the links. Hence $\Delta i_s \to 0$ implies $\Delta v_s \to 0$, i.e., the current-controlled characteristic of the Thévenin equivalent circuit is continuous. The proof of part (b) of Theorem II follows exactly the dual of the above argument.

Part Two: Let us prove Theorem I for a two-node network (see Fig. 9). Let us plot on the (v,i) plane of Fig. 10 the characteristics of the current-controlled tree branch and that of the voltage-controlled link, taking into account the sign conventions defined on Fig. 9. By assumption, the functions g and r are both continuous and have $(-\infty, \infty)$ as domains; therefore their representative curves intersect at least at one point (v,i). We assert that it is the only one: indeed, suppose there were a second one, (v',i'); then the monotonicity of r and g imply, respectively

$$(v' - v)(i' - i) \geqq 0 \quad \text{and} \quad (v' - v)(i' - i) \leqq 0.$$

Hence

$$(v' - v)(i' - i) = 0.$$

Suppose $v' = v$; then since g is a function

$$i = g(-v) = g(-v') = i'.$$

Similarly, if $i' = i$, the fact that r is a function implies $v = v'$. Hence for all possible sources, there is one and only one set of branch voltages and currents that satisfies Kirchhoff's laws and the branch characteristics.

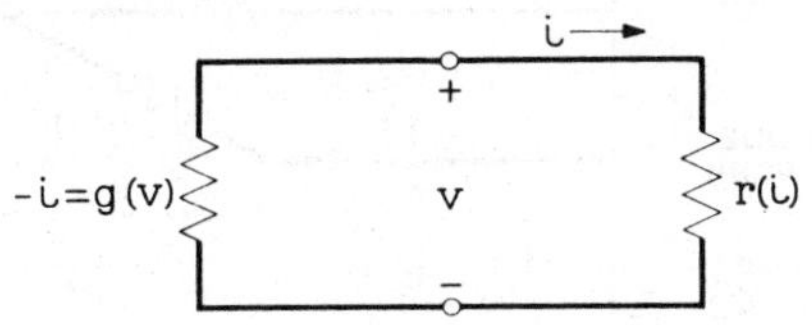

Fig. 9 — Two-node network.

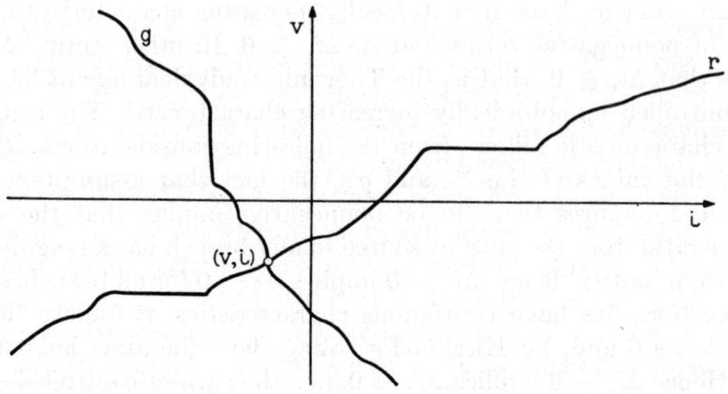

Fig. 10 — Characteristics of current-controlled tree branch and of voltage-controlled link as function of the tree branch current and voltage.

Thus Theorem I is established for a two-node network. The next step in the proof of Theorem I is to show that if it is true for an n-node network it is true for an $(n + 1)$-node network. Consider the n-node network shown in Fig. 11. We shall build out this network into an $(n + 1)$-node network.

Let us first connect the tree branch between node n and node $(n + 1)$, i.e., a current-controlled resistor. (There is no loss of generality in assuming that the numbering of the nodes is such that the branch $(n,n + 1)$ is a tree branch.) It is obvious that, for this network, the existence and uniqueness of the solution holds for all sources. Consequently, from part one of the proof, the input characteristic at any two nodes of this particular $(n + 1)$-node network has the equivalent circuits specified by Theorem II. The next step is to add a link, say between node k and node $(n + 1)$. Since the input characteristic at the node pair $(k,n + 1)$ is as specified in Theorem II (a), the voltage and current in the link are uniquely determined by the reasoning given for the case $n = 2$, and consequently the distribution of voltages and currents in all branches of the network is uniquely determined.

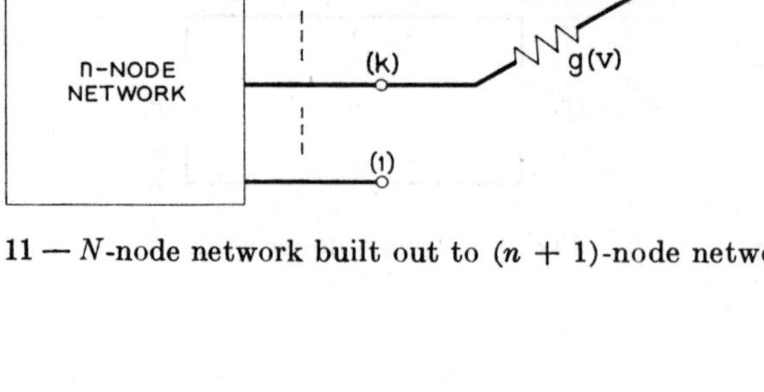

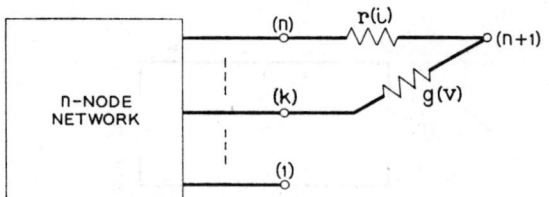

Fig. 11 — N-node network built out to $(n + 1)$-node network.

The process of constructing the $(n + 1)$-node network from the n-node network can be carried out step by step, adding a link at a time. Thus at the end of the process there is one and only one set of branch voltages and currents in the $(n + 1)$-node network that satisfies Kirchhoff's laws and the branch characteristics. Q.E.D.

For the purpose of solving the network differential equations of a general nonlinear *RLC* network it is important to know, for the resistive network case, under what conditions the function which maps the sources, $(\mathbf{E},\mathbf{J})$, into the branch voltages and currents, $(\mathbf{v},\mathbf{i})$, satisfies a Lipschitz condition. It is immediately clear that additional assumptions are required: consider Fig. 12, which shows the characteristic of a current-controlled resistor which fulfills the conditions of Theorem I. In the neighborhood of the operating point A, this resistor may appear to small signals either as an open circuit or a short circuit. Note that the same statement would apply if the resistor were voltage-controlled. It is obvious that under such conditions, the mapping $(\mathbf{E},\mathbf{J}) \rightarrow (\mathbf{v},\mathbf{i})$ will not satisfy a Lipschitz condition. As shown in the following theorem, only weak additional assumptions are required.

Theorem III: Consider a connected network of nonlinear (possibly time-varying) resistors which satisfies conditions (a), (b) and (c) of Theorem I. If, in addition, the following Lipschitz conditions are satisfied: there is a real-valued function $h(R,t)$, defined and positive for $R > 0$ and all t, such that

$$| g_{\alpha\beta}(x,t) - g_{\alpha\beta}(x',t) | \leqq h(R,t) | x - x' |$$

for all links of $\mathfrak{J}$, for all x, x' in $(-R,R)$ and all t and

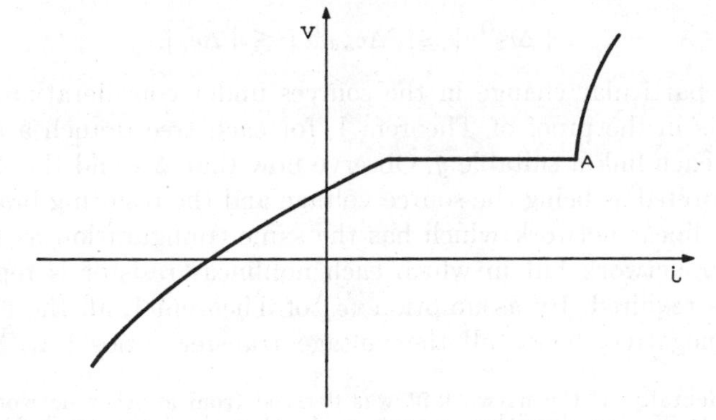

Fig. 12 — Characteristic of current-controlled resistor fulfilling conditions of Theorem I; note unbounded slope at point A.

$$| r_{\alpha\beta}(x,t) - r_{\alpha\beta}(x',t) | \leqq h(R,t) | x - x' |$$

for all branches of $\mathfrak{I}$, for all x,x' in $(-R,R)$ and all t, then the mapping which maps $(\mathbf{E},\mathbf{J})$ *into* $(\mathbf{v},\mathbf{i})$ *satisfies a Lipschitz condition.**

Proof: Consider the effect of a change in the voltage sources **E** on the branch voltages **v** and branch currents **i**. **E** is the vector whose ith component is the output voltage e_i of the source located in the ith branch. In the present $(n + 1)$ node network there are at most $n(n + 1)/2$ branches, hence **E** has at most that many components. Suppose that the change from **E** to $\mathbf{E} + \Delta\mathbf{E}$ is obtained by changing e_i to $e_i + \Delta e_i$ successively with $i = 1, 2, \cdots$. Call α,β the terminals of the first branch and call N_1 the remainder of the network (see Fig. 13). Since the input characteristic of N_1 is monotonically increasing, and since an increase of the voltage across the nonlinear resistance R increases the current through it or keeps it constant, the change in the input voltage of N_1,

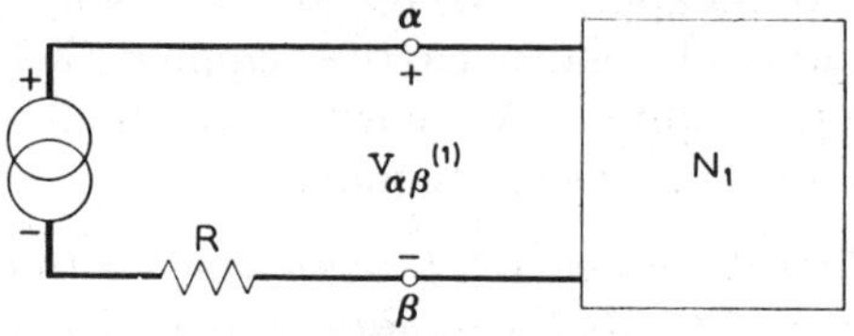

Fig. 13 — Nonlinear resistor R and voltage source in one branch of $(n + 1)$-node network; N_1 represents remainder of network.

$\Delta v_{\alpha\beta}^{(1)}$ due to the change of e_1 to $e_1 + \Delta e_1$ is such that $| \Delta e_1 | \geqq | \Delta v_{\alpha\beta}^{(1)} |$. (The superscript 1 indicates that only the source voltage in the first branch has been changed.) Call $\Delta v_k^{(1)}$ the corresponding change in the kth branch voltage. We assert that

$$| \Delta v_k^{(1)} | \leqq | \Delta v_{\alpha\beta}^{(1)} | \leqq | \Delta e_1 |.$$

For the particular change in the sources under consideration, we may define, as in the proof of Theorem I, for each tree branch a suitable $\tilde{r}$ and for each link a suitable $\tilde{g}$. Observe now that Δe_1 and the $\Delta v_k^{(1)}$ may be interpreted as being the source voltage and the resulting branch voltage of a linear network which has the same configuration as the given nonlinear network but in which each nonlinear resistor is replaced by $\tilde{r}$ or $\tilde{g}$ as required. By assumption (c) of Theorem I, all the $\tilde{r}$'s and $\tilde{g}$'s are nonnegative, hence all the voltage transfer ratios $| \Delta v_k^{(1)}/\Delta e^1 |$ of

* Incidentally, if the network $\mathfrak{N}$ was derived from another network $\mathfrak{N}_A$, by applying to $\mathfrak{N}_A$ the algorithm of Section III, then the mapping $(\mathbf{E},\mathbf{J}) \rightarrow (\mathbf{v},\mathbf{i})$ is one-to-one.

the linear network cannot[11] exceed 1 and the inequality asserted above follows. Thus, for all i's and k's,

$$| \Delta v_k^{(i)} | \leqq | \Delta e_i |.$$

Let Δv_k be the change in the voltage across the kth branch when $\mathbf{E}$ becomes $\mathbf{E} + \Delta\mathbf{E}$. Summing over i, using the triangle inequality, and defining the norm of a vector as the sum of the magnitude of its components, we get

$$| \Delta v_k | \leqq \| \Delta\mathbf{E} \|. \tag{3}$$

Since there are at most $n(n + 1)/2$ branches, we get finally

$$\| \Delta\mathbf{v} \| \leqq [n(n + 1)/2] \| \Delta\mathbf{E} \| \tag{4}$$

where $\Delta\mathbf{v}$ is the change in the branch voltages corresponding to the change of the voltage sources from $\mathbf{E}$ to $\mathbf{E} + \Delta\mathbf{E}$. We next bound the change in the branch currents. Applying (3) to a link and using the Lipschitz condition we find that

$$| \Delta i_k | \leqq h(R,t) \| \Delta\mathbf{E} \| \qquad \text{(for all links)}$$

and since there are at most $n(n - 1)/2$ links and the change in a tree branch current is equal to the change in the sum of currents of the links which belong to its fundamental cut set,

$$| \Delta i_k | \leqq h(R,t)[n(n - 1)/2] \| \Delta\mathbf{E} \| \qquad \text{(for all branches)}.$$

Thus

$$\| \Delta\mathbf{i} \| \leqq h(R,t)[n^2(n^2 - 1)/4] \| \Delta\mathbf{E} \|. \tag{5}$$

The effect of a change in the current sources from $\mathbf{J}$ to $\mathbf{J} + \Delta\mathbf{J}$ is obtained in a dual manner. Since the current transfer ratio may not exceed unity[11] we get

$$| \Delta i_k | \leqq \| \Delta\mathbf{J} \|$$

and

$$\| \Delta\mathbf{i} \| \leqq [n(n + 1)/2] \| \Delta\mathbf{J} \|. \tag{6}$$

This implies

$$| \Delta v_k | \leqq h(R,t) \| \Delta\mathbf{J} \| \qquad \text{(for all tree branches)}$$

and, by Kirchhoff's voltage law,

$$| \Delta v_k | \leqq h(R,t)n \| \Delta\mathbf{J} \| \qquad \text{(for all branches)}.$$

Finally

$$\| \Delta \mathbf{v} \| \leqq h(R,t)[n^2(n + 1)/2] \| \Delta \mathbf{J} \|. \tag{7}$$

Using the usual product topology[17] for both the product spaces of voltage sources and current sources on the one hand and branch voltages and branch currents on the other, and invoking (4) to (7), we conclude that the mapping $(\mathbf{E},\mathbf{J}) \rightarrow (\mathbf{v},\mathbf{i})$ is Lipschitz.

V. NONLINEAR RLC NETWORKS

The previous section required all elements of the network to be of the same kind and to have a monotonically increasing characteristic. In this section both requirements are removed. In addition to independent sources, the network consists of nonlinear (possibly time-varying) resistors, capacitors and inductors and some of the elements are allowed to have characteristics with negative slope.

As a first step let us make one remark. Theorems I, II, and III would still hold if all resistors were montonically decreasing instead of monotonically increasing. In the more complicated situation considered here the same possible choice exists. For example, separable subnetworks of $\mathfrak{N}_{(E)C}$ which contain more than one capacitor could just as well contain monotonically decreasing capacitors. For simplicity, we shall assume that all monotonic elements are increasing.

In order to state the following theorem we need two definitions. A network (or subnetwork) is called a *self-loop* if it consists of a single branch whose end-points are identified: it consists of one branch and one node. A network (or subnetwork) is called an *open branch* if it consists of a single branch whose end-points are not identified: it consists of one branch and two nodes.

Theorem IV: Let $\mathfrak{N}$ be a network of independent sources and nonlinear (possibly time-varying) resistors, capacitors and inductors (without mutual inductance) such that: capacitors of $\mathfrak{N}$ are either charge-controlled or monotonically increasing voltage-controlled; resistors are either voltage-controlled or current-controlled; inductors are either flux-controlled or monotonically increasing current-controlled. It is further assumed that $\mathfrak{N}$ and $\mathfrak{N}_{(E)(J)*}$ are nonseparable and connected. The network $\mathfrak{N}$ is determinate if:

(1) *The capacitor network $S*\mathfrak{N}_{(E)C}$ satisfies the following requirements:*

(a) *Open branches of $S*\mathfrak{N}_{(E)C}$ are charge-controlled and contain all charge-controlled capacitors which are not monotonically increasing.*

(b) *Each subnetwork of $S*\mathfrak{N}_{(E)C}$ which contains more than one element* *has a tree with monotonically increasing charge-controlled tree branches and monotonically increasing voltage-controlled links.*

(*c*) *Self-loops of* $S{*}\mathfrak{N}_{(E)C}$ *are voltage-controlled.*

(2) *The resistive network* $S{*}\mathfrak{N}_{(EC)R}$ *satisfies the following requirements:*

(*a*) *Open branches are current-controlled and contain all current-controlled resistors which are not monotonically increasing.*

(*b*) *Each subnetwork which contains more than one element has a tree with monotonically increasing current-controlled tree branches and monotonically increasing voltage-controlled links.*

(*c*) *Self-loops are voltage-controlled and contain all voltage-controlled resistors which are not monotonically increasing.*

(3) *The inductive network* $S{*}\mathfrak{N}_{(ECR)L}$ *satisfies the following requirements:*

(*a*) *Open branches are current-controlled.*

(*b*) *Each subnetwork which contains more than one element has a tree with monotonically increasing current-controlled tree branches and monotonically increasing flux-controlled branches.*

(*c*) *Self-loops are flux-controlled and contain all flux-controlled inductors that are not monotonically increasing.*

(4) *In any finite interval, and for all time, the characteristics of the network resistors, capacitors and inductors satisfy a Lipschitz condition with respect to the following variables:*

tree branches: capacitors, with respect to q
resistors and inductors, with respect to i
links: capacitors and resistors, with respect to v
inductors, with respect to φ.

Remarks: Note that nonmonotonic voltage-controlled capacitors and current-controlled inductors were excluded from the discussion. Such capacitors and inductors may be included in the discussion provided they fall into the following trivial cases: each nonmonotonic voltage-controlled capacitor is in parallel with a voltage source and each nonmonotonic current-controlled inductor is in series with a current source. In such cases, $\mathfrak{N}_{(E)(J)*}$ is separable unless $\mathfrak{N}$ contains one element only.

The above conditions insure the existence and uniqueness[18] of the solution on some nonvanishing interval $[t_0, t_\alpha)$, where $t_\alpha > t_0$. The length of this interval cannot be specified without further assumptions on the Lipschitz constants $h(R,t)$. This is the well known problem of finite escape time. In particular, if for all branch characteristics the same Lipschitz constant can be used and holds over the whole domain of the characteristic, then the solution exists and is unique on $[t_0, \infty)$ for all regulated **E**'s and **J**'s.

Proof of Theorem IV: Let us denote the voltages and charges of the capacitive branches by $(\mathbf{e}_c, \mathbf{q}_c)$. Similarly, denote the voltages and currents of the resistive branches by $(\mathbf{e}_r, \mathbf{i}_r)$ and fluxes and currents of inductive branches by $(\boldsymbol{\varphi}_L, \mathbf{i}_L)$. Voltage sources and current sources will denoted as usual by $\mathbf{E}$ and $\mathbf{J}$.

We assert that conditions (2) and (4) of Theorem IV imply, first, that the currents and the voltages of the resistive branches at time t, $(\mathbf{e}_r(t),\mathbf{i}_r(t))$ are uniquely determined by the values, *at the same time t*, of the capacitor voltages, the voltage sources, the inductor currents and the current sources, $(\mathbf{e}_c(t),\mathbf{E}(t),\mathbf{i}_L(t),\mathbf{J}(t))$; and, second, that the mappings

$$\mathbf{e}_r(t) = \mathbf{f}_{e_r}(\mathbf{e}_c(t),\mathbf{E}(t),\mathbf{i}_L(t),\mathbf{J}(t),t) \tag{8}$$

$$\mathbf{i}_r(t) = \mathbf{f}_{i_r}(\mathbf{e}_c(t),\mathbf{E}(t),\mathbf{i}_L(t),\mathbf{J}(t),t) \tag{9}$$

satisfy Lipschitz conditions.

Given any set of capacitor voltages $\mathbf{e}_c$ and inductor currents $\mathbf{i}_L$ such that Kirchoff's voltage law is satisfied in each loop formed by capacitors and voltage sources, and such that Kirchoff's current law is satisfied in each cut set formed by inductors and current sources, let us replace each capacitor with a voltage source whose voltage is equal to the voltage of the replaced capacitor and replace each inductor with a current source whose current is equal to the current of the replaced inductor. The network consists now of resistors, current sources and voltage sources only.

Let us use the algorithm of Section III to change the configuration of the sources and to separate the network into its separable parts. Let us denote the resulting network by $\mathfrak{N}^R$ and its sources by $(\mathbf{E}^R,\mathbf{J}^R)$.

The network $\mathfrak{N}^R$ has three sets of subnetworks: (a) connected nonseparable subnetworks which contain sources and two or more resistive branches, (b) subnetworks containing one resistive branch in parallel with a voltage source, and (c) subnetworks containing one resistive branch in parallel with a current source.

Consider the first set of subnetworks. Denote the branch voltages and branch currents of these subnetworks by $(\mathbf{e}_r, \mathbf{i}_r)_1$, and their sources by $(\mathbf{E}_1^R,\mathbf{J}_1^R)$. From conditions (2) and (4) of Theorem IV it follows that each subnetwork contains a tree whose resistive branches are monotonically increasing current-controlled resistors and whose links are monotonically increasing voltage-controlled resistors, and that all elements satisfy Lipschitz conditions. Therefore, from Theorems I and III of Section IV it follows that $(\mathbf{e}_r, \mathbf{i}_r)_1$ are uniquely denfied by $(\mathbf{E}_1^R,\mathbf{J}_1^R)$ and that the mapping $(\mathbf{e}_r(t),\mathbf{i}_r(t)) = \mathbf{f}_r(\mathbf{E}_1^R(t),\mathbf{J}_1^R(t),t)$ satisfies Lipschitz conditions.

The second set of networks corresponds to self-loops of $S*\mathfrak{N}_{(EC)R}$. In each subnetwork, the resistor is voltage-controlled, and hence the current through it is uniquely determined in terms of the voltage source. Since the characteristics satisfy a Lipschitz condition, the mapping from space $\mathbf{E}_2^R$ (the voltage sources) to $(\mathbf{e}_r, \mathbf{i}_r)_2$ (the branch voltages and currents in the subnetworks of the second set) satisfies Lipschitz conditions.

Subnetworks of the third set correspond to open branches of $S*\mathfrak{N}_{(EC)R}$. As each subnetwork contains only one resistor and a current source, the requirement that the branch be current-controlled is enough to insure uniqueness of $(\mathbf{e}_r, \mathbf{i}_r)_3$, the branch voltages and currents of this set, in terms of the corresponding sources $\mathbf{J}_3^R$. From condition (4) it follows that the mapping from $\mathbf{J}_3^R$ to $(\mathbf{e}_r, \mathbf{i}_r)_3$ satisfies the Lipschitz condition.

The voltages of sources $\mathbf{E}^R$ are linear combinations of the voltages $\mathbf{E}$ and $\mathbf{e}_c$, and the currents $\mathbf{J}^R$ are linear combinations of $\mathbf{J}$ and $\mathbf{i}_L$ (see Section IV). From this linearity property and from the properties of the above relations between the voltages and currents of the resistive branches and $(\mathbf{E}^R,\mathbf{J}^R)$ it follows that $(\mathbf{e}_r(t),\mathbf{i}_r(t))$ are uniquely defined by $(\mathbf{e}_c(t),\mathbf{E}(t),\mathbf{i}_L(t),\mathbf{J}(t))$ and that the mappings in (8) and (9) satisfy Lipschitz conditions.

Let us now consider the capacitors of the network $\mathfrak{N}$. Given any set of resistor currents $\mathbf{i}_r$ and inductor currents $\mathbf{i}_L$ such that Kirchoff's current law is satisfied in each cut set formed by resistors, inductors and current sources, let us replace the inductive and resistive branches of $\mathfrak{N}$ by current sources with currents equal to the corresponding currents $\mathbf{i}_L$, $\mathbf{i}_r$. The network consists now of sources and capacitors only. Let us use the algorithm of Section III to change the configuration of the sources and separate the network into its separable parts. The resulting network is denoted by $\mathfrak{N}^C$ and its sources by $\mathbf{E}^C,\mathbf{J}^C$. We are going to establish an analogy between $\mathfrak{N}^C$ and its sources by $\mathbf{E}^C,\mathbf{J}^C$. We are going to establish analogy between $\mathfrak{N}^C$ and $\mathfrak{N}^R$ and use the result just proved for $\mathfrak{N}^R$ to deduce a similar result for $\mathfrak{N}^C$.

$\mathfrak{N}^C$ consists of the three sets of subnetworks which were described in connections with $\mathfrak{N}^R$. Consider the second set of subnetworks of $\mathfrak{N}^C$, which consists of single capacitors in parallel with a voltage source. Except for the trivial case where $\mathfrak{N}$ consists only of a single capacitor in parallel with a voltage source, this set is empty, for otherwise $\mathfrak{N}_{(E)(J)*}$ would be separable. Condition (1) implies that each subnetwork of the first set has a tree, say τ_C, whose tree branches are monotonically increasing and charge-controlled, and whose links are monotonically increasing and voltage-controlled. For each subnetwork of this set, with each fundamental cut set of τ_C defined by a capacitive tree branch, we assign a variable q_i equal to the sum of the charges on all the capacitors of that cut set. For each subnetwork of the third set, we assign a q_i equal to the charge on the capacitor. $\mathbf{q}$ will denote the vector whose components are the q_i's.

The analogy between $\mathfrak{N}^C$ and $\mathfrak{N}^R$ is established in four steps:

(*i*) For $\mathfrak{N}^R$, the Kirchhoff current law applied to the ith cut set associated with a resistive tree branch reads

$$J_i^R = \sum_k i_{ki}$$

where i_{ki} is the current in the kth branch of the ith cut set. The i_{ki} are components of $\mathbf{i}_r$. For $\mathfrak{N}^C$, we have by definition of q_i,

$$q_i = \sum_k q_{ki}$$

where q_{ki} is the charge in the kth branch of the ith cut set. The q_{ki} are components of $\mathbf{q}$.

(*ii*) For both $\mathfrak{N}^R$ and $\mathfrak{N}^C$, the Kirchhoff voltage law holds.

(*iii*) Condition (1) imposes requirements on the topology and element characteristics of $\mathfrak{N}^C$ which are entirely similar to those imposed on $\mathfrak{N}^R$ by condition (2).

(*iv*) Finally, the elements of $\mathfrak{N}^R$ and $\mathfrak{N}^C$ satisfy analogous Lipschitz conditions by condition (4).

Therefore, the variables $(\mathbf{e}_c, \mathbf{q}_c)$ and $(\mathbf{E}^C, \mathbf{q})$ of $\mathfrak{N}^C$ are analogous to the variables $(\mathbf{e}_r, \mathbf{i}_r)$ and $(\mathbf{E}^R, \mathbf{J}^R)$ of $\mathfrak{N}^R$.

Remembering that $\mathbf{E}^C$ is linearly related to $\mathbf{E}$, we conclude that the voltages and charges of the capacitors at time t are uniquely determined by the values, *at the same time* t, of the voltage sources $\mathbf{E}(t)$ and $\mathbf{q}(t)$, and that the mapping

$$\mathbf{e}_c(t) = \mathbf{f}_{e_c}(\mathbf{E}(t), \mathbf{q}(t), t) \tag{10}$$

$$\mathbf{q}_c(t) = \mathbf{f}_{q_c}(\mathbf{E}(t), \mathbf{q}(t), t) \tag{11}$$

satisfies Lipschitz conditions.

Since q_i in any fundamental cut set is equal to the sum of the capacitor charges

$$\frac{dq_i(t)}{dt} = J_i^C(t)$$

where $J_i^C(t)$ is the contribution of the current sources to the ith cut set. As $\mathbf{J}^C$ is a linear combination of $\mathbf{i}_r$, $\mathbf{i}_L$, and $\mathbf{J}$ it follows that

$$\frac{d}{dt}\mathbf{q}(t) = \mathbf{f}_q(\mathbf{i}_r(t), \mathbf{i}_L(t), \mathbf{J}(t)) \tag{12}$$

where $\mathbf{f}_q$ is linear and does not depend explicitly on time.

Let us now consider the inductors of the network $\mathfrak{N}$. Given any set of resistor voltages $\mathbf{e}_r$ and capacitor voltages $\mathbf{e}_c$ such that Kirchoff's voltage law is satisfied in each loop formed by capacitors, resistors and voltage sources, let us replace the capacitor and resistor branches of $\mathfrak{N}$ with voltage sources equal to the corresponding voltages $\mathbf{e}_c$, $\mathbf{e}_r$. The network now consists of sources and inductors only. Let us again use the algorithm of Section III to change the configuration of the sources and separate the network into its separable parts. The resulting network is denoted by $\mathfrak{N}^L$ and its sources by $\mathbf{E}^L$,$\mathbf{J}^L$. As in the case of the resistive network, we are going to use the result previously proved for $\mathfrak{N}^R$ to deduce a similar result for $\mathfrak{N}^L$.

$\mathfrak{N}^L$ consists of the three sets of subnetworks which were described in connection with $\mathfrak{N}^R$. Consider the third set of subnetworks of $\mathfrak{N}^L$, which consists of single inductors in parallel with a current source. Except for the trivial case where $\mathfrak{N}$ consists only of a single inductor in parallel with a current source, this set is empty, for otherwise $\mathfrak{N}_{(E)(J)^*}$ would be separable. Condition (3) implies that each subnetwork of the first set has a tree, say τ_L, whose tree branches are monotonically increasing and current-controlled and whose links are monotonically increasing and flux-controlled. For each subnetwork of the first set, with each fundamental loop of τ_L defined by an inductive link, we assign a variable φ_i equal to the sum of the fluxes of all the inductors of that loop. For each subnetwork of the second set we assign a φ_i equal to the flux of the inductor. $\boldsymbol{\varphi}$ will denote the vector whose components are the φ_i's.

The analogy between $\mathfrak{N}^L$ and $\mathfrak{N}^R$ is established in four steps:

(*i*) For $\mathfrak{N}^R$ the Kirchhoff voltage law applied to the *i*th loop associated with a resistive link reads

$$E_i^R = \sum_k e_{ki}$$

where e_{ki} is the voltage across the *k*th branch of the *i*th loop. The e_{ki} are components of $\mathbf{e}_r$. For $\mathfrak{N}^L$ we have by definition of φ_i,

$$\varphi_i = \sum_k \varphi_{ki}$$

where φ_{ki} is the flux in the *k*th branch of the *i*th loop. The φ_{ki} are components of $\boldsymbol{\varphi}_L$.

(*ii*) For both $\mathfrak{N}^R$ and $\mathfrak{N}^L$ the Kirchhoff current law holds.

(*iii*) Condition (3) imposes requirements on the topology and element characteristics of $\mathfrak{N}^L$ which are entirely similar to those imposed on $\mathfrak{N}^R$ by condition (2).

(*iv*) Finally, the elements of $\mathfrak{N}^R$ and $\mathfrak{N}^L$ satisfy analogous Lipschitz conditions by condition (4).

Therefore, the variables $(\boldsymbol{\varphi}_L, \mathbf{i}_L)$ and $(\boldsymbol{\varphi}, \mathbf{J}^L)$ of $\mathfrak{N}^L$ are analogous to the variables $(\mathbf{e}_r, \mathbf{i}_r)$ and $(\mathbf{E}^R, \mathbf{J}^R)$.

As $\mathbf{J}^L$ is linearly related to $\mathbf{J}$, we conclude that fluxes and currents of the inductors at time t are uniquely determined by the values, at the same time t, of $(\boldsymbol{\varphi}(t), \mathbf{J}(t))$, and that the mappings

$$\boldsymbol{\varphi}_L(t) = \mathbf{f}_{\varphi_L}(\boldsymbol{\varphi}(t), \mathbf{J}(t), t) \tag{13}$$

$$\mathbf{i}_L(t) = \mathbf{f}_{i_L}(\boldsymbol{\varphi}(t), \mathbf{J}(t), t) \tag{14}$$

satisfy Lipschitz conditions.

Since φ_i in each loop is equal to the sum of the fluxes in the loop, it follows from the Kirchhoff voltage law that

$$\frac{d}{dt}\varphi_i(t) = E_i^L(t)$$

where $E_i^R(t)$ is the contribution of the voltage sources in the ith loop. As $\mathbf{E}^L$ is a linear combination of $\mathbf{e}_c$, $\mathbf{e}_r$, and $\mathbf{E}$, it follows that

$$\frac{d}{dt}\boldsymbol{\varphi}(t) = \mathbf{f}_\varphi(\mathbf{e}_c(t), \mathbf{e}_r(t), \mathbf{E}(t)) \tag{15}$$

where $\mathbf{f}_\varphi$ is linear and does not depend explicitly on time.

Any solution of the network requires that (8), (9), (10), (11), (12), (13), (14) and (15) be satisfied simultaneously. It is shown in the following that these equations determine a unique solution.

In (12) and (15) substitute values of $\mathbf{e}_c$, $\mathbf{e}_r$, $\mathbf{i}_r$ and $\mathbf{i}_L$ from (10), (11), (9) and (14). The results are

$$\frac{d}{dt}\mathbf{q} = \mathbf{f}_q[\mathbf{f}_{i_r}(\mathbf{f}_{e_c}(\mathbf{E},\mathbf{q},t),\mathbf{E},\mathbf{f}_{i_L}(\boldsymbol{\varphi},\mathbf{J},t),\mathbf{J},t),\mathbf{f}_{i_L}(\boldsymbol{\varphi},\mathbf{J},t),\mathbf{J}] \tag{16}$$

$$\frac{d}{dt}\boldsymbol{\varphi} = \mathbf{f}_\varphi[\mathbf{f}_{e_c}(\mathbf{E},\mathbf{q},t),\mathbf{f}_{e_r}(\mathbf{f}_{e_c}(\mathbf{E},\mathbf{q},t),\mathbf{E},\mathbf{f}_{i_L}(\boldsymbol{\varphi},\mathbf{J},t),\mathbf{J},t),\mathbf{E}]. \tag{17}$$

Since the right-hand sides of (16) and (17) are compositions of functions satisfying Lipschitz conditions, these equations may be rewritten as

$$\frac{d}{dt}\mathbf{q}(t) = \mathbf{F}_q(\mathbf{E}(t),\mathbf{q}(t),\boldsymbol{\varphi}(t),\mathbf{J}(t),t) \tag{18}$$

$$\frac{d}{dt}\varphi(t) = \mathbf{F}_\varphi(\mathbf{E}(t),\mathbf{q}(t),\varphi(t),\mathbf{J}(t),t) \qquad (19)$$

where $\mathbf{F}_q$ and $\mathbf{F}_\varphi$ satisfy Lipschitz conditions in $\mathbf{q}$ and φ. Therefore, for any $\mathbf{E}(\cdot)$ and $\mathbf{J}(\cdot)$ that are regulated functions of time[13] and for any initial values of φ and $\mathbf{q}$, the differential equations (18) and (19) determine uniquely $\varphi(\cdot)$ and $\mathbf{q}(\cdot)$, and the solutions are continuous functions of time.[18] In terms of $\mathbf{E}(\cdot)$, $\mathbf{J}(\cdot)$, $\varphi(\cdot)$ and $\mathbf{q}(\cdot)$, equations (8), (9), (10), (11), (13) and (14) determine uniquely the currents and voltages of the resistive branches, the voltages and charges of the capacitive branches, and the fluxes and currents of the inductive branches. Therefore the network $\mathfrak{N}$ is determinate. (Incidentally, the proof shows that the state of the network may be represented by $(\mathbf{q}, \varphi)$.)

It is worth indicating an immediate consequence of (18) and (19) and the other circuit relations.

Corollary: If the conditions of Theorem IV are satisfied, $\mathbf{E}$ and $\mathbf{J}$ are continuous functions of time, and all elements depend continuously on time, then $\mathbf{e}_c$, $\mathbf{q}_c$, $\mathbf{e}_r$, $\mathbf{i}_r$, φ_L, $\mathbf{i}_L$ are continuous functions of time; in other words, jump phenomena[1] are excluded.

Corollary: Let the network $\mathfrak{N}$ consist of independent sources, nonlinear (possibly time-dependent) monotonically increasing one-to-one resistors, capacitors and inductors. If the characteristics of all elements Lipschitz conditions as described in condition (4) of Theorem IV, the network $\mathfrak{N}$ is determinate.

Corollary: If branches of a network $\mathfrak{N}$ consist of: (a) voltage sources, current sources; (b) one-to-one monotonically increasing resistors, capacitors, and inductors whose characteristics satisfy condition (4) of Theorem IV; (c) one-to-one two-poles of the types described by the theorems of Section II and which satisfy conditions (*a*), (*b*), (*c*) and (*d*) of these theorems: *then* network $\mathfrak{N}$ is determinate.

Given a physical circuit or device, it may happen that a particular model of the circuit does not satisfy the conditions of Theorem IV. For example, this model $\mathfrak{N}'$ might be such that $S{*}\mathfrak{N}'_{(E)C}$ includes a parallel connection of two charge-controlled capacitors, $D_1(q)$ and $D_2(q)$, with only D_1 monotonically increasing. Under these conditions, it may happen that the current through the parallel combination does not determine uniquely the voltage across it. If, however, the model is changed (call it $\mathfrak{N}''$) and a resistor (or inductor) is inserted in series with D_2, then $S{*}\mathfrak{N}''_{(E)C}$ now includes an open branch D_2, condition (1) of Theorem IV is no longer violated, and $\mathfrak{N}''$ is determinate. Obviously, this idea may be used in the case of inductors and resistors.

Finally, let us conclude this section by a discussion which draws attention to some consequences of the conditions of Theorem IV. Some of the properties considered here will be used in the next section for writing in detail the network equations.

Let $\mathfrak{N}$ be a network which satisfies the conditions of Theorem IV. Denote by r, r_{1-1}, g the sets of resistors which are current-controlled (but not voltage-controlled), one-to-one, and voltage-controlled (and not current-controlled), respectively. Similarly, denote respectively by d, d_{1-1}, c the charge-controlled (and not voltage-controlled), one-to-one, and voltage-controlled capacitors, and by l, Γ_{1-1}, Γ the current-controlled (and not flux-controlled), one-to-one, and flux-controlled inductors.

Let us carry out the following operations:

(a) choose a forest of $\mathfrak{N}_E$

(b) choose a forest of $\mathfrak{N}_{(E)d}$

(c) choose a forest of $\mathfrak{N}_{(Ed)d_{1-1}}$

(d) choose a forest of $\mathfrak{N}_{(Edd_{1-1})r}$

(e) choose a forest of $\mathfrak{N}_{(Edd_{1-1}r)r_{1-1}}$

(f) choose a forest of $\mathfrak{N}_{(Edd_{1-1}rr_{1-1})l}$

(g) choose a forest of $\mathfrak{N}_{(Edd_{1-1}rr_{1-1}l)\Gamma_{1-1}}$.

Since the conditions of Theorem IV are satisfied by $\mathfrak{N}$, it follows that *the union of these forests forms a tree of* $\mathfrak{N}$ which we denote by τ. The construction of this tree is an extension of Bryant's procedure.[9]

This can be proved in the following way: From the conditions of Theorem IV it follows that the union of the forests chosen by (b) and (c) [by (d) and (e), by (f) and (g)] are forests of $\mathfrak{N}_{(E)C}$, [$\mathfrak{N}_{(EC)R}$, and $\mathfrak{N}_{(ECR)L}$, respectively]. Let us add the network's resistors to $\mathfrak{N}_{(ECR)L}$. This is done by splitting nodes and adding the new branches between them. Consider a node which was split, say, to three nodes and a resistor subnetwork connected between these nodes, It is clear that the subtree of this resistive subnetwork completes the forest of $\mathfrak{N}_{(ECR)L}$ for a forest of the $\mathfrak{N}_{(EC)RL}$. We can use the same argument to show that by adding the capacitors and voltage sources we get a forest of the network which includes all branches but the current sources. However, the current sources do not form any cut set and therefore are links of this forest. Thus τ, the union of these forests, is a tree of $\mathfrak{N}$.

From the construction of the tree, the conditions of Theorem IV, and the above discussion it follows that *τ contains all current and charge-controlled elements which are not one-to-one, and all voltage and flux-controlled elements which are not one-to-one are links of this tree.*

Consider the fundamental cut set of $\mathfrak{N}$ defined by an element of set d, a charge-controlled capacitor whose characteristic is not monotonically increasing. By assumption, this capacitor is an open branch of $S{*}\mathfrak{N}_{(E)C}$;

this cut set may not contain capacitors or voltage sources and therefore consists solely of resistors, inductors and current sources. Similar properties exist for fundamental loops defined by the various links. One can exhibit these properties by making a table in which links and tree branches are partitioned according to the types of their elements and properties of their characteristics; for each link and branch the table specifies the type of elements that are allowed to be in the corresponding loop or cut set. As the table is complicated, it is omitted and only some of the more interesting properties are listed below. Here we are going to make use of the rank-order of the elements, $ECRLJ$, defined in Section II:

(*i*) Tree branches with characteristics which are not monotonically increasing are the *highest* ranked elements in their own fundamental cut set. Thus, for example, a charged-controlled nonmonotonically increasing capacitor has a fundamental cut set which may include links which are resistors, inductors and current sources but no other capacitors.

(*ii*) Links with characteristics which are not monotonically increasing are the *lowest* ranked elements in their own fundamental loop. Thus, a fundamental loop defined by a nonmonotonically increasing resistor may have only capacitors or voltage sources in its tree branches.

VI. EQUATIONS FOR RLC NETWORKS

The purpose of this section is to write explicitly the equations of a nonlinear RLC circuit of the type considered in the previous section. Another purpose is to exhibit the similarities and differences between the equations that describe linear networks and those that describe the nonlinear networks under consideration.

To simplify the exposition consider the resistive network of Fig. 14. Call τ_1 the tree formed by the branches 1,2,3, and the voltage source E. If the network were linear, the fundamental cut set equations would read

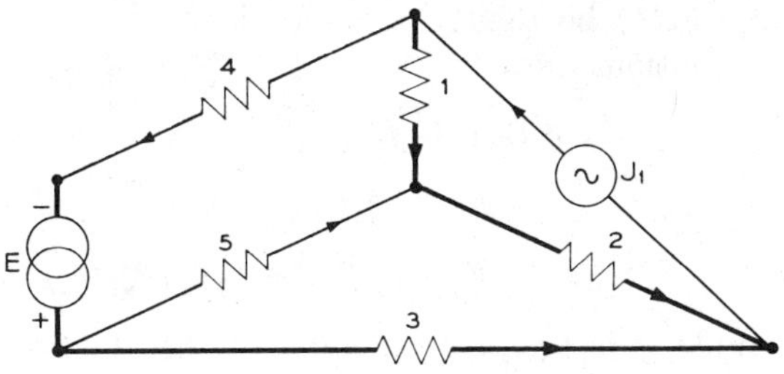

Fig. 14 — Resistive network.

$$\begin{aligned} g_1e_1 + g_4(e_1 + e_2 - e_3 - E) &= J_1 \\ g_4(e_1 + e_2 - e_3 - E) + g_2e_2 - g_5(-e_2 + e_3) &= J_1 \\ -g_4(e_1 + e_2 - e_3 - E) + g_3e_3 + g_5(-e_2 + e_3) &= 0 \end{aligned} \qquad (20)$$

where e_i is the voltage across the ith branch and g_i is the conductance of this branch. In the well known matrix form, the equations become

$$\begin{bmatrix} 1 & 0 & 0 & 1 & 0 \\ 0 & 1 & 0 & 1 & -1 \\ 0 & 0 & 1 & -1 & 1 \end{bmatrix} \begin{bmatrix} g_1 & 0 & 0 & 0 & 0 \\ 0 & g_2 & 0 & 0 & 0 \\ 0 & 0 & g_3 & 0 & 0 \\ 0 & 0 & 0 & g_4 & 0 \\ 0 & 0 & 0 & 0 & g_5 \end{bmatrix} \begin{bmatrix} 1 & 0 & 0 & 0 \\ 0 & 1 & 0 & 0 \\ 0 & 0 & 1 & 0 \\ 1 & 1 & -1 & -1 \\ 0 & -1 & 1 & 0 \end{bmatrix} \begin{bmatrix} e_1 \\ e_2 \\ e_3 \\ E \end{bmatrix} = \begin{bmatrix} J_1 \\ J_1 \\ 0 \end{bmatrix} \qquad (21)$$

or, more generally,

$$\mathbf{\Delta}_{T(R),R} \boldsymbol{G}_R \mathbf{\Delta}'_{T(RE),R} \begin{bmatrix} \mathbf{e} \\ \mathbf{E} \end{bmatrix} + \mathbf{\Delta}_{T(R),L(J)} \mathbf{J} = \mathbf{0} \qquad (22)$$

where **e**, **E** and **J** are column vectors whose components are the tree branch voltages, $\begin{bmatrix} e_1 \\ e_2 \\ e_3 \end{bmatrix}$, the voltage sources, $[E]$, and the current sources, $[J]$, of the network; $\boldsymbol{G}_R$ is the branch admittance matrix. The $\mathbf{\Delta}$'s are appropriate submatrices of the fundamental cut set matrix **Q**. The first subscript of $\mathbf{\Delta}$ denotes the rows and the second subscript denotes the columns of **Q** whose intersection forms the submatrix. Thus $\mathbf{\Delta}_{T(RE),L(R)}$ is a submatrix formed by the intersection of rows corresponding to resistive and voltage source tree branches and columns corresponding to resistive links; $\mathbf{\Delta}_{T(R),R}$ is formed by the intersection of rows corresponding to resistive tree branches and columns corresponding to resistive branches. $\mathbf{\Delta}_{T(R),J}$ is defined similarly. The prime over a matrix indicates transposition. Now, let the resistors become monotonically increasing one-to-one nonlinear resistors. Without loss of generality we can assume these new resistors to be time invariant. Let $\bar{g}_1(\cdot)$, $\bar{g}_2(\cdot)$, $\bar{g}_3(\cdot)$, $\bar{g}_4(\cdot)$ and $\bar{g}_5(\cdot)$ be their characteristics.

The cut set equations are:

$$\begin{aligned} \bar{g}_1(e_1) + \bar{g}_4(e_1 + e_2 - e_3 - E) &= J_1 \\ \bar{g}_4(e_1 + e_2 - e_3 - E) + \bar{g}_2(e_2) - \bar{g}_5(e_3 - e_2) &= J_1 \\ -\bar{g}_4(e_1 + e_2 - e_3 - E) + \bar{g}_5(e_3 - e_2) + \bar{g}_3(e_3) &= 0 \end{aligned} \qquad (23)$$

where, for example, $\bar{g}_1(e_1)$ is now the value of the function $\bar{g}_1$ evaluated at e_1.

The similarity between (20) and (23) suggests a shorthand notation for writing the equations of nonlinear networks. By the product $\boldsymbol{A}*\mathbf{x}$ (where $\boldsymbol{A}$ is a diagonal matrix whose elements are functions $a_i(\cdot)$ and $\mathbf{x}$ is a column vector whose components are $x_1, x_2, \cdots, x_n$), we denote the column vector whose ith component is $a_i(x_i)$, that is, the ith diagonal element of $\boldsymbol{A}$ evaluated at the ith component of $\mathbf{x}$. With this symbolic notation the equations of the network of Fig. 14 can be written (for the nonlinear case) in a form analogous to (22).

$$\boldsymbol{\Delta}_{T(R),R}\boldsymbol{G}_R*\left(\boldsymbol{\Delta}'_{T(RE),R}\begin{bmatrix}\mathbf{e}\\ \mathbf{E}\end{bmatrix}\right) + \boldsymbol{\Delta}_{T(R),L(J)}\mathbf{J} = \mathbf{0}$$

where $\boldsymbol{G}_R$ is the diagonal matrix whose elements are the characteristics $\bar{g}_1, \bar{g}_2, \cdots, \bar{g}_5$ and the $*$ operation must be interpreted as indicated above. $\boldsymbol{G}_R$ will be referred to the branch characteristic matrix. With this symbolic notation, cut set matrices, loop matrices and branch resistance matrices may be used to writing equations of nonlinear networks in the same way as for linear networks.

Let us now assume that the elements of the tree τ_1 of the network of Fig. 14 are monotonically increasing current-controlled but not voltage-controlled and the links are monotonically increasing voltage-controlled but not current-controlled. Since the tree branches are not voltage-controlled, the equations cannot be written in the form of (22). Let $\bar{r}_1(\cdot)$, $\bar{r}_2(\cdot)$ and $\bar{r}_3(\cdot)$ represent the characteristics of the tree branches and i_1, i_2 and i_3 be the currents of the corresponding tree branches. In terms of the tree branch voltages and currents the cut set equations become:

$$\begin{aligned} i_1 + \bar{g}_4(e_1 + e_2 - e_3 - E) &= J_1 \\ i_2 + \bar{g}_4(e_1 + e_2 + e_3 - E) - \bar{g}_5(e_3 - e_2) &= J_1 \\ i_3 - \bar{g}_4(e_1 + e_2 - e_3 - E) + \bar{g}_5(e_2 - e_3) &= 0. \end{aligned} \tag{24}$$

The other set of equations is

$$\begin{aligned} e_1 &= \bar{r}_1(i_1) \\ e_2 &= \bar{r}_2(i_2) \\ e_3 &= \bar{r}_3(i_3) \end{aligned} \tag{25}$$

or symbolically

$$\mathbf{i}_{T(R)} + \boldsymbol{\Delta}_{T(R),L(R)}\boldsymbol{G}_{L(R)}*\left(\boldsymbol{\Delta}'_{T(RE),L(R)}\begin{bmatrix}\mathbf{e}_{T(R)}\\ \mathbf{E}\end{bmatrix}\right) + \boldsymbol{\Delta}_{T(R),L(J)}\mathbf{J} = \mathbf{0}. \tag{26}$$

$$\mathbf{e}_{T(R)} = \boldsymbol{R}_{T(R)}*\mathbf{i}_{T(R)} \tag{27}$$

where $\mathbf{i}_{T(R)}$, $\mathbf{e}_{T(R)}$, $\mathbf{E}$ and $\mathbf{J}$ are the tree currents and voltages, and voltage and current sources respectively. $\boldsymbol{G}$ and $\boldsymbol{R}$ are the link and branch characteristic matrices, and in our example

$$\boldsymbol{G}_{L(R)} = \begin{bmatrix} g_4 & 0 \\ 0 & g_5 \end{bmatrix}, \qquad \boldsymbol{R}_{T(R)} = \begin{bmatrix} \mathcal{R}_1 & 0 & 0 \\ 0 & \mathcal{R}_2 & 0 \\ 0 & 0 & \mathcal{R}_3 \end{bmatrix}$$

The $\boldsymbol{\Delta}$'s are appropriate submatrices of the fundamental cut set matrix $\mathbf{Q}$. The first subscript denotes the rows and the second subscript denotes the columns of $\mathbf{Q}$ whose intersection forms the submatrix. Thus $\boldsymbol{\Delta}_{T(RE),L(R)}$ is a submatrix formed from the intersection of rows corresponding to resistive and voltage source tree branches and columns corresponding to resistive links. $\boldsymbol{\Delta}_{T(R),L(R)}$ and $\boldsymbol{\Delta}_{T(R),L(J)}$ are defined similarly. A comparison of (26), (27) and (22) shows that in the case of current-controlled tree branches and voltage-controlled links which are not one-to-one, we need *both* $\mathbf{i}_{T(R)}$ and $\mathbf{e}_{T(R)}$ for a straightforward writing of the cut set equations and the branch characteristic equations. Either $\mathbf{i}_{T(R)}$ or $\mathbf{e}_{T(R)}$ can be eliminated from the equations. The resulting equations are:

$$\mathbf{i}_{T(R)} + \boldsymbol{\Delta}_{T(R),L(R)} \boldsymbol{G}_{L(R)} * \left(\boldsymbol{\Delta}'_{T(RE),L(R)} \begin{bmatrix} \boldsymbol{R}_{T(R)} * \mathbf{i}_{T(R)} \\ \mathbf{E} \end{bmatrix} \right) + \boldsymbol{\Delta}_{T(R),L(J)} \mathbf{J} = \mathbf{0} \tag{28}$$

Or

$$\mathbf{e}_{T(R)} + \boldsymbol{R}_{T(R)} * \left\{ \boldsymbol{\Delta}_{T(R),L(R)} \boldsymbol{G}_{L(R)} * \left(\boldsymbol{\Delta}'_{T(RE),L(R)} \begin{bmatrix} \mathbf{e}_{T(R)} \\ \mathbf{E} \end{bmatrix} \right) \right\} + \boldsymbol{R}_{T(R)} * (\boldsymbol{\Delta}_{T(R),L(J)} \mathbf{J} = \mathbf{0}. \tag{29}$$

Fundamental loop equations can be written in a similar way using both the voltages and currents of the links $\mathbf{i}_{L(R)}$ and $\mathbf{e}_{L(R)}$. The equations are

$$\mathbf{e}_{L(R)} + \mathbf{1}_{L(R),T(R)} \boldsymbol{R}_{T(R)} * \left(\mathbf{1}'_{L(RJ),T(R)} \begin{bmatrix} \mathbf{i}_{L(R)} \\ \mathbf{J} \end{bmatrix} \right) + \mathbf{1}_{L(R),T(E)} \mathbf{E} = \mathbf{0}$$

$$\mathbf{i}_{L(R)} = \boldsymbol{G}_{L(R)} * \mathbf{e}_{L(R)}$$

where the $\mathbf{1}$'s are appropriate submatrixes of the fundamental tie set matrix $\mathbf{B}$. Similarly to (26) and (27), either $\mathbf{e}_{L(R)}$ or $\mathbf{i}_{L(R)}$ can be eliminated.

We now write the equations for a general *RLC* network (which satisfies the requirements of Theorem IV) by performing the following steps†

† Other systems of variables are possible. For example, one can choose charges and fluxes as above and voltages of resistive links whose loop does not consist of capacitors and voltage sources only.

(*i*) A tree is chosen as explained in Section V.

(*ii*) Variables are chosen. We choose here the charges on the capacitive tree branches, $\mathbf{q}_D$, the currents of the resistive tree branches whose fundamental cut set *does not* consist of inductors and current sources only, $\mathbf{i}_R$, and the fluxes of the inductive links, φ_Γ.

The equations make use of the following characteristic branch matrixes: $\boldsymbol{C}$ and $\boldsymbol{D}$ are diagonal matrixes whose elements are the characteristics of voltage-controlled and charged-controlled capacitors, respectively; $\boldsymbol{G}$ and $\boldsymbol{R}$, those of voltage-controlled and current-controlled resistors, respectively; $\boldsymbol{L}$ and $\boldsymbol{\Gamma}$, those of current-controlled and flux-controlled inductors, respectively. Without loss of generality we can assume that the elements are time-invariant. The equations are

$$\begin{aligned}\frac{d}{dt}\Bigg\{\mathbf{q}_D &+ \mathbf{\Delta}_{T(C),L(C)}\boldsymbol{C}_{L(C)}*\left(\mathbf{\Delta}'_{T(CE),L(C)}\begin{bmatrix}\boldsymbol{D}_{T(C)}*\mathbf{q}_D\\ \mathbf{E}\end{bmatrix}\right)\Bigg\}\\ &+ \mathbf{\Delta}_{T(C),L(\bar{R})}\boldsymbol{G}_{L(\bar{R})}*\left(\mathbf{\Delta}'_{T(C\bar{R}E),L(\bar{R})}\begin{bmatrix}\boldsymbol{D}_{T(L)}*\mathbf{q}_D\\ \boldsymbol{R}_{T(\bar{R})}*\mathbf{i}_R\\ \mathbf{E}\end{bmatrix}\right)\\ &+ \mathbf{\Delta}_{T(C),L(L)}\boldsymbol{\Gamma}_{L(L)}*\varphi_\Gamma + \mathbf{\Delta}_{T(C),L(J)}\mathbf{J} = \mathbf{0}\end{aligned}\tag{30}$$

$$\begin{aligned}\mathbf{i}_R &+ \mathbf{\Delta}_{T(\bar{R}),L(\bar{R})}\boldsymbol{G}_{L(\bar{R})}*\left(\mathbf{\Delta}'_{T(C\bar{R}E),L(\bar{R})}\begin{bmatrix}\boldsymbol{D}_{T(C)}*\mathbf{q}_D\\ \boldsymbol{R}_{T(\bar{R})}*\mathbf{i}_R\\ \mathbf{E}\end{bmatrix}\right)\\ &+ \mathbf{\Delta}_{T(\bar{R}),L(L)}\boldsymbol{\Gamma}_{L(L)}*\varphi_\Gamma + \mathbf{\Delta}_{T(\bar{R}),L(J)}\mathbf{J} = \mathbf{0}\end{aligned}\tag{31}$$

$$\begin{aligned}\frac{d}{dt}\Bigg\{\varphi_\Gamma &+ \mathbf{1}_{L(L),T(L)}\boldsymbol{L}_{T(L)}*\left(\mathbf{1}'_{L(LJ),T(L)}\begin{bmatrix}\boldsymbol{\Gamma}_{L(L)}*\varphi_\Gamma\\ \mathbf{J}\end{bmatrix}\right)\Bigg\}\\ &+ \mathbf{1}_{L(L),T(R_1)}\boldsymbol{R}_{T(R_1)}*\left(\mathbf{1}'_{L(LJ),T(R_1)}\begin{bmatrix}\boldsymbol{\Gamma}_{L(L)}*\varphi_\Gamma\\ \mathbf{J}\end{bmatrix}\right)\\ &+ \mathbf{1}_{L(L),T(C)}\boldsymbol{D}_{T(C)}*\mathbf{q}_D + \mathbf{1}_{L(L),T(\bar{R})}\boldsymbol{R}_{T(\bar{R})}*\mathbf{i}_R\\ &+ \mathbf{1}_{L(L),T(E)}\mathbf{E} = \mathbf{0}\end{aligned}\tag{32}$$

where R_1 is the set of resistive tree branches whose fundamental cut set contains inductive links and current sources only; and $\bar{R}$ is the set which contains all other resistive branches.

The terms in the brackets in (30) and (32) are equal to our state variables $\mathbf{q}$ and φ of Section V. One can write the equations in terms of these variables: the relations between $\mathbf{q}_D$ and $\mathbf{q}$ and φ_Γ and $\boldsymbol{\phi}$ are given by

$$\mathbf{q}_D + \mathbf{\Delta}_{T(C),L(C)} \boldsymbol{C}_{L(C)} * \left(\mathbf{\Delta}'_{T(CE),L(C)} \begin{bmatrix} \boldsymbol{D}_{T(C)} * \mathbf{q}_D \\ \mathbf{E} \end{bmatrix} \right) = \mathbf{q}$$

$$\boldsymbol{\varphi}_\Gamma + \mathbf{1}_{L(L),T(L)} \boldsymbol{L}_{T(L)} * \left(\mathbf{1}'_{L(LJ),T(L)} \begin{bmatrix} \mathbf{\Gamma}_{L(L)} * \boldsymbol{\varphi}_\Gamma \\ \mathbf{J} \end{bmatrix} \right) = \boldsymbol{\varphi}.$$

In summary, the equations of the *RLC* nonlinear network are written in a way which is a generalization of the methods used in linear networks. However, great care must be taken of the fact that some characteristics are representable by functions which do not have inverses. This section indicated a method for tackling the problem. In this section, the equations are written in terms of three sets of variables: $\mathbf{q}_D$, the charges on the capacitive tree branches; $\boldsymbol{\varphi}_L$, the fluxes in the inductive links and $\mathbf{i}_R$, the currents in the resistive tree branches whose fundamental cut sets do not consist of only inductors and current sources. It is interesting to note that (except for the trivial case where $\mathfrak{N}$ consists of a single capacitor in parallel with a voltage source or a single inductor in parallel with a current source) the dimension of the state vector $(\mathbf{q},\boldsymbol{\varphi})$ used above is the same as in the linear case: [number of independent initial conditions] = [number of reactive elements] − [number of independent capacitor-only tie sets] − [number of independent inductor-only cut sets].[7,8,9]

REFERENCES

1. Minorsky, N., *Nonlinear Oscillations*, D. Van Nostrand, 1962, (part IV, especially Chap. 26) p. 614.
2. Duffin, R. J., Nonlinear Networks I, Bull. Am. Math. Soc. **52,** 1946, pp. 836–838.
3. Duffin, R. J., Nonlinear Networks II, Bull. Amer. Math. Soc. **53,** October, 1947, pp. 963–971.
4. Duffin R. J., Nonlinear Network IIb, Bull. Am. Math. Soc., **54,** 1948, pp. 119–127.
5. Birkhoff, G., and Diaz, J. B., Nonlinear Network Problems, Quart. Appl. Math., **13,** Jan., 1956, pp. 431–443.
6. Lock, K., *Coordinate Selection in Numerical Network Analysis*, Electrical Engineering Department, California Institute of Technology, Pasadena, California, 1962.
7. Bers, A., The Degree of Freedom in RLC Networks, IRE Trans. on Circuit Theory, **CT-6,** 1959, pp. 91–95.
8. Bryant, P. R., and Bers, A., The Degree of Freedom in RLC Networks, IRE Trans. on Circuit Theory, **CT-7,** 1960, p. 173.
9. Bryant, P. R., The Explicit Form of Bashkow's A Matrix, IRE Trans. on Circuit Theory, **CT-9,** Sept., 1962, pp. 303–306.
10. Bryant, P. R., The Order of Complexity of Electrical Networks, Proc. IEE, **106C,** June, 1959, pp. 174–188.
11. Weinberg, L., *Network Analysis and Synthesis*, McGraw-Hill, 1962.
12. Seshu, S., and Reed, M. B., *Linear Graphs and Electrical Networks*, Addison-Wesley, 1961.
13. Dieudonné, J., *Foundations of Modern Analysis*, Academic Press, 1960.
14. Apostol, T. M., *Mathematical Analysis*, Addison-Wesley, 1960.

15. Sandberg, I. W., private communication.
16. Branin, F. H., Jr., The Inverse of the Incidence Matrix of a Tree and the Formulation of the Algebraic First-Order Differential Equations of RLC Networks, IEEE Trans. on Circuit Theory, *CT-10*, Dec., 1963, p. 543–544.
17. Simmons, G. F., *Introduction to Topology and Modern Analysis*, McGraw-Hill, New York, 1963.
18. Coddington, E. A., and Levinson, N., *The Theory of Ordinary Differential Equations*, McGraw-Hill, New York, 1955.

Necessary and Sufficient Conditions for the Global Invertibility of Certain Nonlinear Operators That Arise in the Analysis of Networks

IRWIN W. SANDBERG, MEMBER, IEEE

Abstract—For a large class of operators of the form $[F(\cdot)+A]$, in which A is a not necessarily nonsingular real $n \times n$ matrix, and $F(\cdot)$ is a diagonal strictly monotone-increasing mapping of the set of all real n-vectors E^n onto an open subset of E^n, we give necessary and sufficient conditions under which $[F(\cdot)+A]$ possesses a global inverse on E^n. Operators of the type $[F(\cdot)+A]$ frequently arise in the analysis of nonlinear networks and are encountered in other areas as well. In particular, for A the short-circuit conductance matrix of a resistance network, and $F(x)$ the transpose of $(f_1(x_1), f_2(x_2), \cdots, f_n(x_n))$ for all $x \in E^n$ in which the $f_j(\cdot)$ are the usual exponential diode functions, we give a complete solution to the problem of determining whether or not $[F(\cdot)+A]$ possesses an inverse on E^n.

I. Introduction

RECENTLY, a considerable number of results [1]–[3] have been obtained concerning properties of equations of the form

$$F(x) + Ax = B \tag{1}$$

in which A is a real $n \times n$ matrix, $F(\cdot)$ is a mapping of the set of all real n-vectors E^n into itself such that $F(x)$ is equal to the transpose of the row vector $(f_1(x_1), f_2(x_2), \cdots, f_n(x_n))$ for all $x \in E^n$ with each $f_j(\cdot)$ a continuous monotone-nondecreasing mapping of E^1 into E^1, and $B \in E^n$. Those and related results have led to a deeper understanding of the properties of a large and important class of nonlinear networks containing transistors and diodes, and they have also led to some basic results concerning the properties of algorithms for the computation of the solutions of the transient equations of such networks [4].

In [1] it is proved that (1) possesses a unique solution x for each $B \in E^n$ and each $F(\cdot)$ such that all of the $f_j(\cdot)$ are strictly monotone-increasing mappings of E^1 *onto* E^1 if and only if A belongs to the set P_0 of all real square matrices having all principal minors nonnegative.

The functions $f_j(\cdot)$ associated with semiconductor-device models often used in computer simulations are not mappings of E^1 onto E^1. In particular, often for each j, $f_j(\cdot)$ is equal to $a_j[\exp(b_jx_j)-1]$ or $a_j[1-\exp(-b_jx_j)]$ for all $x_j \in E^1$, in which a_j and b_j are positive constants. In that case, $F(\cdot)$ possesses the property that for each j there is a positive constant β_j such that $f_j(\cdot)$ is a strictly monotone-increasing mapping of E^1 onto $(-\beta_j, \infty)$ or $(-\infty, \beta_j)$, and for such $F(\cdot)$ it is known [2] that (1) possesses a unique solution x for each $B \in E^n$ if $A \in P_0$ *and* $\det A \neq 0$.

The previously available results [1]–[3] concerning the existence of solutions of (1) do not deal with cases in which $\det A = 0$ and for some j (or all j) $f_j(\cdot)$ is a strictly monotone-increasing mapping of E^1 onto an open semi-infinite interval. In many applications involving mappings $F(\cdot)$ with such $f_j(\cdot)$, it is known that $\det A \neq 0$. For instance, for the equation of the form $F(x)+Ax=B$ which arises in the study [4] of a large class of algorithms for solving certain systems of differential equations, A is always nonsingular. However, there are important applications in which $\det A = 0$. For example, if A is the short-circuit conductance matrix of a resistor network, then certainly A may be singular.

In this paper we focus attention on mappings $F(\cdot)$ with the property that for some values of j (or for all j) $f_j(\cdot)$ is a strictly monotone-increasing mapping of E^1 onto an open semi-infinite interval, and, for the remaining values of j, $f_j(\cdot)$ is a strictly monotone-increasing mapping of E^1 onto itself. In Section II we prove a theorem according to which for any given such $F(\cdot)$, if $A \in P_0$, and if A satisfies a certain other condition which does not require that $\det A \neq 0$, then (1) possesses a unique solution x for each $B \in E^n$ if and only if a certain set S is empty, and if the set S is not empty, then for any $F(\cdot)$ of a certain type, (1) does not possess a solution for some $B \in E^n$. In particular, Propositions 1 and 3 of Section II-D show that the "certain other condition" concerning A is satisfied whenever A belongs to P_0 and is symmetric (i.e., whenever A is a real symmetric nonnegative-definite matrix) and also whenever A can be written as $P^{-1}Q$ with P and Q real square matrices such that P is strongly column-sum dominant and Q is weakly column-sum dominant.

We state here the following complete result which follows at once from the results of Section II and which is of immediate interest in connection with resistor–diode networks.

For each j, let $f_j(\cdot)$ be a strictly monotone-increasing mapping of E^1 onto either $(-\beta_j, \infty)$ or $(-\infty, \beta_j)$ for some positive constant β_j. Let $f_j(0)=0$ for all j, and for each j let $s_j = 1$ or -1 depending on whether $f_j(\cdot)$ is bounded on $[0, \infty)$ or $(-\infty, 0]$, respectively. Let A be a real symmetric nonnegative-definite $n \times n$ matrix, and let S denote the set

$$\{y: y \in E^n,\ y \neq \theta,\ Ay = \theta, \text{ and } y_js_j \geq 0 \text{ for all } j\}$$

Manuscript received June 8, 1970; revised July 24, 1970.
The author is with the Systems Theory Research Department, Bell Telephone Laboratories, Inc., Murray Hill, N. J. 07974.

Reprinted from *IEEE Transactions on Circuit Theory*, vol. CT-18, pp. 260–263, March 1971.

in which θ is the zero element of E^n. Then (1) possesses a unique solution x for each $B \in E^n$ if and only if S is empty.[1]

It is possible to show that there are convergent algorithms for computing the solution of (1) whenever the hypotheses of the theorem of Section II are satisfied, the set S is empty, and each $f_j(\cdot)$ is twice continuously differentiable on E^1. Indeed this can easily be done by using part of the material of the proof of the theorem and arguments similar to those of [4, sec. 2.3].

II. The Theorem, Three Related Propositions, and Proofs

A. Notation and Definitions

We use the following notation throughout Section II. With n an arbitrary positive integer, the set of all real n-vectors is denoted by E^n, and θ denotes the zero vector of E^n. If $x \in E^n$, then x_j denotes the jth component of x, $\|x\|$ denotes $(\sum_{j=1}^n x_j^2)^{1/2}$, and the inequality $x > \theta$ $(x \geq \theta)$ means that $x_j > 0$ $(x_j \geq 0)$ for all j. For an arbitrary not necessarily square matrix M, M^{tr} denotes the transpose of M.

We say that a real $n \times n$ matrix M is strongly column-sum dominant (weakly column-sum dominant) if and only if $m_{jj} > (\geq) \sum_{i \neq j} |m_{ij}|$ for all j.

Definition 1: The set of all real square matrices M such that all principal minors of M are nonnegative is denoted by P_0.

Definition 2: Let N^n denote the set of all mappings $F(\cdot)$ of E^n into E^n such that $F(x) = (f_1(x_1), f_2(x_2), \cdots, f_n(x_n))^{\text{tr}}$ for all $x \in E^n$, in which 1) for each $j = 1, 2, \cdots, n$ either $f_j(\cdot)$ is a strictly monotone-increasing mapping of E^1 onto E^1, or there is a positive constant β_j such that $f_j(\cdot)$ is a strictly monotone-increasing mapping of E^1 onto either $(-\beta_j, \infty)$ or $(-\infty, \beta_j)$, and 2) $f_j(0) = 0$ for all $j = 1, 2, \cdots, n$.

B. The Theorem

The following theorem is the main result of Section II. Propositions 1, 2, and 3 which follow the proof of the theorem show that a key condition concerning A is satisfied in two important cases.

Theorem 1: Let $F(\cdot) \in N^n$, and for each $j = 1, 2, \cdots, n$ let $s_j = 1, -1$, or 0 depending on whether $f_j(\cdot)$ is bounded on $[0, \infty)$, bounded on $(-\infty, 0]$, or not bounded on either $[0, \infty)$ or $(-\infty, 0]$, respectively. Let A be a real $n \times n$ matrix such that

1) $A \in P_0$;
2) for each diagonal matrix $D = \text{diag}(d_1, d_2, \cdots, d_n)$ with $d_j = 1$ or -1 for all j, there exists a real n-vector p such that $p > \theta$ and $DA^{\text{tr}}Dp \geq \theta$.

Let S denote the set $\{y: y \in E^n, y \neq \theta, Ay = \theta, y_j s_j \geq 0$ for all j such that $|s_j| > 0$, and $y_j = 0$ for all j such that $s_j = 0\}$. Then there exists a unique solution $x \in E^n$ of $F(x) + Ax = B$ for each $B \in E^n$ if and only if S is empty. If S is not empty, then for any $q \in S$ and any $F(\cdot) \in N^n$ for which all of the components of $F(\alpha q)$ are bounded on either $\alpha \in [0, \infty)$ or $\alpha \in (-\infty, 0]$, the equation $F(x) + Ax = B$ does not possess a solution x for some $B \in E^n$.

C. Proof of Theorem 1

Here we prove three lemmas. Lemmas 2 and 3, taken together, prove Theorem 1.

Lemma 1: Let A be a real $n \times n$ matrix. Let $s_j = 1, -1$ or 0 for each $j = 1, 2, \cdots, n$, and let $|s_j| > 0$ for at least one value of j. Let S' denote the set $\{y: y \in E^n, y_j s_j \geq 0$ for all j such that $|s_j| > 0$, and $y_j = 0$ for all j such that $s_j = 0\}$. Then the number

$$\inf_{\substack{y \in S' \\ \|y\| = 1}} \|Ay\| \tag{2}$$

is positive if and only if S' does not contain an element y such that $y \neq \theta$ and $Ay = \theta$.

Proof of Lemma 1: It is obvious that (2) is not positive if there exists an n-vector y belonging to S' such that $y \neq \theta$ and $Ay = \theta$.

Assume now that S' does not contain an element y such that $y \neq \theta$ and $Ay = \theta$. For each j such that $|s_j| > 0$, let u_j denote the n-vector defined by $(u_j)_i = 0$ for all $i \neq j$ and $(u_j)_j = s_j$. An arbitrary $y \in S'$ can be written as $\sum' \alpha_j u_j$ in which $\sum'$ denotes a summation over all j such that $|s_j| > 0$, and the α_j belong to $[0, \infty)$. Therefore,

$$\inf_{\substack{y \in S' \\ \|y\| = 1}} \|Ay\| \tag{3}$$

is equal to

$$\inf_{\alpha \in \mathscr{A}} \|A(\textstyle\sum' \alpha_j u_j)\|$$

in which $\alpha = (\alpha_1, \alpha_2, \cdots, \alpha_n)^{\text{tr}}$ with $\alpha_j = 0$ for all j such that $s_j = 0$, and $\mathscr{A} = \{\alpha = (\alpha_1, \alpha_2, \cdots, \alpha_n)^{\text{tr}}: \alpha \geq \theta, \alpha_j = 0$ for all j such that $s_j = 0$, and $\|\alpha\| = 1\}$. The set $\mathscr{A}$ is closed and bounded. Moreover, on $\alpha \in \mathscr{A}$, $\|A(\sum' \alpha_j u_j)\|$ depends continuously on α and is positive (since S' does not contain an element $y \neq \theta$ such that $Ay = \theta$). Therefore, (3) is positive.

Lemma 2: Let $F(\cdot) \in N^n$, and for each $j = 1, 2, \cdots, n$ let $s_j = 1, -1$, or 0 depending on whether $f_j(\cdot)$ is bounded on $[0, \infty)$, bounded on $(-\infty, 0]$, or not bounded on either $[0, \infty)$ or $(-\infty, 0]$, respectively. Let A be a real $n \times n$ matrix such that

1) $A \in P_0$;
2) there exists a continuous function $\rho(\cdot)$ mapping $[0, \infty)$ into $[0, \infty)$ such that $y \in E^n$, $z \in E^n$, and $F(y) + Ay = z$ imply that $|f_j(y_j)| \leq \rho(\|z\|)$ for all j.

Let S denote the set $\{y: y \in E^n, y \neq \theta, Ay = \theta, y_j s_j \geq 0$ for all j such that $|s_j| > 0$, and $y_j = 0$ for all j such that $s_j = 0\}$. Then there exists a unique solution $x \in E^n$ of $F(x) + Ax = B$ for each $B \in E^n$ if and only if S is empty. If S is not empty, then for any $q \in S$ and any $F(\cdot) \in N^n$ for which all of the elements of $F(\alpha q)$ are bounded on either $\alpha \in [0, \infty)$ or $\alpha \in (-\infty, 0]$, the equation $F(x) + Ax = B$ does not possess a solution x for some $B \in E^n$.

Proof of Lemma 2: We note first that if $s_j = 0$ for all $j = 1, 2, \cdots, n$, then S is empty, and by [1, theorem 3],

[1] When A is the short-circuit conductance matrix of a resistor network, it is frequently possible to determine by inspection whether or not S is empty. An example is presented in the Appendix.

$F(x)+Ax=B$ possesses a unique solution x for each $B \in E^n$. Throughout the remainder of the proof it is assumed that $|s_j|>0$ for at least one value of j.

We shall use the following result due to Palais [5], [6, appendix]. If $R(\cdot)$ is a local homeomorphism on E^n such that $\|R(y)\| \to \infty$ as $\|y\| \to \infty$, then $R(\cdot)$ is a homeomorphism of E^n onto itself.

Since $A \in P_0$ and $F(\cdot) \in N^n$, the equation $F(x)+Ax=B$ possesses at most one solution x for each $B \in E^n$ (see [1]). Let $Q(\cdot)$ be the mapping of E^n into E^n defined by the condition that $Q(y)=F(y)+Ay$ for all $y \in E^n$. In [1] it is proved that $Q(\cdot)$ would be a homeomorphism on E^n if each $f_j(\cdot)$ were a strictly monotone-increasing mapping of E^1 *onto* E^1. It follows that for each $F(\cdot) \in N^n$, $Q(\cdot)$ is a local homeomorphism on E^n.

We now prove that $Q(\cdot)$ is a homeomorphism of E^n onto itself when S is empty. In view of the remarks of the preceding paragraph and the result of Palais cited previously, it suffices to show that $\|Q(y)\| \to \infty$ as $\|y\| \to \infty$ when S is empty. We note that $\|Q(y)\| \to \infty$ as $\|y\| \to \infty$ if for each positive constant k_1, there exists a positive constant k_2 such that $y \in E^n$ and $\|Q(y)\| \le k_1$ imply that $\|y\| \le k_2$.

Assume that S is empty, and let S' denote the set $\{y: y \in E^n, y_j s_j \ge 0$ for all j such that $|s_j|>0$, and $y_j=0$ for all j such that $s_j=0\}$.

Let y and z be arbitrary elements of E^n such that $F(y)+Ay=z$. Since $|f_j(y_j)| \le \rho(\|z\|)$ for all j, we observe that there exist continuous functions $g(\cdot)$ and $g_0(\cdot)$, both mapping $[0, \infty)$ into itself, such that

$$y = -g(\|z\|)s + w + v \tag{4}$$

in which $s=(s_1, s_2, \cdots, s_n)^{tr}$, $w \in S'$, and v satisfies $\|v\| \le g_0(\|z\|)$ and $v_j=0$ for all j such that $|s_j|>0$.[2] Therefore,

$$Aw = g(\|z\|)As - Av + z - F(y)$$

and hence

$$\begin{aligned}\|Aw\| &= \|g(\|z\|)As - Av + z - F(y)\| \\ &\le g(\|z\|)\|As\| + \|Av\| + \|z\| + \|F(y)\| \\ &\le g(\|z\|)\|As\| + ag_0(\|z\|) + \|z\| + n^{1/2}\rho(\|z\|)\end{aligned}$$

in which $a=\sup\{c: \|Aq\| \le c\|q\|, q \in E^n\}$. Let $h(\cdot)$ be the continuous mapping of $[0, \infty)$ into itself defined by $h(\beta) = g(\beta)\|As\| + ag_0(\beta) + \beta + n^{1/2}\rho(\beta)$ for all $\beta \in [0, \infty)$. According to Lemma 1, there exists a positive constant c depending only on A such that $\|Aq\| \ge c\|q\|$ for all $q \in S'$. Thus $\|w\| \le c^{-1}h(\|z\|)$, and, using (4),

$$\|y\| \le n^{1/2}g(\|z\|) + c^{-1}h(\|z\|) + g_0(\|z\|)$$

which shows that for each positive constant k_1 there exists a positive constant k_2 such that $\|z\| \le k_1$ implies that $\|y\| \le k_2$.

Assume now that S is not empty. Let $q \in S$, and let $N^n(q)$ denote the set of all $F(\cdot)$ belonging to N^n with the property that all of the components of $F(\alpha q)$ are bounded on either $\alpha \in [0, \infty)$ or $\alpha \in (-\infty, 0]$.

With $F(\cdot) \in N^n(q)$, suppose that there exists a solution x of $Q(x)=B$ for each $B \in E^n$. Then, since there exists at most one solution of $Q(x)=B$ for each $B \in E^n$, and since $Q(\cdot)$ is a local homeomorphism on E^n, it follows that $Q(\cdot)$ is a homeomorphism of E^n onto E_n. Thus it follows that $Q^{-1}(\cdot)$ exists and is continuous on E^n. In particular, $y=Q^{-1}[Q(y)]$ for all $y \in E^n$. But with $y=\alpha q$ we have $\|Q(y)\|=\|F(\alpha q)\|$, and $\|F(\alpha q)\|$ is bounded on either $\alpha \in [0, \infty)$ or $\alpha \in (-\infty, 0]$. Therefore, the image under $Q^{-1}(\cdot)$ of some bounded subset of E^n is not bounded, which is a contradiction.

Lemma 3: Let $F(\cdot) \in N^n$, and let A be a real $n \times n$ matrix such that

1) $A \in P_0$;
2) for each diagonal matrix $D=\text{diag}(d_1, d_2, \cdots, d_n)$ in which $d_j=1$ or -1 for each j, there exists $p \in E^n$ such that $p>\theta$ and $DA^{tr}Dp \ge \theta$.

Then there exists a positive constant ρ, depending only on A, such that $y \in E^n$, $z \in E^n$, and $F(y)+Ay=z$ imply that $|f_j(y_j)| \le \rho\|z\|$ for all $j=1, 2, \cdots, n$.

Proof of Lemma 3: There are 2^n matrices $D=\text{diag}(d_1, d_2, \cdots, d_n)$ in which $d_j=1$ or -1 for each j, and for each such D there exists a real n-vector p such that $p>\theta$ and $DA^{tr}Dp \ge \theta$. With $q=(1, 1, \cdots, 1)^{tr}$, let $\mathscr{P}$ denote any set of at most 2^n elements of E^n such that $(p-q) \ge \theta$ for all $p \in \mathscr{P}$ and such that for each of the 2^n matrices D there exists a $p \in \mathscr{P}$ such that $DA^{tr}Dp \ge \theta$.

Let $y \in E^n$, $z \in E^n$, and $F(y)+Ay=z$. For at least one of the 2^n matrices D we can write $y=Dw$ with $w_j \ge 0$ for all j, and thus $DF(Dw)+DADw=Dz$. Let $p \in \mathscr{P}$ be such that $DA^{tr}Dp \ge \theta$. Then

$$p^{tr}DF(Dw) + p^{tr}DADw = p^{tr}Dz.$$

Since $p^{tr}DADw \ge 0$, we have $p^{tr}DF(Dw) \le p^{tr}Dz$. Therefore,

$$\sum_{j=1}^{n} p_j|f_j(y_j)| \le \|p\| \cdot \|z\|$$

and hence (since $p_j \ge 1$ for all j) for all j

$$|f_j(y_j)| \le \rho\|z\|$$

in which $\rho=\max_{p \in \mathscr{P}} \|p\|$.

This completes the proof of Theorem 1.

Remark: In an unpublished work, A. N. Willson, Jr., has observed that if $n=3$, if $F(\cdot) \in N^3$ with $f_j(\cdot)$ bounded on $(-\infty, 0]$ for all j, and if

$$A = \begin{bmatrix} 1 & 1 & 1 \\ 0 & 1 & 0 \\ 0 & 1 & 0 \end{bmatrix}$$

then $A \in P_0$, the set S of Theorem 1 is empty, and for some $B \in E^3$ the equation $F(x)+Ax=B$ does not possess a solution x. This shows that Theorem 1 becomes false if condition 2) is omitted. Here, with $D=\text{diag}(1, 1, -1)$, it is obvious that there is no real 3-vector $p>\theta$ such that $DA^{tr}Dp \ge \theta$.

[2] In fact, we may take $g(\cdot)$ and $g_0(\cdot)$ to be the functions defined by $g(\beta)=\max\{\lambda_j: f_j(-s_j\lambda_j)=-s_j\rho(\beta), |s_j|>0\}$ for all $\beta \ge 0$, and with $\bar{v}$ the real n-vector defined by $\bar{v}_j=0$ for all j such that $|s_j|>0$ and $\bar{v}_j=\max\{|\lambda_j|: |f_j(\lambda_j)| =\rho(\beta)\}$ for each j such that $s_j=0$, $g_0(\beta)=\|\bar{v}\|$ for all $\beta \ge 0$.

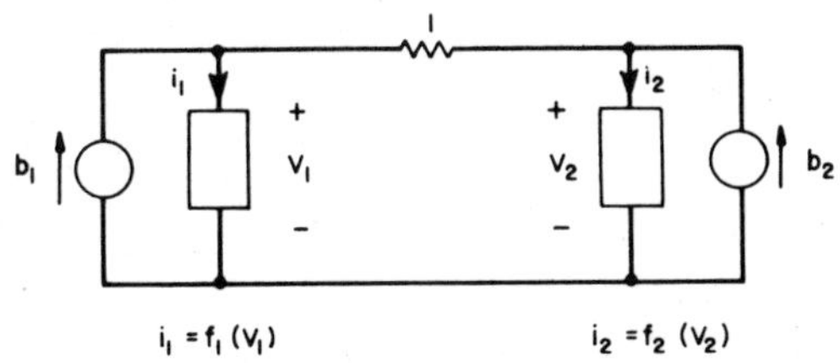

Fig. 1. Circuit for the example.

D. Propositions 1, 2, and 3

Proposition 1: If A is a real symmetric nonnegative-definite $n \times n$ matrix and D is a diagonal matrix of order n with each diagonal element either 1 or -1, then there exists a real n-vector p such that $p > \theta$ and $DADp \geq \theta$.

Proof of Proposition 1: According to a specialization to square matrices of a theorem due Tucker [7], if B and C are real $n \times n$ matrices, then either there exists $u \in E^n$ such that $Bu \geq \theta$ and $Cu \geq \theta$ with at least one component of Bu positive, *or* there exist vectors x and y belonging to E^n such that $y > \theta$, $x \geq \theta$, and $B^{tr}y + C^{tr}x = \theta$, the two alternatives being mutually exclusive. Let M denote the symmetric nonnegative-definite matrix DAD, and let I denote the identity matrix of order n. Suppose that there exists $u \in E^n$ such that $Mu \geq \theta$ and $-u \geq \theta$ with at least one component of Mu positive. Then $u^{tr}Mu \leq 0$. But, with $M^{1/2}$ the symmetric nonnegative-definite square root of M, we have $M^{1/2}u \neq \theta$ (for otherwise we would have $Mu = \theta$), and hence $u^{tr}M^{1/2}M^{1/2}u = u^{tr}Mu > 0$, which is a contradiction. Therefore, by Tucker's theorem, there exist vectors x and y belonging to E^n such that $y > \theta$, $x \geq \theta$, and $My + (-I)x = \theta$.

Proposition 2: If P and Q are real $n \times n$ matrices such that P is strongly column-sum dominant and Q is weakly column-sum dominant, then $P^{-1}Q \in P_0$.

Proposition 2 is proved in [1].

Proposition 3: If P and Q are real $n \times n$ matrices such that P is strongly column-sum dominant and Q is weakly column-sum dominant, and if D is a diagonal matrix of order n with each diagonal element either 1 or -1, then there exists a real n-vector p such that $p > \theta$ and $DQ^{tr}(P^{tr})^{-1}Dp \geq \theta$.

Proof of Proposition 3: With u the n-vector $(1, 1, \cdots, 1)^{tr}$, let $p = DP^{tr}Du$. Clearly, $p > \theta$, and since $u = D(P^{tr})^{-1}Dp$, we have $DQ^{tr}(P^{tr})^{-1}Dp = DQ^{tr}Du \geq \theta$.

Final Remark: The main contribution of this paper is Lemma 2. It is a simple matter to show that it is possible to improve upon Lemma 3 in various ways in order to use Lemma 2 to prove explicit results that are more general than Theorem 1.

Appendix

A Very Simple Example of an Application of Theorem 1

Referring to the circuit of Fig. 1, for $j = 1$ and $j = 2$, let

$$f_j(x_j) = a_j[\exp(b_j x_j) - 1], \qquad \text{for all } x_j \in E^1 \tag{5}$$

or

$$f_j(x_j) = a_j[1 - \exp(-b_j x_j)], \qquad \text{for all } x_j \in E^1 \tag{6}$$

in which a_j and b_j are positive constants; and for $j = 1$ and $j = 2$, let $s_j = -1$ if (5) is satisfied and $s_j = 1$ if (6) is satisfied. With $v = (v_1, v_2)^{tr}$, $B = (b_1, b_2)^{tr}$,

$$G = \begin{bmatrix} 1 & -1 \\ -1 & 1 \end{bmatrix}$$

and $F(x) = (f_1(x_1), f_2(x_2))^{tr}$ for all $x \in E^2$, we have $F(v) + Gv = B$. Here if $s_1 = 1$ and $s_2 = -1$ (or if $s_1 = -1$ and $s_2 = 1$), the set S of Theorem 1 is empty and therefore $F(v) + Gv = B$ possesses a unique solution v for each $B \in E^2$. On the other hand, if $s_1 = s_2 = 1$ (or if $s_1 = s_2 = -1$), then since $G(1, 1)^{tr} = \theta$, S is not empty and, by Theorem 1, there are values of b_1 and b_2 such that $F(v) + Gv = B$ does not have a solution.

References

[1] I. W. Sandberg and A. N. Willson, Jr., "Some theorems on properties of dc equations of nonlinear networks," *Bell Syst. Tech. J.*, vol. 48, Jan. 1969, pp. 1–34.
[2] ——, "Some network-theoretic properties of nonlinear dc transistor networks," *Bell Syst. Tech. J.*, vol. 48, May–June 1969, pp. 1293–1312.
[3] I. W. Sandberg, "Theorems on the analysis of nonlinear transistor networks," *Bell Syst. Tech. J.*, vol. 49, Jan. 1970, pp. 95–114.
[4] ——, "Theorems on the computation of the transient response of nonlinear networks containing transistors and diodes," *Bell Syst. Tech. J.*, vol. 49, Oct. 1970, pp. 1739–1776.
[5] R. S. Palais, "Natural operations on differential forms," *Trans. Amer. Math. Soc.*, vol. 92, no. 1, 1959, pp. 125–141.
[6] C. A. Holzmann and R. Liu, "On the dynamical equations of nonlinear networks with *n*-coupled elements," *Proc. 3rd Ann. Allerton Conf. Circuit and System Theory*, 1965, pp. 536–545.
[7] A. W. Tucker, "Dual systems of homogeneous linear relations," *Linear Inequalities and Related Systems*, H. W. Kuhn and A. W. Tucker, Eds. Princeton, N. J.: Princeton University Press, 1956.

EXISTENCE AND UNIQUENESS OF SOLUTIONS FOR THE EQUATIONS OF NONLINEAR DC NETWORKS*

I. W. SANDBERG AND A. N. WILLSON, JR.†

Abstract. Several theorems are proved concerning the existence and uniqueness of solutions of the equation $AF(x) + Bx = c$, in which A and B are $n \times n$ matrices of real numbers, c is a real n-vector, and $F(x) \equiv [f_1(x_1), \cdots, f_n(x_n)]^T$ is a "diagonal" nonlinear mapping of real n-dimensional Euclidean space E^n into itself. For the case in which the functions f_k are continuous, strictly monotone increasing mappings of E^1 into E^1 (but not necessarily *onto* E^1) and for a large and important class of pairs of matrices (A, B), we give necessary and sufficient conditions for the existence of the global inverse of the mapping $AF(\cdot) + B$. For not-necessarily-monotone f_k (tunnel-diode type functions, for example) we give conditions which guarantee the existence of at least one solution of $AF(x) + Bx = c$ for each $c \in E^n$.

Several significant network-theoretic implications of the results are discussed. For example, we present necessary and sufficient conditions for the existence of the global inverse of the operator $AF(\cdot) + B$ associated with an arbitrary network containing independent sources and nonlinear resistors characterized by arbitrary, continuous, strictly monotone increasing conductance or resistance functions. In particular, the resistors may be described by exponential-type semiconductor diode functions.

1. Introduction. The equation

$$AF(x) + Bx = c \tag{1}$$

is of fundamental importance in the study of nonlinear dc networks. In (1), A and B are $n \times n$ matrices of real numbers, c is a real n-vector, and F, a "diagonal" nonlinear mapping of real n-dimensional Euclidean space E^n into itself, is defined in terms of the functions $f_k : E^1 \to E^1$, by the relation[1] $F(x) = [f_1(x_1), \cdots, f_n(x_n)]^T$ for each $x = (x_1, \cdots, x_n)^T \in E^n$.

This paper is primarily concerned with *resistive* networks; that is, with nonlinear dc networks containing only independent dc voltage and current sources and nonlinear resistors. It will be clear, however, that the results developed here relate to the properties of many networks containing, for example, other passive dc components, such as ideal transformers and gyrators. More explicitly, the main purposes of this paper are (a) to report on a significant generalization of a theorem of [4], and (b) to break entirely new ground by presenting a theorem which asserts the existence of a solution of (1) under certain conditions which do not require that the functions f_k be monotone increasing.

By a *nonlinear resistor* we mean simply a two-terminal electrical element whose voltage and current are related by some not-necessarily-linear function that maps the real line into itself. Thus, for each nonlinear resistor at least one of its two terminal variables, the voltage or the current, can be considered the independent variable.

Nonlinear resistive networks have the canonical form shown in Fig. 1, in which there are n nonlinear resistors, each connected to one of the ports of an

* Received by the editors February 11, 1971, and in revised form September 1, 1971.

† Bell Telephone Laboratories, Incorporated, Murray Hill, New Jersey 07974.

[1] We use the superscript T to denote the transpose of a vector or a matrix.

Reprinted with permission from *SIAM Journal of Applied Mathematics*, vol. 22, pp. 173–186, March 1972.

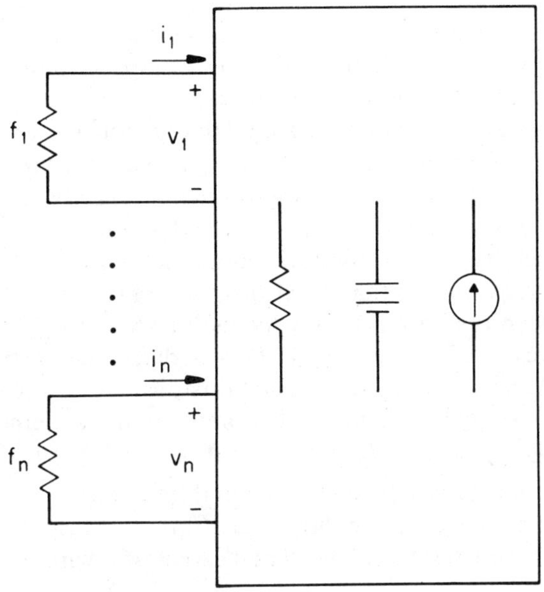

FIG. 1. *Canonical form of a resistive network*

n-port containing the connecting wires, the independent sources, and also any "linear" resistors that might be present in the network. Taking this point of view, it is not difficult to show (by following a development of the type given in [6]) that the network is characterized by an equation having the form of (1). The functions f_k in (1) are the nonlinear conductance or resistance functions for the nonlinear resistors. The components of the vector x are port variables (each component being either a voltage or a current variable; the choice depending upon whether the nonlinear resistor at that port is charactèrized by a conductance function, or by a resistance function, respectively). The matrices A and B, and the vector c, are determined by the n-port. The values of the n-port's linear resistors and the network's topology, alone, determine the matrices A and B; these matrices are not affected by changes in the values of the independent sources. Such changes are expressed by changes in the values of the components of the vector c. In particular, the components of the vector c all become zero when the independent source values all become zero.

We assume that the linear resistors which may be present within the n-port have nonnegative resistance or conductance values. Thus, when the n-port's independent sources are all assigned the value zero, the n-port becomes *passive*. It is not at all surprising, therefore, that the pairs of matrices (A, B) associated through (1) with nonlinear resistive networks possess the following property: for each pair of n-vectors (x, y) satisfying $Ax = By$ it follows that $x^T y \geqq 0$. We call pairs of matrices (A, B) that possess this property, *passive pairs*.

In addition to the class of all nonlinear resistive networks, many nonlinear dc networks containing other passive elements may be described by (1) with a passive pair of matrices (A, B). The existence of a global inverse of the operator $AF(\cdot) + B$ for any such network may also be determined by the methods to be described.

The following notation will be used throughout the paper. For each positive integer n we denote by E^n the n-dimensional Euclidean space, the elements of which are ordered n-tuples of real numbers, which we consider to be column vectors. If $x \in E^n$, then for $k = 1, \cdots, n$, x_k denotes the kth component of x. If $x, y \in E^n$, we denote their inner product by $\langle x, y\rangle = \sum_{k=1}^n x_k y_k$. The norm of each $x \in E^n$ is denoted by $\|x\| = \langle x, x\rangle^{1/2}$. If $x, y \in E^n$, we use the notation $x < y$ to mean that, for $k = 1, \cdots, n$, $x_k < y_k$. Similarly, $x \leqq y$ means that, for $k = 1, \cdots, n$, $x_k \leqq y_k$. In contrast, the inequality $x \leq y$ is used to denote that, for $k = 1, \cdots, n$, $x_k \leqq y_k$, with the additional requirement that strict inequality must hold for at least one such k. Analogous definitions apply for the inequalities $x > y$, $x \geqq y$, and $x \geq y$.

All matrices are taken to be real. If D is a diagonal matrix, then $D > 0$ denotes that each element of the main diagonal is positive. A square matrix A is said to be weakly (strongly) column-sum dominant if its elements a_{ij} satisfy $a_{jj} \geqq \sum_{i \neq j} |a_{ij}|$ $(a_{jj} > \sum_{i \neq j} |a_{ij}|)$. We say that A is weakly (strongly) row-sum dominant if A^T is weakly (strongly) column-sum dominant.

For each $n \times n$ matrix B, we denote by $\mathcal{N}(B)$ the intersection of E^n and the null space of B, that is, $\mathcal{N}(B)$ is the set of all real n-vectors x with the property that[2] $Bx = \theta$.

2. Equations with monotone nonlinearities. Here we study the existence and uniqueness of solutions of (1) under the assumption that the functions f_k specifying the mapping F are continuous and strictly monotone increasing.

DEFINITION. For each positive integer n we denote by $\mathcal{M}^n$ the class of all mappings F of E^n into E^n with the property that there exist, for $k = 1, \cdots, n$, continuous, strictly monotone increasing functions f_k such that for each $x = (x_1, \cdots, x_n)^T \in E^n$, $F(x) = [f_1(x_1), \cdots, f_n(x_n)]^T$.

DEFINITION. For each $F \in \mathcal{M}^n$ we denote by $\mathcal{B}(F)$ the set of all points $x \in E^n$ for which $F(\rho x)$ is bounded as $\rho \to \infty$.

For each $F \in \mathcal{M}^n$ it is clear that the origin θ is a member of $\mathcal{B}(F)$. Moreover, unless each of the functions f_k that specify F has the whole real line as its range, $\mathcal{B}(F)$ will also contain other points. Fortunately, it is a trivial matter to determine the nature of $\mathcal{B}(F)$ for any given $F \in \mathcal{M}^n$: It is well known that the range of any continuous function that maps a connected subset of E^1 into E^1 is also connected. Thus, for every $F \in \mathcal{M}^n$, each of the continuous, strictly monotone increasing functions f_k maps the real line onto an interval (α, β), where $-\infty \leqq \alpha < \beta \leqq \infty$. Four different types of intervals (α, β) are possible. It might be that $\alpha = -\infty$ and β is some real number. In this case the function f_k mapping E^1 onto $(-\infty, \beta)$ has the property that it is bounded on the interval $[0, \infty)$. If both α and β are real numbers, then f_k is bounded on both $(-\infty, 0]$ and on $[0, \infty)$. Similarly, when α is a real number and $\beta = \infty$, then f_k is bounded on $(-\infty, 0]$; and finally, when $\alpha = -\infty$ and $\beta = \infty$, f_k is unbounded on both of the intervals $(-\infty, 0]$ and $[0, \infty)$. Since $F(\rho x)$ is bounded as $\rho \to \infty$ if and only if the same is true for each of the component functions, $f_k(\rho x_k)$, it is clear that $\mathcal{B}(F)$ is the Cartesian product of closed intervals in E^1,

$$\mathcal{B}(F) = I_1(f_1) \times \cdots \times I_n(f_n),$$

[2] We denote the origin in E^n by θ.

where, for $k = 1, \cdots, n$, $I_k(f_k)$ is an interval as specified below: Let f_k map E^1 onto (α, β); then

$$I_k(f_k) = (-\infty, \infty) \qquad \text{if } -\infty < \alpha < \beta < \infty,$$

$$I_k(f_k) = [0, \infty) \qquad \text{if } -\infty = \alpha < \beta < \infty,$$

$$I_k(f_k) = (-\infty, 0] \qquad \text{if } -\infty < \alpha < \beta = \infty,$$

$$I_k(f_k) = \{0\} \qquad \text{if } -\infty = \alpha < \beta = \infty.$$

The concepts of a P_0 matrix and a $\mathscr{W}_0$ pair of matrices have been discussed recently by the authors. See, for example, [5], [6], [7]. For our present purposes the following definitions suffice.

DEFINITION. The set of all real square matrices A with the property that $\det(A + D) \neq 0$ for each diagonal matrix $D > 0$ is called P_0.

DEFINITION. The set of all pairs of real square matrices (A, B) with the property that $\det(AD + B) \neq 0$ for each diagonal matrix $D > 0$ is called $\mathscr{W}_0$.

The following theorem, one of the main results of the paper, is a generalization of Theorem 1 of [4]. It generalizes that theorem in two respects. First, it considers (1) involving a pair of matrices $(A, B) \in \mathscr{W}_0$ rather than an equation involving a single P_0 matrix.[3] Secondly, this theorem allows consideration of *any* continuous, strictly monotone increasing "diagonal" mapping F.[4]

THEOREM 1. *Let $F \in \mathscr{M}^n$ and let (A, B) be a pair of $n \times n$ matrices of real numbers possessing the properties:*

(i)[5] $(A, B) \in \mathscr{W}_0$;

(ii) *for each diagonal matrix $D = \operatorname{diag}(d_1, \cdots, d_n)$ with, for each k, $d_k = +1$ or -1, there exists a real n-vector p such that*

$$DA^Tp \geqq \theta, \quad DB^Tp \geqq \theta \quad \text{and} \quad D(A + B)^Tp > \theta \cdot$$

Then there exists a unique solution of (1) for each $c \in E^n$ if and only if $\mathscr{B}(F) \cap \mathscr{N}(B) = \{\theta\}$. If $\mathscr{B}(F) \cap \mathscr{N}(B) \neq \{\theta\}$, then there exists some $c \in E^n$ such that (1) has no solution.

In the Appendix we give three lemmas which generalize Lemmas 1, 2 and 3 of [4]. Lemmas A2 and A3, together, prove Theorem 1.[6]

Due to the special nature of $\mathscr{B}(F)$, described above, it is a simple matter to determine whether or not any vector $x \in \mathscr{N}(B)$ also satisfies $x \in \mathscr{B}(F)$. The issue is immediately resolved by simply examining the signs of the components of x.

[3] It should be noted, however, that even in the case when (1) reduces to the equation previously considered (i.e., when the matrix A in (1) is the identity matrix), the hypothesis (ii) of our theorem is weaker than the corresponding hypothesis in [4].

[4] Mappings F with some of the f_k bounded on both $(-\infty, 0]$ and $[0, \infty)$ are not considered in [4].

[5] In Theorem 1, property (i) need not be stated explicitly. It is an immediate consequence of Theorem 8 of [7] that property (ii) implies property (i). Alternatively, this implication can easily be derived by using Lemma A3 of the Appendix and the definition of $\mathscr{W}_0$ given above. We choose to mention property (i) here, however, to emphasize that $(A, B) \in \mathscr{W}_0$ is, in fact, being hypothesized. We make explicit use of this property in the theorem's proof; in particular, property (i) is used in Lemma A2 of the Appendix.

[6] Since the class of mappings F considered in Lemma A3 consists of only those $F \in \mathscr{M}^n$ for which $F(\theta) = \theta$, Lemma A2 and Lemma A3, together, prove Theorem 1 only for such mappings F. It is clear, however, that in Theorem 1 the restriction $F(\theta) = \theta$ may be dropped. (The theorem remains the same, for example, if we replace c by $c + AF(\theta)$ in (1).)

For example, if $n = 3$ and if, for $k = 1, 2, 3$, the f_k are the functions whose graphs are shown in Fig. 2, then $\mathscr{B}(F) = (-\infty, \infty) \times (-\infty, 0] \times \{0\}$. Thus, for this example $\mathscr{B}(F) \cap \mathscr{N}(B) = \{\theta\}$ if and only if each real vector $x \neq \theta$ in the null space of B satisfies $x_2 > 0$ or $x_3 \neq 0$. Note, however, that since $x \in \mathscr{N}(B)$ if and only if $-x \in \mathscr{N}(B)$, it thus follows that $\mathscr{B}(F) \cap \mathscr{N}(B) = \{\theta\}$ if and only if each real vector $x \neq \theta$ in the null space of B satisfies $x_3 \neq 0$.

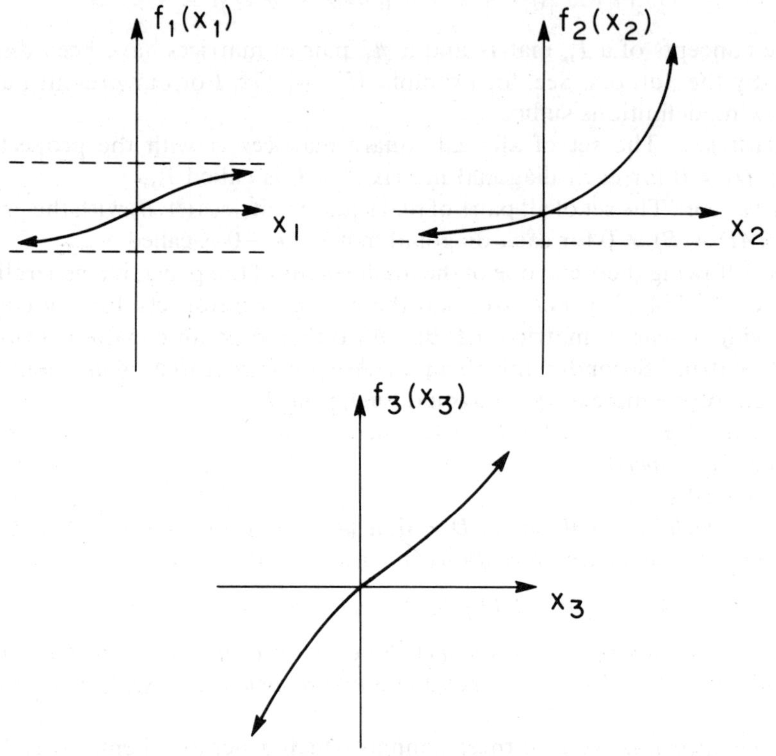

FIG. 2. *Examples of continuous, strictly monotone increasing functions*

When the pair of matrices (A, B) is a passive pair, it follows[7] that $(A, B) \in \mathscr{W}_0$ Theorem 2, which follows, shows that each passive pair (A, B) also possesses property (ii) of Theorem 1. The proof of Theorem 2 makes use of Slater's theorem of the alternative.

SLATER'S THEOREM. *Let A, B, C, and D be given matrices. Then one of the following two statements holds, but never both*:

(i) *There exists x such that*

$$Ax > \theta, \quad Bx \geq \theta, \quad Cx \geqq \theta, \quad Dx = \theta.$$

(ii) *There exist y^1, y^2, y^3, y^4 such that*

$$A^T y^1 + B^T y^2 + C^T y^3 + D^T y^4 = \theta$$

[7] For if $(A, B) \notin \mathscr{W}_0$, there would exist some diagonal matrix $D > 0$ such that $\det(AD + B) = 0$, which implies that for some n-vector $y \neq \theta$, $ADy = -By$ and hence, since (A, B) is a passive pair, $\langle Dy, y \rangle \leqq 0$. But, clearly, for this y, $\langle Dy, y \rangle > 0$.

with either

$$y^1 \geq \theta, \quad y^2 \geqq \theta, \quad y^3 \geqq \theta;$$

or

$$y^1 \geqq \theta, \quad y^2 > \theta, \quad y^3 \geqq \theta.$$

A proof of Slater's theorem is given in [3].

THEOREM 2. *Let (A, B) be a passive pair of real $n \times n$ matrices. Then, for each diagonal matrix $D = \operatorname{diag}(d_1, \cdots, d_n)$ with, for each k, $d_k = +1$ or -1, there exists an n-vector p such that $DA^Tp \geqq \theta$, $DB^Tp \geqq \theta$ and $D(A + B)^Tp > \theta$.*

Proof. It is clear that there exists an n-vector p such that $DA^Tp \geqq \theta$, $DB^Tp \geqq \theta$ and $D(A + B)^Tp > \theta$ if and only if there exists an n-vector p satisfying at least one of the following two sets of inequalities:

(2) $$D(A + B)^Tp > \theta, \quad DA^Tp \geq \theta, \quad DB^Tp \geqq \theta,$$

(3) $$D(A + B)^Tp > \theta, \quad DB^Tp \geq \theta, \quad DA^Tp \geqq \theta.$$

However, by Slater's theorem, there exists an n-vector p satisfying at least one of the sets of inequalities (2), (3) if and only if there is no solution for at least one of the following two equations:

(4a) $$AD(y^1 + y^2) + BD(y^1 + y^3) = \theta$$

with either

(4b) $$y^1 \geq \theta, \quad y^2 \geqq \theta, \quad y^3 \geqq \theta,$$

or

(4c) $$y^1 \geqq \theta, \quad y^2 > \theta, \quad y^3 \geqq \theta;$$

(5a) $$AD(z^1 + z^2) + BD(z^1 + z^3) = \theta$$

with either

(5b) $$z^1 \geq \theta, \quad z^2 \geqq \theta, \quad z^3 \geqq \theta,$$

or

(5c) $$z^1 \geqq \theta, \quad z^2 \geqq \theta, \quad z^3 > \theta.$$

We observe that if both (4a) and (5a) are satisfied under conditions (4c) and (5c), then there exist vectors w^1, w^2 and w^3 such that $w^1 \geqq \theta, w^2 > \theta, w^3 > \theta$, and

$$AD(w^1 + w^2) + BD(w^1 + w^3) = \theta.$$

It follows that there exists an n-vector p satisfying at least one of the sets of inequalities (2), (3) if and only if there exist no vectors y^1, y^2, y^3 such that

(6a) $$AD(y^1 + y^2) + BD(y^1 + y^3) = \theta$$

with either

(6b) $$y^1 \geq \theta, \quad y^2 \geqq \theta, \quad y^3 \geqq \theta,$$

or

(6c) $$y^1 \geqq \theta, \quad y^2 > \theta, \quad y^3 > \theta.$$

But if vectors y^1, y^2, y^3 exist such that either (6b) or (6c) is satisfied, then such vectors would satisfy

$$\langle D(y^1 + y^2), D(y^1 + y^3)\rangle > 0. \tag{7}$$

Such vectors, therefore, could not also satisfy (6a), for (6a) and (7) would then contradict the hypothesis that (A, B) is a passive pair. This completes the proof.

The following corollary of Theorems 1 and 2 provides a convenient criterion for determining whether or not the global inverse of $AF(\cdot) + B$ exists for all cases in which the nonlinear functions f_k which specify F are continuous, strictly monotone increasing mappings of E^1 into E^1, and the pair (A, B) is a passive pair.

COROLLARY 1. *Let $F \in \mathscr{M}^n$ and let (A, B) be a passive pair of real $n \times n$ matrices. Then, there exists a unique solution of* (1) *for each $c \in E^n$ if and only if $\mathscr{B}(F) \cap \mathscr{N}(B) = \{\theta\}$. If $\mathscr{B}(F) \cap \mathscr{N}(B) \neq \{\theta\}$, then there exists some $c \in E^n$ such that* (1) *has no solution.*

It is also possible to show that other classes of pairs of matrices (A, B) possess properties (i) and (ii) of Theorem 1. For example, the following theorem generalizes Propositions 2 and 3 of [4].

THEOREM 3. *Let (A, B) be a pair of $n \times n$ matrices of real numbers with the property that for some nonsingular matrix M, MA and MB are both weakly column-sum dominant matrices and $M(A + B)$ is strongly column-sum dominant. Then, the pair (A, B) possesses properties* (i) *and* (ii) *of Theorem* 1.

Proof. The fact that $(MA, MB) \in \mathscr{W}_0$ follows immediately from Theorem 8 of [7]. Alternatively, it follows directly from the definition of $\mathscr{W}_0$ given earlier, Lemma 1 (below), and the well-known fact that a strongly column-sum dominant matrix is nonsingular.

With $u = (1, \cdots, 1)^T$, and $p = M^T Du$, it follows that $DA^T p \geqq \theta$, $DB^T p \geqq \theta$ and $D(A + B)^T p > \theta$, since any matrix Q is weakly (strongly) row-sum dominant if and only if DQD is weakly (strongly) row-sum dominant.

LEMMA 1. *Let (α, β) be a pair of $n \times n$ matrices of real numbers with the property that α and β are both weakly column-sum dominant, and the matrix $\alpha + \beta$ is strongly column-sum dominant. Then for each diagonal matrix $D > 0$, the matrix $\alpha D + \beta$ is strongly column-sum dominant.*

Proof. Let $D = \operatorname{diag}(d_1, \cdots, d_n) > 0$ be given, and let $j \in \{1, \cdots, n\}$. Either $0 < d_j < 1$ or else $d_j \geqq 1$. If $d_j \geqq 1$, then

$$\begin{aligned}
(\alpha_{jj}d_j + \beta_{jj}) &- \sum_{i \neq j} |\alpha_{ij}d_j + \beta_{ij}| \\
&\geqq (\alpha_{jj} + \beta_{jj}) - \sum_{i \neq j} |\alpha_{ij} + \beta_{ij}| + (d_j - 1)\alpha_{jj} - \sum_{i \neq j} |(d_j - 1)\alpha_{ij}| \\
&= (\alpha_{jj} + \beta_{jj}) - \sum_{i \neq j} |\alpha_{ij} + \beta_{ij}| + (d_j - 1)\left[\alpha_{jj} - \sum_{i \neq j} |\alpha_{ij}|\right] > 0.
\end{aligned}$$

The last inequality is valid since, by hypothesis, $\alpha_{jj} + \beta_{jj} > \sum_{i \neq j} |\alpha_{ij} + \beta_{ij}|$, and $\alpha_{jj} \geqq \sum_{i \neq j} |\alpha_{ij}|$. If $0 < d_j < 1$, then $(\alpha_{jj}d_j + \beta_{jj}) - \sum_{i \neq j} |\alpha_{ij}d_j + \beta_{ij}| > 0$ if (and only if) $(\alpha_{jj} + (1/d_j)\beta_{jj}) - \sum_{i \neq j} |\alpha_{ij} + (1/d_j)\beta_{ij}| > 0$. The last inequality, however, may be verified by using the same technique used in the previous ($d_j \geqq 1$) case. (This time, the hypothesis that $\beta_{jj} \geqq \sum_{i \neq j} |\beta_{ij}|$ is used.)

3. Equations with nonmonotone nonlinearities. The previous section is concerned with equations of the type (1) in which the functions f_k are strictly monotone increasing. For such equations the solution's uniqueness had already been established under the condition that $(A, B) \in \mathscr{W}_0$ (see Theorem 4 of [6]). The main contribution, therefore, was to show how to determine, in cases in which the nonlinear functions f_k were not necessarily *onto* E^1, whether or not a solution *exists* for each $c \in E^n$. We now consider equations of the type (1) in which the nonlinear functions f_k are continuous, but not necessarily monotone. In contrast to the monotone case, we now know at the outset that in some such situations the equations can possess more than one solution even when $(A, B) \in \mathscr{W}_0$. We consider here only the problem of determining conditions under which the equations possess *at least one* solution.

Theorem 4, below, plays an important role in the following development. It shows that if a homeomorphism from E^n onto E^n is modified by the addition of a continuous mapping that does not change the homeomorphism too much (in a certain sense) except, possibly, on some finite region, then the resulting mapping is still at least *onto* E^n.

The following lemma is proved in [1].

LEMMA 2. *Let M be a continuous mapping of E^n into E^n with the property that there exist real numbers $0 < \gamma < 1$ and $k > 0$ such that for every $z \in E^n$ with $\|z\| \geqq k$ it follows that $\|M(z)\| \leqq \gamma\|z\|$. Then for each $y \in E^n$ there exists $x \in E^n$ such that $x + M(x) = y$.*

With the aid of Lemma 2 it is an easy matter to prove the next theorem.

THEOREM 4. *Let U be a homeomorphism of E^n onto E^n, and let V be a continuous mapping of E^n into E^n with the property that there exist real numbers $0 < \gamma < 1$ and $k > 0$ such that for every $q \in E^n$ with $\|q\| \geqq k$ it follows that $\|V(q)\| \leqq \gamma\|U(q)\|$. Then, for each $y \in E^n$ there exists $x \in E^n$ such that $U(x) + V(x) = y$.*

Proof. Since U, and hence U^{-1}, is a homeomorphism of E^n onto E^n, $\|U^{-1}(z)\| \to \infty$ as $\|z\| \to \infty$. Therefore, there exists $k' > 0$ such that $\|U^{-1}(z)\| \geqq k$ whenever $\|z\| \geqq k'$, and hence, $\|V[U^{-1}(z)]\| \leqq \gamma\|z\|$ whenever $\|z\| \geqq k'$. Using Lemma 2, it follows that for each $y \in E^n$ there exists $z \in E^n$ such that $z + V[U^{-1}(z)] = y$. Let $x = U^{-1}(z)$.

Theorem 4 can be used to prove the existence of a solution of (1) in some interesting cases in which the functions f_k entering into the definition of F are not necessarily monotone. Here we consider continuous functions f_k mapping E^1 into E^1 that are strictly monotone increasing for all large arguments – the type of function that characterizes tunnel-diodes, for example.

DEFINITION. For each positive integer n we denote by $\hat{\mathscr{M}}^n$ the class of all mappings F of E^n into E^n with the property that there exist, for $k = 1, \cdots, n$, continuous functions f_k mapping E^1 into E^1, and real numbers $\lambda_k > 0$, with f_k strictly monotone increasing on $\{x_k : |x_k| > \lambda_k\}$, such that for each $x = (x_1, \cdots, x_n)^T \in E^n$, $F(x) = [f_1(x_1), \cdots, f_n(x_n)]^T$.

Obviously $\hat{\mathscr{M}}^n \supset \mathscr{M}^n$. We let the definition of the set $\mathscr{B}(F)$, given in § 2 for $F \in \mathscr{M}^n$, extend to the class $\hat{\mathscr{M}}^n$. All remarks made there concerning the nature of $\mathscr{B}(F)$ hold also for each $F \in \hat{\mathscr{M}}^n$. Using Theorem 1 and Theorem 4, it is not difficult to prove the following theorem.

THEOREM 5. *Let $F \in \hat{\mathscr{M}}^n$ and let (A, B) be a pair of $n \times n$ matrices of real numbers* *possessing properties* (i) *and* (ii) *of Theorem* 1. *Then there exists at least one solution of* (1) *for each* $c \in E^n$ *if* $\mathscr{B}(F) \cap \mathscr{N}(B) = \{\theta\}$.

Proof. For $k = 1, \cdots, n$, let the functions $g_k : E^1 \to E^1$ be defined by

$$g_k(x_k) = f_k(x_k) \qquad \text{for} \quad x_k \leqq -\lambda_k,$$

$$g_k(x_k) = \frac{1}{2\lambda_k}[f_k(\lambda_k) - f_k(-\lambda_k)]x_k + \frac{1}{2}[f_k(\lambda_k) + f_k(-\lambda_k)] \qquad \text{for} \ -\lambda_k < x_k < \lambda_k,$$

$$g_k(x_k) = f_k(x_k) \qquad \text{for} \quad x_k \geqq \lambda_k.$$

Thus, each of the functions g_k is continuous and strictly monotone increasing from E^1 into E^1. Also, each of the functions g_k satisfies $g_k(x_k) = f_k(x_k)$ for all $|x_k| \geqq \lambda_k$. It is therefore clear that if $G(x) \equiv [g_1(x_1), \cdots, g_n(x_n)]^T$, then $\mathscr{B}(G) = \mathscr{B}(F)$. Thus, for each $c \in E^n$, there exists, by Theorem 1, a unique solution of $AG(x) + Bx = c$. We let $U(x) \equiv AG(x) + Bx$. Clearly U is continuous, and we have just shown that U^{-1} exists. That U^{-1} is continuous follows from Theorem 6 of [6]. Thus, U is a homeomorphism of E^n onto E^n. We let $V(x): E^n \to E^n$ be defined by $V(x) = A[F(x) - G(x)]$. V is a continuous mapping of E^n into E^n which, for some $K \geqq 0$ and all sufficiently large $\|x\|$, satisfies $\|V(x)\| \leqq K$. Thus, by Theorem 4, there exists a solution of (1) for each $c \in E^n$.

The following corollary of Theorems 2 and 5 is useful when the pair (A, B) is a passive pair.

COROLLARY 2. *Let* $F \in \hat{\mathscr{M}}^n$ *and let* (A, B) *be a passive pair of real* $n \times n$ *matrices. Then, there exists at least one solution of* (1) *for each* $c \in E^n$ *if* $\mathscr{B}(F) \cap \mathscr{N}(B) = \{\theta\}$.

4. Network-theoretic implications. It seems natural to expect that there exist solutions of equations arising from a problem having physical origin. Whether or not such intuition is warranted, one must certainly distinguish between a physical system and the mathematical model of that system. When constructing the model it is often appropriate to neglect certain attributes of the physical system which appear to be unimportant. As a result, it is quite possible for one to be lead to equations with no solution.

In modeling resistive networks, for example, ideal sources are often used, and nonlinear devices such as diodes are often described by functions whose range is not the entire real line. It is a simple matter to show that such models can yield equations with no solution. It is therefore important that the network theorist have at his disposal methods for determining, at the outset, that there exists a solution (and in many cases, a unique solution) of the network's equations. Since, as discussed in § 1, any resistive network can be described by an equation of type (1) in which (A, B) is a passive pair of matrices, Corollaries 1 and 2 can be useful in this regard.

Often, for networks in which each resistor is described by a continuous, strictly monotone increasing function, the existence of a unique solution for sources having arbitrary values can easily be established by examining the signs of the components of each nonzero real vector in the null space of the matrix B. In case $\mathscr{B}(F) \cap \mathscr{N}(B) \neq \{\theta\}$, it might happen that the network's equation has no solution.

On the other hand, although we know (by Corollary 1) that there exists some $c \in E^n$ such that (1) has no solution, it may or may not be the case that the network's independent sources can be assigned values such that this vector c is obtained in (1).

As a simple illustration of the points made above, consider the network of Fig. 3, in which we assume that the nonlinear resistors R_2, R_3 and R_4 are described by the functions whose graphs are shown in Fig. 4. We wish to determine: *for which types of continuous, strictly monotone increasing, nonlinear resistors R_1 does the network admit a unique solution for all values of the independent sources E and I*?

If R_1 is a voltage-controlled nonlinear resistor, then the equation

$$\begin{bmatrix} 0 & 0 & 0 & 0 \\ 0 & 0 & 0 & -1 \\ 1 & 1 & 0 & 0 \\ 0 & 0 & 1 & 0 \end{bmatrix} \begin{pmatrix} f_1(v_1) \\ f_2(v_2) \\ f_3(v_3) \\ f_4(i_4) \end{pmatrix} + \begin{bmatrix} 1 & -1 & 0 & 0 \\ 0 & 0 & 1 & 0 \\ 0 & 0 & 0 & 0 \\ 0 & 0 & 0 & 1 \end{bmatrix} \begin{pmatrix} v_1 \\ v_2 \\ v_3 \\ i_4 \end{pmatrix} = \begin{pmatrix} E \\ E \\ -I \\ -I \end{pmatrix} \tag{8}$$

describes the network. Clearly,

$$\mathcal{N}(B) = \{x : x_1 = x_2, x_3 = x_4 = 0\}.$$

Now $\mathcal{B}(F) = I_1 \times \cdots \times I_4$, where $I_2 = (-\infty, \infty)$, $I_3 = (-\infty, 0]$ and $I_4 = (-\infty, \infty)$. Thus, we readily determine that $\mathcal{B}(F) \cap \mathcal{N}(B) = \{\theta\}$ if and only if $I_1 = \{0\}$; that is, if and only if the range of f_1 is the entire real line.

In case R_1 is a current-controlled nonlinear resistor, the network is described by the equation (obtained from (8) by interchanging the first columns of the matrices A and B)

$$\begin{bmatrix} 1 & 0 & 0 & 0 \\ 0 & 0 & 0 & -1 \\ 0 & 1 & 0 & 0 \\ 0 & 0 & 1 & 0 \end{bmatrix} \begin{pmatrix} f_1(i_1) \\ f_2(v_2) \\ f_3(v_3) \\ f_4(i_4) \end{pmatrix} + \begin{bmatrix} 0 & -1 & 0 & 0 \\ 0 & 0 & 1 & 0 \\ 1 & 0 & 0 & 0 \\ 0 & 0 & 0 & 1 \end{bmatrix} \begin{pmatrix} i_1 \\ v_2 \\ v_3 \\ i_4 \end{pmatrix} = \begin{pmatrix} E \\ E \\ -I \\ -I \end{pmatrix}$$

In this case, $\mathcal{N}(B) = \{\theta\}$, and hence for *any* continuous, strictly monotone increasing function f_1, $\mathcal{B}(F) \cap \mathcal{N}(B) = \{\theta\}$.

If R_1 is a voltage-controlled nonlinear resistor whose range is not the entire real line, it is clear that the independent sources (in particular the current source I) can be assigned values such that the network admits no solution.

Note that $\mathcal{B}(F) = \{\theta\}$ for any network in which each nonlinear resistor is described by a function that is strictly monotone increasing from the real line *onto* the real line. Hence, in such cases $\mathcal{B}(F) \cap \mathcal{N}(B) = \{\theta\}$, and thus it is clear that there exists a unique solution of the network's equation. This result (the special case of networks containing only resistors that map E^1 *onto* E^1) is a well-known result, proved by R. J. Duffin [2] by an entirely different method.

Although the advances that have been made in recent years concerning the techniques of computation of solutions of circuit equations have been significant in their own right, they have provided very little fundamental *understanding* of "circuits". There is no reason to expect that they should have. Thus, although it is

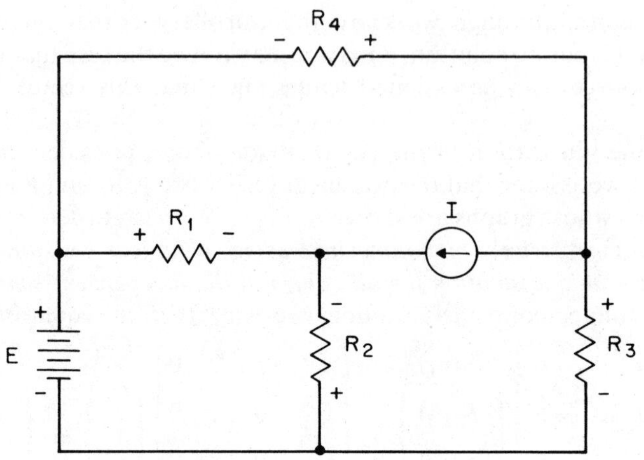

FIG. 3. *Example of a resistive network*

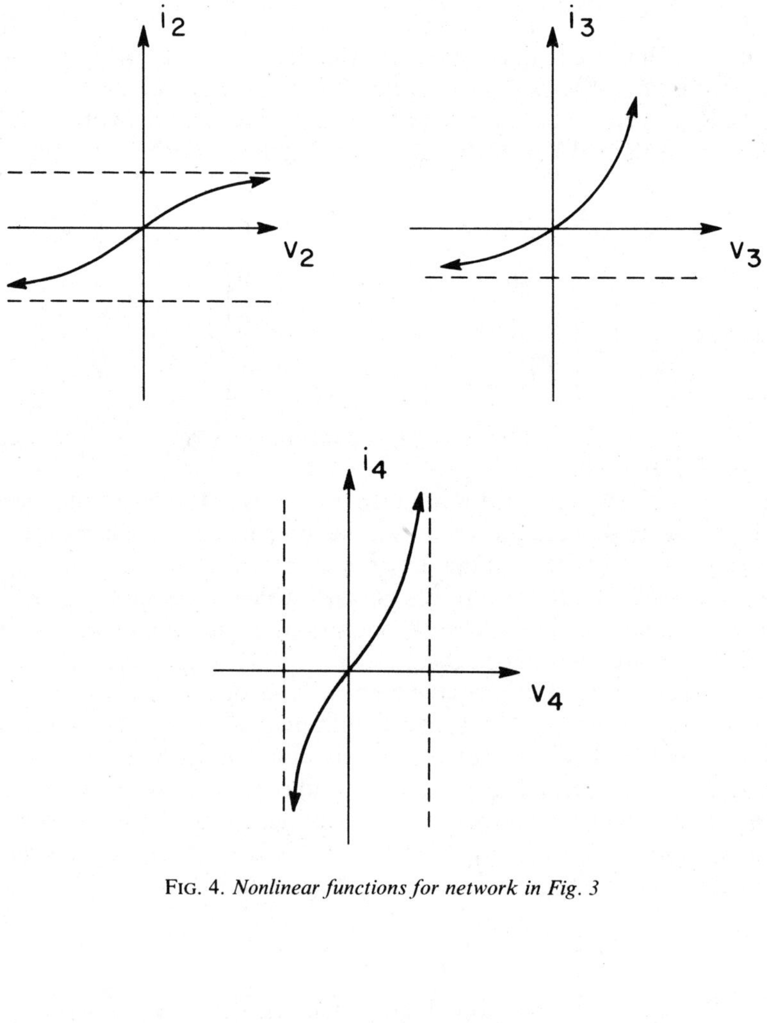

FIG. 4. *Nonlinear functions for network in Fig. 3*

certainly true that the ideas presented here are useful (in that they could be used, say, in a circuit analysis program to make sure that time and money are not wasted on certain pathological problems) it is felt that the full impact of work such as this can only be appreciated when a broader point of view is taken. The primary importance of results such as those presented here lies in the fact that they contribute to a sound analytical basis upon which the theory of nonlinear networks may rest.

Appendix.

LEMMA A1. *Let B be an $n \times n$ matrix of real numbers. Let $R = I_1 \times \cdots \times I_n$, where, for $k = 1, \cdots, n$, I_k is one of the intervals of the real line*: $(-\infty, \infty), (-\infty, 0]$, $[0, \infty), \{0\}$. *Then, unless $R = \{\theta\}$, the number* $\inf\{\|By\| : y \in R, \|y\| = 1\}$ *is positive if and only if $R \cap \mathcal{N}(B) = \{\theta\}$.*

Proof. It is obvious that if $R \cap \mathcal{N}(B) \neq \{\theta\}$ then $\inf\{\|By\| : y \in R, \|y\| = 1\} = 0$.

Assume now that $R \cap \mathcal{N}(B) = \{\theta\}$. This implies that there is no point y in the subset $S \subset R$, $S = \{y : y \in R, \|y\| = 1\}$, for which $\|By\| = 0$. But since R is closed, S is a closed subset of the compact set $\{y : \|y\| = 1\}$. Thus, S is compact. Clearly S is nonempty if $R \neq \{\theta\}$. Therefore, the continuous function $\|By\|$ achieves its greatest lower bound on S for some $y \in S$. For that y we have $\|By\| > 0$.

LEMMA A2. *Let $F \in \mathcal{M}^n$ and let (A, B) be a pair of $n \times n$ matrices of real numbers possessing the properties*:

(i) $(A, B) \in \mathcal{W}_0$;

(ii) *there exists a continuous function ρ mapping $[0, \infty)$ into $[0, \infty)$ such that $AF(y) + By = z$ implies that for $k = 1, \cdots, n$, $|f_k(y_k)| \leqq \rho(\|z\|)$.*

Then there exists a unique solution of (1) *for each $c \in E^n$ if and only if $\mathcal{B}(F) \cap \mathcal{N}(B) = \{\theta\}$. If $\mathcal{B}(F) \cap \mathcal{N}(B) \neq \{\theta\}$, then there exists some $c \in E^n$ such that* (1) *has no solution.*

Proof. We note first that if $\mathcal{B}(F) = \{\theta\}$ then, by Theorem 3 of [6], (1) possesses a unique solution x for each $c \in E^n$. We therefore assume, throughout the remainder of the proof, that $\mathcal{B}(F) \neq \{\theta\}$.

Since $(A, B) \in \mathcal{W}_0$ and $F \in \mathcal{M}^n$, it follows that the mapping $AF(\cdot) + B$ is a local homeomorphism. To prove that this mapping is a homeomorphism of E^n onto E^n, it therefore suffices to show that for each positive constant K_1 there exists a positive constant K_2 such that $\|AF(y) + By\| \leqq K_1$ implies that $\|y\| \leqq K_2$. (See the proof of Lemma 2, in [4].)

Let y and z be arbitrary points in E^n such that $AF(y) + By = z$. Since $|f_k(y_k)| \leqq \rho(\|z\|)$ for all k, we observe that there exist continuous functions g and g_0, both mapping $[0, \infty)$ into itself, such that

$$y = -g(\|z\|)s + w + v, \tag{A.1}$$

in which $s = (s_1, \cdots, s_n)^T$ is the point in $\mathcal{B}(F) \equiv I_1(f_1) \times \cdots \times I_n(f_n)$ defined by, for $k = 1, \cdots, n$,

$$s_k = 1, \qquad \text{if } I_k = [0, \infty),$$

$$s_k = -1, \qquad \text{if } I_k = (-\infty, 0],$$

$$s_k = 0, \qquad \text{if } I_k = \{0\}, \text{ or } (-\infty, \infty),$$

and, in which $w \in \mathscr{B}(F)$ and v satisfies $\|v\| \leqq g_0(\|z\|)$ and $v_k = 0$ for all k such that $|s_k| > 0$.[8]

Therefore,

$$Bw = g(\|z\|)Bs - Bv + z - AF(y),$$

and hence,

$$\|Bw\| \leqq g(\|z\|)\|Bs\| + \|Bv\| + \|z\| + \|AF(y)\|$$

$$\leqq g(\|z\|)\|Bs\| + bg_0(\|z\|) + \|z\| + an^{1/2}\rho(\|z\|),$$

in which $a = \sup\{\xi : \|Ax\| \leqq \xi\|x\|, x \in E^n\}$ and $b = \sup\{\xi : \|Bx\| \leqq \xi\|x\|, x \in E^n\}$. Let h be the continuous mapping of $[0, \infty)$ into itself defined by

$$h(\beta) = g(\beta)\|Bs\| + bg_0(\beta) + \beta + an^{1/2}\rho(\beta)$$

for all $\beta \in [0, \infty)$. Since we assume that $\mathscr{B}(F) \cap \mathscr{N}(B) = \{0\}$, it follows from Lemma A1 that there exists a positive constant γ, depending only on B, such that $\|Bq\| \geqq \gamma\|q\|$ for all $q \in \mathscr{B}(F)$. Thus, $\|w\| \leqq \gamma^{-1}h(\|z\|)$ and, using (A.1),

$$\|y\| \leqq n^{1/2}g(\|z\|) + \gamma^{-1}h(\|z\|) + g_0(\|z\|),$$

which shows that for each positive constant K_1 there exists a positive constant K_2 such that $\|z\| \leqq K_1$ implies that $\|y\| \leqq K_2$.

A proof of the second part of this lemma can be constructed by making very obvious modifications to the corresponding part of the proof of Lemma 2 of [4]. The details are left for the reader.

LEMMA A3. *Let $F \in \mathscr{M}^n$ with $F(0) = 0$. Let (A, B) be a pair of $n \times n$ matrices of real numbers possessing the property: For each diagonal matrix $D = \operatorname{diag}(d_1, \cdots, d_n)$ with, for each k, $d_k = +1$ or -1, there exists a real n-vector p such that $DA^Tp \geqq 0$, $DB^Tp \geqq 0$ and $D(A + B)^Tp > 0$.*

Then there exists a positive constant σ, depending only on the pair (A, B), such that $AF(y) + By = z$ implies that, for $k = 1, \cdots, n$, either $|f_k(y_k)| \leqq \sigma\|z\|$ or $|y_k| \leqq \sigma\|z\|$.

Proof. There are 2^n matrices $D = \operatorname{diag}(d_1, \cdots, d_n)$ in which $d_k = 1$ or -1, and for each such D there exists an n-vector p such that $DA^Tp \geqq 0$, $DB^Tp \geqq 0$ and $D(A + B)^Tp > 0$. Let $\mathscr{P}$ denote any set of at most 2^n elements of E^n such that for each of the 2^n matrices D there exists a $p \in \mathscr{P}$ such that $DA^Tp \geqq 0$, $DB^Tp \geqq 0$ and $D(A + B)^Tp > 0$, and such that for each $k = 1, \cdots, n$,

$$\text{either } (DA^Tp)_k = 0 \quad \text{or else} \quad (DA^Tp)_k \geqq 1,$$

and

$$\text{either } (DB^Tp)_k = 0 \quad \text{or else} \quad (DB^Tp)_k \geqq 1.$$

Let the n-vectors y, z satisfy $AF(y) + By = z$. For at least one of the 2^n matrices D we can write $y = Dw$ with $w \geqq 0$. Thus, $AF(Dw) + BDw = z$. Let $p \in \mathscr{P}$ be such that $DA^Tp \geqq 0$, $DB^Tp \geqq 0$ and $D(A + B)^Tp > 0$. Then,

$$p^TADDF(Dw) + p^TBDw = p^Tz,$$

[8] The same functions g, g_0 specified in the proof of Lemma 2 of [4] may, for example, be used here.

or

$$\sum_{k=1}^{n} [(DA^Tp)_k|f_k(y_k)| + (DB^Tp)_k|y_k|] = p^Tz \leqq \|p\| \cdot \|z\|.$$

Thus, for $k = 1, \cdots, n$,

$$(DA^Tp)_k|f_k(y_k)| + (DB^Tp)_k|y_k| \leqq \|p\| \cdot \|z\| \leqq \sigma\|z\|,$$

in which $\sigma = \max_{p\in\mathscr{P}} \|p\|$.

Therefore, either $|f_k(y_k)| \leqq \sigma\|z\|$, or else $|y_k| \leqq \sigma\|z\|$.

Remark. Given that the conclusion of Lemma A3 holds, it is a trivial matter to demonstrate the existence of a function ρ for which the inequalities in property (ii) of Lemma A2 hold. For example, for $\sigma > 0$, define the continuous, strictly monotone increasing function φ_σ mapping the interval $[0, \infty)$ into itself by $\varphi_\sigma(\tau) = \max\{|f_k(t)| : |t| \leqq \sigma\tau, k = 1, \cdots, n\}$. Then we may define $\rho(\|z\|) = \max\{\varphi_\sigma(\|z\|), \sigma\|z\|\}$.

REFERENCES

[1] V. E. BENEŠ AND I. W. SANDBERG, *Applications of a theorem of Dubrovskii to the periodic responses of nonlinear systems*, Bell System Tech. J., 43 (1964), pp. 2855–2872.

[2] R. J. DUFFIN, *Nonlinear networks. IIa*, Bull. Amer. Math. Soc., 53 (1947), pp. 963–971.

[3] O. L. MANGASARIAN, *Nonlinear Programming*, McGraw-Hill, New York, 1969.

[4] I. W. SANDBERG, *Necessary and sufficient conditions for the global invertibility of certain nonlinear operators that arise in the analysis of networks*, IEEE Trans. Circuit Theory, CT-18 (1971), pp. 260–263.

[5] I. W. SANDBERG AND A. N. WILLSON, JR., *Some theorems on properties of dc equations of nonlinear networks*, Bell System Tech. J., 48 (1969), pp. 1–34.

[6] A. N. WILLSON, JR., *New theorems on the equations of nonlinear dc transistor networks*, Ibid., 49 (1970), pp. 1713–1738.

[7] ———, *A useful generalization of the P_0 matrix concept*, Numer. Math., 17 (1971), pp. 62–70.

NONLINEAR MONOTONE NETWORKS*

CHARLES A. DESOER AND FELIX F. WU†

Abstract. This paper presents some fundamental results on nonlinear networks with uncoupled, monotone increasing, but not necessarily surjective characteristics.

Necessary and sufficient conditions are obtained for the existence and uniqueness of solutions for *strictly increasing* resistive networks and a general class of *increasing* resistive networks. They are also necessary and sufficient for the existence of solutions for *increasing* resistive networks. These conditions are sufficient for the existence of solutions for *eventually strictly increasing* resistive networks, and for a class of *eventually increasing* resistive networks. The conditions are circuit-theoretic and can readily be used as a criterion in design. The dependence of solutions on the inputs is studied and also a bounded-input bounded-solution result is presented. Existence and uniqueness results for monotone RLC networks are obtained by viewing them as combinations of three one-element-kind subnetworks. Finally, two algorithms are given for testing the conditions.

1. Introduction. In this paper we present some fundamental results on nonlinear networks with uncoupled, monotone increasing, but not necessarily surjective characteristics. Necessary and sufficient conditions are obtained for the existence and uniqueness of solutions for *strictly increasing* resistive networks and a general class of *increasing* networks. In such cases, the solution depends continuously on the inputs. These conditions are also necessary and sufficient for the existence of solutions for *increasing* resistive networks. They are sufficient for the existence of solutions for *eventually strictly increasing* resistive networks, and for a class of *eventually increasing* resistive networks. The dependence of these solutions on the inputs is then studied and a bounded-input bounded-solution result is presented. Existence and uniqueness results for monotone RLC networks are obtained by viewing them as combinations of three one-element-kind subnetworks. Finally, two algorithms are given for testing the conditions.

The nature of our necessary and sufficient conditions is circuit-theoretic in the sense that they are checked by considering the network topology and the element characteristics rather than evaluating determinants, or eigenvalues, etc. Moreover, in case one of the conditions fails, the proposed algorithms will pinpoint where the network needs to be modified.

Nonlinear resistive networks were studied by Duffin [1] in 1946. In 1951, Millar [22] points out that there were two dual variational principles associated with solutions (thus generalizing theorems of Maxwell for linear networks); the extremalized quantities were called by him "content" and "co-content". (However, Duffin's proof of existence of a solution was essentially to prove existence of an extremum for the co-content.) Duffin's treatment of existence covered certain cases of nonmonotone resistor characteristics, but he postulated monotonicity in his proof of uniqueness of a solution. His sufficient conditions for existence of a solution are primitive, and far from being necessary conditions.

---

* Received by the editors January 4, 1972, and in revised form March 30, 1973.

† Department of Electrical Engineering and Computer Sciences and the Electronics Research Laboratory, University of California, Berkeley, Calif. 94720. This research was sponsored by the National Science Foundation under Grant GK-10656X1.

Reprinted with permission from *SIAM Journal of Applied Mathematics*, vol. 26, pp. 315–333, March 1974.

In 1960, Minty [15] developed further the methods which have since been called "network programming methods" to study resistive monotone networks. The resistor characteristics are approximated by step functions, and a computational algorithm [19] based on the colored arc lemma is developed to solve the approximate problem exactly. An existence theorem (necessary and sufficient conditions for existence of a solution) is proved by going to the limit as the step-function approximations become closer; the two kinds of degeneracy which prevent existence of a solution are named (i) "unbalanced cycle" which corresponds to a cycle (loop in our language) for which there is no set of possible voltages that sum to zero (hence the Kirchhoff voltage law is violated), and (ii) "unbalanced co-cycle" which is the dual of the first and corresponds to the "infeasible cut" of network programmers. By restricting attention to the case of monotone characteristics, he was able to treat the two extremum principles dually. (For nonmonotone characteristics, one or both of the content/co-content becomes multiple-valued and difficult to treat.)

The present treatment is based on methods of "global analysis" and specifically the "global implicit function theorem" and Brouwer's "invariance of domain" theorem. We consider the "inputs" (independent current and voltage sources) as variables and obtain conditions under which the "input vs. output" function is a homeomorphism. Thus the proof of the existence theorem for a solution appears "less constructive" than the methods of Duffin or Minty, but permits simultaneously the examination of the continuous dependence of output on input, a topic not treated by these two authors. Our conditions (i) and (ii) of Theorem 1 are related to Minty's "unbalanced cycle" and "unbalanced co-cycle".

It should be remarked that Minty's results have been generalized, in the literature, in two significant directions: by Minty [20] to an abstract mathematical structure called a "matroid" or "graphoid" which replaces the "network" of network theory, and has the virtue of exhibiting perfect network duality, and by Rockafellar [21] to the case where the "currents" and "voltage drops" are an arbitrary pair of orthogonal complementary subspaces of R^n, not necessarily derived from any analogue of Kirchhoff's laws. The theorems of the present paper could be proved in the graphoid context with very little rewording of proofs. Rockafellar's treatment cannot easily be compared with that of the present paper, as his existence conditions are not circuit-theoretic. For other relevant literature, we refer the reader to the bibliographies of [14], [15], [20] and [21].

Technological progress has tremendously decreased the price of semiconductor devices, hence made them more important. Since many semiconductor devices have monotone characteristics, networks with monotone nonlinearities have recently received considerable attention. Sandberg and Willson [3]–[6] have made significant advances both theoretically and computationally on networks with strictly increasing nonlinearities. The research reported here was stimulated by the work of Sandberg and Willson and is a generalization of the work of Desoer and Katzenelson.

Proofs of the theorems are included in the text because they improve the understanding of the results. For ease of reference, we state three theorems in Appendix A. In Appendix B, we derive two lemmas which are used in the proof of Theorem 1.

2. Formulation.

2.1. Resistors. In this paper, we define a *resistor* as a two-terminal element that, at any instant time t, is characterized by a *continuous* f which maps the real line $\mathbb{R}$ into itself and $\sigma = f(\rho)$, where either ρ is the branch voltage and σ the branch current, or vice versa. If ρ is the branch voltage, we say that the resistor is *voltage controlled* (v.c.); on the other hand, if ρ is the branch current, we say that the resistor is *current controlled* (c.c.). Such resistors are thus not necessarily linear, not necessarily time-invariant, uncoupled, and either v.c. or c.c. (and possibly both). In the considerations that follow, conditions are examined for fixed t, so that we formulate, for simplicity, as if the resistors were time-invariant.

A resistor is said to be *increasing* if its characteristic f satisfies $f(\rho_2) \geqq f(\rho_1)$ whenever $\rho_2 \geqq \rho_1$; and *strictly increasing* if $f(\rho_2) > f(\rho_1)$ whenever $\rho_2 > \rho_1$. A resistor is said to be *eventually (strictly) increasing* if its characteristic is (strictly) increasing on $\{\rho \in \mathbb{R} \mid |\rho| > M\}$ for some M, which depends on the resistor under consideration.

We say that a resistor is of *type U* if its characteristic has the property that $f(\rho) \to \infty$ as $\rho \to \infty$ and $f(\rho) \to -\infty$ as $\rho \to -\infty$; of *type H* if either (i) $f(\rho) \to -\infty$ as $\rho \to -\infty$ and $|f(\rho)| < B$ for some B as $\rho \to \infty$, or (ii) $|f(\rho)| < B$ for some B as $\rho \to -\infty$ and $f(\rho) \to \infty$ as $\rho \to \infty$; of *type B* if $|f(\rho)| < M$ as $|\rho| \to \infty$ (see Fig. 1).

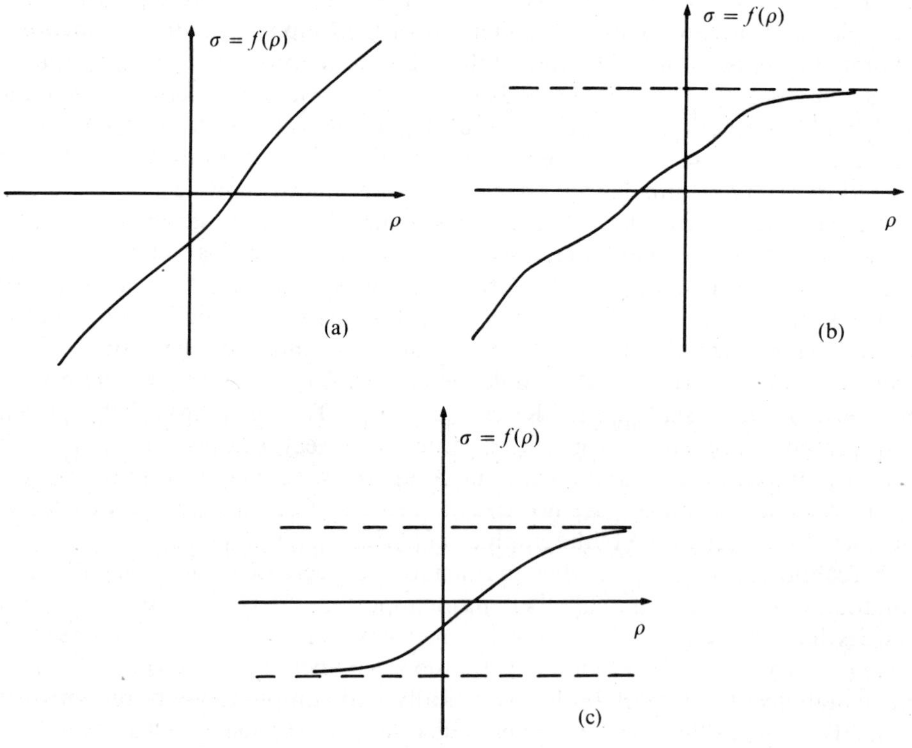

FIG. 1

(a) *Type U resistor: $f(\rho)$ unbounded as ρ grows without bound*

(b) *Type H resistor: $f(\rho)$ bounded on a half real line*

(c) *Type B resistor: $f(\rho)$ bounded on the whole real line*

Clearly the set of all increasing (resp. strictly increasing, eventually increasing, eventually strictly increasing) resistors can be partitioned into type U, type H, and type B resistors.

2.2. Network. Let $\mathcal{N}$ be an interconnection of a finite number of resistors. Without loss of generality, $\mathcal{N}$ is assumed to have a connected and nonseparable graph. If $\mathcal{N}$ inherently has some independent sources, they may be regarded as increasing resistors, or, by source transformation [7, pp. 409–412], be absorbed in the resistive branches.

2.3. Network topology. Let the network variables be partitioned into (v_v, v_c) and (i_v, i_c), where subscripts v and c denote those corresponding to the v.c. resistors and c.c. resistors, respectively.[1] Let us pick a tree which contains the maximum number of v.c. resistors. If $\mathcal{N}$ contains any type H resistor, let the reference direction for that branch be so chosen that $|f(\rho)| < M$ for some M as $\rho \to \infty$. The fundamental loop matrix B and the fundamental cutset matrix Q corresponding to such a choice of tree takes the form[2]:

$$B = \begin{array}{c} \\ vl \\ cl \end{array}\!\!\begin{array}{c} \begin{array}{cccc} vl & cl & vt & ct \end{array} \\ \begin{bmatrix} I & 0 & F_{vv} & 0 \\ 0 & I & F_{vc} & F_{cc} \end{bmatrix} \end{array}$$

$$Q = \begin{array}{c} vt \\ ct \end{array}\!\!\begin{bmatrix} -F_{vv}^T & -F_{vc}^T & I & 0 \\ 0 & -F_{cc}^T & 0 & I \end{bmatrix},$$

where subscript l (resp. t) denotes links (resp. tree branches); hence the double subscript vl, for example, denotes v.c. link resistors. According to this partition, we have

$$\begin{aligned} i_{vl} &= \hat{\imath}_{vl}(v_{vl}), \\ i_{vt} &= \hat{\imath}_{vt}(v_{vt}), \\ v_{cl} &= \hat{v}_{cl}(i_{cl}), \\ v_{ct} &= \hat{v}_{ct}(i_{ct}). \end{aligned} \tag{1}$$

2.4. Independent sources. There are two ways of applying independent sources to a network: namely, pliers entry and soldering iron entry. By *pliers entry* we mean that we enter the network by cutting any branch of the network and connecting the two terminals of a source to the terminals created by the cut. By *soldering iron entry* we mean that we enter the network by connecting the two terminals of a source to any two nodes of the network. Throughout the following, we apply a voltage source only by a pliers entry and a current source only by a soldering iron entry.

2.5. Network equations. Let e_v (resp. e_c) denote the voltage-source vector around fundamental loops defined by v.c. (resp. c.c.) resistive links; j_v (resp. j_c) denotes the current-source vector across a fundamental cutset defined by v.c.

[1] In the case where a resistor is both v.c. and c.c., we can assign it to either class.

[2] Superscript T denotes transpose of a matrix.

(resp. c.c.) resistive tree branches. The Kirchhoff laws are thus expressed by

$$\text{(2a)} \qquad \begin{aligned} i_{ct} - F_{cc}^T i_{cl} &= j_c, \\ v_{vl} + F_{vv} v_{vt} &= e_v, \end{aligned}$$

$$\text{(2b)} \qquad \begin{aligned} v_{cl} + F_{cc} v_{ct} + F_{vc} v_{vt} &= e_c, \\ i_{vt} - F_{vv}^T i_{vl} - F_{vc}^T i_{cl} &= j_v. \end{aligned}$$

Substituting (1) into (2) and eliminating i_{ct} and v_{vl}, we obtain two equations in terms of v_{vt} and i_{cl}:

$$\text{(3)} \qquad \begin{bmatrix} -F_{vv}^T & I & 0 & 0 \\ 0 & 0 & I & F_{cc} \end{bmatrix} \begin{bmatrix} \hat{\imath}_{vl}(-F_{vv} v_{vt} + e_v) \\ \hat{\imath}_{vt}(v_{vt}) \\ \hat{v}_{cl}(i_{cl}) \\ \hat{v}_{ct}(F_{cc}^T i_{cl} + j_c) \end{bmatrix} + \begin{bmatrix} 0 & -F_{vc}^T \\ F_{vc} & 0 \end{bmatrix} \begin{bmatrix} v_{vt} \\ i_{cl} \end{bmatrix} = \begin{bmatrix} j_v \\ e_c \end{bmatrix}.$$

2.6. Notations. To simplify the presentation, we sometimes write (3) in the form:

$$\text{(3')} \qquad C\mathscr{F}(C^T x + u) + Sx = y,$$

where

$$x = \begin{bmatrix} v_{tc} \\ i_{cl} \end{bmatrix} \in \mathbb{R}^m, \qquad y = \begin{bmatrix} j_v \\ e_c \end{bmatrix} \in \mathbb{R}^m,$$

$$u = \begin{bmatrix} e_v \\ 0 \\ 0 \\ j_c \end{bmatrix} \in \mathbb{R}^r, \qquad \mathscr{F} = \begin{bmatrix} \hat{\imath}_{vl} \\ \hat{\imath}_{vt} \\ \hat{v}_{cl} \\ \hat{v}_{ct} \end{bmatrix} : \mathbb{R}^r \to \mathbb{R}^r,$$

$$C = \begin{bmatrix} -F_{vv}^T & I & 0 & 0 \\ 0 & 0 & I & F_{cc} \end{bmatrix} \in \mathbb{R}^{m \times r}, \qquad S = \begin{bmatrix} 0 & -F_{vc}^T \\ F_{vc} & 0 \end{bmatrix} \in \mathbb{R}^{m \times m}.$$

$\mathbb{R}^m$ denotes Euclidean m-space with scalar product $\langle x|y \rangle = \sum_{k=1}^m x_k y_k$ and norm $\|x\| = (\sum_{k=1}^m x_k^2)^{1/2}$.

We sometimes consider the map $G: \mathbb{R}^m \times \mathbb{R}^r \to \mathbb{R}^m$ defined by

$$\text{(4)} \qquad G(x, u) = C\mathscr{F}(C^T x + u) + Sx.$$

Note that S is a real skew-symmetric matrix, C is a matrix of full rank, and $\mathscr{F}$ is a "diagonal" map, i.e., $\mathscr{F}(z) = [f_1(z_1) \cdots f_r(z_r)]^T$. The zero element in $\mathbb{R}^m$ is denoted by θ. The null space of a matrix S is denoted by $\mathscr{N}(S)$, i.e., $\mathscr{N}(S) = \{x | Sx = \theta\}$. $D_1 G(x, u)$ denotes the derivative map of $G(\cdot, u)$ evaluated at x.

Let us define the cone

$$\text{(5)} \qquad \mathscr{B}(\mathscr{F}) \triangleq \{z \in \mathbb{R}^r | \rho \mapsto \|\mathscr{F}(\rho z)\| \text{ is bounded on } [1, \infty)\}.$$

Clearly $z = (z_1, \cdots, z_r) \in \mathscr{B}(\mathscr{F})$ if and only if $z_i = 0$ whenever f_i is of type U; $z_i \geqq 0$ whenever f_i is of type H (Note that by our convention $|f_i(\rho)| < M$ as $\rho \to \infty$), and z_i = any real number whenever f_i is of type B.

By a *solution* of a network, we mean a set of branch voltages and branch currents $\chi = (v, i)$ (including currents in the voltage sources and voltages across the current sources) that satisfy both Kirchhoff laws and the branch characteristics. By an *input*, we mean an independent source vector $\mu = (e_v, e_c, j_v, j_c)$. We say that a solution *depends continuously* on the inputs if and only if considering χ as a function of μ, i.e., $\chi = \hat{\chi}(\mu)$, $\hat{\chi}(\cdot)$ is a continuous function on $\mathbb{R}^r$.

A set of branches belonging to a loop (resp. cutset) in a directed graph is said to be *similarly directed* if we can assign a reference direction to the loop (resp. cutset) such that the direction of each branch in the set agrees with the reference direction of the loop (resp. cutset).

3. Strictly increasing resistive networks. In this section we prove Theorem 1 for strictly increasing resistive networks. We first prove two lemmas, which, taken together, assert that Theorem 1 is true if all resistor characteristics are C^1 (in such a case, the dependence of the unique solution on the inputs is C^1). Then in the proof of Theorem 1 we show that with the aid of two lemmas in Appendix B, the C^1 assumption can be dropped.

THEOREM 1 (Existence, uniqueness, and continuous dependence). *Let $\mathcal{N}$ be a finite network made of* strictly increasing, *continuous, and uncoupled resistors. Then for all independent voltage sources with pliers entries and for all independent current sources with soldering iron entries, the network $\mathcal{N}$ has one and only one solution and this solution depends continuously on the inputs if and only if the following conditions* (i) *and* (ii) *hold*:

(i) *Every loop made of c.c. resistors either contains at least one type U c.c. resistor or if not, then it contains at least two type H c.c. resistors and not all such type H c.c. resistors are similarly directed.*

(ii) *Every cutset made of v.c. resistors either contains at least one type U v.c. resistor or if not, then it contains at least two type H v.c. resistors and not all such type H v.c. resistors are similarly directed.*

Comments. (a) Uniqueness follows directly (by Tellegen's theorem) from the strictly increasing property of the characteristics.

(b) Conditions (i) and (ii) are the duals of each other.

(c) If there is a loop (resp. cutset) of type B c.c. (resp. v.c.) resistors, then condition (i) (resp. condition (ii)) does not hold. If there is a loop (resp. cutset) of type B and type H c.c. (resp. v.c.) resistors, in which all type H resistors are similarly directed, then condition (i) (resp. condition (ii)) does not hold.

(d) Condition (i) (resp. (ii)) is equivalent to: (ia) (resp. (iia)) there is no loop (resp. cutset) of only type B c.c. (resp. v.c.) resistors; (ib) (resp. (iib)) for every type H c.c. (resp. v.c.) resistor b, there is a cutset (resp. loop) containing b, and made of v.c., type U c.c., and type H c.c. (resp. c.c., type U v.c., and type H v.c.) resistors, in which all type H c.c. (resp. v.c.) resistors are similarly directed (by the colored arc lemma (Appendix A)).

LEMMA 1. *Consider equation* (3′):

$$G(x, u) \triangleq C\mathcal{F}(C^T x + u) + Sx = y.$$

Suppose that each f_i is C^1 and strictly increasing. If

$$\{x \in \mathbb{R}^m | C^T x \in \mathcal{B}(\mathcal{F}) \text{ and } x \in \mathcal{N}(S)\} = \{\theta\},$$

then there is a unique C^1*-function* $\Phi:\mathbb{R}^r \times \mathbb{R}^m \to \mathbb{R}^m$, *satisfying* $G(\Phi(u, y), y) = y$, *for all* $y \in \mathbb{R}^m$, $u \in R^r$ *(consequently, for all* $u \in \mathbb{R}^r$, $G(\cdot, u)$ *is a diffeomorphism from* $\mathbb{R}^m$ *onto* $\mathbb{R}^m$*).*

Proof. CLAIM 1. *$D_1G(x, u)$ is nonsingular for all* $x \in \mathbb{R}^m$, $u \in \mathbb{R}^r$. Differentiating (4), $D_1G(x, u) = C[D\mathscr{F}(C^Tx + u)]C^T + S$. Since each component of $\mathscr{F}$ is strictly increasing, $[D\mathscr{F}(C^Tx + u)]$ is diagonal and positive definite for all (x, u). Suppose $D_1G(x, u)$ were singular for some (x, u); then there would be a $\xi \neq \theta$ such that $[D_1G]\xi = \theta$. Note that $\xi \neq \theta$ implies $C^T\xi \neq \theta$ because C is of full rank. Consider $\xi^T[D_1G]\xi = \xi^TC[D\mathscr{F}]C^T\xi = \theta$, which contradicts that $[D\mathscr{F}]$ is positive definite. Therefore, $D_1G(x, u)$ is nonsingular for all (x, u).

CLAIM 2. $\|x\| \to \infty \Rightarrow \|G(x, u)\| \to \infty$ *for all* $u \in \mathbb{R}^r$ *(i.e., for any sequence* $\{x^i\}$ *such that* $\|x^i\| \to \infty$ *implies* $\|G(x^i, u)\| \to \infty$ *for all* $u \in \mathbb{R}^r$*).*

Let $z = C^Tx$, $z \in \mathbb{R}^r$; hence any sequence $\{x^i\}$ defines a sequence $\{z^i\}$ in $\mathbb{R}^r$, where $z^i = C^Tx^i$. Let us partition $\mathbb{R}^r$ into its 2^r orthants; $\{z^i\}$ will accordingly be partitioned into at most 2^r subsequences. Note that C^T contains an identity matrix, thus $\{x^i\} \to \infty$ if and only if $\|z^i\| \to \infty$ and hence at least one of the subsequences is unbounded. We will first consider $\{z|z \in \mathscr{B}(\mathscr{F})\}$, which is a closed convex cone consisting of the union of several orthants, and show that $\|z^i\| \to \infty$ with $z^i \in \mathscr{B}(\mathscr{F})$ implies that the corresponding $\|G(x^i, u)\| \to \infty$. Then we consider each orthant for which $\{z|z \notin \mathscr{B}(\mathscr{F})\}$ and show that the same fact holds.

(a) Consider $\{z|z \in \mathscr{B}(\mathscr{F})$ and $z = C^Tx\}$; for such $x \neq \theta$ we have $x \notin \mathscr{N}(S)$. We are going to show an equivalent condition of $\|x\| \to \infty \Rightarrow \|G(x, u)\| \to \infty$, namely; given any $M > 0$, there exists $N > 0$ such that $\rho > N \Rightarrow \|G(\rho\xi, u)\| > M$ for all $C^T\xi \in \mathscr{B}(\mathscr{F})$ with $\|\xi\| = 1$. Now

$$\|G(\rho\xi, u)\| \geqq \rho\|S\xi\| - \|C\| \cdot \|\mathscr{F}(\rho C^T\xi + u)\|.$$

Since $C^T\xi \in \mathscr{B}(\mathscr{F})$, $\|C\| \cdot \|\mathscr{F}(C^T\xi + u)\|$ is bounded, say by β. Note that $\{x|C^Tx \in \mathscr{B}(\mathscr{F})\}$ is closed and $\{\xi|\|\xi\| = 1\}$ is compact, hence the set $\Sigma \triangleq \{\xi|\|\xi\| = 1$ and $C^T\xi \in \mathscr{B}(\mathscr{F})\}$ is compact. Moreover, $\|S\xi\|$ is a continuous function and $\|S\xi\| > 0$ on the compact set Σ. Therefore,

$$\inf_{\xi \in \Sigma} \|S\xi\| = m > 0.$$

So $\|G(\rho\xi, u)\| \geqq m\rho - \beta$. Thus given any M, if $\rho > (M + \beta)/m$, then $\|G(\rho\xi, u)\| > M$ for all $\xi \in \Sigma$.

(b) Let $\mathscr{O}$ be any orthant of $\mathbb{R}^r$ in which $z = C^Tx \notin \mathscr{B}(\mathscr{F})$. Let $\{z^i\}$, $z^i = C^Tx^i$, be a sequence in $\mathscr{O}$ such that $\|z^i\| \to \infty$.

Case 1. For all j such that $f_j(z_j^i)$ is of type H in the unbounded half or of type U, the sequence $\{|z_j^i|\}_{i=1}^{\infty}$ is bounded. Again, we have

$$\|G(x^i, u)\| \geqq \|Sx^i\| - \|C\| \cdot \|\mathscr{F}(C^Tx^i + u)\|.$$

The second term is bounded in this case. Hence if we show $\|Sx^i\| \to \infty$ then we are done. Observe that $\mathbb{R}^m = \mathscr{N}(S) \oplus \mathscr{N}(S)^\perp$ and the map S restricted to $\mathscr{N}(S)^\perp$ is a bijection of $\mathscr{N}(S)^\perp$ onto $\mathscr{R}(S)$ (see [18, p. 572]). Let $x^i = n^i + p^i$, where $n^i \in \mathscr{N}(S)$ and $p^i \in \mathscr{N}(S)^\perp$. We are going to show by contradiction that the assumption $\|x^i\| \to \infty$ implies that $\|p^i\| \to \infty$, but $Sx^i = Sp^i$ so $\|Sx^i\| \to \infty$ also. Suppose $\|x^i\| \to \infty$ and $\|p^i\|$ is bounded, hence $\|n^i\| \to \infty$. Note that $\mathscr{B}(\mathscr{F})$ is a closed cone and $\{z|z = C^T x \text{ and } x \in \mathcal{N}(S)\}$ is also a closed cone. By assumption the intersection of these two closed cones contains only $\{\theta\}$. Let

$$\delta = \inf \|\xi - \eta\|,$$

where

$$\|\xi\| = 1 \quad \text{and} \quad \xi \in \mathscr{B}(\mathscr{F}),$$
$$\|\eta\| = 1 \quad \text{and} \quad \eta = C^T x, \quad x \in \mathcal{N}(S).$$

We have $\delta > 0$. Hence

$$d_i \triangleq \inf_{\xi \in \mathscr{B}(\mathscr{F})} \|C^T n^i - \xi\| \geqq \frac{\delta}{\sqrt{2}} \|C^T n^i\|$$

and $d_i \to \infty$ as $\|n^i\| \to \infty$. But this requires that at least one of the components of $C^T n^i$ goes to infinity. Clearly this branch variable belongs to a resistor of type H in the unbounded half or of type U. But $z^i = C^T x^i = C^T n^i + C^T p^i$, and since by assumption such $\{z_j^i\}$ are bounded, to compensate we must have $\|C^T p^i\| \to \infty$, hence $\|p^i\| \to \infty$. We have reached the desired contradiction.

Case 2. There is a j such that $|z_j^i| \to \infty$ and $|f_j(z_j^i)| \to \infty$. Recall that z_j^i is either a branch voltage or a branch current and $f_j(z_j^i)$ is the corresponding branch current or branch voltage. Call the branch for which $|z_j^i| \to \infty$ and $|f_j(z_j^i)| \to \infty$ the branch b_j. In the orthant $\mathcal{O}$, each z_j^i, $i = 1, 2, \cdots$, has a definite sign. Let us reassign the reference directions of each type H and of each type U resistor in accordance with the associated sign in the orthant so that if z_k is a branch voltage (resp. branch current) and f_k is a type H or a type U v.c. (resp. c.c.) resistor, the reference direction is so chosen that the branch voltage (resp. branch current) z_k (measured with respect to the new reference direction) is positive whenever z is in the orthant $\mathcal{O}$. Now we have three kinds of branches in the graph, namely (i) type U and type H (v.c. and c.c.) resistors, for which we have assigned directions, (ii) type B c.c. resistors, (iii) type B v.c. resistors. By the colored arc lemma, one of the following alternatives must occur;

Alternative I. There is a loop $\mathscr{L}$ containing b_j, of type U and type H (v.c. and c.c.) resistors, all of which are similarly directed, and of type B c.c. resistors.

Alternative II. There is a cutset $\mathscr{C}$ containing b_j, of type U and type H (v.c. and c.c.) resistors, all of which are similarly directed, and of type B v.c. resistors.

If Alternative I occurs, note that the branch voltage of a type U or a type H v.c. resistor agrees with the reference direction, and the branch voltage of a type U or a type H c.c. resistor either agrees with the reference direction (the direction of its current flow) or, if it is opposite to the reference direction, is bounded. Moreover, the voltage in a type B c.c. resistor is bounded. Hence, in order that KVL be satisfied for $\mathscr{L}$, we have to have an unbounded voltage source e_c to compensate the unbounded voltage in b_j. Dually, if Alternative II occurs, we need an unbounded current source j_v to compensate the unbounded current in b_j. Therefore, in either case, $\|G(x^i, u)\| = \|(j_v, e_c)\| \to \infty$.

Thus, it follows from the global implicit function theorem (Appendix A) that there is a unique C^1-function $\Phi: \mathbb{R}^r \times \mathbb{R}^m \to \mathbb{R}^m$ satisfying $G(\Phi(u, y), u) = y$.

LEMMA 2. *If conditions* (i) *and* (ii) *of Theorem* 1 *hold, then*

$$\mathcal{S} \triangleq \{x \in \mathbb{R}^m | C^T x \in \mathcal{B}(\mathcal{F}) \text{ and } x \in \mathcal{N}(S)\} = \{\theta\}.$$

Proof. Note that $x = (x_1, x_2) \in \mathcal{S}$ if and only if

$$\begin{bmatrix} -F_{vv} \\ I \end{bmatrix} x_1 \in \mathcal{B}\begin{pmatrix} \hat{\imath}_{vl} \\ \hat{\imath}_{vt} \end{pmatrix}, \qquad F_{vc} x_1 = \theta, \tag{6a}$$

$$\begin{bmatrix} I \\ F_{cc}^T \end{bmatrix} x_2 \in \mathcal{B}\begin{pmatrix} \hat{v}_{cl} \\ \hat{v}_{ct} \end{pmatrix}, \qquad F_{vc}^T x_2 = \theta. \tag{6b}$$

We are going to show that if $x = (x_1, x_2) \in \mathcal{S}$, then $x_1 = \theta$ and $x_2 = \theta$. Recall that $x_1 = v_{vt}$ and $x_2 = i_{cl}$.

(a) Let x_2 be a vector satisfying (6b). If we let (θ, x_2) be the link currents of $\mathcal{N}$, then KCL requires that the branch current vector i be

$$i = \begin{bmatrix} I & 0 \\ 0 & I \\ F_{vv}^T & F_{vc}^T \\ 0 & F_{cc}^T \end{bmatrix} \begin{bmatrix} \theta \\ x_2 \end{bmatrix} = \begin{bmatrix} \theta \\ x_2 \\ F_{vc}^T x_2 \\ F_{cc}^T x_2 \end{bmatrix}. \tag{7}$$

Interpreting (6b) in (7), it demands that the branch currents in all v.c. resistors and type U c.c. resistors be zero; and the direction of the actual current flow through any type H c.c. resistor is identical to the preassigned reference direction. We are going to show that these conditions, together with condition (i) of Theorem 1, will force all the branch currents to be zero. As far as KCL is concerned, those v.c. and type U c.c. resistors with zero currents can be removed. It follows from comment (d) (or the Fact in § 7) that condition (i) of Theorem 1 implies that in the remaining network for each type H c.c. resistor b there is a cutset containing b, of similarly-directed type H c.c. resistors. Since the actual current flows in these resistors are the same as their reference directions, KCL requires that all branch currents in type H c.c. resistors be zero. Next remove all type H resistors. In the remaining network which is made of only type B c.c. resistors, there is no loop by condition (i), hence all currents in type B c.c. resistors are also zero. Therefore, $x_2 = \theta$.

(b) Dually one can show that condition (ii) implies that any vector x_1 satisfying (6a) must be zero.

Remark. Lemma 1 and Lemma 2 remain valid if we replace "if" by "if and only if".

Proof of Theorem 1. $\Rightarrow$ is proved by contradiction. Clearly if there is a loop (resp. cutset) of type B c.c. (resp. v.c.) resistors or a loop (resp. cutset) of type H and type B c.c. (resp. v.c.) resistors, in which all type H c.c. (resp. v.c.) resistors are similarly directed, then for some input vector (u, y), KVL (resp. KCL) could not be satisfied.

$\Leftarrow$. We are going to show that for all $u, G(\cdot, u)$ is a homeomorphism from $\mathbb{R}^m$ onto $\mathbb{R}^m$.

CLAIM 1. *$G(\cdot, u)$ is injective.* Suppose not; then for some u, there exists $x \neq \bar{x}$ such that $G(x, u) = G(\bar{x}, u)$. Thus

$$C[\mathscr{F}(C^T x + u) - \mathscr{F}(C^T \bar{x} + u)] + S(x - \bar{x}) = \theta.$$

Premultiply by $[x - \bar{x}]^T$ and obtain

$$\langle C^T(x - \bar{x}) | \mathscr{F}(C^T x + u) - \mathscr{F}(C^T \bar{x} + u)\rangle = \theta.$$

But each component of $\mathscr{F}$ is strictly increasing, hence we reach contradiction.

CLAIM 2. *$G(\cdot, u)$ is surjective.* For a fixed u, given any $\varepsilon > 0$, let us construct, following Lemma B.1 (see Appendix B) for each resistor characteristic f_i in $\mathscr{F}$, a sequence of strictly increasing C^1-functions $\{f_i^k\}_{k=1}^{\infty}$ such that

$$|f_i(\rho) - f_i^k(\rho)| < \frac{\varepsilon}{kr\|C\|} \quad \text{for all } \rho \in \mathbb{R}.$$

Let

$$\mathscr{F}_k = \begin{pmatrix} f_1^k \\ \cdot \\ \cdot \\ f_r^k \end{pmatrix}$$

and

$$G_k(x, u) = C\mathscr{F}_k(C^T x + u) + Sx.$$

Thus

$$\|\mathscr{F}_k(\zeta) - \mathscr{F}(\zeta)\| < \frac{\varepsilon}{k\|C\|} \quad \text{for all } \zeta \in \mathbb{R}^r.$$

Hence

$$\|G_k(x, u) - G(x, u)\| < \varepsilon/k \quad \text{for all } x \in \mathbb{R}^m, \qquad k = 1, 2, 3 \cdots.$$

Note that Lemma 1 and Lemma 2 assert that conditions (i) and (ii) imply that $G_k(\cdot, u)$ is a diffeomorphism (hence, homeomorphism) for each k. It then follows from Lemma B.2 (Appendix B) that $G(\cdot, u)$ is surjective.

CLAIM 3. *$G(\cdot, u)$ is a homeomorphism.* Brouwer's domain invariance theorem [8, pp. 77–78] states that a bijective continuous function is a homeomorphism, hence $G(\cdot, u)$ is a homeomorphism for all $u \in \mathbb{R}^r$. It then follows from the global implicit function theorem that there is a unique continuous function $\Phi: \mathbb{R}^r \times \mathbb{R}^m \to \mathbb{R}^m$ satisfying $G(\Phi(u, y), u) = y$ for all $u \in \mathbb{R}^r$, and for all $y \in \mathbb{R}^m$. Once $x = (v_{vt}, i_{cl})$ is known, all (v, i) will be given by simple substitution into the Kirchhoff laws (2a) and resistor characteristics (1). Finally, the dependence of (v, i) on (u, y) is continuous since (1) and (2a) are all continuous maps.

4. Increasing resistive networks. In this section we allow the resistors to be increasing, but not necessarily strictly increasing. The conditions (i) and (ii) of Theorem 1 are also the necessary and sufficient conditions for the existence of solutions for increasing resistive networks (Theorem 2). With some additional restriction on the topology of the network it is shown in Theorem 3 that these conditions are again necessary and sufficient for the existence *and uniqueness* of solutions for a general class of increasing resistive networks.

THEOREM 2 (Existence). *Let $\mathcal{N}$ be a finite network made of* increasing, *continuous and uncoupled resistors. Then for all independent voltage sources with pliers entries and for all independent current sources with soldering iron entries, the network $\mathcal{N}$ has at least one solution if and only if conditions* (i) *and* (ii) *of Theorem* 1 *hold.*

Proof. $\Rightarrow$. The proof is the same as that of Theorem 1.

$\Leftarrow$. Note that the function f in Lemma B.1 is required only to be increasing, hence in the proof of Theorem 1 the part that $G(\cdot, u)$ is surjective for all u applies here too. However, in the present case, we cannot guarantee uniqueness nor continuous dependence.

THEOREM 3 (Existence, uniqueness, and continuous dependence). *Let $\mathcal{N}$ be a finite network made of* increasing, *continuous, and uncoupled resistors. Suppose that $\mathcal{N}$ satisfies conditions* (U_l) *and* (U_c):

(U_l) *Every loop made of c.c. resistors contains at least one strictly increasing resistor.*

(U_c) *Every cutset made of v.c. resistors contains at least one strictly increasing resistor.*

Under these conditions, for all independent voltage sources with pliers entries and for all independent current sources with soldering iron entries, the network has one and only one solution and this solution depends continuously on the inputs if and only if conditions (i) *and* (ii) *of Theorem* 1 *hold.*

Remarks. (a) Physically condition (U_c) can be explained as follows. Suppose there is a cutset of v.c. resistors in which none of the resistors is strictly increasing. By choosing appropriate voltage sources, we can place the operating point of each resistor in the cutset to be in the interior of an interval where its characteristic is constant. If we change the branch voltages in the cutset by the same sufficiently small amount Δv so that the corresponding current in each resistor remains the same, then we have another solution which is identical to the preceding one except for the branch voltages of the cutset which differ by Δv. Dually for condition (U_l). Therefore, only when conditions (U_l) and (U_c) are satisfied can one expect uniqueness.

(b) If there exists a tree for the network such that all its tree branches are c.c. and all its links are v.c., then clearly conditions (U_l) and (U_c), as well as conditions (i) and (ii) of Theorem 1, are satisfied. Hence the result of Desoer and Katzenelson [2, Thm. I] is a special case of Theorem 3.

Proof of Theorem 3. We need only to show that $G(\cdot, u)$ is injective for all $u \in \mathbb{R}^r$. By contradiction, suppose not. Then there is an input (u, y) for which $G(x, u) = G(\bar{x}, u) = y$ and $x \neq \bar{x}$. Now x and $\bar{x}$ each specify a unique set of branch voltages and branch currents, so $(v, i) \neq (\bar{v}, \bar{i})$, but they satisfy the Kirchhoff laws and the branch characteristics. By the Tellegen theorem, $(v - \bar{v})^T(i - \bar{i}) = 0$, i.e., $\sum_{k=1}^{r} \Delta v_k \Delta i_k = 0$. Since resistor characteristics are increasing, $\Delta v_k \Delta i_k = 0$ for $k = 1, 2, \cdots, r$. Therefore, (a) along every loop in which $\Delta i_i \neq 0$, all $\Delta v_i = 0$, and (b) for every cutset in which $\Delta v_j \neq 0$, all $\Delta i_j = 0$. In case (a) the loop cannot contain a v.c. resistor because for v.c. resistors, $\Delta i_k \neq 0$ implies $\Delta v_k \neq 0$, so (a) can only happen if all resistors in the loop are c.c., However, condition (U_l) requires in such a loop there be a strictly increasing resistor for which $\Delta i_k \neq 0$ implies $\Delta v_k \neq 0$. Hence $\Delta i_k = 0$ for all $k = 1, \cdots, r$. Dually for case (b), hence $\Delta v_k = 0$ for $k = 1, \cdots, r$. This contradicts $x \neq \bar{x}$.

Remark. In Theorem 3, if all resistor characteristics are C^1, then the dependence of the unique solution on the inputs is C^1. Note that, by comparing with Lemmas 1 and 2, all we need to show in this case is that $D_1G(x, u)$ is still nonsingular for all (x, u). Suppose $D_1G(x, u)$ were singular for some $(\underline{x}, \underline{u})$, hence there would exist a $\zeta \neq \theta$ such that

$$C(D\mathscr{F}(C^T\underline{x} + \underline{u}))C^T\zeta + S\zeta = \theta.$$

Now, let us consider the small-signal equivalent circuit $\mathscr{N}_s$ of $\mathscr{N}$ at $(\underline{x}, \underline{u})$. The Kirchhoff laws for $\mathscr{N}_s$ are expressed by

$$C(D\mathscr{F}(C^T\underline{x} + \underline{u}))C^T\Delta x + S\Delta x = \theta,$$

where $\Delta x = (\Delta v_{vt}, \Delta i_{cl})$. However, we have just shown that conditions (U_l) and (U_c) imply that $\Delta x = \theta$, hence $\zeta = \theta$ and we reach a contradiction.

5. Eventually increasing resistive networks. Sandberg and Willson [6] have developed a technique whereby the existence of solutions can be asserted, given only the asymptotic behavior of the characteristics. Applying their technique, we have the following corollary.

COROLLARY 1. (Existence). *Let $\mathscr{N}$ be a finite network made of eventually strictly increasing (resp. eventually increasing), continuous, and uncoupled resistors. Suppose $\mathscr{N}$ satisfies conditions* (i) *and* (ii) *of Theorem* 1 *(resp. conditions* (i) *and* (ii) *of Theorem* 1 *and conditions* (U_l) *and* (U_c) *of Theorem* 3*). Then for all independent voltage sources with pliers entries and for all independent current sources with soldering iron entries, the network $\mathscr{N}$ has at least one solution.*

Proof. There is an $M > 0$ such that for $k = 1, 2, \cdots, r$, $f_k(z_k)$ is (strictly) increasing on $|z_k| > M$. Let us define

$$g_k(z_k) = \begin{cases} f_k(z_k), & |z_k| > M, \\ \dfrac{f_k(M) - f_k(-M)}{2}\dfrac{z_k}{M} + \dfrac{f_k(M) + f_k(-M)}{2}, & |z_k| \leqq M. \end{cases}$$

Let

$$\mathscr{G}(z) = [g_1(z_1), \cdots, g_r(z_r)]^T,$$
$$U(x, u) = C\mathscr{G}(C^Tx + u) + Sx,$$
$$V(x, u) = C[\mathscr{F}(C^Tx + u) - \mathscr{G}(C^Tx + u)].$$

Consider a given u. Let $z \triangleq C^Tx + u$, $z \in \mathbb{R}^r$. Note that $U(\cdot, u)$ is a homeomorphism from $\mathbb{R}^m$ onto $\mathbb{R}^m$, by Theorem 1 (resp. Theorem 3), and $V(\cdot, u)$: $\mathbb{R}^m \to \mathbb{R}^m$ is a continuous map. Now $\|V(x, u)\| \leqq \|C\|\,\|\mathscr{F}(z) - \mathscr{G}(z)\|$. By construction, each component of $[\mathscr{F}(z) - \mathscr{G}(z)]$ satisfies, for $k = 1, \cdots, r$,

$$|f_k(z_k) - g_k(z_k)| \leqq \max_{|z_k| \leqq M} |f_k(z_k) - g_k(z_k)| \triangleq \alpha_k \quad \text{for all } z_k \in \mathbb{R}.$$

Hence $\|V(x, u)\| \leqq \|C\| \cdot \|\alpha\|$ for all $x \in \mathbb{R}^m$, where $\alpha = (\alpha_1 \cdots \alpha_r)^T$. Since $U(\cdot, u)$ is a homeomorphism, by the global inverse function theorem [9], [10], given any $N > 0$ there exists a k such that $\|x\| > k \Rightarrow \|U(x, u)\| \geqq N$. Set $N = 2\|C\| \cdot \|\alpha\|$. We have

$$\|V(x, u)\| \leqq \tfrac{1}{2}\|U(x, u)\| \quad \text{whenever } \|x\| > k.$$

Then apply a theorem of Sandberg and Willson (Appendix A); the corollary is thus proved.

6. Boundedness. For resistive networks, the next basic question, besides the existence and uniqueness of solutions, is the dependence of solutions on the inputs. In the existence and uniqueness Theorems 1 and 3, the results state that the solutions depend continuously on the inputs. We would like to know the dependence of solutions on the inputs for the existence Theorem 2 and Corollary 1. Theorem 4 below asserts that in those cases bounded inputs produce bounded solutions. Not all resistive networks which have continuous characteristics and which have a solution for all inputs have this bounded-input bounded-solution property. Consider a c.c. resistor with the characteristic $v = i \sin^2 i$ connected to an independent voltage source e: for each $e \in \mathbb{R}$, there are infinitely many solutions larger than any prescribed number.

In Theorem 4, we do *not* require that resistor characteristics be increasing, nor even eventually increasing.

THEOREM 4. *Let $\mathcal{N}$ be a finite network made of continuous, and uncoupled resistors. Each resistor is required to be either of type U, or type H, or type B. Suppose that conditions* (i) *and* (ii) *of Theorem* 1 *hold. Suppose that for some independent current sources with soldering iron entries, the network has solutions. Under these conditions, if for some $B_1 < \infty$, the inputs satisfy $|e_k| < B_1$ and $|j_k| < B_1$ for all k, then there exists a $B_2 < \infty$ such that all network solutions satisfy $|v_k| < B_2$ and $|i_k| < B_2$, for all k.*[3]

Proof. Consider the network equation

$$G(x, u) = C\mathcal{F}(C^T x + u) + Sx = y.$$

An equivalent statement of the conclusion of the theorem is that if for some $\bar{B}_1 < \infty$, $\|(u, y)\| \leqq \bar{B}_1$, then there exists a $\bar{B}_2 < \infty$ such that $\|x\| \leqq \bar{B}_2$. Clearly this is true if $\|x\| \to \infty$ implies $\|(u, y)\| \to \infty$. We are going to show this by contradiction. Let $\{x^i\}$ be a sequence with $\|x^i\| \to \infty$, let $\{u^i\}$ and $\{y^i\}$ be two corresponding sequences such that $C\mathcal{F}(C^T x^i + u^i) + Sx^i = y^i$ is satisfied, and suppose $\{(u^i, y^i)\}$ is bounded; hence in particular $\{u^i\}$ is bounded. Lemma 2 states that if conditions (i) and (ii) of Theorem 1 hold then the assumption $\{x | C^T x \in \mathcal{B}(\mathcal{F}) \text{ and } x \in \mathcal{N}(S)\} = \{\theta\}$ of Lemma 1 holds. Note that in Claim 2 of the proof of Lemma 1 we have shown that for any fixed u, $\|x\| \to \infty \Rightarrow \|G(x, u)\| = \|y\| \to \infty$. Observe that in that proof in fact we only require that (i) u remain bounded, (ii) each resistor be either of type U, or type H, or type B. Hence if $\|x^i\| \to \infty$ and $\|u^i\|$ remains bounded, then we must have $\|y^i\| \to \infty$. Thus a contradiction is reached.

7. Monotone RLC networks. The natural framework for considering general nonlinear networks is provided by the differentiable manifold formulation [11], [12]. However, in many special cases, the manifold of configuration space is diffeomorphic to a linear vector space; such networks can then be characterized by a differential equation in normal form.

[3] Here we use e_k (resp. j_k) to denote the magnitude of an independent voltage (resp. current) source; v_k (resp. i_k) to denote the branch voltage (resp. branch current) of a resistor.

An RLC network can be considered as a connection of three one-element-kind subnetworks. Therefore, our results on resistive networks (in fact, on one-element-kind networks) lead directly to Theorem 5 below, which considers increasing RLC networks. Let us first define the class of inductors and capacitors under consideration.

We define a *capacitor* (resp. *inductor*) as a two-terminal element that is characterized by a C^1-function f which maps the real line $\mathbb{R}$ into itself and $\sigma = f(\rho)$, where either ρ is the branch voltage (resp. flux) and σ the stored charge (resp. branch current), or vice versa. If ρ is the branch voltage (resp. flux), we say that the capacitor (resp. inductor) is voltage-controlled (resp. flux-controlled), abbreviated v.c. (resp. ϕ.c.); on the other hand, if ρ is its stored charge (resp. branch current), we say that it is charge-controlled (resp. current-controlled), abbreviated q.c. (resp. c.c.). We define an increasing (resp. strictly increasing, eventually increasing, eventually strictly increasing, type U, type H, type B) capacitor or inductor according to its characteristic, as was done for resistors.

THEOREM 5 (State equations for monotone RLC networks). *Let $\mathcal{N}$ be a finite network made of* increasing, *time-varying, uncoupled resistors, inductors, and capacitors. Thus all characteristics have the form $\sigma = f(\rho, t)$ and we assume that f is C^1 both in ρ and t. Let us derive from $\mathcal{N}$ three subnetworks:*

$\mathcal{N}_L$ (inductive subnetwork): replace by short circuits all elements, except inductors, of $\mathcal{N}$.

$\mathcal{N}_C$ (capacitive subnetwork): remove all elements, except capacitors, of $\mathcal{N}$.

$\mathcal{N}_R$ (resistive subnetwork): replace by short circuits all capacitors, and remove all inductors, of $\mathcal{N}$.

Suppose that in $\mathcal{N}_L$ (resp. $\mathcal{N}_C$, $\mathcal{N}_R$), the conditions (a)–(d) *are satisfied:*

(a) *every loop*[4] *made of c.c. inductors (resp. q.c. capacitors, c.c. resistors) contains at least one which is strictly increasing;*

(b) *every cutset*[5] *made of ϕ.c. inductors (resp. v.c. capacitors, v.c. resistors) contains at least one which is strictly increasing;*

(c) *every loop made of c.c. inductors (resp. q.c. capacitors, c.c. resistors) either contains at least one type U inductor (resp. capacitor, resistor) or if not, then it contains at least two type H inductors (resp. capacitors, resistors) and not all such type H inductors (resp. capacitors, resistors) are similarly directed.*

(d) *every cutset made of ϕ.c. inductors (resp. v.c. capacitors, v.c. resistors) either contains at least one type U inductor (resp. capacitor, resistor) or if not, then it contains at least two type H inductors (resp. capacitors, resistors) and not all such type H inductors (resp. capacitors, resistors) are similarly directed. Suppose that independent sources are all regulated functions* [13, p. 145] *of time. Under these conditions, for all independent voltage sources with pliers entries and for all independent current sources with soldering iron entries, and given any initial time t_0 and any initial conditions, the network $\mathcal{N}$ has one and only one solution on some nonvanishing interval $[t_0, t_\alpha)$.*

Proof. First pick a normal tree and let the subscripts S, R, L (resp. C, G, Γ) correspond to link (resp. tree-branch) capacitors, resistors, and inductors; so that

[4] Self-loop is regarded as a loop.

[5] A cutset may contain only a single branch, in which case, we call it an "open branch".

the fundamental loop matrix takes the form:

$$\begin{bmatrix} I & 0 & 0 & F_{SC} & 0 & 0 \\ 0 & I & 0 & F_{RC} & F_{RG} & 0 \\ 0 & 0 & I & F_{LC} & F_{LG} & F_{L\Gamma} \end{bmatrix}$$

Define a set of state variables as:

$$q = q_C - F_{SC}^T q_S,$$

$$\phi = \phi_L + F_{L\Gamma}\phi_\Gamma.$$

If the following three sets of equations (L), (C) and (R) possess unique solutions for i_L, v_c, i_R and v_G in terms of q, ϕ and t; i.e., $i_L = \tilde{\imath}_L(\phi, t)$, $v_c = \tilde{v}_c(q, t)$, $i_R = \tilde{\imath}_R(\tilde{v}_c(q, t), \tilde{\imath}_L(\phi, t), t)$, and $v_G = \tilde{v}_G(\tilde{v}_c(q, t), \tilde{\imath}_L(\phi, t), t)$, then the network $\mathcal{N}$ is characterized by differential equations in (q, ϕ) (See [17, pp. 61–65]), namely,

$$\dot{q} = F_{RC}^T \tilde{\imath}_R(\tilde{v}_C(q, t), \tilde{\imath}_L(\phi, t), t) + F_{LC}^T \tilde{\imath}_L(\phi, t) + j_C(t),$$

$$\dot{\phi} = -F_{LG}\tilde{v}_G(\tilde{v}_C(q, t), \tilde{\imath}_L(\phi, t), t) - F_{LC}\tilde{v}_C(q, t) + e_L(t).$$

$$[I \quad F_{L\Gamma}]\begin{bmatrix} \phi_L \\ \phi_\Gamma \end{bmatrix} = \phi,$$

(L) $$[-F_{L\Gamma}^T \quad I]\begin{bmatrix} i_L \\ i_\Gamma \end{bmatrix} = j_L(t),$$

$$f_L(\phi_L, \phi_\Gamma, i_L, i_\Gamma, t) = \theta;$$

$$[I \quad F_{SC}]\begin{bmatrix} v_S \\ v_C \end{bmatrix} = e_S(t),$$

(C) $$[-F_{SC}^T \quad I]\begin{bmatrix} q_S \\ q_C \end{bmatrix} = q;$$

$$f_C(v_S, v_C, q_S, q_C, t) = \theta;$$

$$[I \quad F_{RG}]\begin{bmatrix} v_R \\ v_G \end{bmatrix} = F_{RC}v_C + e_R(t),$$

(R) $$[-F_{RG}^T \quad I]\begin{bmatrix} i_R \\ i_G \end{bmatrix} = F_{RL}^T i_L + j_G(t),$$

$$f_R(v_R, v_G, i_R, i_G, t) = \theta;$$

where we use f_L (resp. f_C, f_R) to denote inductor (resp. capacitor, resistor) characteristics. Note that if we consider the right-hand sides as inputs, the first two sets of equations of (L) (resp. (C), (R)) are precisely KVL and KCL for $\mathcal{N}_L$ (resp. $\mathcal{N}_C, \mathcal{N}_R$). Therefore, by the remark following the proof of Theorem 3,[6] conditions (a)–(d) on $\mathcal{N}_L$ (resp. $\mathcal{N}_C$) imply that i_L (resp. v_c) is uniquely determined by $(\phi, j_L(t), t)$ (resp. $(q, e_s(t), t)$)[7] and the dependence is C^1. Thus, $i_L = \tilde{\imath}_L(\phi, t)$ (resp. $v_c = \tilde{v}_c(q, t)$), where

[6] In the time-varying case, we will have $C\mathcal{F}(C^T x + u, t) + Sx = y$. By assumption $\mathcal{F}$ is C^1 in t. In applying the global implicit function theorem, consider $G(x, u, t): \mathbb{R}^m \times \mathbb{R}^{r+1} \to \mathbb{R}^m$.

[7] Note that the solutions exist even if ϕ.c. self-loops (resp. c.c. open branches) of $\mathcal{N}_L$ (resp. $\mathcal{N}_C$) are not increasing; and v.c. self-loops and c.c. open branches of $\mathcal{N}_R$ are not increasing.

$\tilde{\imath}_L$ (resp. $\tilde{v}_c$) is C^1 in ϕ (resp. q). Moreover, since $j_L(t)$ (resp. $e_s(t)$) is a regulated function, $\tilde{\imath}_L$ (resp. $\tilde{v}_c$) is a regulated function of t. Conditions (a)–(d) on $\mathcal{N}_R$ imply that $i_R = \tilde{\imath}_R(v_c, i_L, t)$ and $v_G = \tilde{v}_G(v_c, i_L, t)$, where $\tilde{\imath}_R$ and $\tilde{v}_G$ are C^1 in (v_c, i_L) and regulated in t. Hence $i_R = \tilde{\imath}_R(\tilde{v}_c(q, t), \tilde{\imath}_L(\phi, t), t) \triangleq i_R(q, \phi, t)$ and $v_G = \tilde{v}_G(\tilde{v}_c(q, t), \tilde{\imath}_L(\phi, t), t) \triangleq \tilde{v}_G(q, \phi, t)$, where $\tilde{\imath}_R$ and $\tilde{v}_G$ are C^1 in (q, ϕ) and regulated in t. The theorem then follows from the fundamental theorem of differential equations [13, pp. 285–289].

Remark. In comparison with the result of Desoer and Katzenelson [2, Thm. IV], we note that their circuit-theoretic conditions are sufficient for conditions (a)–(d); we require that characteristics be differentiable, however.

8. Algorithms. We propose two efficient algorithms for checking conditions (i) and (ii) of Theorem 1. First we present an immediate consequence of the colored arc lemma.

Fact. Condition (i) (resp. (ii)) of Theorem 1 holds if and only if, after removing all v.c. and type U c.c. resistors (resp. replacing by short circuits all c.c. and type U v.c. resistors),

(A) every loop (resp. cutset) contains at least one type H c.c. (resp. v.c.) resistor;

(B) for every type H c.c. (resp. v.c.) resistor b, there is a similarly-directed cutset (resp. loop) of type H c.c. (resp. v.c.) resistors, containing b.

ALGORITHM 1. (For checking condition (i)).

Step 1. Remove all v.c. and type U c.c. resistors from $\mathcal{N}$. Remove all "open branches". Call the resultant network $\mathcal{N}_0$; set $i = 0$.

Step 2. If $\mathcal{N}_i$ has a type H c.c. resistor go to (4), else go to (3).

Step 3. If $\mathcal{N}_i$ has a loop, output: *condition* (i) *is not satisfied*. (There is a loop in $\mathcal{N}$ of type B c.c. resistors.) Else output: *condition* (i) *holds*. (This conclusion follows in view of the foregoing Fact.)

Step 4. Pick a type H c.c. resistor b, directed from, say, node n_0 to node n_1. Set $V_1 = \{n_1\}$, set $k = 1$.

Step 5. If there is a type H c.c. resistor directed from some node in V_k to a node t not in V_k, go to (6); otherwise, if there is a type B c.c. resistor connecting some node in V_k and a node t not in V_k, go go (6), else go to (7).

Step 6. If $t = n_0$, output: *condition* (i) *is not satisfied*. (There is a loop of type H and type B c.c. resistors, in which all type H resistors are similarly directed.) Else set $V_{k+1} = V_k \cup \{t\}$, set $k = k + 1$, go to (5).

Step 7. Remove all resistors which have only one terminal node in V_k. Call the resultant network $\mathcal{N}_{i+1}$, set $i = i + 1$, go to (2). (There is a cutset, in $\mathcal{N}_0$, containing b, of similarly-directed type H c.c. resistors. We may remove them from further consideration.)

ALGORITHM 2 (For checking condition (ii)).

Step 1. Replace all c.c. resistors and type U v.c. resistors in $\mathcal{N}$ by short circuits and identify any two nodes connected by a short circuit. Remove all self-loops. Call the resultant network $\mathcal{N}_0$; set $i = 0$.

Step 2. If $\mathcal{N}_i$ has a type H v.c. resistor, go to (4); else go to (3).

Step 3. If $\mathcal{N}_i$ has a type B v.c. resistor, output: *condition* (ii) *is not satisfied*. (There is a cutset in $\mathcal{N}$ of type B v.c. resistors.) Else output: *condition* (ii) *holds*.

(This conclusion follows in view of the foregoing Fact.)

Step 4. Pick a type H v.c. resistor b, directed from, say, node n_0 to node n_1; set $k = 1$.

Step 5. If there is a type H v.c. resistor directed from node n_k to some node t, go to (6). Else output: *condition* (ii) *is not satisfied.* (There is a type H v.c. resistor which is not in a loop, in $\mathcal{N}_0$, of similarly-directed type H v.c. resistors. This violates condition (B).)

Step 6. If $t \neq n_j$ for $0 \leqq j < k$, set $n_{k+1} = t$, $k = k + 1$; go to (5); else go to (7).

Step 7. Identify node $n_j, n_{j+1}, \cdots, n_k, t$ and remove all self-loops. (There is a loop in $\mathcal{N}_0$ of similarly-directed type H v.c. resistors. We may disregard them from further consideration, i.e., shrink them down into a node.) If $j > 1$, set $k = j$ and go to (5); else call the resultant network $\mathcal{N}_{i+1}$, set $i = i + 1$, go to (2). (The loop contains b; we have to start again.)

Appendix A.

PALAIS GLOBAL IMPLICIT FUNCTION THEOREM [9], [14]. *Let* $G:\mathbb{R}^m \times \mathbb{R}^r \to \mathbb{R}^m$ *be continuous (resp.* C^k, $1 \leqq k \leqq \infty$*). Given* $G(x, u) = y$ *there exists a unique continuous (resp.* C^k*) function* $\Phi:\mathbb{R}^m \times \mathbb{R}^r \to \mathbb{R}^m$ *such that* $G(\Phi(y, u), u) = y$ *for all* $u \in \mathbb{R}^r$, $y \in \mathbb{R}^m$ *if and only if*

(i) $G(\cdot, u)$ *is a local homeomorphism (resp. local diffeomorphism*[8]*) from* $\mathbb{R}^m$ *onto* $\mathbb{R}^m$ *for all* $x \in \mathbb{R}^m$, $u \in \mathbb{R}^r$;

(ii) *for all fixed* $u \in \mathbb{R}^r$, $\|G(x, u)\| \to \infty$ *whenever* $\|x\| \to \infty$.

MINTY'S COLORED ARC LEMMA [15], [16]. *Let* $\mathcal{N}$ *be a directed graph whose branches are partitioned into three sets (or colored with three colors) A, B, and C, and let* $b \in B$. *Then there exists one and only one of the following:*

(i) *There is a loop, containing b, of branches in A and B only, in which all branches of B are similarly directed.*

(ii) *There is a cutset, containing b, of branches in B and C only, in which all branches of B are similarly directed.*

SANDBERG AND WILLSON'S THEOREM [6]. *Let U be a homeomorphism from* $\mathbb{R}^m$ *onto* $\mathbb{R}^m$, *and let V be a continuous map from* $\mathbb{R}^m$ *into* $\mathbb{R}^m$ *with the property that there exist real numbers* $0 < c < 1$ *and* $M > 0$ *such that for all* $\|x\| > M$,

$$\|V(x)\| \leqq c\|U(x)\|.$$

Then for each $y \in \mathbb{R}^m$ *there exists at least one* $x \in \mathbb{R}^m$ *such that* $U(x) + V(x) = y$.

Appendix B.

LEMMA B1. *Let* $f: \mathbb{R} \to \mathbb{R}$ *be a continuous and increasing function. Then given any* $\varepsilon > 0$ *there exists a strictly increasing* C^1*-function* $f^\varepsilon: \mathbb{R} \to \mathbb{R}$ *such that*

$$|f(\rho) - f^\varepsilon(\rho)| < \varepsilon \quad \textit{for all } \rho \in \mathbb{R}.$$

Proof. The proof is by construction. First assume that f is continuous and *strictly increasing*. Consider the compact interval $I_n = [n, n + 1]$, n an integer. f is uniformly continuous on I_n, hence there exists a δ_n such that $|\rho_1 - \rho_2| < \delta_n$

[8] By the inverse function theorem, $G(\cdot, u)$ is a local diffeomorphism from $\mathbb{R}^m$ onto $\mathbb{R}^m$ for all $x \in \mathbb{R}^m$, $u \in \mathbb{R}^r$ if and only if $D_1G(x, u)$ is nonsingular for all $x \in \mathbb{R}^m$, $u \in \mathbb{R}^r$.

$\Rightarrow |f(\rho_1) - f(\rho_2)| < \varepsilon/4$ for all $\rho_1, \rho_2 \in I_n$. Without loss of generality, we may take δ_n to be the inverse of a positive integer. Now construct a piecewise linear function $h(\rho)$ on I_n such that $h(\rho) = f(\rho)$ for $\rho = n, n + \delta_n, \cdots n + 1$, and linear between any two consecutive points. Repeating the construction for all n, we obtain a piecewise linear, strictly increasing function $h: \mathbb{R} \to \mathbb{R}$ and $|h(\rho) - f(\rho)| \leqq \varepsilon/4$ for all $\rho \in \mathbb{R}$. Now we round off the corners of h by circular arcs with radius $1/(8\delta_n)$ for corners inside I_n and $(8 \min\{\delta_n, \delta_{n+1}\})^{-1}$ for corners at n. The result is a strictly increasing C^1-function f^ε such that $|f(\rho) - f^\varepsilon(\rho)| < \varepsilon/2$ for all $\rho \in \mathbb{R}$.

Suppose now that f is only increasing. Then the piecewise linear approximation h constructed above may have line segments with zero slope. So it needs modification to make h a piecewise linear, strictly increasing function h_1.

Case 1. $h(\rho) = c$ *on* $[\alpha, \beta]$. Let ρ_1, ρ_2 be the points where $f(\rho_1) = c - \varepsilon/4$ and $f(\rho_2) = c + \varepsilon/4$.[9] Let $h_1(\rho)$ on $[\rho_1, \rho_2]$ be the straight line connecting $(\rho_1, c - \varepsilon/4)$ and $(\rho_2, c + \varepsilon/4)$.

Case 2. $h(\rho) = c$ *on* $[\alpha, \infty)$. Let ρ_1 be the point where $h(\rho_1) = c - \varepsilon/4$. Let $h_1(\rho)$ on $[\rho_1, \infty)$ be $h_1(\rho) = c - (\varepsilon/4)e^{-\lambda(\rho - \rho_1)}$, where $\lambda = (4/\varepsilon)h(\rho_1)$.

Case 3. $h(\rho) = c$ *on* $(-\infty, \infty)$. Let $h_1(\rho) = (\varepsilon/4)\tanh \rho + c$. Clearly $|h_1(\rho) - f(\rho)| < \varepsilon/2$ for all $\rho \in \mathbb{R}$. Rounding off corners of h_1, we obtain f^ε and $|f^\varepsilon(\rho) - f(\rho)| < \varepsilon$ for all $\rho \in \mathbb{R}$.

LEMMA B2. *Let* $\{\mathscr{F}_k\}$ *be a sequence of homeomorphisms from* $\mathbb{R}^n$ *onto* $\mathbb{R}^n$, *and* $\mathscr{F}: \mathbb{R}^n \to \mathbb{R}^n$ *be continuous. If for a given* $\varepsilon > 0$, $\|\mathscr{F}_k(x) - \mathscr{F}(x)\| < \varepsilon/k$ *for all* $x \in \mathbb{R}^n$, $k = 1, 2, \cdots$, *then* $\mathscr{F}$ *is a surjective map and the inverse image under* $\mathscr{F}$ *of any bounded set is a bounded set.*

Proof. (i) First we show that $\mathscr{F}^{-1}(B)$ is bounded whenever $B \subseteq \mathbb{R}^n$ is bounded. Suppose not. Then there exists an unbounded sequence $\{\xi_i\} \to \infty$ and $\|\mathscr{F}(\xi_i)\| < M$ for $i = 1, 2, \cdots$. Since $\|\mathscr{F}_k(\xi_i) - \mathscr{F}(\xi_i)\| < \varepsilon/k$ for all i, $\|\mathscr{F}_k(\xi_i)\| < \|\mathscr{F}(\xi_i)\| + \varepsilon/k < M + \varepsilon/k$. But $\mathscr{F}_k$ is a homeomorphism, from the global inverse function theorem, and so $\|\xi_i\| \to \infty \Rightarrow \|\mathscr{F}_k)\| \to \infty$. We reach contradiction.

(ii) We shall show that given any $y \in \mathbb{R}^n$, there exists an $x \in \mathbb{R}^n$ such that $\mathscr{F}(x) = y$. Let x_k be the points satisfying $\mathscr{F}_k(x_k) = y$. Consider the sequence $\{x_k\}$. We claim that it is bounded. Since for all positive integers k, $\|\mathscr{F}(x_k) - \mathscr{F}_k(x_k)\| < \varepsilon/k$, we have $\|\mathscr{F}(x_k)\| < \|y\| + \varepsilon/k$. Hence $\{x_k\}$ is in the inverse image under $\mathscr{F}$ of a bounded set, this it is bounded by (i). Therefore, $\{x_k\}$ has a convergent subsequence, $\{x_{k_i}\}$, say converging to $\bar{x}$. Now we claim $\mathscr{F}(\bar{x}) = y$. Indeed,

$$\begin{aligned}\|y - \mathscr{F}(\bar{x})\| &= \|\mathscr{F}_{k_i}(x_{k_i}) - \mathscr{F}(\bar{x})\| \\ &\leqq \|\mathscr{F}_{k_i}(x_{k_i}) - \mathscr{F}(x_{k_i})\| + \|\mathscr{F}(x_{k_i}) - \mathscr{F}(\bar{x})\| \\ &\leqq \varepsilon/k_i + \|\mathscr{F}(x_{k_i}) - \mathscr{F}(\bar{x})\|.\end{aligned}$$

Since $\mathscr{F}$ is continuous, and $x_{k_i} \to \bar{x}$ as $k_i \to \infty$, the right-hand side can be made as small as we please by picking k_i large enough, hence $\|y - \mathscr{F}(\bar{x})\| = 0$, i.e., $y = \mathscr{F}(\bar{x})$.

Acknowledgment. The authors would like to thank L. T. Kou for his helpful discussions and suggestions on § 8.

[9] If such points cannot be found, reduce $\varepsilon/4$ to $\varepsilon/8$, and so on.

REFERENCES

[1] R. J. DUFFIN, *Nonlinear networks, I, II, IIb*, Bull. Amer. Math. Soc., *I*: 52 (1946), pp. 836–838; *II*: 53 (1947), pp. 963–971; *IIb*: 54 (1948), pp. 119–127.

[2] C. A. DESOER AND J. KATZENELSON, *Nonlinear RLC networks*, Bell Systems Tech. J., 44 (1965), pp. 161–198.

[3] I. W. SANDBERG AND A. N. WILLSON, JR., *Some theorems on properties of DC equations of nonlinear networks*, Ibid., 28 (1969), pp. 1–34.

[4] A. N. WILLSON, JR., *New theorems on the equations of nonlinear DC transistor networks*, Ibid., 49 (1970), pp. 1713–1738.

[5] I. W. SANDBERG, *Necessary and sufficient conditions for the global invertibility of certain nonlinear operators that arise in the analysis of networks*, IEEE Trans. Circuit Theory, CT-18 (1971), pp. 260–263.

[6] I. W. SANDBERG AND A. N. WILLSON, JR., *Existence and uniqueness of solutions for the equations of nonlinear DC equations*, this Journal, 22 (1972), pp. 173–186.

[7] C. A. DESOER AND E. S. KUH, *Basic Circuit Theory*, McGraw-Hill, New York, 1969.

[8] J. T. SCHWARTZ, *Nonlinear Functional Analysis*, Gordon and Breach, New York, 1969.

[9] R. S. PALAIS, *Natural operations on differential forms*, Trans. Amer. Math. Soc., 92 (1959), pp. 125–141.

[10] F. F. WU AND C. A. DESOER, *Global inverse function theorem*, IEEE Trans. Circuit Theory, CT-19 (1972), pp. 199–201.

[11] S. SMALE, *On the mathematical foundations of electrical circuit theory*, to be published.

[12] C. A. DESOER AND F. F. WU, *Trajectories of nonlinear RLC networks: A geometric approach*, Special Issue on Nonlinear Circuits, IEEE Trans. Circuit Theory, CT-19 (1972).

[13] J. DIEUDONNÉ, *Foundations of Modern Analysis*, Academic Press, New York, 1969.

[14] E. S. KUH AND I. N. HAJJ, *Nonlinear circuit theory: Resistive networks*, Proc. IEEE, 59 (1971), pp. 340–355.

[15] G. J. MINTY, *Monotone networks*, Proc. Royal Soc. London Ser. A, 257 (1960), pp. 194–212.

[16] C. BERGE AND A. GHOUILA-HOURI, *Programming, Games and Transportation Networks*, John Wiley, New York, 1962.

[17] T. E. STERN, *Theory of Nonlinear Networks and Systems*, Addison-Wesley, Reading, Mass., 1965.

[18] L. A. ZADEH AND C. A. DESOER, *Linear System Theory*, McGraw-Hill, New York, 1963.

[19] G. J. MINTY, *Solving steady-state nonlinear networks of monotone elements*, IRE Trans. Circuit Theory, CT-8 (1961), pp. 99–104.

[20] ———, *On the axiomatic foundations of the theories of direct linear graphs, electrical networks and network programming*, J. Math. Mech., 15 (1966), pp. 485–520.

[21] R. T. ROCKAFELLAR, *Convex programming and systems of elementary monotonic relations*, J. Math. Anal. Appl., 19 (1967), pp. 543–564.

[22] W. MILLAR, *Some general theorems for nonlinear systems possessing resistance*, Philos. Mag., 42 (1951), pp. 1150–1160.

On the Solutions of Equations for Nonlinear Resistive Networks

By A. N. WILLSON, JR.

(Manuscript received December 13, 1967)

Several theorems are proved concerning the solutions of equations that arise in the study of resistive nonlinear electrical networks. The first, an existence and uniqueness theorem, applies to equations describing an interesting class of networks which includes certain active and nonreciprocal networks for which the existence and uniqueness of solutions has not previously been established. A method of computing bounds on the location of the solutions is given, and two iterative techniques are presented for computing the solutions. It is proved that the iterative techniques converge for a subclass of the equations which also includes equations describing certain active and nonreciprocal networks. Finally, the rate of convergence of the iterative techniques is compared with that of another well-known iterative technique and some practical computational aspects are pointed out. Computations for two example problems, not reported here, were carried out to show the practicality of applying these iterative techniques to the equations of specific networks.

I. INTRODUCTION

In this paper we consider the solution of the equation

$$F(x) + Ax = B \tag{1}$$

where $x \equiv \begin{bmatrix} x_1 \\ \vdots \\ x_n \end{bmatrix}$ is a point in the n-dimensional Euclidean space E^n,

$F(x) \equiv \begin{bmatrix} f_1(x_1) \\ \vdots \\ f_n(x_n) \end{bmatrix}$ is a nonlinear function mapping E^n into E^n, A is an

$n \times n$ matrix of real numbers, and $B \equiv \begin{bmatrix} b_1 \\ \vdots \\ b_n \end{bmatrix}$ is an arbitrary point in E^n. We prove (Theorem 1) that there is a unique solution of (1) if:

(*i*) Each f_i is a strictly monotone increasing function mapping E^1 onto E^1,

and

(*ii*) The elements a_{ij} of the matrix A satisfy the inequality

$$a_{ii} \geq \sum_{\substack{j=1 \\ j \neq i}}^{n} | a_{ij} |, \quad \text{for} \quad i = 1, \cdots, n.$$

We then demonstrate a straightforward method of computing bounds on the location of this solution. Finally, we present two iterative techniques for computing the solution; and prove (Theorem 3) that the two additional assumptions:

(*iii*) Either all of the functions f_i are convex, or else all f_i are concave,

and

(*iv*) $a_{ij} \leqq 0$ if $i \neq j$,

are sufficient to guarantee that the iterations converge to the solution.

Equations of type (1) occur often in the study of nonlinear electrical networks. For example, if a linear n-port containing resistors, independent sources, and dependent sources has a two-terminal device whose V vs I curve is specified by $I_i = f_i(V_i)$, for $i = 1, \cdots, n$, connected across each port, then the port voltages may often be expressed as the solution of an equation of type (1). In this case the matrix A will be the y-parameter matrix of the n-port, the constant vector B will account for the presence of the independent sources, and the components of the vector x will be the desired port voltages.

II. ACTIVE AND NONRECIPROCAL n-PORTS

In case the n-port of the above example contains no dependent sources and the functions f_i satisfy condition (*i*) above, the existence and uniqueness of a solution of (1) follows immediately from the well-known result of R. J. Duffin.[1] In fact, with the additional assumption that the slope of each f_i is bounded by positive constants the computational technique of J. Katzenelson and L. H. Seitelman may be used to compute the solution.[2] This computational technique is based upon a theorem of I. W. Sandberg which relies upon the contraction-mapping fixed point theorem.[3]

Sandberg's theorem may, in fact, be used to prove the existence and uniqueness of a solution of (1), and to construct a convergent iteration process for computing this solution, whenever the matrix A is positive semidefinite* and the slope of each f_i is bounded by positive constants. Other theorems which do not require that the slopes of each strictly monotone increasing f_i be bounded by positive constants also exist. (For example, see Ref. 4.) These theorems guarantee existence and uniqueness of a solution of (1) whenever A is positive semidefinite but do not specify a procedure for computing it.

Suppose, however, that the matrix A is not positive semidefinite; that is, suppose the n-port in the above example is active. Then the above results no longer apply. It may often happen that the matrix A is not positive semidefinite but still satisfies condition (*ii*) above. The matrix

$$A = \begin{bmatrix} 1 & 1 \\ 5 & 7 \end{bmatrix},$$

for example, has this property. It is interesting to notice that in this case the matrix A will necessarily also be nonsymmetric (the corresponding n-port will be nonreciprocal). This follows from the fact that for symmetric matries A, condition (*ii*) implies that A is a dominant matrix[5] which, in turn, implies that A is positive semidefinite. It is for this class of active nonreciprocal n-ports that our work provides entirely new results. Even for the passive case, however, notice that our computational techniques do not require that the slopes of the functions f_i be bounded. Also, there is reason to believe that for certain problems our iteration schemes may converge more rapidly than the ones based upon the contraction mapping theorem. More is said about this in Section VII.

III. EXISTENCE AND UNIQUENESS

Before proving the existence and uniqueness theorem we first prove a lemma which is used many times in this and the following section.

Lemma 1: Let the $n \times n$ matrix A of real numbers satisfy condition (ii) of Section I. For $j = 1, \cdots, n$ let p_j denote the jth component of $p \in E^n$. Let $k \in \{1, \cdots, n\}$ be chosen such that $|p_k| = \max \{|p_j| : j = 1, \cdots, n\}$. Then,

---

* The $n \times n$ matrix A is said to be positive semidefinite if $\langle x, Ax \rangle \geq 0$ for all x in E^n.

$$p_k > 0 \Rightarrow \sum_{j=1}^{n} a_{kj}p_j \geqq 0,$$

and

$$p_k < 0 \Rightarrow \sum_{j=1}^{n} a_{kj}p_j \leqq 0.$$

Proof:

$$a_{kk} \mid p_k \mid \geqq \sum_{\substack{j=1 \\ j\neq k}}^{n} \mid a_{kj} \mid \cdot \mid p_k \mid \geqq \sum_{\substack{j=1 \\ j\neq k}}^{n} \mid a_{kj}p_j \mid \geqq \left| \sum_{\substack{j=1 \\ j\neq k}}^{n} a_{kj}p_j \right|.$$

Thus,

$$a_{kk} \mid p_k \mid \geqq \pm \sum_{\substack{j=1 \\ j\neq k}}^{n} a_{kj}p_j .$$

But then,

$$p_k > 0 \Rightarrow a_{kk}p_k \geqq -\sum_{\substack{j=1 \\ j\neq k}}^{n} a_{kj}p_j \Rightarrow \sum_{j=1}^{n} a_{kj}p_j \geqq 0,$$

and,

$$p_k < 0 \Rightarrow -a_{kk}p_k \geqq \sum_{\substack{j=1 \\ j\neq k}}^{n} a_{kj}p_j \Rightarrow \sum_{j=1}^{n} a_{kj}p_j \leqq 0. \quad \square$$

Theorem 1: There exists a unique solution of (1) whenever conditions (i) and (ii) of Section I are satisfied.

Proof: We first prove that if a solution exists it is unique. Let x^1 and x^2 be solutions of (1). Then,

$$F(x^2) - F(x^1) = A(x^1 - x^2).$$

For $j = 1, \cdots, n$ let x_j^1 and x_j^2 denote the jth components of x^1 and x^2, respectively, and choose $k \in \{1, \cdots, n\}$ such that

$$\mid x_k^1 - x_k^2 \mid = \max \{\mid x_j^1 - x_j^2 \mid : j = 1, \cdots, n\}.$$

If $x_k^1 > x_k^2$ then, by Lemma 1,

$$f_k(x_k^2) - f_k(x_k^1) = \sum_{j=1}^{n} a_{kj}(x_j^1 - x_j^2) \geqq 0.$$

If $x_k^1 < x_k^2$ then, by Lemma 1,

$$f_k(x_k^2) - f_k(x_k^1) = \sum_{j=1}^{n} a_{kj}(x_j^1 - x_j^2) \leqq 0.$$

Both of these conclusions contradict the fact that f_k is strictly monotone increasing. Thus, $x_k^1 = x_k^2$ and hence $x_j^1 = x_j^2$ for $j = 1, \cdots, n$. That is, the solution of (1) is unique, if it exists.

We prove existence of a solution by induction. For $k = 1, \cdots, n$ let

$$F_k(x) \equiv \begin{bmatrix} f_1(x_1) \\ \vdots \\ f_k(x_k) \end{bmatrix}, \qquad A_k \equiv \begin{bmatrix} a_{11} & \cdots & a_{1k} \\ \cdots & \cdots & \cdots \\ a_{k1} & \cdots & a_{kk} \end{bmatrix}, \qquad B_k \equiv \begin{bmatrix} b_1 \\ \vdots \\ b_k \end{bmatrix}.$$

Clearly, the matrix A_k satisfies condition (*ii*) of Section I. Also, it is clear that there exists a unique solution* of $F_1(x) + A_1x = B_1$ for every strictly monotone increasing function f_1 mapping E^1 onto E^1.

Assume that there exists a solution of $F_k(x) + A_kx = B_k$ for arbitrary strictly monotone increasing functions f_i, $i = 1, \cdots, k$ mapping E^1 onto E^1. Then, for every real number x_{k+1}, the equation

$$F_k(x) + A_kx + \begin{bmatrix} a_{1,k+1} \\ \vdots \\ a_{k,k+1} \end{bmatrix} x_{k+1} = B_k$$

has a (unique) solution; since for $i = 1, \cdots, k$ the function $f_i(x_i) + a_{i,k+1}x_{k+1}$ is strictly monotone increasing from E^1 onto E^1. Let the components of this solution be denoted by $x_i = m_i(x_{k+1})$ for $i = 1, \cdots, k$. We have thus defined k functions m_i on E^1.

We now prove that for every $x_{k+1}^1, x_{k+1}^2 \in E^1$,

$$| x_{k+1}^2 - x_{k+1}^1 | \geqq | m_j(x_{k+1}^2) - m_j(x_{k+1}^1) |, \quad \text{for } j = 1, \cdots, k. \tag{2}$$

This inequality, incidentally, implies that each m_i is continuous.

Let $x_{k+1}^1, x_{k+1}^2 \in E^1$ and choose $l \in \{1, \cdots, k\}$ such that

$$| m_l(x_{k+1}^2) - m_l(x_{k+1}^1) |$$
$$= \max \{| m_j(x_{k+1}^2) - m_j(x_{k+1}^1) | : j = 1, \cdots, k\}.$$

Assume that $| m_l(x_{k+1}^2) - m_l(x_{k+1}^1) | > | x_{k+1}^2 - x_{k+1}^1 |$. Clearly, then, $m_l(x_{k+1}^2) - m_l(x_{k+1}^1) \neq 0$. If $m_l(x_{k+1}^2) - m_l(x_{k+1}^1) > 0$ then,

$$f_l[m_l(x_{k+1}^2)] - f_l[m_l(x_{k+1}^1)] > 0.$$

---

* We take the liberty of using the same symbol x to denote points in any of the spaces E^k, $1 \leq k \leq n$. No confusion should arise since the subscripts on F and A will make our choice clear.

Also, since the matrix A_{k+1} satisfies condition (*ii*) of Section I, letting

$$p_j = m_j(x_{k+1}^2) - m_j(x_{k+1}^1), \quad \text{for} \quad j = 1, \cdots, k,$$

$$p_{k+1} = x_{k+1}^2 - x_{k+1}^1 ,$$

we have, by Lemma 1,

$$\sum_{j=1}^{k} a_{lj}[m_j(x_{k+1}^2) - m_j(x_{k+1}^1)] + a_{l,k+1}(x_{k+1}^2 - x_{k+1}^1) \geqq 0.$$

Thus,

$$f_l[m_l(x_{k+1}^2)] + \sum_{j=1}^{k} a_{lj}m_j(x_{k+1}^2) + a_{l,k+1}x_{k+1}^2$$

$$> f_l[m_l(x_{k+1}^1)] + \sum_{j=1}^{k} a_{lj}m_j(x_{k+1}^1) + a_{l,k+1}x_{k+1}^1 , \qquad (3)$$

which is a contradiction since the quantity on each side of this inequality is equal to b_l. If $m_l(x_{k+1}^2) - m_l(x_{k+1}^1) < 0$ then,

$$f_l[m_l(x_{k+1}^2)] - f_l[m_l(x_{k+1}^1)] < 0.$$

By applying Lemma 1 again, as above, one arrives again at (3) with $>$ replaced by $<$. This is also a contradiction. Thus, we must have

$$| x_{k+1}^2 - x_{k+1}^1 | \geqq | m_l(x_{k+1}^2) - m_l(x_{k+1}^1) |,$$

and hence (2) is proved

Now, consider the function

$$\sum_{j=1}^{k} a_{k+1,j}m_j(x_{k+1}) + a_{k+1,k+1}x_{k+1} . \qquad (4)$$

Let x_{k+1}^1, $x_{k+1}^2 \in E^1$ with $x_{k+1}^1 < x_{k+1}^2$. Then,

$$a_{k+1,k+1} \geqq \sum_{j=1}^{k} | a_{k+1,j} |$$

implies

$$a_{k+1,k+1}(x_{k+1}^2 - x_{k+1}^1)$$

$$= a_{k+1,k+1} | x_{k+1}^2 - x_{k+1}^1 | \geqq \sum_{j=1}^{k} (| a_{k+1,j} | \cdot | x_{k+1}^2 - x_{k+1}^1 |).$$

But, using (2),

$$a_{k+1,k+1}(x_{k+1}^2 - x_{k+1}^1) \geqq \sum_{j=1}^{k} | a_{k+1,j}[m_j(x_{k+1}^2) - m_j(x_{k+1}^1)] |$$

$$\geqq -\sum_{j=1}^{k} a_{k+1,j}[m_j(x_{k+1}^2) - m_j(x_{k+1}^1)],$$

which implies

$$\sum_{j=1}^{k} a_{k+1,j}m_j(x_{k+1}^1) + a_{k+1,k+1}x_{k+1}^1 \leqq \sum_{j=1}^{k} a_{k+1,j}m_j(x_{k+1}^2) + a_{k+1,k+1}x_{k+1}^2 .$$

That is, the function (4) is monotone increasing. Clearly (4) is continuous. It follows, therefore, that if f_{k+1} is a strictly monotone increasing function mapping E^1 onto E^1, then so is the function

$$f_{k+1}(x_{k+1}) + \sum_{j=1}^{k} a_{k+1,j}m_j(x_{k+1}) + a_{k+1,k+1}x_{k+1} .$$

Thus, there exists a unique solution of the equation

$$f_{k+1}(x_{k+1}) + \sum_{j=1}^{k} a_{k+1,j}m_j(x_{k+1}) + a_{k+1,k+1}x_{k+1} = b_{k+1} .$$

If x_{k+1}^0 denotes this solution then

$$x^0 \equiv \begin{bmatrix} m_1(x_{k+1}^0) \\ \vdots \\ m_k(x_{k+1}^0) \\ x_{k+1}^0 \end{bmatrix}$$

is the (unique) solution of

$$F_{k+1}(x) + A_{k+1}x = B_{k+1} .$$

Thus, we have proved that there exists a unique solution of (1). □

IV. BOUNDS ON THE SOLUTION

Having established the existence and uniqueness of a solution of (1) a natural question to arise is: What can one say about the location of this solution? It turns out that we can say quite a bit (again assuming that conditions (*i*) and (*ii*) of Section I are satisfied). One can, in fact, with little effort (compared with the effort required, in general, to actually compute the solution) determine a finite region R in E^n, in which the solution must lie. This region is the cartesian product of finite intervals $I_i \subset E^1$, for $i = 1, \cdots, n$, each of which has the property that if

$$x^0 \equiv \begin{bmatrix} x_1^0 \\ \vdots \\ x_n^0 \end{bmatrix}$$

is the solution of (1) then $x_i^0 \varepsilon I_i$ and, as

$$\sum_{\substack{j=1 \\ j \neq i}}^{n} | a_{ij} | \rightarrow 0,$$

the length of I_i, $l(I_i) \rightarrow 0$. Thus, when the off-diagonal elements of A are small, the region R will also be small.

In many applications it may be sufficient to know only that there exists a unique solution of (1) and to know the region R in which it must lie. If, however, one actually does want to compute the solution by some iterative technique, the knowledge of R should be useful in determining a starting point for the iteration. In fact, it will be shown that if the point x^* is the solution of

$$F(x) + \text{diag}\,[a_{11}, \cdots, a_{nn}]\, x = B, \tag{5}$$

then x^* is also in R and thus might be a reasonable starting point for an iterative computation of x^0.

The computation of bounds for the solution of (1) proceeds in two steps. First, one solves each of the equations

$$f_i(x_i) = b_i, \quad \text{for} \quad i = 1, \cdots, n. \tag{6}$$

Letting α_i denote the solutions of (6), and defining

$$\alpha = \max \{| \alpha_i | : i = 1, \cdots, n\},$$

$$B' = \begin{bmatrix} \sum_{\substack{j=1 \\ j \neq 1}}^{n} | a_{1j} | \\ \vdots \\ \sum_{\substack{j=1 \\ j \neq n}}^{n} | a_{nj} | \end{bmatrix},$$

one then solves each of the equations

$$F(x) + \text{diag}\,[a_{11}, \cdots, a_{nn}]\, x = B - \alpha B', \tag{7a}$$

$$F(x) + \text{diag}\,[a_{11}, \cdots, a_{nn}]\, x = B + \alpha B'. \tag{7b}$$

Denoting the solutions of (7a) and (7b) by

$$\eta \equiv \begin{bmatrix} \eta_1 \\ \vdots \\ \eta_n \end{bmatrix} \quad \text{and} \quad \xi \equiv \begin{bmatrix} \xi_1 \\ \vdots \\ \xi_n \end{bmatrix},$$

respectively, one has $R = I_1 \times \cdots \times I_n$, where

$$I_i = [\eta_i , \xi_i], \quad \text{for} \quad i = 1, \cdots , n.$$

It is clear from the fact that each component of the vector $\alpha B'$ is a nonnegative number and from the monotone nature of the left-hand sides of (7) that x^*, the solution of (5), is (as claimed) always in R. It is also clear that, for $i = 1, \cdots , n$, as

$$\sum_{\substack{j=1 \\ j \neq i}}^{n} | a_{ij} | \to 0,$$

the ith components of both $B - \alpha B'$ and $B + \alpha B'$ approach b_i, and hence $\eta_i \to x_i^*$ and $\xi_i \to x_i^*$. Thus, $l(I_i) \to 0$. We now prove that the solution of (1) is in R.

Theorem 2: If R is constructed as described above, then the solution of (1) is contained in R whenever conditions (i) and (ii) of Section I are satisfied.

Proof: Let x^0 be the solution of (1) and let $k \in \{1, \cdots , n\}$ be chosen such that $| x_k^0 | = \max\{| x_i^0 |: i = 1, \cdots , n\}$. Then, by Lemma 1, if $x_k^0 > 0$, $\sum_{j=1}^{n} a_{kj} x_j^0 \geqq 0$ and hence,

$$0 = f_k(x_k^0) + \sum_{j=1}^{n} a_{kj} x_j^0 - b_k \geqq f_k(x_k^0) - b_k$$

or $f_k(x_k^0) \leqq b_k$.

Thus, because of the monotonicity of f_k,

$$| x_k^0 | = x_k^0 \leqq \alpha_k \leqq \alpha,$$

and hence $| x_i^0 | \leqq \alpha$ for $i = 1, \cdots , n$. Similarly, by Lemma 1, if $x_k^0 < 0$, $\sum_{j=1}^{n} a_{kj} x_j^0 \leqq 0$ and hence,

$$f_k(x_k^0) \geqq b_k ,$$

and thus

$$| x_k^0 | = -x_k^0 \leqq -\alpha_k \leqq \alpha,$$

and hence $| x_i^0 | \leqq \alpha$, for $i = 1, \cdots, n$. Thus, in any case, $| x_i^0 | \leqq \alpha$, for $i = 1, \cdots, n$.

Now, for all x with $| x_j | \leqq \alpha$ for $j = 1, \cdots, n$, and for each $i \in \{1, \cdots, n\}$ we have,

$$\alpha \sum_{\substack{j=1 \\ j \neq i}}^{n} | a_{ij} | = \sum_{\substack{j=1 \\ j \neq i}}^{n} (| a_{ij} | \alpha) \geqq \sum_{\substack{j=1 \\ j \neq i}}^{n} | a_{ij} x_j |,$$

which implies

$$\alpha \sum_{\substack{j=1 \\ j \neq i}}^{n} | a_{ij} | \geqq \sum_{\substack{j=1 \\ j \neq i}}^{n} a_{ij} x_j ,$$

and

$$-\alpha \sum_{\substack{j=1 \\ j \neq i}}^{n} | a_{ij} | \leqq \sum_{\substack{j=1 \\ j \neq i}}^{n} a_{ij} x_j .$$

Thus,

$$a_{ii} x_i - \alpha \sum_{\substack{j=1 \\ j \neq i}}^{n} | a_{ij} | \leqq \sum_{j=1}^{n} a_{ij} x_j \leqq a_{ii} x_i + \alpha \sum_{\substack{j=1 \\ j \neq i}}^{n} | a_{ij} |.$$

In particular, for $x = x^0$, we have

$$f_i(x_i^0) + a_{ii} x_i^0 - \alpha \sum_{\substack{j=1 \\ j \neq i}}^{n} | a_{ij} | \leqq b_i \leqq f_i(x_i^0) + a_{ii} x_i^0 + \alpha \sum_{\substack{j=1 \\ j \neq i}}^{n} | a_{ij} |.$$

Comparing this result with (7) we have, as a consequence of the monotonicity of the functions on the left-hand sides of (7),

$$\eta_i \leqq x_i^0 \leqq \xi_i , \quad \text{for} \quad i = 1, \cdots, n.$$

Hence, $x^0 \in R$. □

Since in the above proof it was shown that $| x_i^0 | \leqq \alpha$ for $i = 1, \cdots, n$ it might seem to some readers that the intervals I_i might be reduced in length if we simply define them to be: $I_i = [-\alpha, \alpha] \cap [\eta_i , \xi_i]$. This, however, is unnecessary since it is easily shown that $-\alpha \leqq \eta_i \leqq \xi_i \leqq \alpha$, for $i = 1, \cdots, n$.

V. EXAMPLE

We now give an example of the use of the above method for the computation of solution bounds. Consider the equation

$$\begin{bmatrix} f_1(x_1) \\ f_2(x_2) \end{bmatrix} + \begin{bmatrix} 5 & 4 \\ -3 & 4 \end{bmatrix} \begin{bmatrix} x_1 \\ x_2 \end{bmatrix} = \begin{bmatrix} 15 \\ 13 \end{bmatrix},$$

where f_1 and f_2 are defined by

$$f_1(x_1) = \begin{cases} 4^{x_1} - 2, & x_1 \geqq 0 \\ \frac{1}{10}x_1 - 1, & x_1 < 0, \end{cases}$$

and

$$f_2(x_2) = \begin{cases} x_2 + 9, & x_2 \geqq 3 \\ 4x_2\,, & -3 < x_2 < 3 \\ x_2 - 9, & x_2 \leqq -3. \end{cases}$$

Figure 1 shows the graphs of f_1 and f_2. Since we know that the region R will be small if the off-diagonal terms of A are small enough, we have intentionally chosen an example in which these terms are rather large.

The computation of α by solving (6) may be done by inspection for this example. One finds that $4^{\alpha_1} = 17$ implies that α_1 is slightly greater than 2, and since $\alpha_2 = 4$ we have $\alpha = 4$. Using this result in (7) one readily computes

$$\eta = \begin{bmatrix} 0 \\ 0.125 \end{bmatrix}, \qquad \xi \approx \begin{bmatrix} 2.23 \\ 3.2 \end{bmatrix}.$$

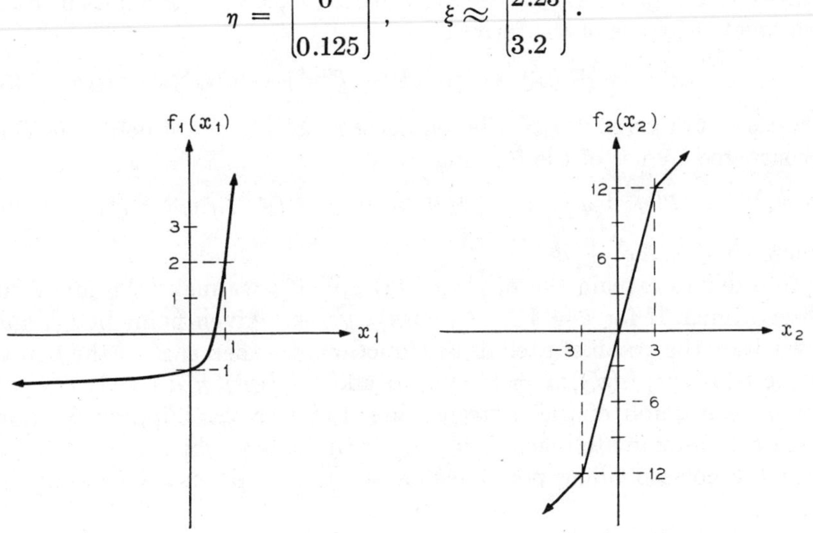

Fig. 1 — The nonlinear functions f_1 and f_2 for the example.

Actually, it is easily verified that the solution of this example is $x^0 = \begin{pmatrix} 1 \\ 2 \end{pmatrix}$.

VI. COMPUTATION OF THE SOLUTION

For $i = 1, \cdots, n$ we denote by $f_i'(x_i)$ the right-hand derivative of f_i at the point $x_i \in E^1$. For each $x \in E^n$ we denote by $F'(x)$ the following matrix:

$$F'(x) = \text{diag}\,[f_1'(x_1), \cdots, f_n'(x_n)].$$

It is easy to prove that if F satisfies condition (*i*) of Section I then $F'(x)$ exists for all $x \in E^n$. Also, it is clear that each element of the main diagonal of $F'(x)$ is nonnegative for all $x \in E^n$. (Each element is in fact positive if, in addition, F satisfies condition (*iii*) of Section I.) Finally, we note that

$$F^{-1}(y) \equiv \begin{bmatrix} f_1^{-1}(y_1) \\ \vdots \\ f_n^{-1}(y_n) \end{bmatrix}$$

is defined for all $y \in E^n$, assuming again that F satisfies condition (*i*) of Section *I*.

The following two iteration schemes are proposed for the computation of the solution of (1):

Scheme 1: For given $x^1 \in E^n$ the sequence $x^1, x^2, x^3, \cdots$ of points in E^n is constructed by use of the formula

$$x^{k+1} = [F'(x^k) + A]^{-1}(B - F(x^k) + F'(x^k)x^k). \tag{8}$$

Scheme 2: For given $x^1 \in E^n$ the sequence $x^1, x^2, x^3, \cdots$ of points in E^n is constructed by use of the formula

$$x^{k+1} = [F'(F^{-1}(y^k)) + A]^{-1}(B - y^k + F'(F^{-1}(y^k))F^{-1}(y^k)), \tag{9}$$

where $y^k = -Ax^k + B$.

In order to explain the origin of (8) and (9) we make the following observations: If for $i = 1, \cdots, n$ (x_i^k, y_i^k) is a given point in E^2, and if we draw the graph of each of the functions f_i, then each of the points in the sets $\{(x_i^k, f_i(x_i^k)): i = 1, \cdots, n\}$ and $\{(f_i^{-1}(y_i^k), y_i^k): i = 1, \cdots, n\}$ lies on the graph of the corresponding function f_i. Suppose we now replace (approximate) each f_i by the straight line which is tangent to it at the corresponding point in one of the above sets.* Choosing the

---

* Our definition of *tangent* coincides with the usual one, except that the right-hand derivative is used at those points where the derivative fails to exist.

first set of points we approximate F by

$$\hat{F}(x) \equiv F'(x^k)x + F(x^k) - F'(x^k)x^k.$$

Choosing the second set gives

$$\tilde{F}(x) \equiv F'(F^{-1}(y^k))x + y^k - F'(F^{-1}(y^k))F^{-1}(y^k).$$

If we now define $y^k = -Ax^k + B$ and compute the solution of the equation

$$\hat{F}(x) + Ax = B$$

and call it x^{k+1}, we obtain (8). Calling x^{k+1} the solution of

$$\tilde{F}(x) + Ax = B,$$

yields (9).

The above remarks have a very meaningful interpretation for problems arising from nonlinear electrical networks of the type described in Section I. Iteration Scheme 1 implements the following procedure: Given the vector x^k of port voltages for the linear n-port, replace each two-terminal nonlinear device with a linear Thévenin's "equivalent" circuit whose V vs I curve is a straight line, tangent to the given curve at the point $(x_i^k, f_i(x_i^k))$. Compute the port voltages in the resulting linear network to obtain x^{k+1}.

Iteration Scheme 2 has a similar interpretation; this time, however, the vector of port currents, $y^k = -Ax^k + B$, is used to determine the linear equivalent circuit replacing the nonlinear devices at each step.

In view of the above remarks it is apparent that if one has some facility for solving linear network problems (a computer program, for example) then it might easily be adapted to solve many nonlinear problems as well.

We finally remark that the use of the first iteration scheme is, in essence, the same as using the Newton-Raphson technique to compute the root of (1).

We now prove a theorem which specifies conditions which are sufficient to ensure convergence of each of the above iteration schemes. We emphasize, however, that these iteration schemes will converge for many problems in which the conditions of the theorem are not satisfied—especially if a good enough starting point is provided.

In the following we denote the origin in E^n by θ and, for the points $x, y \in E^n$, the notation $x \leqq y$ means $x_i \leqq y_i$ for $i = 1, \cdots, n$. The relations $x < y$, $x \geqq y$, $x > y$ are defined similarly. We also make use of the concept of a *matrix of monotone kind*.[6] The matrix A is said to be of monotone kind if $x \in E^n$, $Ax \geqq \theta \Rightarrow x \geqq \theta$. It is easy to show that A is of monotone kind if and only if A^{-1} contains only nonnegative elements. It is also easy to show that if A is of monotone kind and $x, y \in E^n$ with $Ax \leqq y$, then $x \leqq A^{-1}y$. Ref. 6 shows that if the strict inequality $>$ holds in condition (*ii*) of Section I, then conditions (*ii*) and (*iv*) are sufficient to ensure that A is of monotone kind.

Theorem 3: For an arbitrary starting point x^1, both of the above iteration schemes will converge to the solution of (1) if conditions (i) through (iv) of Section I are satisfied.

Proof: We give here only the proof for the second iteration scheme, assuming that all of the functions f_i are convex. The other three cases are quite similar and it will be apparent to the reader how this proof may easily be modified to take care of them.*

We first remark that the iteration scheme is well defined. The fact that for every $y^k \in E^n$, $F'(F^{-1}(y^k))$ is a diagonal matrix containing all positive numbers on the main diagonal, and the fact that A satisfies conditions (*ii*) and (*iv*) of Section I, assures us that the matrix $[F'(F^{-1}(y^k)) + A]$ is nonsingular (it is, in fact, of monotone kind—see Ref. 6, p. 376).

Let x^1 be an arbitrary point in E^n. Then, since for $i = 1, \cdots, n$ and $k = 2, 3, 4, \cdots$ each of the points (x_i^k, y_i^k) lies on some straight line, tangent to the corresponding function f_i, and since each f_i is strictly monotone increasing and convex, we have that $F^{-1}(y^k) \leqq x^k$ for $k = 2, 3, 4, \cdots$. We now show that $F^{-1}(y^k) \leqq x^k$ implies that $x^{k+1} \leqq x^k$. Obviously,

$$F'(F^{-1}(y^k))(x^k - F^{-1}(y^k)) \geqq \theta.$$

But, by definition, $Ax^k + y^k - B = \theta$; hence,

$$F'(F^{-1}(y^k))(x^k - F^{-1}(y^k)) + Ax^k + y^k - B \geqq \theta,$$

which implies

$$[F'(F^{-1}(y^k)) + A]x^k \geqq B - y^k + F'(F^{-1}(y^k))F^{-1}(y^k).$$

But then, since $[F'(F^{-1}(y^k)) + A]$ is a matrix of monotone kind,

---

* After this manuscript had been completed, the author became aware of J. S. Vandergraft's paper (Ref. 7). With a certain amount of reformulation, the (monotone) convergence of the first iteration scheme, when all f_i are convex, can be shown to follow, in essence, from his Theorem 5.1.

$$x^k \geqq [F'(F^{-1}(y^k)) + A]^{-1}(B - y^k + F'(F^{-1}(y^k))F^{-1}(y^k)),$$

or, $x^k \geqq x^{k+1}$. Thus, the sequence $x^2, x^3, x^4, \cdots$ has the property

$$x^2 \geqq x^3 \geqq x^4 \geqq \cdots.$$

We now show that for $k = 2, 3, 4, \cdots, x^k \geqq x^0$, where x^0 is the solution of (1). For each x^k, $k = 2, 3, 4, \cdots$, there is some point $p \in E^n$ ($p \equiv F^{-1}(y^{k-1})$) such that

$$Ax^k - B = F'(p)p - F'(p)x^k - F(p). \tag{10}$$

Furthermore, from the convexity of each f_i, it is clear that for every pair of points $q^1, q^2 \in E^n$,

$$F(q^1) \geqq F(q^2) + F'(q^2)(q^1 - q^2).$$

In particular,

$$F(x^0) \geqq F(p) + F'(p)(x^0 - p).$$

Hence,

$$F'(p)p - F(p) + F(x^0) \geqq F'(p)x^0$$

which implies

$$F'(p)(p - x^k) - F(p) + F(x^0) \geqq F'(p)(x^0 - x^k).$$

Using (10) we have, therefore,

$$Ax^k - B + F(x^0) \geqq F'(p)(x^0 - x^k).$$

But, $F(x^0) = -Ax^0 + B$, hence

$$A(x^k - x^0) \geqq F'(p)(x^0 - x^k)$$

or,

$$[F'(p) + A](x^k - x^0) \geqq \theta.$$

But then, since $[F'(p) + A]$ is of monotone kind, $x^k - x^0 \geqq \theta$, or $x^k \geqq x^0$. Thus, we have shown that each sequence $x_i^2, x_i^3, x_i^4, \cdots$ is a bounded monotone sequence and hence the sequence $x^2, x^3, x^4, \cdots$ converges to some point x^* in E^n. We now prove that $x^* = x^0$; that is, we show that x^* satisfies (1).

Let $y^* = -Ax^* + B$. Then, as $k \to \infty$, $x^k \to x^*$ and $y^k \to y^*$. Thus, $F^{-1}(y^k) \to F^{-1}(y^*)$ and each element of the matrix $F'(F^{-1}(y^k))$ approaches the corresponding element of $F'(F^{-1}(y^*))$. Now, from (9), we have

$$Ax^{k+1} + F'(F^{-1}(y^k))x^{k+1} = Ax^k + F'(F^{-1}(y^k))F^{-1}(y^k)$$

which implies

$$F'(F^{-1}(y^k))(F^{-1}(y^k) - x^{k+1}) = A(x^{k+1} - x^k)$$

and hence

$$F'(F^{-1}(y^k))(F^{-1}(y^k) - x^* = A(x^{k+1} - x^k) - F'(F^{-1}(y^k))(x^* - x^{k+1}).$$

But as $k \to \infty$, $(x^{k+1} - x^k) \to \theta$ and hence $A(x^{k+1} - x^k) \to \theta$; also, $(x^* - x^{k+1}) \to \theta$ and hence $F'(F^{-1}(y^k))(x^* - x^{k+1}) \to F'(F^{-1}(y^*))\theta = \theta$. Thus, as $k \to \infty$,

$$F'(F^{-1}(y^k))(F^{-1}(y^k) - x^*) \to \theta$$

which implies

$$F^{-1}(y^k) - x^* \to \theta$$

or

$$F^{-1}(y^k) \to x^*$$

and therefore

$$y^k \to F(x^*).$$

Hence, $y^* = F(x^*)$, and thus,

$$F(x^*) + Ax^* = B.$$

Thus, the iteration converges to the solution of (1). □

Although Theorem 3 states that both of our iteration schemes will converge for the same class of problems, only one of the schemes might converge for some problems for which all of the conditions (*i*) through (*iv*) of Section I are not satisfied. Also, for some problems a prior knowledge of the region in which the solution lies might dictate the choice of one iteration scheme over the other. For example, if it is known that some of the functions f_i are quite steep in the neighborhood of the solution then perhaps F^{-1} may be evaluated in this region more accurately than F. In this case Scheme 2 might be preferred to Scheme 1.

VII. SPEED OF CONVERGENCE

Section II mentions that in certain situations our iteration schemes may converge to the solution of (1) more rapidly than those based upon the contraction-mapping fixed point theorem. To illustrate this property we have chosen to compare the rate of convergence of Sandberg's iteration scheme to that of our schemes.[3]

If we define the operator G mapping E^n into E^n by

$$G(x) \equiv F(x) + Ax,$$

then, as a special case of Sandberg's Theorem I, we have the result: If there are positive constants k_1 and k_2 such that

$$\langle G(x) - G(y), x - y \rangle \geqq k_1 \, || x - y ||^2, \tag{11}$$

and

$$|| G(x) - G(y) ||^2 \leqq k_2 \, || x - y ||^2, \tag{12}$$

for all $x, y \,\varepsilon\, E^n$, then there is a unique solution of (1) and the solution is given by $\lim_{k \to \infty} x^k$, where x^1 is an arbitrary point in E^n, and

$$x^{k+1} = \frac{k_1}{k_2} [B - G(x^k)] + x^k,$$

for $k = 1, 2, 3, \cdots$. The proof of this theorem consists of showing that the mapping

$$H(x) \equiv \frac{k_1}{k_2} [B - G(x)] + x$$

is a contraction.

It is interesting to observe that if the inequalities (11) and (12) are satisfied then positive constants k_3 and k_4 exist, such that

$$\langle G(x) - G(y), x - y \rangle \leqq k_3 \, || x - y ||^2, \tag{13}$$

and

$$|| G(x) - G(y) ||^2 \geqq k_4 \, || x - y ||^2, \tag{14}$$

for all $x, y \,\varepsilon\, E^n$. In fact, a simple application of the Schwarz inequality to (11) and (12) yields (13) and (14) with $k_3 = (k_2)^{\frac{1}{2}}$ and $k_4 = k_1^2$. Now (13) and (14) are of the same form as (11) and (12), except that the inequalities are reversed. Thus, if one uses (13) and (14) in the proof of Sandberg's theorem, reversing all inequalities, one obtains:

$$|| H(x) - H(y) ||^2 \geqq K \, || x - y ||^2,$$

where,

$$K = 1 - 2(k_1^2/k_2)^{\frac{1}{2}} + (k_1^2/k_2)^2.$$

It is readily seen that if $4k_1^2 < k_2$, then K is positive. If we let x^0 denote the solution of (1), and hence $H(x^0) = x^0$, we have, for $k = 1, 2, \cdots$,

$$|| x^{k+1} - x^0 ||^2 = || H(x^k) - H(x^0) ||^2 \geqq K \, || x^k - x^0 ||^2.$$

Thus, $(K)^{\frac{1}{2}}$ represents, in this case, a lower bound on the rate of convergence of the iteration scheme. It is true that $(K)^{\frac{1}{2}}$ is always in the interval (0,1), for indeed Sandberg has proved that the sequence x^k does converge to x^0. However, as k_1 becomes small, and as k_2 becomes large, K approaches 1 and the sequence converges quite slowly. For (1) the largest value that may be used for k_1 and the smallest value that may be used for k_2 will many times be dictated by the positive constants which are bounds on the slopes of the functions f_i. If, for example, the slopes of the f_i become so large for large x_i, and so small for large negative x_i that one must choose $k_1 = 10^{-1}$ and $k_2 = 10^2$, then one easily computes $(K)^{\frac{1}{2}} \approx 0.99$. Thus, no matter how close any iterate is to the solution, the next iterate will be no more than about one percent closer.

It is of course true that Sandberg's iteration scheme is applicable to a much more general class of problems than we consider in this paper. If, however, for any problem to which it is applied, the constants k_1 and k_2 must be restricted such that k_1/k_2 is quite small, then the rate of convergence will always be adversely affected. In the Katzenelson-Seitelman application of Sandberg's iteration scheme, their "heuristic refinement" (see Ref. 2) attempts to overcome this difficulty.

Although the classes of equations to which our iteration schemes and the Katzenelson-Seitelman algorithm may be applied are not identical, in those cases where both techniques may be used the advantage that our schemes offer is now clear. From (8) and (9) one easily obtains

$$x^{k+1} - x^0 = [F'(x^k) + A]^{-1}(F(x^0) - F(x^k) - F'(x^k)(x^0 - x^k)),$$

and

$$x^{k+1} - x^0 =$$
$$[F'(F^{-1}(y^k)) + A]^{-1} (F(x^0) - y^k - F'(F^{-1}(y^k))(x^0 - F^{-1}(y^k))),$$

respectively. These equations show that $\| x^{k+1} - x^0 \|$ will be small (even if $\| x^k - x^0 \|$ is rather large) so long as for $i = 1, \cdots, n$,

$$\frac{f_i(x_i^0) - f_i(x_i^k)}{x_i^0 - x_i^k} \approx f_i'(x_i^k),$$

for Scheme 1, or

$$\frac{f_i(x_i^0) - y_i^k}{x_i^0 - f_i^{-1}(y_i^k)} \approx f_i'(f_i^{-1}(y_i^k)),$$

for Scheme 2. That is, as soon as the kth iterate comes close enough to the solution that each of the functions f_i is approximately linear, the rate of convergence of our iterations becomes quite rapid. In fact, the rate of convergence increases without bound as the iterates approach the solution. It is also clear that if each of the functions f_i is piecewise linear then our iterations will converge in a finite number of steps.

From the standpoint of computational efficiency it is, of course, the amount of time required to compute an approximate solution that is the major concern. For those problems to which both our iteration schemes and the Katzenelson-Seitelman algorithm may be applied, it can happen that our methods might still be slower than theirs even in the case when the convergence rate of our methods is faster. This can happen because, for some problems, the equation with which we are concerned may be of a higher order than theirs, and also because we must compute the inverse of a matrix at each iteration step. On the other hand, it is clear that for many problems, even from the standpoint of total computation time, our techniques will be more efficient.

VIII. ACKNOWLEDGMENT

The author is grateful to I. W. Sandberg for encouragement and many helpful conversations.

REFERENCES

1. Duffin, R. J., "Nonlinear Networks II," Bull. Amer. Math. Soc., *53* (October 1947), pp. 963–971.
2. Katzenelson, J. and Seitelman, L. H., "An Iterative Method for Solution of Networks of Nonlinear Monotone Resistors," IEEE Trans. Circuit Theory, *CT-13,* No. 3 (September 1966), pp. 317–323.
3. Sandberg, I. W., "On the Properties of Some Systems That Distort Signals-I," B.S.T.J., *42,* No. 5 (September 1963), pp. 2033–2046.
4. Minty, G. J., "Two Theorems on Nonlinear Functional Equations in Hilbert Space," Bull. Amer. Math. Soc., *69,* No. 5 (September 1963), pp. 691–692.
5. Weinberg, L., *Network Analysis and Synthesis,* New York: McGraw-Hill, 1962.
6. Collatz, L., *Functional Analysis and Numerical Mathematics* (tr. from German), New York: Academic Press, 1966.
7. Vandergraft, J. S., "Newton's Method for Convex Operators in Partially Ordered Spaces," SIAM J. Numerical Anal., *4,* No. 3 (September 1967), pp. 406–432.

A THEORY OF NONLINEAR NETWORKS—I*

By

R. K. BRAYTON (*International Business Machines Corporation, Yorktown Heights, N. Y.*)
and J. K. MOSER (*Courant Institute of Mathematical Sciences, New York University*)

Abstract. This report describes a new approach to nonlinear RLC-networks which is based on the fact that the system of differential equations for such networks has the special form

$$L(i)\frac{di}{dt} = \frac{\partial P(i, v)}{\partial i}, \qquad C(v)\frac{dv}{dt} = -\frac{\partial P(i, v)}{\partial v}.$$

The function, $P(i, v)$, called the mixed potential function, can be used to construct Liapounov-type functions to prove stability under certain conditions. Several theorems on the stability of circuits are derived and examples are given to illustrate the results. A procedure is given to construct the mixed potential function directly from the circuit. The concepts of a complete set of mixed variables and a complete circuit are defined.

Introduction. A. In the extensive theory of electrical circuits many impressive advances have led to a powerful tool for the engineer and the designer. For a wide class of problems one is able to construct a circuit with required properties using a rather complete theory which is available in several textbooks (see, e.g., [1], [2]). Most of these theories are based on the linear differential equations of electrical circuits. However, in recent times many engineering problems have led to the study of nonlinear networks which cannot appropriately be approximated by linear equations. Typical examples in this direction are the so-called flip-flop circuits which have several equilibrium states. Since a linear circuit obviously admits only one equilibrium, a flip-flop circuit can only be described by nonlinear differential equations. The main difference between such circuits and linear ones lies in the nonmonotone character of the voltage-current relations for the resistors. It will be a main point in the following to admit such "negative resistors".

B. The electrical circuits considered in this paper are general RLC-circuits in which any or all of the elements may be nonlinear. One of the purposes of this paper is to show that the differential equations of such electrical circuits have a special form which has its ultimate basis in the conservation laws of Kirchhoff. It will be derived that under very general assumptions the differential equations have the form

$$\begin{aligned} L_\rho \frac{di_\rho}{dt} &= \frac{\partial P}{\partial i_\rho}, \qquad (\rho = 1, \cdots, r), \\ C_\sigma \frac{dv_\sigma}{dt} &= -\frac{\partial P}{\partial v_\sigma}, \qquad (\sigma = r+1, \cdots, r+s), \end{aligned} \tag{1}$$

*Received May 29, 1963. The results reported in this paper were obtained in the course of research jointly sponsored by IBM and the Air Force Office of Scientific Research, Contract AF49(638)-1139.

Reprinted with permission from *Quarterly of Applied Mathematics*, vol. 22, pp. 1–33, April 1964.

where the i_ρ represent the currents in the inductors and v_σ the voltages across the capacitors. The function $P(i, v)$ describes the physical properties of the resistive part of the circuit. Since it has the dimension of voltage times current, it will be called a potential function. This function can be formed additively from potential functions of the single elements similar to the way that the Hamiltonian is formed in particle dynamics from the potential energy and the kinetic energy of the different particles. However, it should be observed that equations (1) do not represent a Hamiltonian system since the latter describes nondissipative motion while in equations (1) the potential P contains dissipative terms. Also, the transformation properties of the above equations are different from Hamiltonian equations in that equations (1) preserve their form under coordinate transformations which leave the indefinite metric

$$-\sum_{\rho=1}^{r} L_\rho(di_\rho)^2 + \sum_{\sigma=1+r}^{s+r} C_\sigma(dv_\sigma)^2 \tag{2}$$

invariant.

C. A geometrical interpretation of the special form of equations (1) is the following. We consider a box containing an electrical circuit with only resistive elements. There are n pairs of terminals on the box which are connected internally to the electrical circuit. To measure the external electrical properties of this box we connect each terminal to either a current source of prescribed current i_ρ ($\rho = 1, \cdots, r$) or a voltage source of prescribed voltage v_σ ($\sigma = r + 1, \cdots, r + s = n$). Under natural compatibility assumptions for the arrangement of these sources, an equilibrium state (i_ν, v_ν) ($\nu = 1, \cdots, n$) will be attained, i.e., the missing quantities v_ρ ($\rho = 1, \cdots, r$) and i_σ ($\sigma = r + 1, \cdots, r + s$) will be determined. In other words the $2n$ voltages and currents satisfy n relations which define an n-dimensional surface in $2n$-dimensional space. We call this surface the characteristic surface Σ of the box. In fact, if $n = 1$, Σ is a curve usually called the voltage-current characteristic for an element or a circuit. The result that the equations have the form (1) can be expressed compactly by the identity,

$$\sum_{\nu=1}^{n} di_\nu \wedge dv_\nu = 0, \tag{3}$$

i.e., this two-dimensional differential form in the sense of Cartan [3] vanishes identically on the surface Σ. This fact will be explained and proved in section 13, part II.

D. It is also the purpose of this study to draw some conclusions concerning the solutions of the differential equations (1) from their special form. To show that such implications can be expected, consider, for instance, an RC-circuit (i.e., a circuit without inductors or $r = 0$ in (1)). In this case, the quadratic form (2) is positive definite and can be used as a metric

$$(ds)^2 = \sum_{\sigma=1}^{s} C_\sigma(dv_\sigma)^2.$$

One verifies immediately that in this case $P(i, v)$ decreases along solutions of (1) since

$$\frac{dP}{dt} = \sum_{\sigma=1}^{s} \frac{\partial P}{\partial v_\sigma}\frac{dv_\sigma}{dt} = -\left(\frac{ds}{dt}\right)^2$$

which is negative except at the equilibrium points. This implies that all solutions of an RC-circuit approach equilibrium states for $t \to \infty$ even if the resistors are negative in some regions. Of course, some natural assumptions have to be added and these will be found in section 8.

Especially in case a circuit contains negative resistors is it of interest to find criteria which guarantee that the solutions approach the equilibria as time increases and therefore do not oscillate. We saw that this is generally the case for RC-circuits and similarly for RL-circuits. On the other hand, RLC-circuits certainly will admit oscillations in general even in the linear case. But one would expect a nonoscillatory behavior of circuits in which the inductance—or a quantity of the dimension L/R^2C—is sufficiently small. Such criteria for nonoscillation will be derived in section 8. The main idea is to associate with the differential equation another metric which is positive and so find a function P^* which decreases along the solutions.

Such criteria are especially valuable for large circuits which contain many loops. It is usually hard to judge intuitively whether the presence of many loops may lead to oscillatory behavior. In section 9 we discuss an example of an arbitrarily large ladder network containing nonlinear elements, which demonstrates that our criteria are the best possible in general.

In section 20, part II, similar methods are used to establish the existence of periodic solutions for periodically excited nonlinear circuits. This result can be considered as an extension of a theorem of R. Duffin [4].

This paper is divided into two parts. The more important part is the first which leads to the main results rather directly without containing all the detailed proofs and refinements. The second part contains several additional results as well as detailed proofs complementing part I.

Originally, this work started with the study of some nonlinear circuits proposed by Goto and others [5]. Some preliminary investigations in this direction have been published earlier (see [6, 7]). In this paper, we present these ideas in a more systematic fashion in the hope that it will be useful to the theoretically inclined electrical engineers as well as mathematicians.

1. Complete sets of variables for a network. A network is an idealized concept in circuit theory which can be defined as a set of points, called nodes, and a set of connecting lines, called branches. It is irrelevant whether nodes and branches lie in a plane or whether the branches can be realized by straight lines. It is essential, however, that every branch connect exactly two nodes. Such a network is frequently called a graph. Actually, for applications other natural restrictions—like connectedness of the graph—could be imposed which, however, we will not need.

In each branch labeled by $\mu = 1, \cdots, b$ we specify a direction arbitrarily, indicating it by an arrow ("directed" graph). Accordingly, we distinguish the two connected nodes as initial and end nodes. The current flow in such a network is completely described by giving the amount of current i_μ flowing in the direction of the arrow; that means i_μ is negative if the flow is against the specified direction and positive otherwise. Similarly, we associate with each branch a voltage v_μ with a specified sign by taking the voltage level at the end node minus the voltage level at the initial node of the branch.

The $2b$ variables i_μ, v_μ $(\mu = 1, \cdots, b)$ are restricted by the well-known Kirchhoff laws. The node law expresses that the currents arriving at any node (taken with proper sign) add up to zero which we write symbolically in the form

$$\sum_{\text{node}} \pm i_\mu = 0. \tag{1.1}$$

Kirchhoff's loop law expresses that the voltage drop over any loop (closed chain of branches) is zero, or

$$\sum_{\text{loop}} \pm v_\mu = 0. \tag{1.2}$$

Another way of describing this loop law is that to every node one can assign a voltage level such that v_μ equals the difference of the defined voltage levels between the end node and initial node.*

In the investigation of the circuit dynamics it will be of first importance to know how restrictive Kirchhoff's laws are. They form a set of linear equations, and we study first which of the currents and voltages can be chosen independently. More precisely, we call a set of variables $i_1, \cdots, i_r, v_{r+1}, \cdots, v_{r+s}$** "*complete*" if they can be chosen *independently* without leading to a violation of Kirchhoff's laws and if they determine in each branch *at least one* of the two variables, the current or the voltage. The problem is to describe a complete set of variables for a given graph.

This can be done in several ways. For instance, for $r = 0$ the answer is well-known. Choose in the graph a maximal tree τ, i.e., a subgraph which does not contain any loops and cannot be enlarged as a tree. It is clear that the t voltages, $v_1, \cdots, v_t$, on this tree can be assigned arbitrarily without interfering with Kirchhoff's loop law since the tree does not contain any loops. The branches not contained in the tree are called "links", and it remains to determine the voltages in the links. Since τ is maximal, such a link added to τ forms a loop, and therefore the link voltage is expressible in terms of the tree voltages $v_1, \cdots, v_t$ by (1.2). This proves that *the voltages in a maximal tree form a complete set of variables,* $(r = 0, s = t)$.

Similarly, it is well-known how to choose a complete set of currents. Let τ be a maximal tree and $\mathcal{L}$ its l links. Decomposing the graph into independent loops, one finds readily that *the link currents of a maximal tree form a complete set of variables,* $(r = l, s = 0)$. See Guillemin [2].

Finally, we construct a complete set of variables $i_1, \cdots, i_r, v_{r+1}, \cdots, v_{r+s}$ in the "mixed" case where $rs > 0$. For this purpose we begin with a maximal tree τ (with a corresponding set of links $\mathcal{L}$) and choose a subtree τ' in τ. With $\mathcal{L}'$ (links of τ') we denote all branches which connect two nodes of τ' and make a loop with branches of τ' only.† Thus $\mathcal{L}'$ is contained in $\mathcal{L}$ and τ' is a maximal tree of the graph $\tau' + \mathcal{L}'$. The number of branches in τ', $\mathcal{L}'$ will be denoted by t', l' respectively. Then, *the currents* $i_1, \cdots, i_r$ *in the branches of* $\mathcal{L} - \mathcal{L}'$ *together with the voltages* $v_{r+1}, \cdots, v_{r+s}$ *in* τ' *form a complete set of variables,* $(r = l - l', s = t')$.

This is easily seen. Since τ' is a maximal tree of $\tau' + \mathcal{L}'$, the voltages $v_{r+1}, \cdots, v_{r+s}$ are independent and determine all voltages in $\tau' + \mathcal{L}'$. Also, the currents $i_1, \cdots, i_r$—being link currents—are independent, and it remains to be shown that the currents in all branches outside $\tau' + \mathcal{L}'$ are determined too. We form the r loops through the links $\mathcal{L} - \mathcal{L}'$ whose loop currents are $i_1, \cdots, i_r$. Recalling that *all* l loop currents determine all branch currents and the fact that the loops through the l' links of $\mathcal{L}'$ belong entirely

*A more analytic description of these laws with a connection matrix of the graph will be found in section 12, part II.

**We always reserve the freedom to relabel the branches and choose here the branches corresponding to the given set of variables as the first ones.

†It could happen that τ' is not connected and a branch connecting two nodes in different components of τ' would not form a loop within τ'. These branches are to be excluded from $\mathcal{L}'$.

to $\tau' + \mathfrak{L}'$, it follows that all branch currents outside $\tau' + \mathfrak{L}'$ are independent of the l' loop currents and hence are already determined by the $r = l - l'$ loop currents $i_1, \cdots, i_r$.

One sees that the mixed case of a complete set contains, in particular, the "pure" cases. If τ' is empty, one has $s = 0$ and $r = l$, i.e., the case of a pure set of currents; if $\tau' = \tau$, one has $r = l - l' = 0$, $s = t$, which gives a pure set of t voltages.

We remark that the link currents in $\mathfrak{L}'$ and the voltages in $\tau - \tau'$ give rise to another complete set of variables.

In the mixed case the whole graph $\mathfrak{N}$ is broken up into $\mathfrak{N}_v = \tau' + \mathfrak{L}'$, those branches in which the voltages can be determined from $v_{r+1}, \cdots, v_{r+s}$ by Kirchhoff's loop law, and the remaining branches $\mathfrak{N}_i$ in which the currents can be determined from $i_1, \cdots, i_r$ by Kirchhoff's node law.

2. Network theorems. The two theorems in this section are derived using only the geometry of the network and Kirchhoff's laws. They do not depend on the types of elements in the branches. Theorem 1 is known as Tellegen's theorem [8], but it is stated here in a form which has a geometrical interpretation.

We consider a directed network with b branches and n nodes. The set of branch currents $i = (i_1, \cdots, i_b)$ and the set of branch voltages $v = (v_1, \cdots, v_b)$ are vectors in the b-dimensional Euclidean vector space $\mathcal{E}_b$. The inner product is defined for two vectors $x, y \in \mathcal{E}_b$ as $(x, y) = \sum_{\mu=1}^{b} x_\mu y_\mu$. Let $\mathcal{J}$ be the set of all vectors in $\mathcal{E}_b$ such that if $x \in \mathcal{J}$ and the components of x are taken as the branch currents of the directed network, then Kirchhoff's node law, $\sum_{\text{node}} \pm x_\mu = 0$, must hold at every node. Similarly, we let $\mathcal{V}$ denote the set of all vectors in $\mathcal{E}_b$ such that if $x \in \mathcal{V}$ and the components of x are taken as the branch voltages, then Kirchhoff's loop law, $\sum_{\text{loop}} \pm x_\mu = 0$, should be satisfied for every loop. It is obvious that $\mathcal{J}$ and $\mathcal{V}$ are subspaces of $\mathcal{E}_b$ since they are defined through linear equations.

Theorem 1. If $i \in \mathcal{J}$ and $v \in \mathcal{V}$, then $(i, v) = 0$, i.e., $\mathcal{J}$ and $\mathcal{V}$ are orthogonal subspaces of $\mathcal{E}_b$.

Proof. Since v satisfies Kirchhoff's loop law, there exists a set of node voltages $(V_1, \cdots, V_n)$ such that v_μ is the difference between the voltages of the end node and the initial node (see previous section). Let the current flowing from node k to node l be denoted by i_{kl} which is taken to be zero if there is no connecting branch. Thus, if the μth branch connects nodes k and l, we have

$$v_\mu i_\mu = (V_k - V_l)i_{kl} = (V_l - V_k)i_{lk}^{*},$$

and because of this symmetry in k and l, we can express the inner product (i, v) as a free sum, i.e.,

$$(i, v) = \sum_{\mu=1}^{b} i_\mu v_\mu = \tfrac{1}{2} \sum_{k,l} (V_k - V_l) i_{kl},$$

or

$$(i, v) = \tfrac{1}{2}\Big[\sum_k V_k\Big(\sum_l i_{kl}\Big) - \sum_l V_l\Big(\sum_k i_{kl}\Big)\Big].$$

However,

$$\sum_l i_{kl} = \sum_{\text{node } k} \pm i_\mu = 0 \quad \text{and} \quad \sum_k i_{kl} = \sum_{\text{node } l} \pm i_\mu = 0$$

so that $(i, v) = 0$.

*There is no restriction in assuming that at most one branch connects the same two nodes.

We remark that $\mathcal{J}$ and $\mathcal{V}$ are not only orthogonal subspaces, but that they even span $\mathcal{E}_b$. The simple proof of this fact is found in section 12, part II, although it is used in the next theorem.

The next theorem is similar but not equivalent to the first theorem and leads directly to one of the main results of this paper. Let Γ denote a one-dimensional curve in $\mathcal{E}_b$ with projections on $\mathcal{J}$ and $\mathcal{V}$ denoted by i and v, respectively.

Theorem 2.

$$\int_\Gamma \sum_{\mu=1}^{b} v_\mu \, di_\mu = \int_\Gamma \sum_{\mu=1}^{b} i_\mu \, dv_\mu = 0.$$

Proof. Since $di = (di_1, \cdots, di_b)$ is the limit of the difference of two vectors in $\mathcal{J}$ and since $\mathcal{J}$ is a subspace, then $di \in \mathcal{J}$ and by theorem 1

$$(v, di) = \sum_{\mu=1}^{b} v_\mu \, di_\mu = 0.$$

Integrating along Γ we obtain

$$\int_\Gamma \sum_{\mu=1}^{b} v_\mu \, di_\mu = 0,$$

and integration by parts yields

$$\int_\Gamma \sum_{\mu=1}^{b} v_\mu \, di_\mu = (i, v) \Big|_\Gamma - \int_\Gamma \sum_{\mu=1}^{b} i_\mu \, dv_\mu .$$

Since $(i, v) = 0$ by theorem 1, we have

$$\int_\Gamma \sum_{\mu=1}^{b} i_\mu \, dv_\mu = 0.$$

3. Nonlinear elements. So far we have only discussed facts which depend on network concepts, and we have seen that Kirchhoff's laws impose certain restrictions among the branch voltages and among the branch currents. On the other hand, physical properties of the elements in the branches lead to further restrictions which relate branch currents to branch voltages, and it is our purpose here to discuss the nature of these relations.

We consider elements which are two terminal devices and restrict our discussion to purely resistive, inductive or capacitive elements.

The name "resistor" usually refers to a linear passive device which has a resistance R such that the current and voltage at its terminals are related by $v = -Ri$.* A more general concept is obtained by considering a resistor as a continuous function such that the relation $f(i, v) = 0$ holds. This defines a continuous curve in the (i, v)-plane which we will call the characteristic of the resistor. From $f(i, v) = 0$ we could solve for i or v as a function of the other. It is not necessary to require that either such function be single valued, and, in fact, the more interesting cases are when this is not true. For example, the tunnel diode [9] is a nonlinear resistor which has the characteristic shown in figure 1, and clearly v as a function of i is not single valued.

It will also not be necessary to require that the characteristic pass through the origin,

*The negative sign is chosen here in order to be consistent with the convention adopted in Section 1 on the direction of positive current and positive voltage.

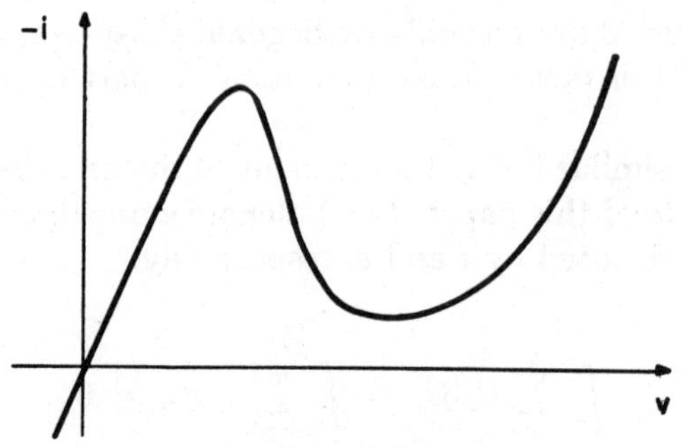

FIG. 1. Voltage-Current Characteristic for a Tunnel Diode

i.e., that the element be passive. A generator (voltage or current) is therefore a special type of resistor for which the characteristic is parallel to one of the axes.

There are some assumptions which will be made about resistors. It will be assumed that there exists $B > 0$ such that for $|i|$, $|v| > B$ the characteristic of the resistor lies in the first and third quadrants and is monotone increasing there. We make this assumption in order to guarantee that the equilibrium problem can be solved, i.e., that we can solve the circuit equations under steady state conditions. A proof of this statement under less stringent assumptions will be found in section 14, part II.

For completeness we discuss the well-known laws for inductors and capacitors which we also allow to be nonlinear. An inductor is a function relating the magnetic flux linkages to the current, i.e.,

$$\phi = -f(i).$$

In terms of voltage and current

$$v = \frac{d\phi}{dt} = -f'(i)\frac{di}{dt} = -L(i)\frac{di}{dt},$$

where $L(i)$ is the inductance and is non-negative. Similarly, a capacitor is a function relating the charge and the voltage, i.e.,

$$q = -f(v).$$

Differentiating, we obtain

$$i = \frac{dq}{dt} = -f'(v)\frac{dv}{dt} = -C(v)\frac{dv}{dt},$$

where $C(v)$ is the capacitance and is non-negative.

We remark that mutual inductance can be handled by simply changing to vector notation where $L(i)$ would be a symmetric matrix.

4. The form of the equations. The general RLC circuit can be thought of as a resistive circuit (R-circuit) with n ports to which either an inductor or a capacitor is attached (Fig. 2). We want to derive the differential equations describing the dynamical behavior of such a circuit.

We assume that the resistors are of the type discussed in the previous section so that the equilibrium problem can be solved. This means that if we know the currents denoted by $i^* = (i_1, \cdots, i_r)$ in all the inductors and the voltages denoted by $v^* = (v_{r+1}, \cdots, v_{r+s})$

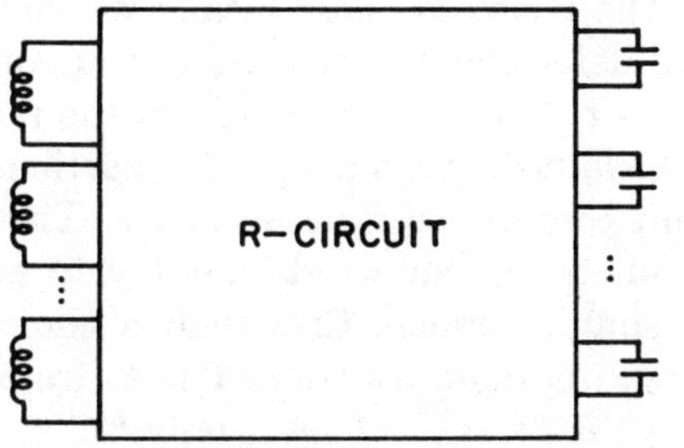

FIG. 2. General RLC Circuit

across all the capacitors, the problem of determining the voltage and current in any branch of the circuit can be solved from some implicit equations, i.e.,

$$\begin{aligned} v_\mu &= f_\mu(i^*, v^*), \qquad \mu = 1, \cdots, b, \\ i_\mu &= g_\mu(i^*, v^*), \qquad \mu = 1, \cdots, b. \end{aligned} \tag{4.1}$$

In general, there may be several solutions for v_μ, i_μ leading to multiple-valued functions f_μ, g_μ with branch points, cusps, etc.

Of particular interest is the voltage across any inductor and the current through any capacitor given by

$$\begin{aligned} v_\rho &= f_\rho(i^*, v^*), \qquad \rho = 1, \cdots, r, \\ i_\sigma &= g_\sigma(i^*, v^*), \qquad \sigma = r + 1, \cdots, r + s. \end{aligned} \tag{4.2}$$

On the other hand, the dynamical laws of inductors and capacitors, as discussed in section 3, require

$$\begin{aligned} v_\rho &= -L_\rho(i_\rho) \frac{di_\rho}{dt}, \qquad \rho = 1, \cdots, r, \\ i_\sigma &= -C_\sigma(v_\sigma) \frac{dv_\sigma}{dt}, \qquad \sigma = r + 1, \cdots, r + s. \end{aligned} \tag{4.3}$$

Combining (4.2) and (4.3), we obtain the desired differential equations,

$$\begin{aligned} L_\rho(i_\rho) \frac{di_\rho}{dt} &= -f_\rho(i^*, v^*), \qquad \rho = 1, \cdots, r, \\ C_\sigma(v_\sigma) \frac{dv_\sigma}{dt} &= -g_\sigma(i^*, v^*), \qquad \sigma = r + 1, \cdots, r + s \end{aligned} \tag{4.4}$$

which give di^*/dt and dv^*/dt explicitly in terms of i^* and v^*.

It is our aim to show that the functions on the right-hand side of (4.4) can be derived from one function.

For this purpose we make use of theorem 2

$$\int_\Gamma \sum_{\mu=1}^{b} v_\mu \, di_\mu = 0, \tag{4.5}$$

where μ ranges over *all* branches. For Γ we will now choose any curve from a fixed state to a variable one in such a manner that along Γ the voltages and currents satisfy the relations characteristic for the resistors. This means we choose $i^*(s)$, $v^*(s)$ arbitrarily as continuous functions and determine the remaining components i_μ, v_μ for $\mu > r + s$ and v_ρ, i_σ ($\rho = 1, \cdots, r; \sigma = r + 1, \cdots, r + s$) from the relations (4.1). Although the f_μ, g_μ are possibly multiple-valued, we make, by a particular choice, the $v_\mu(s)$, $i_\mu(s)$, $v_\rho(s)$, $i_\sigma(s)$ single-valued and continuous functions of s. This is, in general, possible if one excludes pathological functions, but we do not try to give a precise discussion of necessary restrictions and simply assume that such a choice of continuous functions exists. Since $\mathcal{J}$ and $\mathcal{V}$ span $\mathcal{E}_b$, this defines a curve Γ in $\mathcal{E}_b$ from a fixed point to a variable point determined by $i_1, \cdots, i_r, v_{r+1}, \cdots, v_{r+s}$ only.

With Γ chosen in the manner specified, we now make the obvious but important observation that the integral,

$$\int_\Gamma \sum_{\mu > r+s} v_\mu \, di_\mu ,$$

taken *not* over all branches as in (4.5) but only over all resistor branches, depends only on the end points of Γ. This simply follows from the fact that in a resistor v_μ depends on i_μ only, i.e., that a resistor relates the current and voltage in one and the same branch only.

We write (4.5) in the form

$$\int_\Gamma \sum_{\rho=1}^{r} v_\rho \, di_\rho + \int_\Gamma \sum_{\sigma=r+1}^{r+s} v_\sigma \, di_\sigma + \int_\Gamma \sum_{\mu > r+s} v_\mu \, di_\mu = 0 \tag{4.6}$$

or, integrating the second line integral by parts,

$$\int_\Gamma \sum_{\rho=1}^{r} v_\rho \, di_\rho - \int_\Gamma \sum_{\sigma=r+1}^{r+s} i_\sigma \, dv_\sigma + P = 0, \tag{4.7}$$

where

$$P = \int_\Gamma \sum_{\mu > r+s} v_\mu \, di_\mu + \sum_{\sigma=r+1}^{r+s} i_\sigma v_\sigma \Big|_\Gamma \tag{4.8}$$

is a function depending only on the end points of Γ. In other words, P is a function of the variable end point of Γ which, in turn, depends only on the variables $i_1, \cdots, i_r, v_{r+1}, \cdots, v_{r+s}$, i.e., $P = P(i^*, v^*)$. It is also only defined up to a constant which depends on the choice of the fixed initial point of Γ. From (4.7) we read off

$$\begin{aligned} v_\rho &= -\frac{\partial P}{\partial i_\rho}, \qquad \rho = 1, \cdots, r, \\ i_\sigma &= +\frac{\partial P}{\partial v_\sigma}, \qquad \sigma = r + 1, \cdots, r + s \end{aligned} \tag{4.9}$$

which, with (4.2) and (4.4), gives the desired differential equations

$$\begin{aligned} L_\rho(i_\rho) \frac{di_\rho}{dt} &= \frac{\partial P}{\partial i_\rho}, \qquad \rho = 1, \cdots, r, \\ C_\sigma(v_\sigma) \frac{dv_\sigma}{dt} &= -\frac{\partial P}{\partial v_\sigma}, \qquad \sigma = r + 1, \cdots, r + s. \end{aligned} \tag{4.10}$$

These equations put into evidence that the right-hand sides are derived from one single function P. The computation of P required solving implicit equations and therefore is still complicated. But, we will show in the next section how to construct and interpret P in simple cases.

An immediate consequence of the equation (4.10) is that for linear circuits the equilibrium equations can be written with a symmetric R (resistance) matrix. This fact is discussed in section 15, part II, and is closely related to the reciprocity theorem.

5. Construction of P. We want to describe in some detail how the function P, which we will call the *mixed potential*, can be constructed directly from the circuit. For this purpose, we use the definition of P according to (4.8),

$$P(i^*, v^*) = \int_\Gamma \sum_{\mu > r+s} v_\mu \, di_\mu + \sum_{\sigma=r+1}^{r+s} i_\sigma v_\sigma \Big|_\Gamma , \tag{5.1}$$

where $i^* = (i_1, \cdots, i_r)$ is the set of currents through the inductors and $v^* = (v_{r+1}, \cdots, v_{r+s})$ is the set of voltages across the capacitors. Since v_μ depends only on i_μ for $\mu > r + s$, P can be written as

$$P(i^*, v^*) = \sum_{\mu > r+s} \int_\Gamma v_\mu \, di_\mu + \sum_{\sigma=r+1}^{r+s} i_\sigma v_\sigma \Big|_\Gamma . \tag{5.2}$$

The integral $\int_\Gamma v_\mu \, di_\mu$ is, of course, well-defined as a line integral even if v_μ cannot be written as a single-valued function of i_μ. Taken as a line integral, the path of integration is along the characteristic of the resistor, as can be easily seen from the definition of Γ. We give this integral a special name, the *current potential* of the element in the branch labeled by μ. Similarly, the line integral $\int_\Gamma i_\mu \, dv_\mu$ will be called the *voltage potential*, and it is easily seen that

$$\int_\Gamma i_\mu \, dv_\mu + \int_\Gamma v_\mu \, di_\mu = i_\mu v_\mu \Big|_\Gamma .^* \tag{5.3}$$

The current or voltage potential has a simple interpretation if the graph of a resistor can be expressed as a single-valued function of one of the variables. For example, if i_μ is a single-valued function of v_μ, then the voltage potential is an ordinary integral and consequently is the shaded area shown in figure 3 assuming that the initial fixed point of the path was at $v_\mu = 0$.

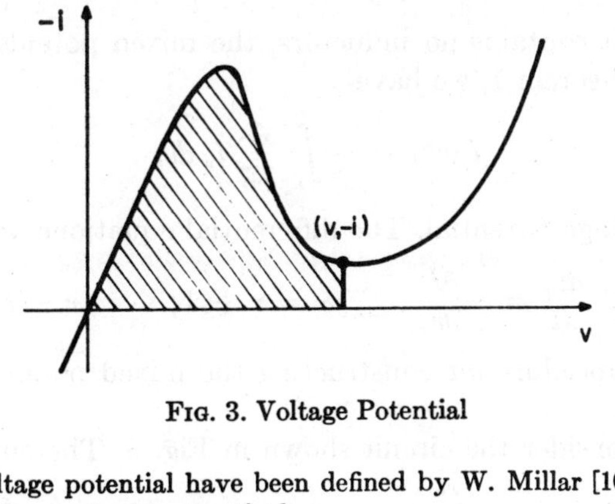

FIG. 3. Voltage Potential

*The current and voltage potential have been defined by W. Millar [10] and C. Cherry [11] who call them the contents and cocontents respectively.

We shall speak of the current (voltage) potential of a network as the sum of the current (voltage) potentials of its resistors.

It is useful to make some remarks about the sign of a resistor characteristic. The current induced through a linear resistor R is in the direction of the voltage drop, i.e., the induced current flows from $+$ to $-$ (Fig. 4). Now suppose that the direction assigned

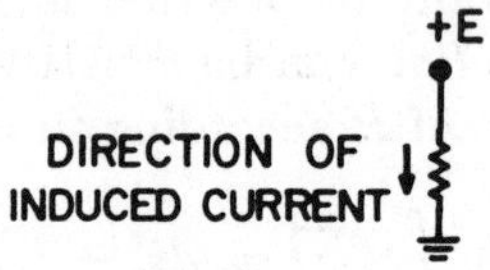

FIG. 4. Direction of Induced Current Through a Resistor

to this branch of the directed network is from E to ground. Then, according to the convention adopted in section 1 on the direction of positive current and voltage, the branch current i_μ is equal to i, the induced current, whereas the branch voltage $v_\mu = -E$. We find that $v_\mu = -Ri_\mu$ and the current potential is $-\frac{1}{2}Ri_\mu^2$. We obtain this even if the opposite direction were assigned to this branch because then $v_\mu = E$, $i_\mu = -i$, and again $v_\mu = -Ri_\mu$. For similar reasons, the current potential of a battery E is $+Ei_\mu$.

Now we may state the procedure for constructing the mixed potential directly from the circuit:

(1) determine the current potential for each resistor;
(2) determine the product $i_\sigma v_\sigma$ for each capacitor;
(3) form the sum of these terms and express it in terms of i^*, v^*.

We show now that the concept of a mixed potential contains the voltage potential and the current potential as special cases. If the circuit contains no capacitors, the second sum in the mixed potential is absent, and hence the mixed potential reduces to

$$P(i^*) = \int_\Gamma \sum_{\mu > r+s} v_\mu \, di_\mu ,$$

which is the current potential. In this case, the differential equations take the simple form

$$L_\rho \frac{di_\rho}{dt} = \frac{\partial P}{\partial i_\rho}, \qquad \rho = 1, \cdots, r.$$

Similarly, if the circuit contains no inductors, the mixed potential contains a term for each branch, and by theorem 1, we have

$$P(v^*) = -\int_\Gamma \sum_{\sigma=1}^{s} i_\mu \, dv_\mu ,$$

i.e., $-P(v^*)$ is the voltage potential. The differential equations are

$$C_\sigma \frac{dv_\sigma}{dt} = -\frac{\partial P}{\partial v_\sigma}, \qquad \sigma = r+1, \cdots, r+s.$$

To illustrate the procedure for constructing the mixed potential, we consider a few examples.

Example 1. We consider the circuit shown in Fig. 5. The current potential of R is

$$\int_0^i (-Ri) \, di = -\tfrac{1}{2}Ri^2,$$

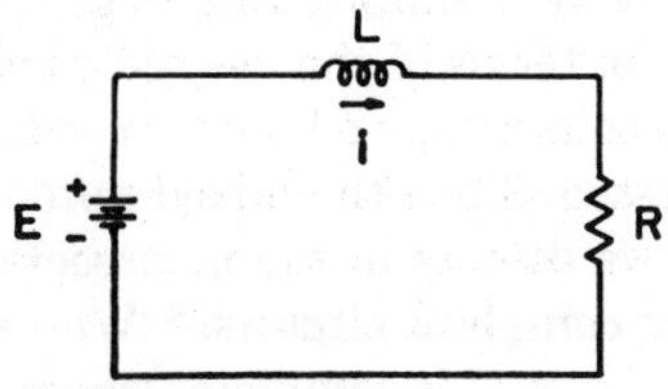

FIG. 5. Series RL Circuit

and the current potential of the battery is

$$\int_0^i E\,di = Ei.$$

Since no capacitors are present, we have

$$P(i) = Ei - \tfrac{1}{2}Ri^2,$$

which gives the desired differential equations by (4.10)

$$L\frac{di}{dt} = \frac{\partial P}{\partial i} = E - Ri.$$

Example 2. Next we consider the tunnel diode circuit shown in Fig. 6. The current

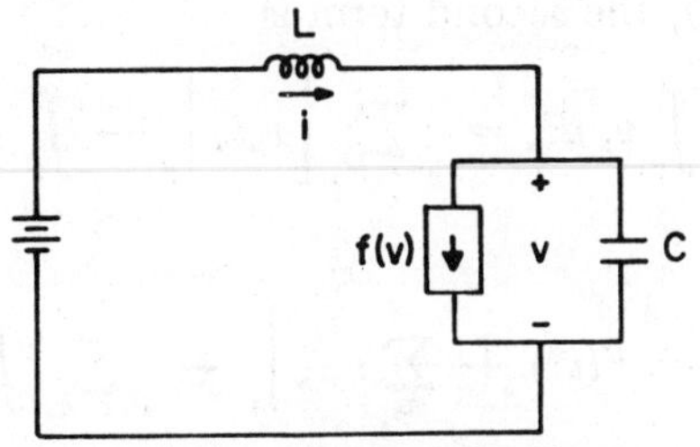

FIG. 6. Tunnel-Diode Circuit

through the box in the direction shown is given by the nonlinear function $f(v)$ shown in Fig. 1. According to (5.3) the current potential of the resistor is

$$-\int_\Gamma v\,d(f(v)) = -vf(v) + \int_0^v f(v)\,dv.$$

The current through the capacitor is $i - f(v)$ and, since it is in the opposite direction of v, the $i_\rho v_\rho$ product is $-(i - f(v))v$. Thus, the mixed potential is

$$P(i, v) = Ei + \int_0^v f(v)\,dv - iv,$$

which gives the differential equations

$$L\frac{di}{dt} = \frac{\partial P}{\partial i} = E - v,$$

$$C\frac{dv}{dt} = -\frac{\partial P}{\partial v} = i - f(v).$$

Of course, these examples are very simple, and, in general, it may be difficult although not impossible to express v_μ, i_μ in terms of the desired variables $i_1, \cdots, i_r, v_{r+1}, \cdots, v_{r+s}$. However, if these variables are complete, we know according to the discussion of section 1 that v_μ, i_μ can be simply expressed. Then the mixed potential function has a particularly simple and useful form which we discuss in the next section.

6. The mixed potential for complete circuits. We call a circuit *complete* if the set of variables $i_1, \cdots, i_r, v_{r+1}, \cdots, v_{r+s}$ is complete, where $i_1, \cdots, i_r$ denote the currents through the inductors and $v_{r+1}, \cdots, v_{r+s}$ denote the voltages across the capacitors. It is our purpose here to discuss the form that the mixed potential takes for the class of complete circuits and to give a simple procedure for its construction.

Recall (Section 1) that a complete circuit could be split into two subnetworks $\mathfrak{N}_i$ and $\mathfrak{N}_v$ such that the branch currents of $\mathfrak{N}_i$ and the branch voltages of $\mathfrak{N}_v$ are known from the complete set of variables. In particular, $\mathfrak{N}_v$ contains all the capacitors and $\mathfrak{N}_i$ the inductors. According to (4.8), the mixed potential is*

$$
\begin{aligned}
P(i^*, v^*) &= \sum_{\mu>t} \int_\Gamma v_\mu \, di_\mu + \sum_{\sigma=r+1}^{t} v_\sigma i_\sigma \Big|_\Gamma \\
&= \sum_{\mu>t,\mathfrak{N}_i} \int_\Gamma v_\mu \, di_\mu + \sum_{\mu>t,\mathfrak{N}_v} \int_\Gamma v_\mu \, di_\mu + \sum_{\sigma=r+1}^{t} v_\sigma i_\sigma \Big|_\Gamma . \qquad (6.1)
\end{aligned}
$$

The first term is simply the current potential of $\mathfrak{N}_i$ and, since i^* determines all the branch currents in $\mathfrak{N}_i$, we can express this term as a function of i^*, which we denote by $F(i^*)$. According to (5.3), the second term is

$$\sum_{\mu>t,\mathfrak{N}_v} \int_\Gamma v_\mu \, di_\mu = \sum_{\mu>t,\mathfrak{N}_v} \left[v_\mu i_\mu \Big|_\Gamma - \int_\Gamma i_\mu \, dv_\mu \right],$$

and therefore we have

$$P(i^*, v^*) = F(i^*) + \sum_{\mathfrak{N}_v} v_\mu i_\mu \Big|_\Gamma - \sum_{\mu>t,\mathfrak{N}_v} \int_\Gamma i_\mu \, dv_\mu .^{**}$$

The last term is simply the voltage potential of $\mathfrak{N}_v$ and, since v^* determines all branch voltages of $\mathfrak{N}_v$, we denote this by $G(v^*)$ and we have

$$P(i^*, v^*) = F(i^*) - G(v^*) + \sum_{\mathfrak{N}_v} v_\mu i_\mu \,|_\Gamma . \qquad (6.2)$$

It remains to be shown that the last term of (6.2) can be expressed in terms of the complete set of variables $i_1, \cdots, i_r, v_{r+1}, \cdots, v_{r+s}$ alone, which is not obvious since in every branch only one of the variables i_μ, v_μ is known.

Lemma. There exists an $r \times s$ matrix $\gamma = (\gamma_{\rho\sigma})$ with $\gamma_{\rho\sigma} = +1, -1, 0$ such that

$$\sum_{\mathfrak{N}_v} v_\mu i_\mu = \sum_{\rho=1,\sigma=1}^{r,s} \gamma_{\rho\sigma} i_\rho v_{r+\sigma} = (i^*, \gamma v^*). \qquad (6.3)$$

Proof. We draw attention to the set $\mathfrak{N}_0$ of n_0 nodes at which branches of $\mathfrak{N}_i$ and $\mathfrak{N}_v$ come together. Since the currents in $\mathfrak{N}_i$ can be expressed in terms of $i_1, \cdots, i_r$, we can determine the currents j_ν, $(\nu = 1, \cdots, n_0)$, at the nodes of $\mathfrak{N}_0$ flowing from $\mathfrak{N}_i$ to $\mathfrak{N}_v$.

*For brevity we denote $r + s$ by t.

**Since the branches labeled by $\sigma = r + 1, \cdots, r + s$ containing the capacitors belong to $\mathfrak{N}_v$, the second term contains the last sum of (6.1).

From Kirchhoff's node law it follows that

$$\sum_{\nu=1}^{n_0} j_\nu = 0. \tag{6.4}$$

Similarly, since we can express the voltage difference between any two nodes of $\mathfrak{N}_0$ in terms of $v_{r+1}, \cdots, v_{r+s}$, we can assign a voltage level w_ν to every node in $\mathfrak{N}_0$. It is clear that the sum

$$\sum_{\nu=1}^{n_0} j_\nu w_\nu$$

can be expressed in terms of $i_1, \cdots, i_r, v_{r+1}, \cdots, v_{r+s}$. The proof will be complete if we identify this sum with the given one (up to the sign). For this purpose we dissect the graph $\mathfrak{N}$ at the nodes $\mathfrak{N}_0$ and replace $\mathfrak{N}_i$ by artificial branches from the first node to the νth node ($\nu = 2, \cdots, n_0$) (figure 7). We assign the currents j_ν and voltages $w_\nu - w_1$

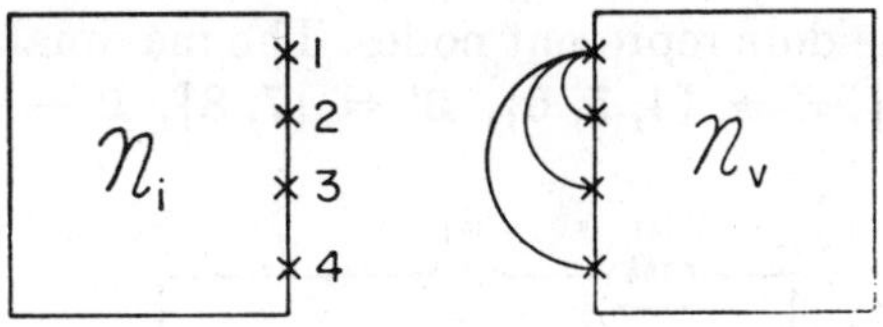

FIG. 7. The Network $\mathfrak{N}$ Dissected at the Nodes $\mathfrak{N}_0$

($\nu = 2, \cdots, n_0$) to these artificial branches. For the new graph consisting of the new branches and $\mathfrak{N}_v$, we have by theorem 1

$$\sum_{\mathfrak{N}_v} i_\mu v_\mu + \sum_{\nu=2}^{n_0} j_\nu(w_\nu - w_1) = 0 \tag{6.5}$$

or, by (6.4),

$$\sum_{\mathfrak{N}_v} i_\mu v_\mu = -\sum_{\nu=1}^{n_0} j_\nu w_\nu . \tag{6.6}$$

This shows that the sum on the left is expressed in terms of $j_1, \cdots, j_{n_0}, w_1, \cdots, w_{n_0}$ which are in turn expressible in terms of i^*, v^*. In fact, the j_ν depend linearly on i^* and the w_ν depend linearly on v^*. Hence, with some constants $\gamma_{\rho\sigma}$ we have

$$\sum_{\mathfrak{N}_v} i_\mu v_\mu = \sum \gamma_{\rho\sigma} i_\rho v_{r+\sigma} .$$

This proves the lemma except for the verification that $\gamma_{\rho\sigma} = \pm 1, 0$ which is left to section 13, part II.

Moreover, we have found an interpretation of the term $\sum_{\mathfrak{N}_v} i_\mu v_\mu$; it is the sum of the product of the current j_ν passing from $\mathfrak{N}_i$ to $\mathfrak{N}_v$ and the voltage level w_ν of the node summed over the nodes of $\mathfrak{N}_0$.

Combining (6.2) and (6.3) gives us the final form for the mixed potential in terms of the variables i^*, v^* only, i.e.,

$$P(i^*, v^*) = F(i^*) - G(v^*) + (i^*, \gamma v^*). \tag{6.7}$$

The above formula already gives us a procedure for constructing the mixed potential from a complete circuit; *the mixed potential is the sum of the current potential for* $\mathfrak{N}_i$, *the voltage potential for* $\mathfrak{N}_v$, *and a term* $(i^*, \gamma v^*)$ *which is constructed as outlined in the lemma.* We remark that our reason for calling $P(i^*, v^*)$ the *mixed* potential is a consequence of the form of (6.7).

There is a more direct and simpler method for constructing the cross-product term $(i^*, \gamma v^*)$ and we state this without proof. Take any link $\rho \in \mathfrak{L} - \mathfrak{L}'$ and consider the loop Λ_ρ which it determines. The branches of Λ_ρ other than the link branch are branches of the maximal tree τ of which some may be branches of τ'. The desired term is the sum of products of the loop current i_ρ and the voltage v_σ of the branches of τ' in Λ_ρ summed over all links of $\mathfrak{L} - \mathfrak{L}'$, i.e.,

$$(i^*, \gamma v^*) = \sum_{\rho=1}^{r} i_\rho \sum_{\Lambda_\rho \cap \tau'} \pm v_\sigma .$$

Now we consider a fairly complicated example of a complete circuit to illustrate the the procedure for constructing the mixed potential.

Example 3. Considered only as a graph, the circuit in Fig. 8 becomes the graph shown in Fig. 9, where the dots represent nodes. The maximal tree τ consists of branches $\{4, 5, 6, 9, 10, 11, 12, 13\}$, $\tau' = \{4, 5, 6\}$, $\mathfrak{L}' = \{7, 8\}$, $\mathfrak{L} - \mathfrak{L}' = \{1, 2, 3\}$ and we see

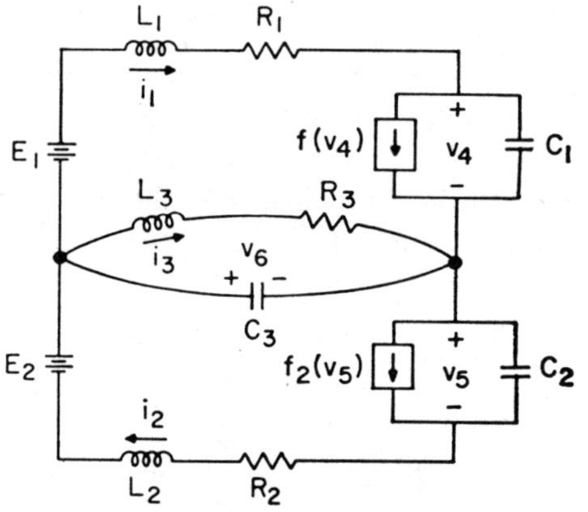

FIG. 8. Twin Tunnel-Diode Circuit

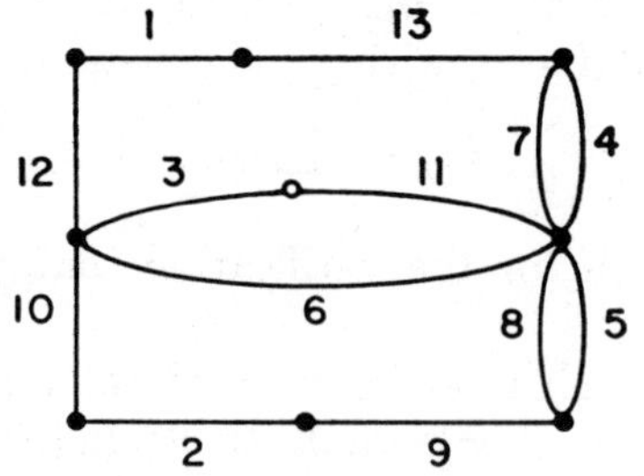

FIG. 9. The Graph of the Circuit of Figure 8

that i_1, i_2, i_3, v_4, v_5, v_6 form a complete set of mixed variables. We construct the current potential of $\mathfrak{N}_i = \{1, 2, 3, 9, 10, 11, 12, 13\}$ where terms come only from the resistive branches $\{9, 10, 11, 12, 13\}$ and are listed in order:

$$F(i_1, i_2, i_3) = -\tfrac{1}{2}R_2 i_2^2 + E_2 i_2 - \tfrac{1}{2}R_3 i_3^2 + E_1 i_1 - \tfrac{1}{2}R_1 i_1^2 . \qquad (6.8)$$

Similarly, the current potential of $\mathfrak{N}_v = \{4, 5, 6, 7, 8\}$ is due to the resistive branches $\{7, 8\}$ and is

$$G(v_4, v_5, v_6) = -\int_0^{v_4} f_1(v)\,dv - \int_0^{v_5} f_2(v)\,dv. \tag{6.9}$$

It remains to determine $(i^*, \gamma v^*)$ which we do by considering the loops $\Lambda_1 = \{1, 13, 4, 6, 12\}$, $\Lambda_2 = \{2, 10, 6, 5, 9\}$ and $\Lambda_3 = \{3, 11, 6\}$ and their intersections with τ'. We obtain the correct sign by determining which way the loop currents i_ρ ($\rho = 1, 2, 3$) flow through the voltages v_σ ($\sigma = 4, 5, 6$) of τ' which gives us

$$(i^*, \gamma v^*) = i_1(-v_4 + v_6) + i_2(-v_6 - v_5) + i_3(v_6). \tag{6.10}$$

Combining (6.8), (6.9), and (6.10) according to (6.7), we obtain the mixed potential for this circuit, i.e.,

$$P(i^*, v^*) = E_1 i_1 + E_2 i_2 - \tfrac{1}{2}R_1 i_1^2 - \tfrac{1}{2}R_2 i_2^2 - \tfrac{1}{2}R_3 i_3^2$$
$$+ \int_0^{v_4} f_1(v)\,dv + \int_0^{v_5} f_2(v)\,dv - i_1 v_4 + i_1 v_6 - i_2 v_6 - i_2 v_5 + i_3 v_6 .$$

It is easily verified that the equations

$$L_1 \frac{di_1}{dt} = E_1 - R_1 i_1 - v_4 + v_6 = \frac{\partial P}{\partial i_1},$$

$$L_2 \frac{di_2}{dt} = E_2 - R_2 i_2 - v_6 - v_5 = \frac{\partial P}{\partial i_2},$$

$$L_3 \frac{di_3}{dt} = -R_3 i_3 + v_6 = \frac{\partial P}{\partial i_3},$$

$$C_1 \frac{dv_4}{dt} = i_1 - f_1(v_4) = -\frac{\partial P}{\partial v_4},$$

$$C_2 \frac{dv_5}{dt} = i_2 - f_2(v_5) = -\frac{\partial P}{\partial v_5},$$

$$C_3 \frac{dv_6}{dt} = -i_1 + i_2 - i_3 = -\frac{\partial P}{\partial v_6}$$

are the correct ones.

7. Limit situations. Although a complete circuit can, in some sense, be considered as typical, nevertheless it is sometimes more appropriate to neglect inductances or capacitances which are of minor importance. However, this may lead to circuits which are not complete. To see this we consider the complete circuit shown in Fig. 10. This

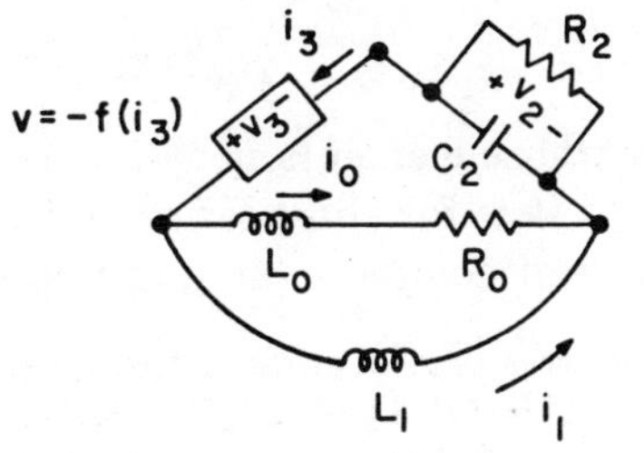

FIG. 10. A Complete Circuit When $L_0 \neq 0$

circuit is obviously complete since the currents of the lower branches determine the current $i_3 = (i_0 + i_1)$ in the upper branch. The mixed potential is easily computed to be

$$P(i_0, i_1, v_2) = -\tfrac{1}{2}R_0 i_0^2 - \int_0^{i_0+i_1} f(i)\, di + \frac{1}{2}\frac{v_2^2}{R_2} + v_2(i_0 + i_1), \tag{7.1}$$

which leads to the following differential equations.

$$L_0 \frac{di_0}{dt} = v_2 - f(i_0 + i_1) - R_0 i_0 = \frac{\partial P}{\partial i_0},$$

$$L_1 \frac{di_1}{dt} = -f(i_0 + i_1) + v_2 = \frac{\partial P}{\partial i_1},$$

$$C_2 \frac{dv_2}{dt} = -(i_0 + i_1) - \frac{v_2}{R_2} = -\frac{\partial P}{\partial v_2}.$$

However, if we let $L_0 \to 0$, then i_0 is no longer an independent variable and must be eliminated using $\partial P/\partial i_0 = 0$ or

$$v_2 = f(i_0 + i_1) + R_0 i_0 .$$

This leads to

$$v_2 + R_0 i_1 = f(i_0 + i_1) + R_0(i_0 + i_1),$$

or

$$i_0 + i_1 = h(v_2 + R_0 i_1),$$

and substituting this into the mixed potential, we obtain the new mixed potential in terms of the independent variables, i_1, v_2

$$Q(i_1, v_2) = -\tfrac{1}{2}R_0(h(v_2 + R_0 i_1) - i_1)^2 - \int_0^{h(v_2+R_0 i_1)} f(i)\, di + \frac{1}{2}\frac{v_2^2}{R_2} + v_2 h(v_2 + R_0 i_1). \tag{7.2}$$

This shows that the concept of a mixed potential is still meaningful,* but cannot be interpreted as easily as in the complete case.

It is obvious that one can add inductors in series and capacitors in parallel to a circuit to make it complete and then consider the original circuit as a limiting case of the new circuit so obtained. This procedure can be justified on physical grounds since these reactances are present anyway as parasitic elements.

Of course, the mixed potential of a noncomplete circuit is not necessarily so complicated as in our example. In determining stability criteria in the next section we will only require that the mixed potential have the form

$$P(i^*, v^*) = -A(i^*) + B(v^*) + (i^*, Dv^*), \tag{7.3}$$

where D is a constant matrix whose elements need not be ± 1, 0. Obviously this is the case for a complete circuit, and it is for this reason and the fact that all other circuits can be considered as limit cases of complete circuits that we have considered complete

*It is remarkable that the derivatives (e.g. with respect to v_2) of Q and P agree where $\partial P/\partial i_0 = 0$. This follows from the identity $\partial Q/\partial v_2 = \partial P/\partial v_2 + \partial P/\partial i_0\, \partial i_0/\partial v_2$ and the observation that the last term vanishes for $\partial P/\partial i_0 = 0$. In other words instead of eliminating i_0 from the differential equations one can eliminate it from the potential function.

circuits in such detail. We emphasize, however, that completeness is certainly not necessary for the mixed potential to have the form of (7.3).

8. Nonoscillating circuits. We wish to consider circuits without time-varying elements and to establish criteria which rule out the existence of self-sustained oscillations. For a linear circuit with only positive resistors, no oscillations can occur if there is dissipation, but this is not true of nonlinear circuits as, for example, the van der Pol circuit. This question has practical significance in the design of computer circuits, control systems, etc., in which are utilized nonlinear elements and, in such cases, one wants to establish overall stability requirements. We shall see in this section how one can use the mixed potential to fulfill such requirements.

We have a system of differential equations which we write in vector form*

$$-Jx' = \partial P(x)/\partial x, \tag{8.1}$$

where

$$x = \begin{pmatrix} i \\ v \end{pmatrix},$$

$$J = \begin{bmatrix} -L & 0 \\ 0 & C \end{bmatrix},$$

which is, in general**, an indefinite symmetric matrix, and $\partial P/\partial x$ denotes the gradient of P. We note that the stationary points of $P(x)$, i.e., where $\partial P(x)/\partial x = 0$, are exactly the equilibria of (8.1) and we want to discuss conditions under which *all* solutions approach these equilibria as $t \to \infty$. In particular, this would exclude the existence of periodic solutions if there are only a finite number of equilibria.

For this purpose, we differentiate $P(x)$ along the solutions of (8.1) and find

$$\frac{dP(x)}{dt} = (x', P_x) = -(x', Jx'). \tag{8.2}$$

If the circuit does not contain any inductors, then we have

$$\frac{dP}{dt} = -(x', Cx')$$

since $J = C$ and obviously P decreases along solutions except at equilibria. Following Liapounov's ideas, one could derive from this, assuming, in addition, that $P(x) \to \infty$ as $|x| \to \infty$, that every solution tends to one of the equilibrium points as $t \to \infty$. Similarly, if there are no capacitors in the circuit, then

$$\frac{d(-P)}{dt} = -(x', Lx'),$$

and assuming that $-P(x) \to \infty$ as $|x| \to \infty$, we have asymptotic stability again.

It is intuitively clear that if L or C is sufficiently small, then a similar result should hold also. Just how large L or C can become without causing oscillations will be answered in the following investigation.

We first ask whether one can describe the system (8.1) by another pair J^*, P^*† in

*Here x' denotes dx/dt.

**If no inductors are present, J is, of course, positive definite.

†The notation J^* should not be confused with the adjoint of the matrix J.

place of J, P such that

$$-J^*x' = P^*_x(x), \tag{8.3}$$

and such that J^* is positive definite. Since $-x' = J^{-1}P_x(x)$, a necessary and sufficient condition for a new pair J^*, P^* describing the differential equations in the form (8.3) is

$$J^*J^{-1}P_x = P^*_x . \tag{8.4}$$

Our aim is to find a pair J^*, P^* such that (x', J^*x') is positive definite and $P^*(x) \to \infty$ for $|x| \to \infty$.

We note first if (J_1, P_1), (J_2, P_2) are two pairs describing (8.1), then so are

$$(\alpha J_1 + \beta J_2, \alpha P_1 + \beta P_2), \tag{8.5}$$

which gives considerable freedom in constructing other pairs.

To find one nontrivial pair other than (J, P) we observe that if M is any constant *symmetric* matrix, then the pair

$$J^* = P_{xx}MJ, \qquad P^* = \tfrac{1}{2}(P_x, MP_x),$$

is a possible choice. This is easily seen since

$$P^*_x = (P_{xx}M)P_x$$

and therefore

$$J^*J^{-1}P_x = P_{xx}MJJ^{-1}P_x = P^*_x$$

which by (8.4) implies (8.3). By superposition we obtain the more general pairs

$$J^* = (\lambda I + P_{xx}M)J, \qquad P^* = \lambda P + \tfrac{1}{2}(P_x, MP_x) \tag{8.6}$$

where M ranges over all constant symmetric matrices and λ is an arbitrary constant.

Having made these observations, we shall now prove some theorems which depend on the mixed potential having the form

$$P(i, v) = -A(i) + B(v) + (i, \gamma v), \tag{8.7}$$

where γ is a constant matrix *not* necessarily with elements ± 1, 0.

We first consider a "semilinear" case, i.e.,

$$P(i, v) = -\tfrac{1}{2}(i, Ai) + B(v) + (i, \gamma v - a), \tag{8.8}$$

where A is a constant symmetric matrix and a is a constant vector so that the first r equations in (4.10) are linear. In case the circuit is complete, this means that all the nonlinear resistors are in the subnetwork $\mathfrak{N}_v$. In theorems 3 and 4 which follow, we can allow nonlinear inductors, i.e.,

$$L = L(i),$$

and nonlinear capacitors, i.e.,

$$C = C(v).$$

However, it is necessary to assume that $L(i)$ and $C(v)$ are symmetric, positive definite, and their least eigenvalues are bounded away from zero.

Theorem 3. If A is positive definite, $B(v) + |\gamma v| \to \infty$ as $|v| \to \infty$, and

$$\| L^{1/2}(i)A^{-1}\gamma C^{-1/2}(v) \| \le 1 - \delta, \qquad \delta > 0,^* \tag{8.9}$$

*The notation $\| K \|^2$ of a matrix K denotes $\max_{|x|=1} (Kx, Kx)$.

for all i, v, then all solutions of (8.1) tend to the set of equilibrium points as $t \to \infty$.

Proof. We choose M and λ in (8.6) as follows:

$$M = \begin{bmatrix} 2A^{-1} & 0 \\ 0 & 0 \end{bmatrix}, \qquad \lambda = 1.$$

Then

$$J^* = J + P_{xx}MJ = \begin{bmatrix} L & 0 \\ -2\gamma^T A^{-1} L & C \end{bmatrix},^*$$

and

$$(x', J^*x') = (y, y) - 2(z, C^{-1/2}\gamma^T A^{-1} L^{1/2} y) + (z, z),$$

where

$$y = L^{1/2}\frac{di}{dt} \quad \text{and} \quad z = C^{1/2}\frac{dv}{dt}.$$

With $K = L^{1/2}A^{-1}\gamma C^{-1/2}$, we have

$$(x', J^*x') = (y - Kz, y - Kz) + (z, z) - (Kz, Kz),$$

and since $\| K \| \leq 1 - \delta$, then

$$(x', J^*x') \geq | y - Kz |^2 + \delta | z |^2 \geq 0,$$

which is zero if and only if $di/dt = dv/dt = 0$. Thus

$$P^* = P + (P_i , A^{-1}P_i)$$

is monotone decreasing except at equilibria and it remains to be proved that $P^*(x) \to \infty$ as $| x | \to \infty$.

To show this we rewrite P in the form

$$P(\alpha, v) = -\tfrac{1}{2}(\alpha, A^{-1}\alpha) + U(v),$$

where $\alpha = P_i = -Ai + \gamma v - a$ and

$$U(v) = \tfrac{1}{2}[(a - \gamma v), A^{-1}(a - \gamma v)] + B(v). \tag{8.10}$$

Then P^* becomes

$$P^*(\alpha, v) = \tfrac{1}{2}(\alpha, A^{-1}\alpha) + U(v).$$

Since A is positive definite and $B(v) + | \gamma v | \to \infty$ as $| v | \to \infty$, it is clear also that $U(v) \to \infty$ as $| v | \to \infty$. It remains to be shown that $| \alpha | + | v | \to \infty$ as $| i | + | v | \to \infty$. For this purpose we consider the matrix

$$S = \begin{bmatrix} -A & \gamma \\ 0 & I \end{bmatrix},$$

which is nonsingular and gives rise to the transformation

$$\begin{bmatrix} \alpha + a \\ \omega \end{bmatrix} = \begin{bmatrix} -A & \gamma \\ 0 & I \end{bmatrix} \begin{bmatrix} i \\ v \end{bmatrix},$$

*The notation γ^T denotes the transpose of the matrix γ.

or

$$y = Sx.$$

Hence $(y, y) = (Sx, Sx) \geq 0$. However, S is nonsingular and its least eigenvalue is bounded away from zero so that $(Sx, Sx) > 0$ for $x \neq 0$. Since $(Sx, Sx) \geq \lambda_1(x, x)$ where λ_1 is the smallest eigenvalue of $S^T S$, then we have

$$(y, y) = (Sx, Sx) \geq \lambda_1(x, x),$$

where $\lambda_1 > 0$. This gives

$$|\alpha + a|^2 + |\omega|^2 \geq \lambda_1(|i|^2 + |v|^2),$$

which implies $|\alpha| + |v| \to \infty$ for $|i| + |v| \to \infty$. By applying a well-known theorem of Liapounov (see LaSalle and Lefschetz [12]), we conclude that every solution of (8.1) approaches the set of equilibrium points as $t \to \infty$.

Next we consider the other semilinear case, i.e.,

$$P(i, v) = -A(i) + \tfrac{1}{2}(v, Bv) + (v, \gamma^T i + b), \tag{8.11}$$

where B is a constant symmetric matrix and b is a constant vector so that the last s equations of (4.10) are linear. In case the circuit is complete, this means that all the nonlinear resistors are in the subnetwork $\mathfrak{N}_i$.

Theorem 4. If B is positive definite, $A(i) + |\gamma^T i| \to \infty$ as $|i| \to \infty$, and

$$\| C^{1/2}(v)B^{-1}\gamma^T L^{-1/2}(i) \| \leq 1 - \delta, \qquad \delta > 0, \tag{8.12}$$

for all i, v, then all solutions of (8.1) tend to the set of equilibrium points as $t \to \infty$.

Proof. Since the proof is similar to the proof of theorem 3, we will only indicate it. We choose M and λ in (8.6) as follows:

$$M = \begin{bmatrix} 0 & 0 \\ 0 & 2B^{-1} \end{bmatrix}, \qquad \lambda = -1.$$

Then it is easy to show that $P^* = -P + (P_v, B^{-1}P_v)$ is monotone decreasing along solutions of (8.1). To show that $P^*(i, v) \to \infty$ as $|i| + |v| \to \infty$, we write P^* in the form

$$P^*(i, v) = \tfrac{1}{2}(P_v, B^{-1}P_v) + W(i),$$

where

$$W(i) = \tfrac{1}{2}[(b + \gamma^T i), B^{-1}(b + \gamma^T i)] + A(i), \tag{8.13}$$

and proceed as in the proof of theorem 3.

We remark that it is not obvious, in general, when the first two conditions of theorems 3 and 4 hold. In section 19, part II, conditions on the network and the resistors are given which are equivalent to these conditions if the circuit is complete.

The next theorem does not require semilinearity but does require that the matrices L and C are constant symmetric and positive definite. The mixed potential has the form (8.7) where nonlinearities may occur in both $A(i)$ and $B(v)$. For this theorem we construct the function

$$P^*(i, v) = \left(\frac{\mu_1 - \mu_2}{2}\right)P(i, v) + \tfrac{1}{2}(P_i, L^{-1}P_i) + \tfrac{1}{2}(P_v, C^{-1}P_v), \tag{8.14}$$

where μ_1 is the smallest eigenvalue of the matrix $L^{-1/2}A_{ii}(i)L^{-1/2}$ for all i and μ_2 is the smallest eigenvalue of $C^{-1/2}B_{vv}(v)C^{-1/2}$ for all v. We shall use $\mu(M)$ to denote the smallest eigenvalue of a symmetric matrix M.

Theorem 5. If

$$\mu(L^{-1/2}A_{ii}(i)L^{-1/2}) + \mu(C^{-1/2}B_{vv}(v)C^{-1/2}) \geq \delta, \qquad \delta > 0, \tag{8.15}$$

for all i, v and

$$P^*(i, v) \to \infty \quad \text{as} \quad |i| + |v| \to \infty, \tag{8.16}$$

where $P^*(i, v)$ is given by (8.14), then all solutions of (8.1) approach the equilibrium solutions as $t \to \infty$.

It is difficult, in general, to replace (8.16) by simple conditions on $P(i, v)$. We leave it therefore in the form given since it can be checked directly.

Proof. For M and λ in (8.6) we choose

$$M = \begin{bmatrix} L^{-1} & 0 \\ 0 & C^{-1} \end{bmatrix} \quad \text{and} \quad \lambda = \frac{\mu_1 - \mu_2}{2}.$$

Then

$$J^* = \begin{bmatrix} A_{ii} & \gamma \\ -\gamma^T & B_{vv} \end{bmatrix} + \lambda \begin{bmatrix} -L & 0 \\ 0 & C \end{bmatrix}.$$

With $z = L^{1/2}\,(di/dt)$ and $w = C^{1/2}\,(dv/dt)$ we have

$$\begin{aligned}(x', J^*x') &= (z, L^{-1/2}A_{ii}L^{-1/2}z) + (w, C^{-1/2}B_{vv}C^{-1/2}w) + \lambda((w, w) - (z, z)) \\ &\geq (\mu_1 - \lambda)(z, z) + (\mu_2 + \lambda)(w, w) \\ &\geq \tfrac{1}{2}(\mu_1 + \mu_2)[(z, z) + (w, w)] \geq 0.\end{aligned}$$

Since $\mu_1 + \mu_2 \geq \delta > 0$ by assumption, then (x', J^*x') is positive definite and equal to zero if and only if $di/dt = dv/dt = 0$. Thus, $P^*(i, v)$, given by (8.14), is monotone decreasing except at the equilibria.

In summary, we have three theorems which give sufficient conditions for asymptotic stability in the large. Theorems 3 and 4 give conditions which depend on the graph of the network as given by the matrix γ but are independent of the nonlinearities. On the other hand, theorem 5 gives conditions which depend on the nonlinearities but are independent of the graph γ of the network.

9. Example. In this section we consider a large ladder network shown in the following figure, and we wish to apply theorem 3 to find conditions for nonoscillation. Then, in order to demonstrate that these criteria are sharp, we will choose a particular nonlinear element for this circuit and exhibit an exact periodic solution.

We will see that the conditions which cause this oscillation can be chosen very close to the nonoscillatory conditions and also, in one case, that the point separating oscillation and nonoscillation approaches zero as the number of loops in the circuit becomes large.

The circuit considered is shown in Fig. 11. The nonlinear element is given by $g(v)$ which is the current through the element in the direction shown.

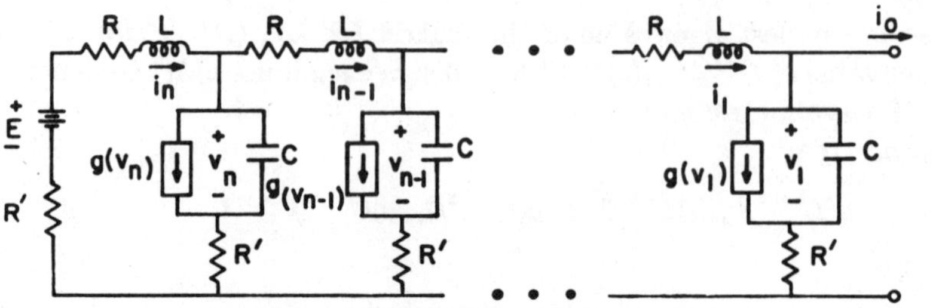

FIG. 11. Large Nonlinear Ladder Circuit

Assuming that $v_{n+1} = E$ and $i_0 = 0$, the mixed potential for the circuit is

$$P(i, v) = -\tfrac{1}{2}R'i_n^2 + \sum_{k=1}^{n}\left[-\tfrac{1}{2}Ri_k^2 - \tfrac{1}{2}R'(i_{k-1} - i_k)^2 + i_k(v_{k+1} - v_k) + \int_0^{v_k} g(v)\,dv\right],$$

or written in vector notation

$$P(i, v) = -\tfrac{1}{2}R(i, i) - \tfrac{1}{2}R'(\alpha i, \alpha i) + (i, \gamma v) + B(v) - (a, i),$$

where α is an $n + 1$ by n matrix and γ is an n by n matrix given by

$$\alpha = \begin{pmatrix} -1 & 0 & \cdots & \cdots & 0 \\ 1 & -1 & \ddots & & \vdots \\ 0 & 1 & -1 & \ddots & \vdots \\ \vdots & \ddots & 1 & \ddots & \vdots \\ \vdots & & \ddots & -1 & 0 \\ \vdots & & & 1 & -1 \\ 0 & \cdots & \cdots & 0 & 1 \end{pmatrix}, \quad \gamma = \begin{pmatrix} -1 & 1 & 0 & \cdots & 0 \\ 0 & -1 & 1 & \ddots & \vdots \\ \vdots & \ddots & -1 & 1 & \vdots \\ \vdots & & \ddots & \ddots & 0 \\ \vdots & & & \ddots & 1 \\ 0 & \cdots & \cdots & 0 & -1 \end{pmatrix}, \quad a = \begin{pmatrix} 0 \\ \vdots \\ \vdots \\ 0 \\ -E \end{pmatrix},$$

and

$$B(v) = \sum_{k=1}^{n} \int_0^{v_k} g(v)\,dv.$$

In applying theorem 3 we shall consider two cases separately: $R' = 0$ and $R = 0$. With $R' = 0$, we want to show first that the circuit has only a finite number of equilibrium solutions. The equilibrium equations are

$$\frac{v_{k-1} - 2v_k + v_{k+1}}{R} = g(v_k), \qquad k = 2, \cdots, n,$$

$$\frac{v_2 - v_1}{R} = g(v_1),$$

where $v_{n+1} = E$. The solution is found graphically as shown in Fig. 12, where, for example, we have chosen $g(v)$ to be the characteristic for a tunnel diode. We see that there are at most three intersections and hence the number of equilibria is at most 3^n.

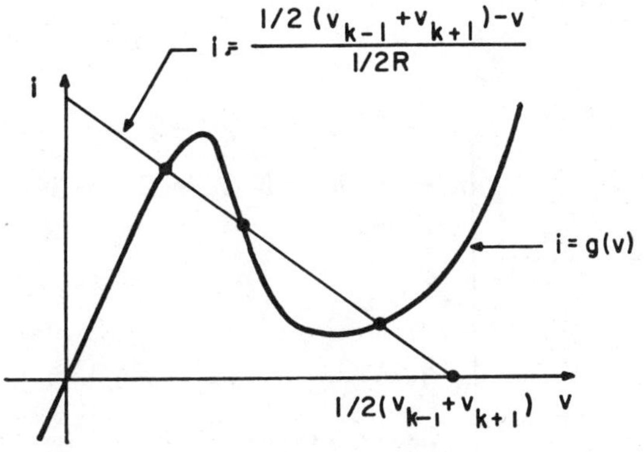

FIG. 12. Equilibrium Solutions for the kth Circuit

In checking the conditions of theorem 3, it is clear that $B(v) \to \infty$ as $|v| \to \infty$ and since $A = RI$,* then A is positive definite. We now compute the norm of the matrix $K = L^{1/2}A^{-1}\gamma C^{-1/2}$ or in this case $K = (L^{1/2}/RC^{1/2})\gamma$. Computing the norm of γ, we find $\|\gamma\|^2 \leq 4$ and therefore

$$\|K\|^2 \leq \frac{4L}{R^2C}.$$

Hence, from theorem 3, $\|K\|^2 < 1$ for nonoscillation is satisfied if

$$\frac{L}{R^2C} < \frac{1}{4}. \tag{9.1}$$

For the case $R = 0$, the equilibrium equations are of the form

$$\frac{E - R' \sum_{l \neq k} i_l - v_k}{2R'} = g(v_k), \qquad k = 1, \cdots, n,$$

and it is clear again that there are at most 3^n equilibrium solutions. In this case the matrix A is

$$A = R'\alpha^T\alpha = R' \begin{pmatrix} 2 & -1 & 0 & \cdots & \cdots & 0 \\ -1 & 2 & -1 & \ddots & & \vdots \\ 0 & \ddots & \ddots & \ddots & \ddots & \vdots \\ \vdots & \ddots & \ddots & \ddots & \ddots & 0 \\ \vdots & & \ddots & -1 & 2 & -1 \\ 0 & \cdots & \cdots & 0 & -1 & 2 \end{pmatrix},$$

which is positive definite since α has full rank. One computes

---

* I denotes the identity matrix.

$$R'(-\gamma^{-1}A)^{-1} = -R'A^{-1}\gamma$$

$$= \frac{1}{(n+1)}\begin{bmatrix} n & -1 & \cdots & \cdots & \cdots & -1 \\ n-1 & n-1 & -2 & \cdots & \cdots & -2 \\ n-2 & n-2 & n-2 & -3 & \cdots & -3 \\ \vdots & & & \ddots & \ddots & \vdots \\ 3 & \cdots & 3 & 2-n & & 2-n \\ 2 & \cdots & \cdots & 2 & & 1-n \\ 1 & \cdots & \cdots & \cdots & & 1 \end{bmatrix}$$

In order to estimate the norm of $K = L^{1/2}A^{-1}\gamma C^{-1/2}$, it is necessary to estimate the norm of $-R'A^{-1}\gamma$. We compute

$$\begin{aligned}(-R'A^{-1}\gamma x, -R'A^{-1}\gamma x) &= \frac{1}{(n+1)^2}\sum_{l=1}^{n}\left[-l\sum_{k=l+1}^{n} x_k + (n-l+1)\sum_{k=1}^{l} x_k\right]^2 \\ &\le \frac{n}{(n+1)^2}\sum_{l=1}^{n}\left[\sum_{k=1}^{l}(n-l+1)^2x_k^2 + \sum_{k=l+1}^{n} l^2x_k^2\right] \\ &\le \frac{n}{(n+1)^2}\sum_{l=1}^{n}[(n-l+1)^2 + l^2]\sum_{k=1}^{n} x_k^2 \\ &\le \frac{n}{(n+1)^2}\left(\sum_{k=1}^{n} x_k\right)\sum_{l=1}^{n}(n+1)^2 \\ &\le n^2(x, x).\end{aligned}$$

Thus $\| -R'A^{-1}\gamma \|^2 \le n^2$.

According to theorem 3, we are forced to require

$$\frac{L}{R'^2C} < \frac{1}{n^2} \tag{9.2}$$

to ensure $\| K \|^2 < 1$, which is sufficient for nonoscillation. In contrast to the previous criterion for $R' = 0$, this condition is more and more stringent as n increases. That this is not only due to our estimates will be seen from the following consideration.

We now consider a particular nonlinearity for $g(v)$ which is the piecewise linear function shown in Fig. 13. Our purpose is to find a periodic solution such that the magnitude

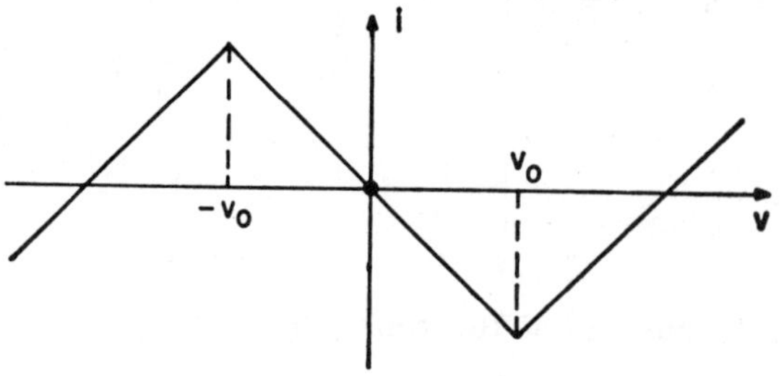

FIG. 13. A Special Choice for $g(v)$

of the voltage across any diode is less than v_0. Then we can replace $g(v)$ by $-Gv$ for $|v| < v_0$ where G is a positive constant, and we have the following differential equations to solve:

$$L\frac{di_k}{dt} = -Ri_k + R'(i_{k+1} - 2i_k + i_{k+1}) - (v_k - v_{k+1}),$$

$$C\frac{dv_k}{dt} = -(i_{k-1} - i_k) + Gv_k, \qquad k = 1, \cdots, n, \tag{9.3}$$

with the boundary conditions $v_{n+1} = E = 0$, $i_0 = i_{n+1} = 0$.

Assuming the solution is of the form

$$i_k = ae^{\alpha t}\sin k\Delta,$$

$$v_k = be^{\alpha t}\cos(k - \tfrac{1}{2})\Delta, \tag{9.4}$$

we find from the boundary conditions that

$$\cos(n + \tfrac{1}{2})\Delta = 0$$

or

$$\Delta = \Delta_\nu = \left(\frac{\nu + \frac{1}{2}}{n + \frac{1}{2}}\right)\pi, \qquad \nu = 0, \cdots, n - 1. \tag{9.5}$$

Note that $\nu = n$ is excluded because this corresponds to the identically zero solution. We denote $2\sin\Delta_\nu/2$ by λ_ν and substituting (9.4) into (9.3) we find the following simultaneous equations for a and b:

$$L a\alpha = -Ra - \lambda_\nu^2 R' a + b\lambda_\nu,$$
$$Cb\alpha = -\lambda_\nu a + Gb. \tag{9.6}$$

This has a nontrivial solution if and only if the determinant of the coefficients of a and b is zero, i.e.,

$$\alpha^2 + \left[\frac{R}{L} + \frac{R'}{L}\lambda_\nu^2 - \frac{G}{C}\right]\alpha + \frac{1}{LC}[\lambda_\nu^2 - G(R + R'\lambda_\nu^2)] = 0.$$

Oscillations occur if α is purely imaginary and nonzero, which means

$$\frac{R}{L} + \frac{R'}{L}\lambda_\nu^2 = \frac{G}{C}, \tag{9.7}$$

and

$$\lambda_\nu^2 > G(R + R'\lambda_\nu^2). \tag{9.8}$$

Eliminating G, these conditions become

$$\frac{R}{\lambda_\nu} + R'\lambda_\nu \leq \sqrt{\frac{L}{C}}. \tag{9.9}$$

The left-hand side is plotted as a function of λ in Fig. 14.

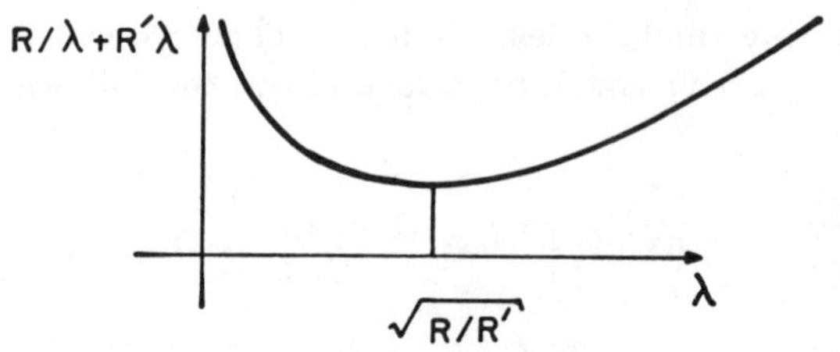

FIG. 14. $R/\lambda + R'\lambda$ versus λ

We also see that the condition $|v| \le |v_0|$ is easily satisfied; for purely imaginary α, this becomes $|b| \le v_0$, and we are at liberty to choose b since equation (9.6) is homogeneous.

For $R' = 0$ condition (9.9) becomes

$$\frac{L}{R^2C} > \frac{1}{\lambda_\nu^2}, \tag{9.10}$$

and, for the purposes of this example, we want to choose ν such that λ_ν is maximum. This leads to the choice $\nu = n - 1$ so that

$$\lambda_{n-1}^2 = 4\sin^2 \frac{n - 0.5}{n + 0.5}\frac{\pi}{2} \approx 4 - \frac{4\pi^2}{(2n+1)^2} \quad \text{for large } n,$$

and (9.10) becomes

$$\frac{L}{R^2C} > \frac{1}{4}\left[1 - \frac{\pi^2}{(2n+1)^2}\right]^{-1} \quad \text{for large } n. \tag{9.11}$$

Thus, under this condition, an oscillatory solution exists and is given by (9.4).

In the case $R = 0$ equation (9.9) becomes

$$\frac{L}{R'^2C} > \lambda_\nu^2, \tag{9.12}$$

and we make λ_ν^2 a minimum by choosing $\nu = 0$. Then

$$\lambda_\nu^2 = 4\sin^2 \frac{1}{(2n+1)}\left(\frac{\pi}{2}\right) \approx \frac{\pi^2}{(2n+1)^2} \quad \text{for large } n,$$

and condition (9.12) becomes

$$\frac{L}{R'^2C} > \frac{\pi^2}{(2n+1)^2} \quad \text{for } n \text{ large.} \tag{9.13}$$

Thus, under this condition, an oscillatory solution exists and is given by (9.4).

In summary, we have shown in the case $R' = 0$ that under the condition

$$\frac{L}{R^2C} < \frac{1}{4},$$

no oscillations can exist, but if

$$\frac{L}{R^2C} > \frac{1}{4}\left[1 - \frac{\pi^2}{(2n+1)^2}\right]^{-1} \quad \text{for large } n,$$

and if we choose $g(v)$ appropriately, then a periodic solution actually does exist. This shows that the criterion of theorem 3 is fairly sharp, and, in fact, it is the best possible in general.

In the case $R = 0$, no oscillations can occur if

$$\frac{L}{R'^2C} < \frac{1}{n^2},$$

but if

$$\frac{L}{R'^2C} > \frac{\pi^2}{(2n+1)^2} \quad \text{for large } n,$$

and if $g(v)$ is chosen appropriately, then a periodic solution exists. We also see in this case that the tendency for this circuit to oscillate increases as the number of loops increases.

This discussion suggests the investigation of the continuous analog of (9.3) which corresponds to a nonlinear transmission line. If x in $0 \leq x \leq 1$ is a variable along the line, the equations take the form

$$L\frac{\partial i}{\partial t} = -Ri + R'\frac{\partial^2 i}{\partial x^2} + \frac{\partial v}{\partial x},$$

$$C\frac{\partial v}{\partial t} = \frac{\partial i}{\partial x} + g(v),$$

with the boundary conditions

$$i(t, 0) = 0, \qquad v(t, 1) = E.$$

Here R, R', L, C represent some appropriate densities of resistance, inductance, and capacitance. For $R = R' = 0 = g(v)$, one obtains the equations for a lossless transmission line. As in the discrete case, one can attempt to describe the solutions for large values of t. For instance, the function P would become an integral

$$\int_0^1 \left[-\frac{R}{2}i^2 + R'i_x^2 + iv_x + G(v(t, x))\right] dx, \qquad G(v) = \int_0^v g(v)\, dv.$$

However, the detailed discussion is out of place here. We just mention that for $R = 0$ and appropriate $g(v)$, E, there is a continuum of equilibrium solutions, namely any function $v(x)$, for which

$$-\frac{v}{R'} + g(v) = \text{constant} = -\frac{E}{R'} + g(E)$$

holds, gives rise to an equilibrium solution. If $-v/R' + g(v) = \text{constant}$ has several roots, v can be a step function taking on these roots in arbitrary intervals.

10. Solutions near an equilibrium. The question of the stability of an equilibrium solution can be studied by two methods. One is the standard method of investigating the structure of the variational equations, and the other is a method which uses the functions P^* of section 8. In this section we discuss both of these methods separately.

A. We study the solutions of the system

$$L \frac{di}{dt} = \frac{\partial P}{\partial i}, \qquad C \frac{dv}{dt} = -\frac{\partial P}{\partial v} \tag{10.1}$$

near an equilibrium $i = i_0$, $v = v_0$. For this purpose, we replace the above equations by the linearized system (variational equations). This system will be simplified by introducing

$$x = [L(i_0)]^{1/2}(i - i_0) \qquad y = [C(v_0)]^{1/2}(v - v_0).$$

Then

$$\frac{dx}{dt} = Ax - By, \qquad \frac{dy}{dt} = B^T y + Dy, \tag{10.2}$$

where

$$A = L^{-1/2}P_{ii}L^{-1/2}, \qquad B = -L^{-1/2}P_{iv}C^{-1/2}, \qquad D = -C^{-1/2}P_{vv}C^{-1/2}$$

at $i = i_0$, $v = v_0$. Here, A and D are symmetric matrices.

The properties of the matrix

$$M = \begin{bmatrix} A & -B \\ B^T & D \end{bmatrix} \tag{10.3}$$

are most easily discussed in terms of the indefinite bilinear form

$$-(x, x') + (y, y'). \tag{10.4}$$

Combining x, y into one vector

$$z = \begin{pmatrix} x \\ y \end{pmatrix}, \qquad z' = \begin{pmatrix} x' \\ y' \end{pmatrix},$$

(10.4) can be written as

$$(z, Jz') = -(x, x') + (y, y'), \tag{10.5}$$

where

$$J = \begin{bmatrix} -I & 0 \\ 0 & I \end{bmatrix}.$$

The matrix M of (10.3) is then symmetric with respect to the form (z, Jz'), i.e.,

$$\begin{aligned} (Mz, Jz') &= (Ax - By, -x') + (Bx + Dy, y') \\ &= -(Ax, x') + (By, x') + (Bx, y') + (Dy, y') \\ &= (Mz', Jz). \end{aligned}$$

It is well-known that the behavior of the solutions of the system (10.2) is described by the eigenvalues of the matrix M. For instance, the stability of the equilibrium is ensured if all the eigenvalues of M have negative real parts. We wish to study which restrictions the symmetry properties of M impose on its eigenvalues. A matrix which is symmetric with respect to a definite form has only real eigenvalues. Since the form (10.5) is indefinite, such a statement does not hold for M. However, we will prove the following lemma.

Lemma. If z is a (complex) eigenvector of M corresponding to a nonreal eigenvalue λ, then

$$(z, J\bar{z}) = -|x|^2 + |y|^2 = 0.$$

Proof. From $Mz = \lambda z$ follows

$$(Mz, J\bar{z}) = \lambda(z, J\bar{z}) = \lambda(|y|^2 - |x|^2),$$

and since M is symmetric with respect to (10.5), this expression must be real. Hence,

$$(\mathrm{Im}\,\lambda)(-|x|^2 + |y|^2) = 0,$$

which proves the lemma.

For such matrices as (10.3) the condition of section 8, theorem 3,

$$\|K\| < 1,$$

takes the form

$$\|A^{-1}B\| < 1. \tag{10.6}$$

We want to investigate the restrictions on the eigenvalues of M imposed by condition (10.6).

Theorem 6. If in the matrix M of (10.3) the matrices A, D are symmetric, $-A$ is positive definite and (10.6) holds, then the eigenvalues of M lie either in the left-half plane or on the real line.

Remark. Note that D can be negative definite since the only restriction on D is symmetry. Therefore, one cannot expect that the equilibrium is stable, in general. In fact, in case the nonlinear differential equations have several equilibria, some of them must be unstable and give rise to matrices M with positive eigenvalues.

Since the eigenvalues are not purely imaginary, periodic solutions are excluded. Moreover, the solutions escaping from the unstable equilibria are not oscillatory.

Proof. If λ is not real and if

$$z = \begin{pmatrix} x \\ y \end{pmatrix} \neq 0,$$

is the corresponding complex eigenvector, then according to the lemma

$$|x| = |y|,$$

and from $Mz = \lambda z$ follows

$$(A - \lambda I)x = By,$$

or

$$(I - \lambda A^{-1})x = A^{-1}By.$$

If we normalize $|x| = |y| = 1$, we have

$$\| (I - \lambda A^{-1})x \|^2 = \theta < 1,$$

where $\theta = \| A^{-1}B \|^2$. This gives

$$|x|^2 - 2(\mathrm{Re}\,\lambda)(\bar{x}, A^{-1}x) + |\lambda|^2 |A^{-1}x|^2 = \theta,$$

or

$$|\lambda|^2 |A^{-1}x|^2 - 2(\mathrm{Re}\,\lambda)(\bar{x}, A^{-1}x) = \theta - 1 < 0. \tag{10.7}$$

Dropping the first term, one finds

$$(\mathrm{Re}\,\lambda) < 0,$$

since $(\bar{x}, A^{-1}x)$ is negative by assumption.

More precisely, for each eigenvector of M, (10.7) defines a circle in the left-half λ-plane, and the corresponding eigenvalue must lie on this circle or on the real axis. (See Fig. 15).

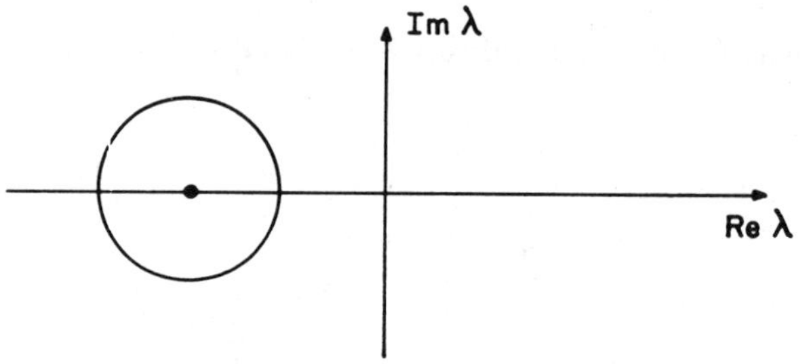

FIG. 15. Circle Containing Eigenvalues of M

The condition of theorem 4 leads to a similar restriction on the eigenvalues.

Theorem 7. If $\mu(A)$ and $\mu(D)$ denote the maximal eigenvalues of the symmetric matrices A, D and if

$$\mu(A) + \mu(D) < 0,$$

then all nonreal eigenvalues of M lie in $\mathrm{Re}\,\lambda < 0$.

Proof. With the above notation we have

$$Ax - By = \lambda x,$$

$$B^T\bar{x} + D\bar{y} = \bar{\lambda}\bar{y},$$

if λ is a nonreal eigenvalue. Multiplying the first equation by $\bar{x}$ and the second by y and adding, we find

$$(\bar{x}, Ax) + (y, D\bar{y}) = \lambda |x|^2 + \bar{\lambda} |y|^2.$$

From the lemma $|x| = |y|$ and normalizing $|x| = |y| = 1$, we have

$$(\bar{x}, Ax) + (y, D\bar{y}) = 2\,\mathrm{Re}\,\lambda.$$

By assumption, the left-hand side remains negative, namely,

$$0 > \mu(A) + \mu(D) \geq 2 \operatorname{Re} \lambda,$$

which proves the theorem.

The linearized equations also show that the conditions of theorems 3, 4, and 5 are sharp. For instance, suppose that x and y in equations (10.2) are scalar. Then M has two eigenvalues which under the condition

$$\text{trace} \quad M = A + D = 0,$$

are both imaginary, and hence the equations have only oscillatory solutions. The condition of theorem 5 for nonoscillation is simply

$$A + D < 0.$$

B. The stability of an equilibrium can also be decided by studying one of the functions P^* of section 8. It is easily verified that in all cases the stationary points of P^* coincide with the equilibrium solutions.

From a well-known theorem due to Dirichlet and Lagrange (and later exploited by Liapounov), if a function exists which decreases with time along the solutions except the equilibrium solutions, then an equilibrium solution is stable if and only if the function has a local minimum there. For a precise formulation of this statement and its proof we refer to the book of Chetayev [13]. Thus, we have

Theorem 8. Under the conditions of theorem 3, 4, or 5, the equilibria coincide with the stationary points of P^* and the local minima of P^* coincide with the stable equilibrium solutions.

In the semilinear case we can discuss the stability of the equilibrium solutions in terms of a function of the voltages only or of the currents only. For instance, in theorem 3 we had by (8.10)

$$P^*(i, v) = \tfrac{1}{2}(P_i \, , A^{-1}P_i) + U(v),$$

and it is obvious that the stationary points are given by

$$P_i = 0 \quad \text{and} \quad U_v = 0.$$

Since the first relation is linear and hence trivial, the equilibrium solutions can be obtained as stationary points of $U(v)$. Moreover, since A is positive definite, the local minima of $U(v)$ are the stable equilibria.

The advantage of this approach is that frequently it is much easier to determine the minima of a function than it is to determine the eigenvalues of the linearized equations.

Bibliography

[1] W. Bode, *Network analysis and feedback amplifier design,* D. Van Nostrand Co., Inc., Princeton, N. J., 1945

[2] E. A. Guillemin, *Introductory circuit theory,* J. Wiley and Sons, Inc., N. Y., 1958

[3] E. J. Cartan, *Leçons sur les invariants integraux,* A. Hermann et Fils, Paris, 1922

[4] R. Duffin, *Nonlinear networks III,* Bull. Amer. Math. Soc. **54,** (1948) 119

[5] E. Goto et al., *Esaki diode high-speed logical circuits,* IRE Trans. on Electronic Computers *EC-9* (1960) 25

[6] J. Moser, *Bistable systems of differential equations,* Proc. of the Rome Symposium, Provisional International Computation Centre, Birkhäuser Verlag, 1960, pp. 320–329

[7] J. Moser, *Bistable systems with applications to tunnel diodes*, IBM J. Res. Dev. **5** (1961) 226

[8] B. D. H. Tellegen, *A general network theorem with applications*, Phillips Research Reports **7** (1952) 259

[9] L. Esaki, *New phenomenon in narrow Ge p-n junctions*, Phys. Rev. **109** (1958) 603

[10] W. Millar, *Some general theorems for nonlinear systems possessing resistance*, Phil. Mag. **42** (1951) 1150

[11] C. Cherry, *Some general theorems for nonlinear systems possessing reactance*, Phil. Mag. **42** (1951) 1161

[12] J. LaSalle and S. Lefschetz, *Stability by Liapunov's direct method with applications*, Academic Press, New York, London, 1961, p. 66

[13] N. G. Chetayev, *Stability of motion*, Moscow, 1946

On the Dynamic Equations of a Class of Nonlinear RLC Networks

L. O. CHUA, MEMBER, IEEE, AND R. A. ROHRER, MEMBER, IEEE

***Abstract*—A method for obtaining the dynamic equations for a broad class of driven nonlinear networks is presented. The parametric approach to element value characterization leads to a mathematical description for any *unicursal* network element. The parametric representation allows the mathematical description of any RLC network which contains such elements and independent sources by means of a set of coupled algebraic-differential equations. The conditions under which these governing equations can be reformulated in the mathematically convenient normal form are given with the explicit means for doing so. Finally, simple methods are presented for revising the original network model so that the normal form exists over the entire dynamic space under mild restrictions.**

I. Introduction

THE ANALYSIS of nonlinear networks may be broken into three distinct steps: 1) the mathematical characterization of the network; 2) qualitative analysis of the network equations; and 3) exact—or approximate—solution of the network behavior. The past two decades have witnessed much work on the analysis of networks containing *monotone* nonlinearities [1]–[4]. Moreover, recently there have appeared several treatments capable of dealing with networks containing elements the characteristic curves of which are *singled-valued* functions of one terminal variable [5]–[7]. Each of these approaches has of necessity limited the class of networks being characterized in step 1) so as not to render impotent the qualitative analysis of step 2). This paper is concerned only with analysis step 1) for a broad class of nonlinear networks: the mathematical characterization is obtained for that class of nonlinear networks containing *unicursal* elements.

Manuscript received October 24, 1964; revised June 12, 1965. The work reported here was originally supported by the Air Force Office of Scientific Research Contract No. AF 49(638)–1383 while the authors were both affiliated with the Coordinated Science Lab., University of Illinois, Urbana, and a WAS-NRC postdoctoral fellowship to the second author. This work was also supported in part by National Science Foundation Grant No. GK–26, Purdue University, Lafayette, Ind.

L. O. Chua is with the School of Electrical Engineering, Purdue University, West Lafayette, Ind.

R. A. Rohrer is with the Dept. of Applied Analysis, State University of New York, Stonybrook, N. Y.

Many networks or network elements of practical import must be characterized by multivalued characteristic curves. Consider, for example, the quiescent characteristic curve of a single tunnel diode in Fig. 1(a), and observe that the series connection (with appropriate biasing) of two such devices yields widely different composite characteristic curves: Fig. 1(b), a multivalued function of each terminal variable; Fig. 1(c), a disconnected curve which is also multivalued; Fig. 1(d), a disconnected curve with an isolated point. The characteristic curve of Fig. 1(b) represents the quiescent model for many practical flip-flop circuits [8]; the curve of Fig. 1(c) and (d) might be obtained inadvertantly in an attempt to construct a flip-flop. Whereas the composite characteristic curves remain monotone for interconnections of those elements originally specified by monotone characteristic curves;

Reprinted from *IEEE Transactions on Circuit Theory*, vol. CT-12, pp. 475–489, December 1965.

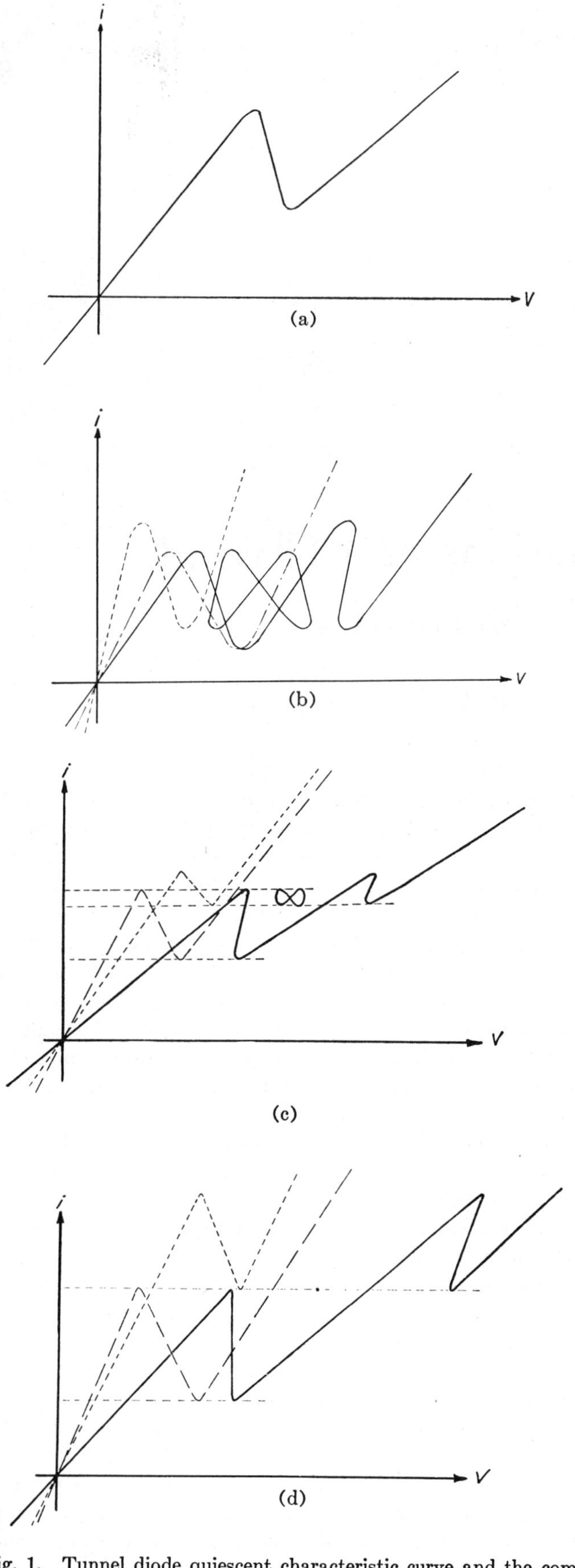

Fig. 1. Tunnel diode quiescent characteristic curve and the composite quiescent characteristic of the interconnection of two such devices. (a) V–I characteristic of a typical tunnel diode. (b) A multivalued curve. (c) A disconnected curve. (d) A curve with an isolated point.

beyond this class a much more exotic behavior results. It is too ambitious at present to deal with a higher class of elements which possess the desirable property that interconnections of elements remain in the class; consequently, in the remainder of the paper, attention is confined to *unicursal* elements to be defined in Section II.

II. Definition and Representation of Unicursal Elements

The class of *unicursal* two-terminal resistance, inductance, or capacitance elements consists of all those which can be defined by a unicursal characteristic curve.

A *unicursal curve* is a subset C of the Euclidean plane E^2 defined by

$$C = \{(w, z) \in E^2 : w = w(\xi), z = z(\xi) \text{ and } \xi \in [a, b]\}, \quad (1)$$

where $w(\xi)$ and $z(\xi)$ are *continuous* functions of the parameter ξ, for which the domain $[a, b]$ may coincide with the *entire* real line; and where both $w(\xi)$ and $z(\xi)$ are functions of *bounded variation* for all *finite* values of a' and b', where $a \leq a' < b' \leq b$.

This definition reduces to the standard definition of an arc if the domain $[a, b]$ is bounded [9], in which case the curve is rectifiable. Hence, a unicursal curve is rectifiable over any bounded domain, but as the curve approaches infinity, the length of the curve approaches infinity. In more intuitive terms, any curve which can be traced over *any bounded* domain of the plane in a *single, continuous,* and *finite* stroke is a unicursal curve.

The above definition of unicursal curve leads naturally to the class of elements which it can represent. Any two-terminal element which can be characterized by a *unicursal curve* is called a *unicursal element.* In particular, a unicursal element may be a resistance, inductance, or capacitance.

It is clear from the definition of unicursal curve that each of the three unicursal element types is amenable to a parametric representation,

$$v = v(\xi) \quad \text{and} \quad i = i(\xi) \quad (2a)$$

for a unicursal resistance;

$$\varphi = \varphi(\xi), i = i(\xi), \text{and } v = \frac{d\varphi}{d\xi}\frac{d\xi}{dt} \quad (2b)$$

for a unicursal inductance;[1] and

$$q = q(\xi), v = v(\xi) \text{ and } i = \frac{dq}{d\xi}\frac{d\xi}{dt} \quad (2c)$$

for a unicursal capacitance.[1] There are many ways to

[1] It is interesting to observe here that since any function of bounded variation has a unique finite derivative almost everywhere [10], the functions $d\varphi/d\xi$ and $dq/d\xi$ are well defined except perhaps on a set of measure zero. One can, of course, avoid this possibility by requiring φ and q to be differentiable functions of ξ. However, in general, this requirement is not necessary provided one defines clearly the domain of existence of the functions $d\varphi/d\xi$ and $dq/d\xi$. It is actually necessary to impose the stronger requirement that the curves $w(\xi)$ and $z(\xi)$ be absolutely continuous in order to avoid infrequent but possible pathological behavior.

find the parametric functions representing a unicursal element. If the element is characterized by a function of one terminal variable, then the independent variable itself can be chosen as the parameter and the representation becomes the conventional one. If the curve is multivalued, then one convenient method is to choose the arc-length as the parameter [9].

The following extension of the concept of *state function* [11] usually associated with network elements gives an introductory example of the utility of the parametric approach. One can define, for example, the *content* and *co-content* of a unicursal resistance from the point (v_1, i_1) to the point (v_2, i_2) on the curve to be given by the *Stieltjes* integrals [9]

$$R(s_2, s_1) \equiv \int_{s_1}^{s_2} v(s)\, di(s) \tag{3a}$$

and

$$R^*(s_2, s_1) \equiv \int_{s_1}^{s_2} i(s)\, dv(s), \tag{3b}$$

respectively, where s_1 is the arc-length to (v_1, i_1) and s_2 is the arc-length to (v_2, i_2). A useful relation exists between these two state functions,

$$R(s_2, s_1) + R^*(s_2, s_1) = v(s_2)i(s_2) - v(s_1)i(s_1); \tag{4}$$

a generalization of the *Legendre* transformation [11], (4) follows immediately from the integration-by-parts theorem for Stieltjes integrals. Similar extensions can be made for the *energy* and *co-energy* of unicursal inductances or capacitances.

III. Governing Equations in Parametric Form

In this section a set of governing equations is obtained for any nonlinear RLC network containing unicursal elements and sources. To this end, let $\mathfrak{N}$ be an n-node *connected* RLC network containing b-unicursal elements. For complete generality, each element is assumed to be accompanied by a series voltage source and a parallel current source as is shown in Fig. 2. The sources may be either independent sources or controlled sources, where the latter may be a nonlinear function of several controlling variables.[2] In the oriented graph for $\mathfrak{N}$, each network element (including its sources) is considered to be represented by only one edge of the graph, and the orientation of each edge k is consistent with the reference direction for current I_k as is shown in Fig. 2.

It is notationally convenient to label the network elements in the following order: first, all of the resistance branches; then, all of the inductance branches; finally, all of the capacitance branches. If $\mathfrak{N}$ contains b_r resistances, b_l inductances, and b_c capacitances, then the network elements corresponding to the numbers 1 to b_r

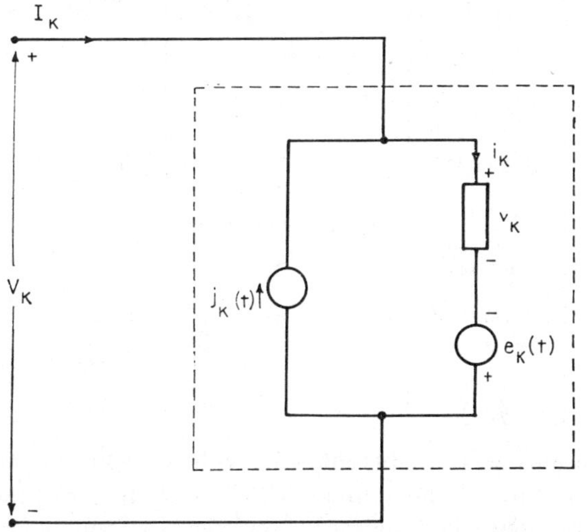

Fig. 2. A typical nonlinear two-terminal element; each element is accompanied by a series voltage source and a parallel current source in the general branch.

are resistances, those for $b_r + 1$ to $b_r + b_l$ are inductances, and those for $b_r + b_l + 1$ to b are capacitances. Now, let $x_1, x_2, \cdots x_b$ be the b-real variables parametrizing the b-unicursal element characteristics and define the following b-dimensional column vectors

$$\mathbf{i}[\mathbf{x}(t)] \equiv \begin{bmatrix} i_1[x_1(t)] \\ \vdots \\ i_{b_r}[x_{b_r}(t)] \\ i_{(b_r+1)}[x_{b_r+1}(t)] \\ \vdots \\ i_{(b_r+b_l)}[x_{(b_r+b_l)}(t)] \\ \dfrac{dq_{(b_r+b_l+1)}}{dx_{(b_r+b_l+1)}} \cdot \dfrac{dx_{(b_r+b_l+1)}}{dt} \\ \vdots \\ \dfrac{dq_b}{dx_b} \cdot \dfrac{dx_b}{dt} \end{bmatrix}, \tag{5a}$$

$$\mathbf{v}[\mathbf{x}(t)] \equiv \begin{bmatrix} v_1[x_1(t)] \\ \vdots \\ v_{b_r}[x_{b_r}(t)] \\ \dfrac{d\varphi_{(b_r+1)}}{dx_{(b_r+1)}} \cdot \dfrac{dx_{(b_r+1)}}{dt} \\ \vdots \\ \dfrac{d\varphi_{(b_r+b_l)}}{dx_{(b_r+b_l)}} \cdot \dfrac{dx_{(b_r+b_l)}}{dt} \\ v_{(b_r+b_l+1)}[x_{(b_r+b_l+1)}(t)] \\ \vdots \\ v_b[x_b(t)] \end{bmatrix} \tag{5b}$$

(representing the element currents and voltages, respectively);

[2] It is in this manner that a nonlinear element of more than two terminals can be handled by the present approach; it is well-known that any multiterminal element governed solely by algebraic equations can be modeled by means of two-terminal elements of the same type, controlled sources, and (perhaps) independent sources.

$$\mathbf{j}(t) \equiv \begin{bmatrix} j_1(t) \\ \vdots \\ j_{b_r}(t) \\ j_{(b_r+1)}(t) \\ \vdots \\ j_{(b_r+b_l)}(t) \\ j_{(b_r+b_l+1)}(t) \\ \vdots \\ j_b(t) \end{bmatrix} \qquad (5c) \qquad \mathbf{e}(t) \equiv \begin{bmatrix} e_1(t) \\ \vdots \\ e_{b_r}(t) \\ e_{(b_r+1)}(t) \\ \vdots \\ e_{(b_r+b_l)}(t) \\ e_{(b_r+b_l+1)}(t) \\ \vdots \\ e_b(t) \end{bmatrix} \qquad (5d)$$

(representing the current and voltage sources, respectively). Any of the sources $j_k(t)$ or $e_k(t)$ in (5c) and (5d) may be controlled sources; however, for convenience of notation, these sources are shown as explicit functions of time.

Then the set of dynamic equilibrium equations for $\mathfrak{N}$ is given by

$$\mathbf{Ci}[\mathbf{x}(t)] = \mathbf{Cj}(t) \qquad (6a)$$

and

$$\mathbf{Bv}[\mathbf{x}(t)] = \mathbf{Be}(t), \qquad (6b)$$

where the $(n - 1) \times b$ matrix $\mathbf{C}$ is the fundamental cut-set matrix corresponding to *any* tree of $\mathfrak{N}$ and the $[b - (n - 1)] \times b$ matrix $\mathbf{B}$ is the associated fundamental circuit matrix [12]. Equations (6) constitute a system of b equations in the b parametric unknowns $x_1, x_2, \cdots, x_b$, and, hence, are the desired set of *governing equations in parametric form*. In general, some of the b equations are *nonlinear algebraic equations* while the remainder are *first-order nonlinear differential equations*. Observe that these equations are undefined at those points where $d\varphi_i/dx_i$ and dq_k/dx_k do not exist.

As an illustrative example of an RLC network governed by parametric equations, consider the simple single loop RLC network shown in Fig. 3(a), where the resistance, inductance, and capacitance characteristics are shown in Figs. 3(b), 3(c), and 3(d), respectively. The $v - i$ characteristic of the resistance R is the four-cusped hypocycloid with parametric representation

$$v_1 = \cos^3 x_1 \quad \text{and} \quad i_1 = \sin^3 x_1. \qquad (7)$$

The $\varphi - i$ characteristic of the inductance L is the ellipse with parametric representation

$$\varphi_2 = \cos x_2 \quad \text{and} \quad i_2 = 2 \sin x_2. \qquad (8)$$

The $q - v$ characteristic of the capacitance C is the cycloid with parametric representation

$$q_3 = x_3 - \sin x_3 \quad \text{and} \quad v_3 = 1 - \cos x_3. \qquad (9)$$

The *KVL* and *KCL* equations are

$$v_1(x_1) + \frac{d\varphi_2(x_2)}{dx_2}\cdot\frac{dx_2}{dt} + v_3(x_3) = 0, \qquad (10a)$$

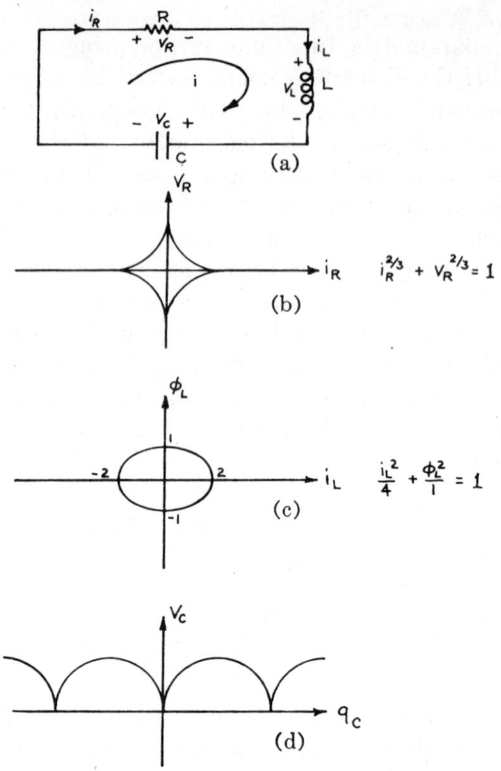

Fig. 3. A single loop RLC network.

$$i_1(x_1) = i_2(x_2), \qquad (10b)$$

and

$$\frac{dq_3(x_3)}{dx_3}\cdot\frac{dx_3}{dt} = i_2(x_2). \qquad (10c)$$

The substitution of (7)–(9) into (10) yields the desired set of dynamic equilibrium equations in parametric form

$$\sin x_2 \cdot \frac{dx_2}{dt} = \cos^3 x_1 - \cos x_3 + 1, \qquad (11a)$$

$$\sin^3 x_1 = 2 \sin x_2, \qquad (11b)$$

and

$$(1 - \cos x_3)\cdot\frac{dx_3}{dt} = 2 \sin x_2. \qquad (11c)$$

Equations (11) constitute the complete set of governing equations for the given network. This example is indicative of the utility of the parametric approach: some characteristic curves represent multi-valued functions of both terminal variables, some even have discontinuous derivatives.

IV. Parametric Representation of Nonlinear RLC Networks in Normal Form

The set of governing equations formulated in Section III is in general a system of nonlinear algebraic and first-order differential equations. However, the governing equations would be more tractable if they could be reduced to some canonic system containing all of the pertinent information on the dynamic behavior of the network.

Therefore, it is usually desirable to formulate the dynamic equations in the form $\dot{\mathbf{x}} = \mathbf{f}(\mathbf{x}, t)$, henceforth referred to as the *normal form* [13]. Many important results concerning the behavior of solutions of ordinary differential equations have been enunciated in terms of the normal form; consequently, it has the proper orientation for analysis steps two and three mentioned in Section I. Networks have been found for which the dynamic equations cannot be expressed in normal form [7]; hence, only a subclass of the networks containing unicursal elements can be represented in the normal form. In the sequel, the normal form is said to *exist* for a given network $\mathfrak{N}$ if the dynamic equations of $\mathfrak{N}$ can be expressed in the form

$$\dot{\mathbf{x}} = \mathbf{f}(\mathbf{x}, t), \tag{12}$$

where the vector $\mathbf{x}$ represents a *complete set of dynamically independent variables* [14], henceforth referred to as the dynamic variables.[3]

A. Definitions and Notation

In the following, any term, definition, or symbol which is not explained is understood to be employed in the same manner as given by Bryant [14].

Definition 1: Normal Tree

Any tree of a network which is chosen according to the set of rules given by Bryant is defined to be a normal tree. If a resistance (inductance, capacitance) belongs to the normal tree $\mathfrak{T}$, then it is called a tree-resistance (inductance, capacitance); whereas, if a resistance (inducttance, capacitance) does not belong to $\mathfrak{T}$, then it is called a link-resistance (inductance, capacitance).

Definition 2: Partitioning and Labelling of Network Elements

Let $\mathfrak{T}$ be any normal tree of a network $\mathfrak{N}$ and let $\mathfrak{L}$ be the corresponding set of links. Partition the set of network elements into six disjoint subsets, $\alpha, \beta, \gamma, \delta, \epsilon, \zeta$, where α is the set of all link-capacitances, β is the set of all link-resistances, γ is the set of all link-inductances, δ is the set of all tree-capacitances, ϵ is the set of all tree-resistances, and ζ is the set of all tree-inductances. Next label the M_C link-capacitances in subset α from 1 to M_C, label the $M_{CR} - M_C$ link-resistances in subset β from $M_C + 1$ to M_{CR}, label the $M - M_{CR}$ link-inductances in subset γ from $M_{CR} + 1$ to M, label the $N - S_C$ tree-capacitances in subset δ from $M + 1$ to $M + N - S_C$, label the $S_C - S_{CR}$ tree-resistances in subset ϵ from $M + N - S_C + 1$ to $M + N - S_{CR}$, and label the $S_{CR} - 1$ tree-inductances in subset ζ from $M + N - S_{CR} + 1$ to $M + N - 1$.

[3] In modern usage the vector $\mathbf{x}$ would represent a set of state variables only if the normal form equations possessed a unique solution [15]. The analysis presented in this paper is addressed to networks the models of which, at least, may not generate unique solutions. Thus, the term dynamic variable is introduced; in those instances where the solution is unique, the dynamic variables become the state variables a posteriori.

In the sequel, each network element is assumed to be specified in its parametric form with the parameter taken to be the arc-length, or any other parameter having a continuous one-to-one correspondence with the arc-length. The domain of each parametric function is assumed to be the entire real line R^1, but the range may be any connected subset of R^1.

Definition 3: Element Parametrization

Each link-capacitance C_{α_i} is specified by $q_{\alpha_i} = Q_{\alpha_i}(x_{\alpha_i})$ and $v_{\alpha_i} = V_{\alpha_i}(x_{\alpha_i})$, each link-resistance R_{β_i} is specified by $i_{\beta_i} = I_{\beta_i}(x_{\beta_i})$ and $v_{\beta_i} = V_{\beta_i}(x_{\beta_i})$, each link-inductance L_{γ_i} is specified by $i_{\gamma_i} = I_{\gamma_i}(x_{\gamma_i})$ and $\varphi_{\gamma_i} = \Phi_{\gamma_i}(x_{\gamma_i})$, each tree-capacitance C_{δ_i} is specified by $v_{\delta_i} = V_{\delta_i}(x_{\delta_i})$ and $q_{\delta_i} = Q_{\delta_i}(x_{\delta_i})$, each tree-resistance R_{ϵ_i} is specified by $v_{\epsilon_i} = V_{\epsilon_i}(x_{\epsilon_i})$ and $i_{\epsilon_i} = I_{\epsilon_i}(x_{\epsilon_i})$, and each tree-inductance L_{ζ_i} is specified by $\varphi_{\zeta_i} = \Phi_{\zeta_i}(x_{\zeta_i})$ and $i_{\zeta_i} = I_{\zeta_i}(x_{\zeta_i})$.

Definition 4: Element Parameter Vectors

The network element parameters in the subset $\alpha, \beta, \gamma, \delta, \epsilon$, and ζ are defined by vectors of the following form:

$$\mathbf{x}_\mu \equiv \begin{bmatrix} x_{\mu_1} \\ x_{\mu_2} \\ \vdots \\ x_{\mu_\mu} \end{bmatrix}, \quad \mathbf{I}_\mu(\mathbf{x}_\mu) \equiv \begin{bmatrix} I_{\mu_1}(x_{\mu_1}) \\ I_{\mu_2}(x_{\mu_2}) \\ \vdots \\ I_{\mu_\mu}(x_{\mu_\mu}) \end{bmatrix}, \text{ and } \mathbf{V}_\mu(\mathbf{x}_\mu) \equiv \begin{bmatrix} V_{\mu_1}(x_{\mu_1}) \\ V_{\mu_2}(x_{\mu_2}) \\ \vdots \\ V_{\mu_\mu}(x_{\mu_\mu}) \end{bmatrix},$$

where μ is a characteristic index which may take on any of the letters $\alpha, \beta, \gamma, \delta, \epsilon$, and ζ when appropriate. In particular, $\alpha_1 = 1, \cdots, \alpha_\alpha = M_C$; $\beta_1 = M_C + 1, \cdots, \beta_\beta = M_{CR}$; $\gamma_1 = M_{CR} + 1, \cdots \gamma_\gamma = M$; $\delta_1 = M + 1, \cdots, \delta_\delta = M + N - S_C$; $\epsilon_1 = M + N - S_C + 1, \cdots, \epsilon_\epsilon = M + N - S_{CR}$; $\zeta_1 = M + N - S_{CR} + 1, \cdots, \zeta_\zeta = M + N - 1$.

B. Governing Equations in Partitioned Form

Now, let each edge of the network graph for $\mathfrak{N}$ represent a composite element as given in Fig. 2 and let $\mathbf{B}$ and $\mathbf{C}$ be the fundamental circuit matrix and the fundamental cut-set matrix, respectively, associated with the normal tree $\mathfrak{T}$ [12]. Then in view of the previous notation the governing equations can be written in the following partitioned form:

$$\begin{bmatrix} \mathbf{V}_\alpha(\mathbf{x}_\alpha) \\ \mathbf{V}_\beta(\mathbf{x}_\beta) \\ \mathbf{V}_\gamma(\mathbf{x}_\gamma) \end{bmatrix} = - \begin{bmatrix} \mathbf{F}_{\alpha\delta} & \mathbf{O} & \mathbf{O} \\ \mathbf{F}_{\beta\delta} & \mathbf{F}_{\beta\epsilon} & \mathbf{O} \\ \mathbf{F}_{\gamma\delta} & \mathbf{F}_{\gamma\epsilon} & \mathbf{F}_{\gamma\zeta} \end{bmatrix} \begin{bmatrix} \mathbf{V}_\delta(\mathbf{x}_\delta) \\ \mathbf{V}_\epsilon(\mathbf{x}_\epsilon) \\ \mathbf{V}_\zeta(\mathbf{x}_\zeta) \end{bmatrix} + \begin{bmatrix} \mathbf{E}_\alpha \\ \mathbf{E}_\beta \\ \mathbf{E}_\gamma \end{bmatrix} \tag{13}$$

and

$$\begin{bmatrix} \mathbf{I}_\delta(\mathbf{x}_\delta) \\ \mathbf{I}_\epsilon(\mathbf{x}_\epsilon) \\ \mathbf{I}_\zeta(\mathbf{x}_\zeta) \end{bmatrix} = \begin{bmatrix} \mathbf{F}'_{\alpha\delta} & \mathbf{F}'_{\beta\delta} & \mathbf{F}'_{\gamma\delta} \\ \mathbf{O} & \mathbf{F}'_{\beta\epsilon} & \mathbf{F}'_{\gamma\epsilon} \\ \mathbf{O} & \mathbf{O} & \mathbf{F}'_{\gamma\zeta} \end{bmatrix} \begin{bmatrix} \mathbf{I}_\alpha(\mathbf{x}_\alpha) \\ \mathbf{I}_\beta(\mathbf{x}_\beta) \\ \mathbf{I}_\gamma(\mathbf{x}_\gamma) \end{bmatrix} + \begin{bmatrix} \mathbf{T}_\delta \\ \mathbf{T}_\epsilon \\ \mathbf{T}_\zeta \end{bmatrix}, \tag{14}$$

where the submatrices $\mathbf{F}_{\alpha\delta}$, $\mathbf{F}_{\beta\delta}$, $\mathbf{F}_{\beta\epsilon}$, $\mathbf{F}_{\gamma\delta}$, $\mathbf{F}_{\gamma\epsilon}$, and $\mathbf{F}_{\gamma\zeta}$ and the vectors $\mathbf{E}_\alpha$, $\mathbf{E}_\beta$, $\mathbf{E}_\gamma$, $\mathbf{T}_\delta$, $\mathbf{T}_\epsilon$, and $\mathbf{T}_\zeta$ are as defined by Bryant [14] and where the primes denote transposition of the associated submatrices. It is shown in the sequel that $\mathbf{x}_\gamma$ and $\mathbf{x}_\delta$ together constitute the set of *dynamic variables* which describe the dynamic behavior of the nonlinear network with *unicursal* elements. It is convenient to denote the Euclidean $\mathbf{x}_\gamma$-space by R^γ and the Euclidean $\mathbf{x}_\delta$-space by R^δ. Then, clearly, the *dynamic space* (i.e., the state space when state concepts are applicable) is simply the Cartesian product [9] of R^γ and R^δ: dynamic space $R^{\gamma+\delta} = R^\gamma \times R^\delta$. The problem now becomes that of reducing (13) and (14) to the desired normal form. It is necessary, therefore, to express the nondynamic variables $\mathbf{x}_\alpha$, $\mathbf{x}_\beta$, $\mathbf{x}_\epsilon$, and $\mathbf{x}_\zeta$ in terms of the dynamic variables $\mathbf{x}_\gamma$ and $\mathbf{x}_\delta$. The first step to this end is to consider the possibility of eliminating the variables associated with the resistances $\mathbf{x}_\beta$ and $\mathbf{x}_\epsilon$. The equations involving these variables are from (13) and (14),

$$\mathbf{V}_\beta(\mathbf{x}_\beta) + \mathbf{F}_{\beta\epsilon}\mathbf{V}_\epsilon(\mathbf{x}_\epsilon) = -\mathbf{F}_{\beta\delta}\mathbf{V}_\delta(\mathbf{x}_\delta) + \mathbf{E}_\beta \tag{15}$$

and

$$-\mathbf{F}'_{\beta\epsilon}\mathbf{I}_\beta(\mathbf{x}_\beta) + \mathbf{I}_\epsilon(\mathbf{x}_\epsilon) = \mathbf{F}'_{\gamma\epsilon}\mathbf{I}_\gamma(\mathbf{x}_\gamma) + \mathbf{T}_\epsilon. \tag{16}$$

It is convenient to write (15) and (16) in the form of a *nonlinear* transformation from the Euclidean k-space R^k into itself, where $k = k_\beta + k_\epsilon$ ($k_\beta = M_{CR} - M_C$ = numbers of link-resistances and $k_\epsilon = S_C - S_{CR}$ = number of tree-resistances). This transformation takes the form

$$\begin{aligned} f_1(x_{\beta_1}, x_{\beta_2}, \cdots, x_{\beta_k}, x_{\epsilon_1}, x_{\epsilon_2}, \cdots, x_{\epsilon_{k_\epsilon}}) &= y_1 \\ \vdots \qquad\qquad & \quad \vdots \\ f_{k_\beta}(x_{\beta_1}, x_{\beta_2}, \cdots, x_{\beta_{k_\beta}}, x_{\epsilon_1}, x_{\epsilon_2}, \cdots, x_{\epsilon_{k_\epsilon}}) &= y_{k_\beta} \\ \vdots \qquad\qquad & \quad \vdots \\ f_k(x_{\beta_1}, x_{\beta_2}, \cdots, x_{\beta_{k_\beta}}, x_{\epsilon_1}, x_{\epsilon_2}, \cdots, x_{\epsilon_{k_\epsilon}}) &= y_k, \end{aligned} \tag{17}$$

or in operator form

$$\mathbf{F}(\mathbf{x}_{\beta\epsilon}) = \mathbf{y}, \tag{18}$$

where the vector $\mathbf{x}_{\beta\epsilon} = (\mathbf{x}'_\beta, \mathbf{x}'_\epsilon)'$ denotes the k nondynamic variables on the left side of (15) and (16) and the vector $\mathbf{y}$ denotes the expressions on the right side of (15) and (16), which are functions of dynamic variables $\mathbf{x}_\gamma$ and $\mathbf{x}_\delta$ and excitations $\mathbf{E}_\beta$ and $\mathbf{T}_\epsilon$.

C. Existence of the Normal Form for Time-Invariant Nonlinear Networks

The most general of several other existence theorems on the normal form to be presented in the sequel, the *main existence theorem*, requires a mathematical constraint which seems to have no equivalent topological counterpart.[4] The proof of the main existence theorem is given in detail because this theorem forms the basis for the later theorems, which, strictly speaking, are but corollaries of it.

Theorem 1: Main Existence Theorem

Let $\mathfrak{N}$ be any time-invariant, nonlinear unicursal RLC network containing neither controlled sources nor mutual inductances and let $\mathfrak{I}$ be some normal tree of $\mathfrak{N}$. Moreover, let the following hypotheses be satisfied.

1) The nonlinear transformation $\mathbf{F}$ is a *one-to-one* mapping [16] from a set $D \subset R^k$ *onto* a set $D^* \subset R^k$;
2) Each link-capacitance is voltage-controlled;[5]
3) Each tree-inductance is current-controlled;[5]
4) Each tree-capacitance and link-inductance is characterized by a unicursal curve (elements which are multivalued functions of both terminal variables are allowed);
5) The parametric representations of all elements are differentiable everywhere and, in particular, each link-capacitance or tree-inductance has a parametric representation which has a nonzero derivative everywhere.

Then the normal form exists over some subset $S \subset R^{\gamma+\delta}$, where S depends in general on the time t (through the source functions $\mathbf{E}_\beta$ and $\mathbf{T}_\epsilon$). The normal form is given explicitly in (44) and (45) and the set S is defined in (46). Conversely, if a network $\mathfrak{N}$ admits of a normal form representation with dynamic variables $\mathbf{x}_\gamma$ and $\mathbf{x}_\delta$ then there necessarily exists a normal tree in $\mathfrak{N}$ such that hypotheses 1) to 5) are satisfied.

Proof: Hypothesis 1) is a necessary and sufficient condition for the existence of an inverse transformation [16] $\mathbf{x}_{\beta\epsilon} = \mathbf{G}(\mathbf{y})$, from D^* *onto* D. Now the vector $\mathbf{y}$ is a function of the dynamic variables $\mathbf{x}_\gamma$ and $\mathbf{x}_\delta$ as well as the sources (independent time functions) $\mathbf{E}_\beta$ and $\mathbf{T}_\epsilon$, $\mathbf{y} = \mathbf{Y}(\mathbf{x}_\gamma, \mathbf{x}_\delta, \mathbf{E}_\beta, \mathbf{T}_\epsilon)$. Hence, $\mathbf{x}_{\beta\epsilon}$ is simply the composition [9] of $\mathbf{G}$ and $\mathbf{Y}$ (denoted by $\mathbf{G} \circ \mathbf{Y}$) and has the form

$$\mathbf{x}_{\beta\epsilon} = \mathbf{G} \circ \mathbf{Y}(\mathbf{x}_\gamma, \mathbf{x}_\delta, \mathbf{E}_\beta, \mathbf{T}_\epsilon), \tag{19}$$

where (19) is defined on the set

$$S^* = \{(\mathbf{x}_\gamma, \mathbf{x}_\delta) \in R^{\gamma+\delta} : \mathbf{Y}(\mathbf{x}_\gamma, \mathbf{x}_\delta, \mathbf{E}_\beta, \mathbf{T}_\epsilon) \in D^*\}. \tag{20}$$

Expression (19) also takes the form

$$\mathbf{x}_\beta = \mathbf{X}_\beta(\mathbf{x}_\gamma, \mathbf{x}_\delta, \mathbf{E}_\beta, \mathbf{T}_\epsilon) \tag{21}$$

and

$$\mathbf{x}_\epsilon = \mathbf{X}_\epsilon(\mathbf{x}_\gamma, \mathbf{x}_\delta, \mathbf{E}_\beta, \mathbf{T}_\epsilon), \tag{22}$$

where (21) and (22) are defined on the set $S^* \subset R^{\gamma+\delta}$. The next step is to express $\mathbf{x}_\alpha$ in terms of the dynamic variables. The relation involving $\mathbf{x}_\alpha$ is found from (13) to be

$$\mathbf{V}_\alpha(\mathbf{x}_\alpha) = -\mathbf{F}_{\alpha\delta}\mathbf{V}_\delta(\mathbf{x}_\delta) + \mathbf{E}_\alpha. \tag{23}$$

[4] By "equivalent" is meant a set of *necessary and sufficient* conditions on network topology which relates to the mathematical constraint; no such set has been forthcoming.

[5] A unicursal element is said to be ρ-controlled if ρ can be taken to be the independent variable specifying the characteristic and if ρ ranges over the entire real line.

Equation (23) consists of M_C scalar equations, and the problem here is to take the inverse of each function $V_{\alpha_i}(x_{\alpha_i})$ and "compose" it with the right side of (23) in a manner similar to that employed in the resistance case. The necessary and sufficient condition for the inverse function of $V_{\alpha_i}(x_{\alpha_i})$ to exist is that hypothesis 2) be satisfied. Moreover, since the capacitance voltage V_{α_i} ranges over the entire real line by definition, the domain of the inverse function $V_{\alpha i}^{-1}$ is the entire real line. Hence, one can write

$$\mathbf{x}_\alpha = \mathbf{V}_\alpha^{-1}(\mathbf{x}_\delta, \mathbf{E}_\alpha), \tag{24}$$

where (24) is defined in the entire dynamic space $R^{\gamma+\delta}$. Moreover, in view of hypothesis 5), the inverse function $V_{\alpha i}^{-1}$ is differentiable everywhere and, hence, $\mathbf{x}_\alpha$ given by (24) is a differentiable function of its arguments. Similarly, one obtains from (14) the equations

$$\mathbf{I}_\zeta(\mathbf{x}_\zeta) = \mathbf{F}'_{\gamma\zeta}\mathbf{I}_\gamma(\mathbf{x}_\gamma) + \mathbf{T}_\zeta \tag{25}$$

and

$$\mathbf{x}_\zeta = \mathbf{I}_\zeta^{-1}(\mathbf{x}_\gamma, \mathbf{T}_\zeta), \tag{26}$$

where (26) is defined over the entire dynamic space $R^{\gamma+\delta}$ and is a differentiable function of its arguments. The voltages $\mathbf{V}_\zeta(\mathbf{x}_\zeta)$ across the tree-inductances are given by

$$\mathbf{V}_\zeta(\mathbf{x}_\zeta) = \dot{\varphi}_\zeta(\mathbf{x}_\zeta), \tag{27}$$

where $\dot{\varphi}_\zeta(\mathbf{x}_\zeta)$ is an $(S_{CR} - 1) \times 1$ vector, the elements of which are the time derivatives of $\varphi_\zeta(\mathbf{x}_\zeta)$. If one "composes" $\varphi_\zeta(\mathbf{x}_\zeta)$ with the inverse function vector $\mathbf{x}_\zeta$ as defined by (26), the composition becomes a function of $\mathbf{x}_\gamma$ and is defined in the entire dynamic space $R^{\gamma+\delta}$. Moreover, since (26) is differentiable, so is the composition. Hence, by the chain rule, (27) becomes

$$\mathbf{V}_\zeta(\mathbf{x}_\zeta) = \mathbf{\Phi}_\zeta^*(\mathbf{x}_\gamma)\dot{\mathbf{x}}_\gamma + \frac{\partial}{\partial t}\varphi_\zeta, \tag{28}$$

where $\mathbf{\Phi}_\zeta^*$ is an $(S_{CR} - 1) \times (M - M_{CR})$ Jacobian matrix the jkth element of which is given by

$$[\mathbf{\Phi}_\zeta^*(\mathbf{x}_\gamma)]_{jk} = \frac{\partial}{\partial x_{\gamma_k}}[\varphi_{\zeta_j} \circ I_{\zeta j}^{-1}(\mathbf{x}_\gamma, \mathbf{T}_\zeta)] \tag{29}$$

(the "composition" on the right is a function of $\mathbf{x}_\gamma$ and $\mathbf{T}_\zeta$), where $\dot{\mathbf{x}}_\gamma$ is an $(M - M_{CR}) \times 1$ vector the elements of which are the time derivatives of $\mathbf{x}_\gamma$ and $(\partial/\partial t)\varphi_\zeta$ is an $(S_{CR} - 1) \times 1$ vector the elements of which are the partial derivatives of $\varphi_\zeta \circ \mathbf{I}_\zeta^{-1}(\mathbf{x}_\gamma, \mathbf{T}_\zeta)$ with respect to time. Similarly, for the currents $\mathbf{I}_\alpha(\mathbf{x}_\alpha)$ in the link-capacitances, one obtains

$$\mathbf{I}_\alpha(\mathbf{x}_\alpha) = \mathbf{Q}_\alpha^*(\mathbf{x}_\delta)\dot{\mathbf{x}}_\delta + \frac{\partial}{\partial t}\mathbf{Q}_\alpha, \tag{30}$$

where $\mathbf{Q}_\alpha^*(\mathbf{x}_\delta)$ is an $M_C \times (N - S_C)$ Jacobian matrix the jkth element of which is given by

$$[\mathbf{Q}_\alpha^*(\mathbf{x}_\delta)]_{jk} = \frac{\partial}{\partial x_{\delta_k}}[Q_{\alpha_j} \circ V_{\alpha j}^{-1}(\mathbf{x}_\delta, \mathbf{E}_\alpha)], \tag{31}$$

(the "composition" on the right is a function of $\mathbf{x}_\delta$ and $\mathbf{E}_\alpha$), where $\dot{\mathbf{x}}_\delta$ is an $(N - S_C) \times 1$ vector the elements of which are the time derivatives of $\mathbf{x}_\delta$ and $(\partial/\partial t)\mathbf{Q}_\alpha$ is an $M_C \times 1$ vector the elements of which are the partial derivatives of $\mathbf{Q}_\alpha \circ \mathbf{V}_\alpha^{-1}(\mathbf{x}_\delta, \mathbf{E}_\alpha)$ with respect to time. The voltages $\mathbf{V}_\gamma(\mathbf{x}_\gamma)$ across the link inductances are given by

$$\mathbf{V}_\gamma(\mathbf{x}_\gamma) = \mathbf{\Phi}_\gamma^*(\mathbf{x}_\gamma)\dot{\mathbf{x}}_\gamma, \tag{32}$$

where $\mathbf{\Phi}_\gamma^*(\mathbf{x}_\gamma)$ is an $(M - M_{CR}) \times (M - M_{CR})$ diagonal Jacobian matrix, each diagonal element of which is given by

$$[\mathbf{\Phi}_\gamma^*(\mathbf{x}_\gamma)]_{ii} = \frac{d\varphi_{\gamma_i}}{dx_{\gamma_i}} \tag{33}$$

and $\dot{\mathbf{x}}_\gamma$ is an $(M - M_{CR}) \times 1$ vector the elements of which are the time derivatives of $\mathbf{x}_\gamma$. In view of hypotheses 4) and 5) and the previous assumption that the domain of each parametric function is the entire real line, (32) is defined in the entire dynamic space $R^{\gamma+\delta}$. Similarly, the currents $\mathbf{I}_\delta(\mathbf{x}_\delta)$ in the tree-capacitances are given by

$$\mathbf{I}_\delta(\mathbf{x}_\delta) = \mathbf{Q}_\delta^*(\mathbf{x}_\delta)\dot{\mathbf{x}}_\delta, \tag{34}$$

where $\mathbf{Q}_\delta^*(\mathbf{x}_\delta)$ is an $(N - S_C) \times (N - S_C)$ diagonal Jacobian matrix, the diagonal elements of which are given by

$$[\mathbf{Q}_\delta^*(\mathbf{x}_\delta)]_{ii} = \frac{dQ_{\delta_i}}{dx_{\delta_i}} \tag{35}$$

and $\dot{\mathbf{x}}_\delta$ is an $(N - S_C) \times 1$ vector, the elements of which are the time derivatives of $\mathbf{x}_\delta$. In view of hypotheses 4) and 5) and the previous assumption that the domain of each parametric function is the entire real line, (34) is defined in the entire dynamic space $R^{\gamma+\delta}$. Now, the substitution of (32), (22), and (28) into (13), for $\mathbf{V}_\gamma(\mathbf{x}_\gamma)$, yields

$$\begin{aligned}\mathbf{\Phi}_\gamma^*(\mathbf{x}_\gamma)\dot{\mathbf{x}}_\gamma &= -\mathbf{F}_{\gamma\delta}\mathbf{V}_\delta(\mathbf{x}_\delta) - \mathbf{F}_{\gamma\epsilon}[\mathbf{V}_\epsilon \circ \mathbf{x}_\epsilon(\mathbf{x}_\gamma, \mathbf{x}_\delta, \mathbf{E}_\beta, \mathbf{T}_\epsilon)]\\ &\quad - \mathbf{F}_{\gamma\zeta}\mathbf{\Phi}_\zeta^*(\mathbf{x}_\gamma)\dot{\mathbf{x}}_\gamma - \mathbf{F}_{\gamma\zeta}\frac{\partial}{\partial t}\varphi_\zeta + \mathbf{E}_\gamma.\end{aligned} \tag{36}$$

Similarly, the substitution of (34), (30), and (21) into (14), for $\mathbf{I}_\delta(\mathbf{x}_\delta)$, yields

$$\begin{aligned}\mathbf{Q}_\delta^*(\mathbf{x}_\delta)\dot{\mathbf{x}}_\delta &= \mathbf{F}'_{\alpha\delta}\mathbf{Q}_\alpha^*(\mathbf{x}_\delta)\dot{\mathbf{x}}_\delta + \mathbf{F}'_{\alpha\delta}\frac{\partial}{\partial t}\mathbf{Q}_\alpha\\ &\quad + \mathbf{F}'_{\beta\delta}[\mathbf{I}_\beta \circ \mathbf{x}_\beta(\mathbf{x}_\gamma, \mathbf{x}_\delta, \mathbf{E}_\beta, \mathbf{T}_\epsilon)] + \mathbf{F}'_{\gamma\delta}\mathbf{I}_\gamma(\mathbf{x}_\gamma) + \mathbf{T}_\delta.\end{aligned} \tag{37}$$

Equation (36) can be recast in the form

$$\mathbf{M}_\gamma\dot{\mathbf{x}}_\gamma = \mathbf{K}_\gamma, \tag{38}$$

where

$$\mathbf{M}_\gamma = \mathbf{\Phi}_\gamma^*(\mathbf{x}_\gamma) + \mathbf{F}_{\gamma\zeta}\mathbf{\Phi}_\zeta^*(\mathbf{x}_\gamma) \tag{39}$$

and

$$\begin{aligned}\mathbf{K}_\gamma &= -\mathbf{F}_{\gamma\delta}\mathbf{V}_\delta(\mathbf{x}_\delta) - \mathbf{F}_{\gamma\epsilon}[\mathbf{V}_\epsilon \circ \mathbf{x}_\epsilon(\mathbf{x}_\gamma, \mathbf{x}_\delta, \mathbf{E}_\beta, \mathbf{T}_\epsilon)]\\ &\quad - \mathbf{F}_{\gamma\zeta}\frac{\partial}{\partial t}\varphi_\zeta + \mathbf{E}_\gamma.\end{aligned} \tag{40}$$

Similarly, (37) can be written in the form

$$\mathbf{M}_\delta \dot{\mathbf{x}}_\delta = \mathbf{K}_\delta, \tag{41}$$

where

$$\mathbf{M}_\delta = \mathbf{Q}_\delta^*(\mathbf{x}_\delta) - \mathbf{F}'_{\alpha\delta}\mathbf{Q}_\alpha^*(\mathbf{x}_\delta) \tag{42}$$

and

$$\mathbf{K}_\delta = \mathbf{F}'_{\alpha\delta}\frac{\partial}{\partial t}\mathbf{Q}_\alpha + \mathbf{F}'_{\beta\delta}[\mathbf{I}_\beta \circ \mathbf{x}_\beta(\mathbf{x}_\gamma, \mathbf{x}_\delta, \mathbf{E}_\beta, \mathbf{T}_\epsilon)] + \mathbf{F}'_{\gamma\delta}\mathbf{I}_\gamma(\mathbf{x}_\gamma) + \mathbf{T}_\delta. \tag{43}$$

Finally, upon taking the appropriate inverses in (38) and (41), one obtains the desired normal form

$$\dot{\mathbf{x}}_\gamma = \mathbf{M}_\gamma^{-1}\mathbf{K}_\gamma \tag{44}$$

and

$$\dot{\mathbf{x}}_\delta = \mathbf{M}_\delta^{-1}\mathbf{K}_\delta; \tag{45}$$

these equations are defined at all points in S^* where the matrices $\mathbf{M}_\gamma$ and $\mathbf{M}_\delta$ are nonsingular. Thus, the normal form is defined in the set

$$S = S^* - S^{**}, \tag{46}$$

where S^* is as defined in (20) and

$$S^{**} = \{(\mathbf{x}_\gamma, \mathbf{x}_\delta) \in R^{\gamma+\delta} : \mathbf{M}_\gamma \text{ or } \mathbf{M}_\delta \text{ is singular}\}. \tag{47}$$

Observe that the set S is in general a function of time (through the source functions) and that S is nonempty except in pathological cases. The proof of the first part of the theorem is concluded. Conversely, if a network $\mathfrak{N}$ admits of a normal form representation with dynamic variables $\mathbf{x}_\gamma$ and $\mathbf{x}_\delta$, then there must exist at least one normal tree with the aforementioned associated dynamic variables. A retracing of the steps involved in the previous proof makes clear the fact that hypotheses 1) to 5) must hold for at least one normal tree. Q.E.D.

Some remarks concerning the main existence theorem are in order. First, observe that with respect to a normal tree $\mathfrak{I}$ hypotheses 1) to 5) are necessary and sufficient conditions for the existence of normal form equations. Consequently, the phenomenon that the normal form may not be defined in the entire dynamic space is a characteristic feature of nonlinear networks and *nothing* can be done about it—it is a property inherent in the network. Of course, it might very well be that such a network is not physical; but this consideration is not valid here, since the present point of view is to derive the most general condition whereby an arbitrary network, physical or nonphysical, may be represented by a single system of first-order differential equations over a suitably defined region of the dynamic space. Observe, also, that hypotheses 2) to 5) of Theorem 1 can be readily examined by inspection and only hypothesis 1) requires more detailed consideration; this investigation is undertaken in Section IV-D.

In many cases $\mathbf{F}$ is a one-to-one mapping from R^k onto R^k, in which case the following corollary holds.

Corollary 1: Global Representation

If the nonlinear transformation $\mathbf{F}$ is a *one-to-one* mapping of the entire Euclidean space R^k onto R^k, then the normal form given by (44) and (45) is defined in the set $S = R^{\gamma+\delta} - S^{**}$, where S^{**} is as defined in (47).

Proof: Since $D^* = R^k$, the set S^* in (20) becomes the dynamic space $R^{\gamma+\delta}$; hence, in view of (46), $S = R^{\gamma+\delta} - S^{**}$.

Theorem 1 is somewhat restrictive for practical applications since it allows only two-terminal elements and many practical devices have more than two terminals. Multi-terminal devices can be modeled with two-terminal elements and controlled sources, a situation which motivates the inclusion of the following corollary on controlled sources.

Corollary 2: Controlled Sources

The main existence theorem remains valid when controlled sources are present provided that the following additional constraints are satisfied. Let S_i be any controlled source in $\mathfrak{N}$ and let $\psi = \{p_1, p_2, \cdots, p_\rho\}$ be any subset of network elements in $\mathfrak{N}$ belonging to the subsets α, β, ϵ, and ζ. Then the following additional hypotheses are required.

1) That S_i does not depend on the voltage across any inductance or the current through any capacitance;
2) Whenever S_i depends on the voltages across the elements in ψ or the currents through the elements in ψ (i.e., $S_i = g[\psi_{p_1}, \psi_{p_2}, \cdots, \psi_{p_\rho}]$), where ψ_{p_i} may be either the voltage or current associated with element p_i, then S_i is not directly connected to the elements in ψ;
3) If S_i is connected to a resistance, then it does not depend on the voltage or current in another resistance.

Proof: The proof relies on the demonstration that the nondynamic variables $\mathbf{x}_\alpha$, $\mathbf{x}_\beta$, $\mathbf{x}_\epsilon$, and $\mathbf{x}_\zeta$ can be expressed in terms of the dynamic variables $\mathbf{x}_\gamma$ and $\mathbf{x}_\delta$. First, observe that when controlled sources are present, the explicit relations of (21), (22), (24), and (26) become implicit functions of some nondynamic variables because the sources $\mathbf{E}_\alpha$, $\mathbf{E}_\beta$, $\mathbf{T}_\epsilon$, and $\mathbf{T}_\zeta$ may now be functions of nondynamic variables. However, in view of conditions 1), 2), and 3) mentioned above, each of the nondynamic variables which appears in $\mathbf{E}_\alpha$, $\mathbf{E}_\beta$, and $\mathbf{T}_\epsilon$, and $\mathbf{T}_\zeta$ can again be expressed in terms of the dynamic variables $\mathbf{x}_\gamma$ and $\mathbf{x}_\delta$ simply by substituting (21), (22), (24), and (26) in place of the nondynamic variables. This substitution immediately eliminates all nondynamic variables because the aforementioned three additional constraints assure that no chain-dependence is possible. Q.E.D.

Observe that condition 2) of Corollary 2 can be replaced by the less stringent condition that no chain-dependence among the controlled sources be allowed; i.e., the substitution of (21), (22), (24), and (26) for the nondynamic variables which appear in the source terms does not result in a "vicious circle." This situation may occur, for example, when a controlled source S_A is connected to element A and depends on the voltage of element B, while another controlled source S_B is connected to element B and depends on the voltage of element A. There are other examples of more complicated chain dependence among several controlled sources, but these situations are quite uncommon. In fact, most practical circuits of interest cannot be modeled in such a way that the circuit model exhibits nontrivial chain-dependence among the controlled sources. Observe, finally, that the controlled sources may be nonlinear and may depend on several controlling quantities; hence, under Corollary 2, a wide class of electronic and solid state circuits is admissible to this study.

D. Existence Theorems in Terms of Topological Constraints

From the point of view of the network theorist, it is desirable to express the constraints on the existence of the normal form in terms of the nature of the characteristic curves of the network elements and the network topology. For such a characterization to be possible, however, one must reduce the generality of Theorem 1 by imposing additional topological constraints to assure the mathematical constraints. In this sense, the following theorems are more restrictive, but still general enough to handle a large class of networks. Here, the main existence Theorem 1 is specialized to theorems, the hypotheses of which are stated strictly in terms of the nature of the characteristic curves of the network elements and the topological constraints imposed by the network. Although these theorems are less general, they may be more useful because the existence of the normal form can be ascertained from them by inspection of the network.

In the sequel, a particular resistance subnetwork plays an important role; hence, it merits a special definition.

Definition: Resistance Subnetwork $\mathfrak{N}_{\beta\epsilon}$

Given a network $\mathfrak{N}$, let all capacitances and voltage sources in $\mathfrak{N}$ be replaced by short circuits and all inductances and current sources in $\mathfrak{N}$ be replaced by open circuits. The resulting network $\mathfrak{N}^*$ is a purely resistance network which may contain more than one separate part. Replace by open circuits all resistances of $\mathfrak{N}^*$ which do not form closed loops with at least one other resistance in $\mathfrak{N}^*$ and remove any isolated nodes which result. The resulting network is defined to be the resistance subnetwork $\mathfrak{N}_{\beta\epsilon}$.

Observe that the resistance subnetwork $\mathfrak{N}_{\beta\epsilon}$ consists of in general a subset of the resistances in $\mathfrak{N}$. In particular, any resistance in parallel with a capacitance or in series with an inductance does not belong to $\mathfrak{N}_{\beta\epsilon}$ as well as those resistances which do not form closed loops with others.

The importance of the resistance subnetwork $\mathfrak{N}_{\beta\epsilon}$ is readily inferred from the following theorems.

Theorem 2 [1], [6]: Monotone-Increasing Resistances

If all unicursal resistances in $\mathfrak{N}_{\beta\epsilon}$ have strictly monotone-increasing curves which tend to $+\infty$ $(-\infty)$ when the independent variables tend to $+\infty$ $(-\infty)$, then the nonlinear transformation $\mathbf{F}$ is a one-to-one mapping from the entire Euclidean space R^k onto R^k.

Proof: From the definition of $\mathbf{F}_{\beta\epsilon}$ [14], it is clear that the nonzero elements of $\mathbf{F}_{\beta\epsilon}$ correspond to the resistances in $\mathfrak{N}_{\beta\epsilon}$. Hence, since a resistance network which contains strictly monotone-increasing elements, which tend to $+\infty$ $(-\infty)$ as the independent variables tend to $+\infty$ $(-\infty)$, has one and only one solution [1], [6] it follows that $\mathbf{F}$ is a one-to-one mapping from R^k onto R^k.
Q.E.D.

Corollary 3: Monotone-Decreasing Resistances

Theorem 2 remains valid if all resistances in $\mathfrak{N}_{\beta\epsilon}$ have strictly monotone-decreasing curves.

Proof: The proof is identical to that given previously for the opposite case. Q.E.D.

Corollary 4: RC Circuits

Let $\mathfrak{I}$ be a normal tree of $\mathfrak{N}$. If for each link resistance R_{β_i}, the elements in $\mathfrak{I}$ forming a fundamental circuit with R_{β_i} consists of only capacitances, then the transformation $\mathbf{F}$ is a one-to-one mapping from R^k *onto* R^k.

Proof: The previous hypothesis guarantees that $\mathfrak{N}_{\beta\epsilon}$ is the empty set, and, hence, the hypothesis of *Theorem* 2 is satisfied automatically. Q.E.D.

Observe that Theorem 2 and Corollaries 3 and 4 impose conditions which can be determined by inspection. Hence, if a network satisfies the hypothesis of either Theorem 2 or Corollary 3 or 4, then, by Theorem 1, the existence of the normal form can be ascertained by inspection. Observe also that Theorem 2 and Corollaries 3 and 4 are only sufficient conditions. If a network does not satisfy the hypothesis of Theorem 2 or of Corollary 3 or 4, it may still satisfy hypothesis 1) of Theorem 1. One such possibility occurs when the transformation $\mathbf{F}$ can be inverted by simple algebraic substitution and elimination, in which case one need not even bother finding a global inversion theorem. At the worst, one must check to ascertain whether $\mathbf{F}$ is one-to-one, but only for the resistance subnetwork $\mathfrak{N}_{\beta\epsilon}$—a relatively simpler network than the original network.

It is possible to state a theorem which allows the

determination of whether the normal form exists over the entire dynamic space by inspection. Again, this ability is obtained only with the imposition of additional constraints; multi-valued elements are excluded and certain topological constraints are introduced. An unexpected bonus is that when these constraints are imposed, the time-invariance requirement can be relaxed.

Theorem 3: Global Characteristic Representation

Let $\mathfrak{N}$ be any time-variable, nonlinear unicursal RLC network containing neither controlled sources nor mutual inductances. Let the following hypotheses be satisfied.

1) There exists a tree $\mathfrak{T}$ in $\mathfrak{N}$ containing all of the capacitances in $\mathfrak{N}$ but none of the inductances in $\mathfrak{N}$ (this statement is equivalent to the requirement that there be no capacitance-only circuits and no inductance-only cut-sets);

2) All link-resistances are voltage-controlled;

3) All tree-resistances are current-controlled;

4) All capacitances are characterized by differentiable, charge-controlled functions;

5) All inductances are characterized by differentiable, flux-controlled functions;

6) All resistances belonging to $\mathfrak{N}_{\beta\epsilon}$ are characterized by strictly monotone-increasing functions which map $(-\infty, \infty)$ onto $(-\infty, \infty)$ for all time t.

Then the normal form exists over the entire dynamic space $R^{\gamma+\delta}$ for all time t (for all possible values of current and voltage sources) and is given explicitly by

$$\dot{\mathbf{x}}_\gamma = [\mathbf{\Phi}^*_\gamma(\mathbf{x}_\gamma)]^{-1}\{-\mathbf{F}_{\gamma\delta}\mathbf{V}_\delta(\mathbf{x}_\delta) - \mathbf{F}_{\gamma\epsilon}[\mathbf{V}_\epsilon \circ \mathbf{x}_\epsilon(\mathbf{x}_\gamma, \mathbf{x}_\delta, \mathbf{E}_\beta, \mathbf{T}_\epsilon)] + \mathbf{E}_\gamma\} \tag{48}$$

and

$$\dot{\mathbf{x}}_\delta = [\mathbf{Q}^*_\delta(\mathbf{x}_\delta)]^{-1}\{\mathbf{F}'_{\gamma\delta}\mathbf{I}_\gamma(\mathbf{x}_\gamma) + \mathbf{F}'_{\beta\delta}[\mathbf{I}_\beta \circ \mathbf{x}_\beta(\mathbf{x}_\gamma, \mathbf{x}_\delta, \mathbf{E}_\beta, \mathbf{T}_\epsilon)] + \mathbf{T}_\delta\}. \tag{49}$$

Proof: Hypothesis 1) is equivalent to $\mathbf{F}_{\alpha\delta} = \mathbf{F}_{\gamma\zeta} = \mathbf{0}$; hence, (36) and (37) reduce to (48) and (49), respectively. The parametric functions $\Phi_{\gamma i}(x_{\gamma i})$ and $Q_{\delta i}(x_{\delta i})$ are strictly monotone-increasing in view of hypotheses 4) and 5). Moreover, since $\mathbf{\Phi}^*_\gamma$ and $\mathbf{Q}^*_\delta$ are diagonal matrices, they must be positive-definite over the entire dynamic space $R^{\gamma+\delta}$; hence, $[\mathbf{\Phi}^*_\gamma(\mathbf{x}_\gamma)]^{-1}$ and $[\mathbf{Q}^*_\delta(\mathbf{x}_\delta)]^{-1}$ exist throughout $R^{\gamma+\delta}$. Now, in view of Theorem 2, hypothesis 6) implies that the nonlinear transformation $\mathbf{F}$ is a one-to-one mapping from R^k onto R^k, and, hence, the normal form is defined over the entire dynamic space $R^{\gamma+\delta}$. Finally, observe that if each element is allowed to be time-variable, the proof goes through as before without difficulty; thus, the theorem is proved. Q.E.D.

E. Physical Considerations in the Existence of Normal Form Equations

It may happen that a given nonlinear network or its mathematical model is not amenable to the theorems of the foregoing sections. When such a case arises, one may examine the model to ascertain whether perhaps an oversimplification in modeling has actually rendered a given physical network mathematically untractable. If, on the other hand, one is adamant about his mathematical model, there is nothing that can be done about the elusiveness of the normal form. There are, however, a number of assumptions which may be *consistent with physical reality* that can be used to assure the existence of normal form equations; these are outlined as follows.

The following techniques of individual element augmentation may be considered as a prologue to the general procedure.

Method 1: Technique to Satisfy Corollary 4

The hypothesis of *Corollary 4* is equivalent to the requirement that $\mathbf{F}_{\beta\epsilon} = \mathbf{0}$. There are at least two methods, both of which may be consistent with physical reality, to alter a given network so that $\mathbf{F}_{\beta\epsilon} = \mathbf{0}$.

1a) One may consider a small linear inductance to be connected in series with any link-resistance which prevents $\mathbf{F}_{\beta\epsilon}$ from being zero. This augmentation immediately converts the "troublesome" link-resistance into a tree-resistance; therefore, $\mathbf{F}_{\beta\epsilon} = \mathbf{0}$ in the augmented network. The linear inductance can be chosen to be small enough so as to simulate the lead inductance which is actually present in practice.

1b) One may alternately consider a small linear capacitance to be connected in parallel with any tree-resistance which prevents $\mathbf{F}_{\beta\epsilon}$ from being zero. This augmentation immediately converts the "troublesome" tree-resistance into a link-resistance; therefore, $\mathbf{F}_{\beta\epsilon} = \mathbf{0}$ in the augmented network. The linear capacitance can be chosen to be small enough so as to simulate the stray parasitic capacitance which is actually present in practice.

Method 2: Technique to Guarantee $\mathbf{F}_{\alpha\delta} = \mathbf{0}$

Since the presence of capacitance-only circuits is solely responsible for the nonvanishing of the matrix $\mathbf{F}_{\alpha\delta}$, it is clear that if one considers a small linear resistance to be connected in series with one of the capacitances in each capacitance-only circuit, the matrix $\mathbf{F}_{\alpha\delta}$ becomes zero. The linear resistance can be chosen to be small enough so as to simulate the lead resistance which is actually present in practice.

Method 3: Technique to Guarantee $\mathbf{F}_{\gamma\zeta} = \mathbf{0}$

Since the presence of inductance-only cut-sets is solely responsible for the nonvanishing of the matrix $\mathbf{F}_{\gamma\zeta}$, it is clear that if one considers a large linear resistance to be connected in parallel with one of the inductances in each inductance-only cut-set, the matrix $\mathbf{F}_{\gamma\xi}$ becomes zero. The linear resistance can be chosen to be large enough so as to simulate the insulator resistance which is actually present in practice.

These methods of augmentation lead naturally to the following theorem on the existence of the normal form equations.

Theorem 4: Network Augmentation for Normal Form

Let $\mathfrak{N}$ be any time-invariant, nonlinear RLC network containing neither controlled sources nor mutual inductances. Let the characteristic curve of each element represent a continuous function of bounded variation [9] of at least one of its two terminal variables. Moreover, let each inductance and capacitance characteristic curve be differentiable. Then, if necessary, under the augmentation of the network with certain linear elements (the values of which may be chosen so as to simulate reasonable physical conditions), the normal form can be made to exist over the *entire* dynamic space.

Proof: It suffices to have a constructive procedure so that any network which satisfies the hypotheses of Theorem 4 can be made to satisfy the hypotheses of Theorem 3 by means of the appropriate augmentation. The crucial step consists of decomposing each function which is not a one-to-one mapping into the sum of a strictly monotone-increasing function and a strictly monotone-decreasing function. This decomposition is always possible since the functions are of bounded variation [9]. Each element in decomposition becomes either two similar elements in series or two similar elements in parallel. Hence, the resulting network consists solely of elements characterized by either continuous strictly monotone-increasing or continuous strictly monotone-decreasing curves. The second step involves the application of the three augmentation methods already discussed, adding linear elements until $\mathbf{F}_{\beta\epsilon} = \mathbf{0}$, $\mathbf{F}_{\alpha\delta} = \mathbf{0}$, and $\mathbf{F}_{\gamma\zeta} = \mathbf{0}$. Since $\mathbf{F}_{\beta\epsilon} = \mathbf{0}$ implies that $\mathfrak{N}_{\beta\epsilon}$ is the empty set, the augmented network satisfies all the hypotheses of Theorem 3; hence, the normal form equations for the augmented network must exist over the entire dynamic space.

F. Two Illustrative Examples

Consider first the network of Fig. 4(a) which has the following element characterizations: $i_1 = I_1(v_1)$ and $i_2 = I_2(v_2) = v_2$ for resistances R_1 and R_2, respectively; $i_k = I_k(\varphi_k)$, $k = 3, 4, 5, 6$, for the four inductances; $v_k = V_k(q_k)$, $k = 7, 8, 9, 10$, for the four capacitances; and $v_{11} = V_{11}(i_{11}) = i_{11}^3 - i_{11}$ for resistance R_{11} and $v_{12} = V_{12}(i_{12})$ for resistance R_{12} (note: the usual ordering of network elements is abandoned here to single out the most interesting elements). The functions $I_1(v_1)$, $I_k(\varphi_k)$, $V_k(q_k)$, and $V_{12}(i_{12})$ are all assumed to be differentiable functions of their respective arguments, but none of these functions is assumed to be monotonic. The critical resistance subnetwork $\mathfrak{N}_{\beta\epsilon}$ consists only of resistances R_2 and R_{11} as is shown in Fig. 4(b). Since the resistance R_{11} is characterized by a nonmonotonic curve, the hypotheses of Theorems 2 and 3 as well as those of Corollaries 3 and 4 are not satisfied; it is therefore necessary to resort to the *Main Existence Theorem.* Since all elements are characterized by functions, it is expedient to simply choose v_1, v_2, φ_3, φ_4, φ_5, φ_6, q_7, q_8, q_9, q_{10}, i_{11}, and i_{12} as the parameters. Since the network has neither capacitance-only circuits nor inductance-only cut-sets, $\mathbf{F}_{\alpha\delta} = \mathbf{F}_{\gamma\xi} = \mathbf{0}$;

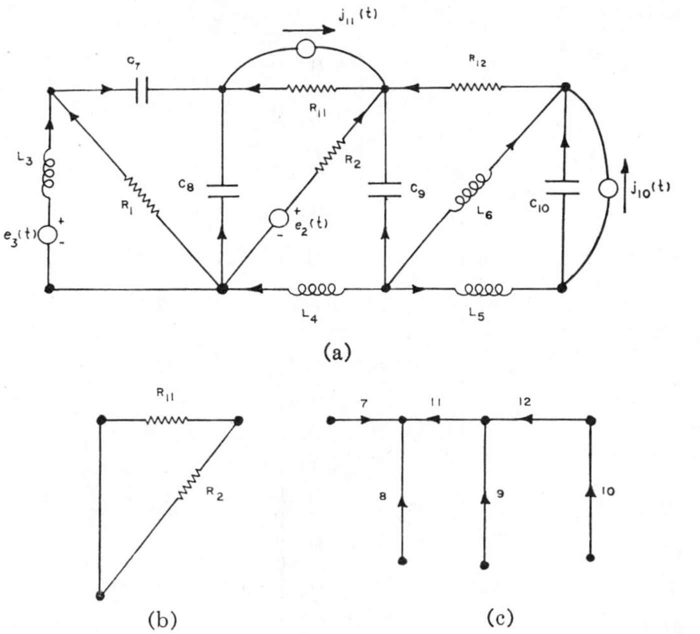

Fig. 4. Example of a nonlinear RLC network to be represented in normal form.

hence, the normal form given by (44) and (45) reduces to the following:

$$\dot{\boldsymbol{\varphi}}_\gamma = -\mathbf{F}_{\gamma\delta}\mathbf{V}_\delta(\mathbf{q}_\delta) - \mathbf{F}_{\gamma\epsilon}[\mathbf{V}_\epsilon \circ \mathbf{x}_\epsilon(\boldsymbol{\varphi}_\gamma, \mathbf{q}_\delta, \mathbf{E}_\beta, \mathbf{T}_\epsilon)] + \mathbf{E}_\gamma \tag{50}$$

and

$$\dot{\mathbf{q}}_\delta = \mathbf{F}'_{\gamma\delta}\mathbf{I}_\gamma(\boldsymbol{\varphi}_\gamma) + \mathbf{F}'_{\beta\delta}[\mathbf{I}_\beta \circ \mathbf{x}_\beta(\boldsymbol{\varphi}_\gamma, \mathbf{q}_\delta, \mathbf{E}_\beta, \mathbf{T}_\epsilon)] + \mathbf{T}_\delta, \tag{51}$$

where $\mathbf{x}_\epsilon$ and $\mathbf{x}_\beta$ are defined in (21) and (22) which represents the inverse nonlinear operator $\mathbf{G}$. The various submatrices of $\mathbf{F}$ can be found from the choice of normal tree exhibited in Fig. 4(c). The fundamental circuit matrix $\mathbf{B}$ is

$$\mathbf{B} = \left[\begin{array}{cccccc|cccc|cc} 1&0&0&0&0&0 & 1&-1&0&0 & 0&0 \\ 0&1&0&0&0&0 & 0&-1&0&0 & 1&0 \\ \hline 0&0&1&0&0&0 & 1&-1&0&0 & 0&0 \\ 0&0&0&1&0&0 & 0&1&-1&0 & -1&0 \\ 0&0&0&0&1&0 & 0&0&-1&1 & 0&1 \\ 0&0&0&0&0&1 & 0&0&-1&0 & 0&1 \end{array}\right] \tag{52}$$

which leads to the following partitioning:

$$\mathbf{F}_{\beta\delta} = \begin{bmatrix} 1 & -1 & 0 & 0 \\ 0 & -1 & 0 & 0 \end{bmatrix},$$

$$\mathbf{F}_{\beta\epsilon} = \begin{bmatrix} 0 & 0 \\ 1 & 0 \end{bmatrix},$$

$$\mathbf{F}_{\gamma\delta} = \begin{bmatrix} 1 & -1 & 0 & 0 \\ 0 & 1 & -1 & 0 \\ 0 & 0 & -1 & 1 \\ 0 & 0 & -1 & 0 \end{bmatrix},$$

and

$$\mathbf{F}_{\gamma\epsilon} = \begin{bmatrix} 0 & 0 \\ -1 & 0 \\ 0 & 1 \\ 0 & 1 \end{bmatrix};$$

$$\mathbf{E}_\beta = \begin{bmatrix} 0 \\ e_2(t) \end{bmatrix},$$

$$\mathbf{T}_\epsilon = \begin{bmatrix} j_{11}(t) \\ 0 \end{bmatrix},$$

$$\mathbf{E}_\gamma = \begin{bmatrix} e_3(t) \\ 0 \\ 0 \\ 0 \end{bmatrix},$$

and

$$\mathbf{T}_\delta = \begin{bmatrix} 0 \\ 0 \\ 0 \\ -j_{10}(t) \end{bmatrix};$$

$$-\mathbf{F}_{\gamma\delta}\mathbf{V}_\delta(\mathbf{q}_\delta) = \begin{bmatrix} -V_7(q_7) + V_8(q_8) \\ -V_8(q_8) + V_9(q_9) \\ V_9(q_9) - V_{10}(q_{10}) \\ V_9(q_9) \end{bmatrix},$$

$$\mathbf{F}'_{\gamma\delta}\mathbf{I}_\gamma(\varphi_\gamma) = \begin{bmatrix} I_3(\varphi_3) \\ -I_3(\varphi_3) + I_4(\varphi_4) \\ -I_4(\varphi_4) - I_5(\varphi_5) - I_6(\varphi_6) \\ I_5(\varphi_5) \end{bmatrix},$$

$$-\mathbf{F}_{\gamma\epsilon}(\mathbf{V}_\epsilon \circ \mathbf{x}_\epsilon) = \begin{bmatrix} 0 \\ V_{11} \circ i_{11} \\ -V_{12} \circ i_{12} \\ -V_{12} \circ i_{12} \end{bmatrix},$$

$$\mathbf{F}'_{\beta\delta}(\mathbf{I}_\beta \circ \mathbf{x}_\beta) = \begin{bmatrix} I_1 \circ v_1 \\ -I_1 \circ v_1 - I_2 \circ v_2 \\ 0 \\ 0 \end{bmatrix},$$

where

$$\mathbf{x}_\beta \equiv \begin{bmatrix} v_1 \\ v_2 \end{bmatrix}, \quad \mathbf{x}_\epsilon \equiv \begin{bmatrix} i_{11} \\ i_{12} \end{bmatrix} \quad \text{and} \quad \mathbf{x}_{\beta\epsilon} \equiv \begin{bmatrix} \mathbf{x}_\beta \\ \mathbf{x}_\epsilon \end{bmatrix}.$$

It remains to invert the expressions for $\mathbf{x}_\beta$ and $\mathbf{x}_\epsilon$,

$$\mathbf{v}_\beta + \mathbf{F}_{\beta\epsilon}\mathbf{V}_\epsilon(\mathbf{i}_\epsilon) = \begin{bmatrix} v_1 \\ v_2 + V_{11}(i_{11}) \end{bmatrix},$$

$$-\mathbf{F}'_{\beta\epsilon}\mathbf{I}_\beta(\mathbf{v}_\beta) + \mathbf{i}_\epsilon = \begin{bmatrix} i_{11} - I_2(v_2) \\ i_{12} \end{bmatrix}. \tag{53}$$

Hence, the nonlinear transformation of (18) becomes

$$\mathbf{F}(\mathbf{x}_{\beta\epsilon}) = \begin{bmatrix} v_1 \\ v_2 + (i_{11}^3 - i_{11}) \\ i_{11} - v_2 \\ i_{12} \end{bmatrix} = \begin{bmatrix} y_1 \\ y_2 \\ y_3 \\ y_4 \end{bmatrix} \equiv \mathbf{y}. \tag{54}$$

Although the resistance R_2 is described by a non-monotonic curve, $\mathbf{F}(\mathbf{x}_{\beta\epsilon})$ is a one-to-one operator from R^4 *onto* R^4 and, consequently, possesses an inverse

$$\mathbf{x}_{\beta\epsilon} = \begin{bmatrix} v_1 \\ v_2 \\ i_{11} \\ i_{12} \end{bmatrix} = \begin{bmatrix} y_1 \\ -y_3 + (y_2 + y_3)^{1/3} \\ (y_2 + y_3)^{1/3} \\ y_4 \end{bmatrix}. \tag{55}$$

Hence, all of the hypotheses of Theorem 1 are satisfied and the normal form exists over the *entire* dynamic space. The vector $\mathbf{y}$ is

$$\mathbf{y} \equiv \begin{bmatrix} -\mathbf{F}_{\beta\delta}\mathbf{V}_\delta(\mathbf{q}_\delta) + \mathbf{E}_\beta \\ \mathbf{F}'_{\gamma\epsilon}\mathbf{I}_\gamma(\varphi_\gamma) + \mathbf{T}_\epsilon \end{bmatrix} = \begin{bmatrix} -V_7(q_7) + V_8(q_8) \\ V_8(q_8) + \zeta(t) \\ -I_4(\varphi_4) + j_{11}(t) \\ I_5(\varphi_5) + I_6(\varphi_6) \end{bmatrix}, \tag{56}$$

which follows from (15) and (16). Hence, from (54)–(56),

$$\mathbf{x}_\beta(\varphi_\gamma, \mathbf{q}_\delta, \mathbf{E}_\beta, \mathbf{T}_\epsilon) \equiv \begin{bmatrix} v_1 \\ v_2 \end{bmatrix}$$

$$= \begin{bmatrix} -V_7(q_7) + V_8(q_8) \\ I_4(\varphi_4) + j_{11}(t) + [V_8(q_8) + e_2(t) - I_4(\varphi_4) + j_{11}(t)]^{1/3} \end{bmatrix} \tag{57}$$

and

$$\mathbf{x}_\epsilon(\varphi_\gamma, q_\delta, E_\beta, T_\epsilon) \equiv \begin{bmatrix} i_{11} \\ i_{12} \end{bmatrix}$$

$$= \begin{bmatrix} [V_8(q_8) + e_2(t) - I_4(\varphi_4) + j_{11}(t)]^{1/3} \\ I_5(\varphi_5) + I_6(\varphi_6) \end{bmatrix}. \tag{58}$$

The desired normal form equations follow immediately when the expressions are substituted into (50) and (51),

$$\dot{\varphi}_3 = -V_7(q_7) + V_8(q_8) + e_3(t)$$

$$\dot{\varphi}_4 = V_9(q_9) + e_2(t) - I_4(\varphi_4) + j_{11}(t) - [V_8(q_8) + e_2(t) - I_4(\varphi_4) + j_{11}(t)]^{1/3}$$

$$
\begin{aligned}
\dot{\varphi}_5 &= V_9(q_9) - V_{10}(q_{10}) - V_{12} \circ [I_5(\varphi_5) + I_6(\varphi_6)] \\
\dot{\varphi}_6 &= V_9(q_9) - V_{12} \circ [I_5(\varphi_5) + I_6(\varphi_6)] \\
\dot{q}_7 &= I_3(\varphi_3) + I_1 \circ [V_8(q_8) - V_7(q_7)] \\
\dot{q}_8 &= -I_3(\varphi_3) - I_1 \circ [V_8(q_8) - V_7(q_7)] - j_{11}(t) \\
&\quad - [V_8(q_8) + e_2(t) - I_4(\varphi_4) + j_{11}(t)]^{1/3} \\
\dot{q}_9 &= -I_4(\varphi_4) - I_5(\varphi_5) - I_6(\varphi_6) \\
\dot{q}_{10} &= I_5(\varphi_5) - j_{10}(t).
\end{aligned} \tag{59}
$$

This example clearly points out the significance of Theorem 1; it does not satisfy the hypotheses of Theorems 2 or 3 or Corollaries 3 or 4, but it does satisfy the hypotheses of Theorem 1. Furthermore, the example shows that the normal form can exist over the entire dynamic space even if the resistances in $\mathfrak{N}_{\beta\epsilon}$ are not characterized by monotonic curves. One should not infer, however, that the monotonicity requirements of Theorems 2 and 3 and Corollary 3 are too restrictive in general; the tightness ("almost necessity") of such conditions and the method of parasitic element argumentation is illustrated in the following example.

Consider the network shown in Fig. 5(a), where R_1 is a nonlinear resistance characterized by

$$v_1 = i_1^3. \tag{60}$$

The problem here is to determine whether the normal form exists over the entire dynamic space. Since (60) gives v_1 as a single-valued function of i_1, there is no need for parametrization and i_1 and i_2 are taken to be the parameters x_1 and x_2, respectively. The first analysis step requires the choice of a normal tree. There are but two possibilities: C and R_1 form the normal tree; or C and R_2 form the normal tree. In either case, the mesh formed by one (link-) resistance necessarily contains also the other (tree-) resistance; hence, $\mathbf{F}_{\beta\epsilon} \neq \mathbf{0}$ and $\mathfrak{N}_{\beta\epsilon}$ consists of R_1 and R_2. Hypothesis 6) of Theorem 3 is not satisfied since the resistance R_1 is a monotone-increasing function but the resistance R_2 is a monotone-decreasing function. Neither is the hypothesis of Corollary 4 satisfied. Consequently, one must resort to Theorem 1. Since hypotheses 2) to 5) of Theorem 1 are satisfied (by inspection), only hypothesis 1) need be examined. Equation (15) becomes

$$i_1^3 - 3i_2 = v \tag{61}$$

and (16) becomes

$$i_1 - i_2 = 0; \tag{62}$$

the transformation, (18), becomes

$$\mathbf{F}(\mathrm{i}) = \begin{bmatrix} f_1(i_1, i_2) \\ f_2(i_1, i_2) \end{bmatrix}, \tag{63}$$

where

$$f_1(i_1, i_2) = i_1^3 - 3i_2 \tag{64}$$

and

$$f_2(i_1, i_2) = i_1 - i_2. \tag{65}$$

It is easy to see that the transformation $\mathbf{F}$ is not one-to-one, for $\mathbf{F}$ maps the two points (1, 1) and $(-2, -2)$ into the same point $(-2, 0)$. Since there are only two normal trees, neither of which satisfy hypothesis 1) of Theorem 1, the normal form does not exist over the entire dynamic space.

For this simple network one can alternately come to the same conclusion by writing down the following equations:

$$v_1^{1/3} = -1/3v_2; \tag{66}$$

$$v = v_1 - 3v_1^{1/3}; \tag{67}$$

and

$$\frac{dv}{dt} = -v_1^{1/3}. \tag{68}$$

In order for the normal form to exist over the entire dynamic space it is necessary and sufficient to be able to solve for v_1 in terms of v in (67). But this solution is clearly impossible since $v(0) = v(3\sqrt{3})$ implies that v is not a single-valued function of v_1. Hence, one can more simply arrive at the same conclusion as obtained earlier by the use of Theorem 1. Of course, for this simple network the use of Theorem 1 is not necessary, but for more complicated networks it becomes a useful tool of

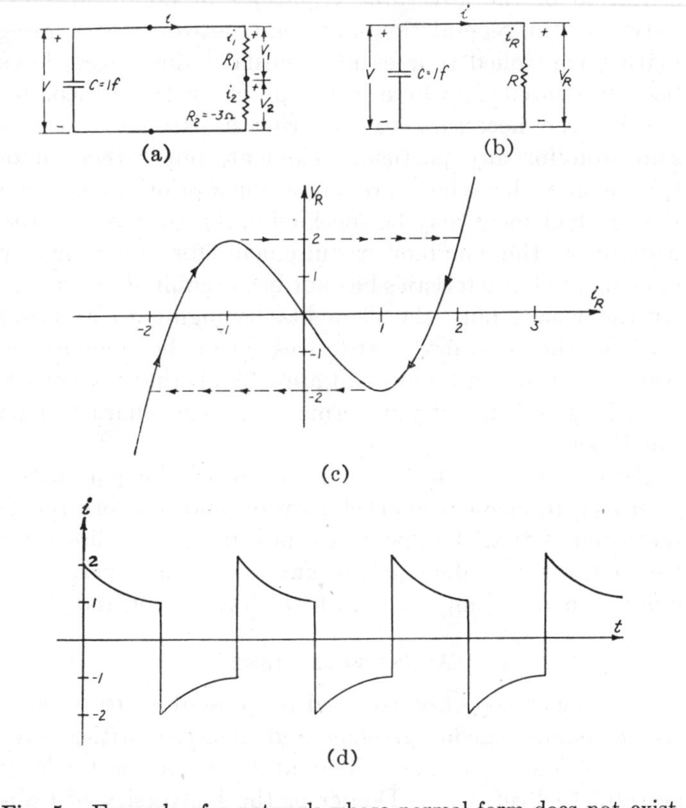

Fig. 5. Example of a network whose normal form does not exist.

analysis. Observe further that this example clearly points out the fact that hypothesis 6) of Theorem 3 is a reasonably weak condition since there are networks, that of the present example being one, for which this hypothesis is a necessary as well as a sufficient condition for the existence of the normal form throughout the dynamic space. Moreover, to show that hypothesis 2) of Theorem 3 is also a reasonably weak condition, one need only replace the network of Fig. 5(a) by its equivalent network in Fig. 5(b), where the composite resistance is now specified by

$$v_R = i_R^3 - 3i_R. \tag{69}$$

The characteristic curve which represents this element is shown in Fig. 5(c), and the element is seen to be current-controlled. The network response is readily found to be the periodic waveform shown in Fig. 5(d). The appearance of abrupt jumps in the waveform is a phenomenon characteristic of all astable multivibrators. Since this jump phenomenon occurs, it is clear that the normal form does not exist over the entire dynamic space for no single differential equation in normal form can admit of a discontinuous solution [13]. Observe that hypothesis 3) of Theorem 3 is not satisfied by the network of Fig. 5(b), and, hence, this hypothesis is not too restrictive in general. Finally, to complete the illustration, the network of Fig. 5(b) is next augmented by a small linear inductance in accordance with Theorem 4; the resulting network is shown in Fig. 6(a). The normal form now exists over the entire dynamic space:

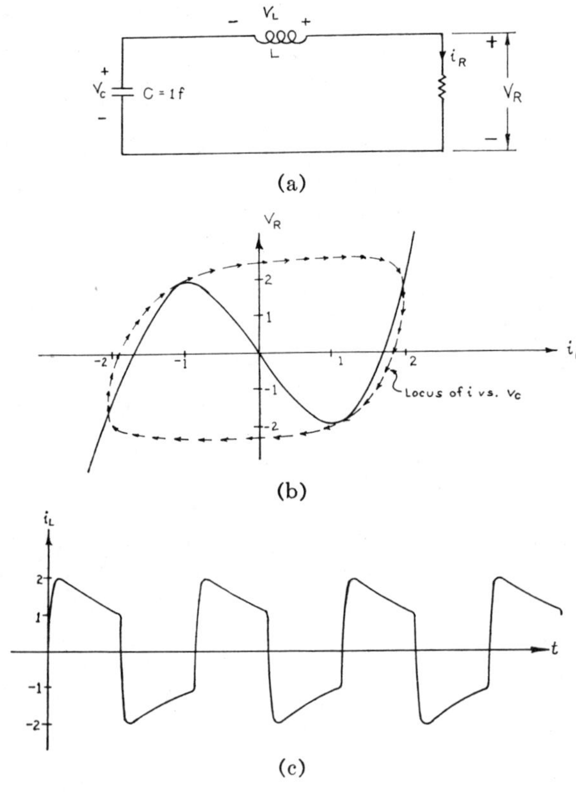

Fig. 6. Example illustrating the existence of normal form by the addition of parasitic elements.

$$\frac{dv_c}{dt} = -i_L \tag{70}$$

and

$$\frac{di_L}{dt} = \frac{1}{L}(v_c - i_L^3 + 3i_L). \tag{71}$$

Since the inductance L is very small its effect is only significant during the switching instant; hence, the solution of (70) and (71) can be found by standard graphical procedures [17]. The graphical solution is sketched in Fig. 6(b) and the solution is given in Fig. 6(c). Figure 6(c) is almost identical to Fig. 5(d) with the significant exception that no jump phenomenon appears in the former. The solution curve now represents a continuous function of time, as it should because the network may now be described by the normal form (70) and (71), the solution of which must necessarily be continuous.

V. Conclusion

The parametric approach has been seen to be a useful concept for the representation of a broad class of nonlinear elements. In particular, when multivalued elements are present the parametric approach represents the only method presently available for the formulation of the dynamic equations which govern network behavior. More specifically, any network containing only unicursal elements and sources (including controlled sources) admits of a parametric representation.

Several theorems regarding the normal form representation of the governing equations of nonlinear RLC networks (in general time-invariant networks containing neither controlled sources nor mutual inductances) have been presented. Nowhere in the proofs of these theorems has it been necessary to construct a parametric representation for any particular element, but rather all of the theorems have been proven on the a priori assumption that each element may be specified in its parametric form. Moreover, the common requirement for monotonicity of element characteristics has not been required in general. At the worst, only the elements belonging to a subset $\mathfrak{N}_{\beta\epsilon}$ of the network resistances need be monotonic. Another feature of the present approach is that the normal form is given explicitly in terms of element characteristic functions.

Perhaps the greatest single feature of the parametric approach to element characterization and network representation is that it appears to indicate future directions which must be taken in the analysis of networks even more general than those which have appeared here.

Acknowledgment

This paper has only attained its present state through the constant encouragement and incisive criticism of many of our colleagues. The authors are particularly grateful to Prof. C. A. Desoer of the University of California, Berkeley; Dr. J. Katzenelson of the Bell Telephone Laboratories, Murray Hill, N. J.; Profs. J. B. Cruz, Jr. and M. E. Van Valkenburg of the University of Illinois, Urbana; Profs. L. V. Auth and R. W. Liu of the University of Notre Dame, Ind.; Prof. D. C. Youla of the Polytechnic Institute of Brooklyn, N. Y.; and Profs. D. R. Anderson and B. J. Leon of Purdue University, Lafayette, Ind.

References

[1] R. J. Duffin, "Nonlinear networks (I, II, and IIb)," *Bull. Amer. Math. Soc.*, vols. 52, 53 and 54, pp. 836–838, 963–971, and 119–127; 1946, 1947 and 1948.
[2] G. Birkhoff and J. B. Diaz, "Nonlinear network problems," *Quart. Appl. Math.*, vol. 13, pp. 431–443, January 1956.
[3] G. J. Minty, "Solving steady-state nonlinear networks of 'monotone' elements," *IRE Trans. on Circuit Theory*, vol. CT-8, pp. 99–104, June 1961.
[4] R. Liu and L. V. Auth, "Qualitative synthesis of nonlinear networks," 1963 *Proc. First Annual Allerton Conf. on Circuit and System Theory*, pp. 330–343.
[5] R. K. Brayton and J. K. Moser, "A theory of nonlinear networks," *Quart. Appl. Math.*, vol. 12, pp. 1–33, 81–104, April 1964.
[6] C. A. Desoer and J. Katzenelson, "Nonlinear RLC Networks," *Bell Sys. Tech. J.*, pp. 161–198, January 1965.
[7] T. E. Stern, "On the equations of nonlinear networks," to be published in *IEEE Trans. on Circuit Theory*, March 1966.
[8] J. K. Moser, "Bistable systems of differential equations with applications to tunnel diode circuits," *IBM J. Res. and Dev.*, vol. 5, p. 2, July 1961.
[9] T. M. Apostol, *Mathematical Analysis*. Reading, Mass.: Addison-Wesley, 1956.
[10] E. J. McShane, *Integration*, sec. 34. Princeton, N. J.: Princeton University Press, 1944.
[11] P. Penfield, Jr., *Frequency Power Formulas*. New York: Wiley, 1960.
[12] S. Seshu and M. B. Reed, *Linear Graphs and Electrical Networks*. Reading, Mass.: Addison-Wesley, pp. 88–114, 1961.
[13] E. A. Coddington and N. Levinson, *Theory of Ordinary Differential Equations*. New York: McGraw-Hill, pp. 21–22, 1955.
[14] P. R. Bryant, "The order of complexity of electrical networks," *Proc. IEE (London)*, Monograph 335E, vol. 106 C, pp. 174–188, June 1959.
[15] L. A. Zadeh and C. A. Desoer, *Linear System Theory—The State Space Approach*. New York: McGraw-Hill, 1963.
[16] W. Fulks, *Advanced Calculus*. New York: McGraw-Hill, 1956.
[17] H. J. Zimmermann and S. J. Mason, *Electronic Circuit Theory*. New York: Wiley, 1959.

Some Theorems on Properties of DC Equations of Nonlinear Networks

By I. W. SANDBERG and A. N. WILLSON, JR.

(Manuscript received July 3, 1968)

Several results are presented concerning the equation $F(x) + Ax = B$ (with $F(\cdot)$ a "diagonal" nonlinear mapping of real Euclidean n-space E^n into itself, and A a real $n \times n$ matrix) which plays a central role in the dc analysis of transistor networks. In particular, we give necessary and sufficient conditions on A such that the equation possesses a unique solution x for each real n-vector B and each strictly monotone increasing $F(\cdot)$ that maps E^n onto itself.

There are several direct circuit-theoretic implications of the results. For example, we show that if the short-circuit admittance matrix G of the linear portion of the dc model of a transistor network satisfies a certain dominance condition, then the network cannot be bistable. Therefore, a fundamental restriction on the G matrix of an interesting class of switching circuits is that it must violate the dominance condition.

I. INTRODUCTION

For each positive integer n let $\mathfrak{F}^n$ denote that collection of mappings of the real n-dimensional Euclidean space E^n onto itself, defined by: $F \varepsilon \mathfrak{F}^n$ if and only if there exist, for $i = 1, \cdots, n$, strictly monotone increasing functions f_i mapping E^1 onto E^1 such that, for each $x \equiv (x_1, \cdots, x_n)^t \varepsilon E^n$, $F(x) = (f_1(x_1), \cdots, f_n(x_n))^t$.

The main purpose of this paper is to report on some results concerning properties of the equation

$$F(x) + Ax = B, \tag{1}$$

where A is an $n \times n$ matrix of real numbers, F maps E^n into E^n, and $B \in E^n$. In particular, a condition to be satisfied by A is given which is both necessary and sufficient to guarantee that for each $F \in \mathfrak{F}^n$ and each $B \in E^n$ there exists a unique solution of equation (1).

We also study the problem of obtaining bounds on the solution of equation (1). These bounds show that (if $F \in \mathfrak{F}^n$ and our condition on A is satisfied) the solution depends continuously on B. The bounds are often of use in computing the solution by standard iteration methods such as the Newton-Raphson method. By appealing to a theorem of R. S. Palais it is shown that the bounds can also be used to obtain a theorem essentially the same as, but somewhat weaker than, our principal result.

Several results can be found in the literature which specify sufficient conditions for the existence of a unique solution of equation (1). For example, if A is positive semidefinite then a special case of a theorem of Ref. 1 guarantees the existence of a unique solution of equation (1) for all those $F \in \mathfrak{F}^n$ which have the property that the slope of each f_i is bounded from above and below by positive constants, and for all $B \in E^n$. This theorem also specifies that a certain iteration scheme will always converge to the solution.

A theorem of G. J. Minty[2], when applied to equation (1), also implies essentially the same result. The boundedness condition on the slopes of the functions f_i is not required by Minty's theorem. On the other hand, Minty's theorem does not provide a procedure for computing the solution of equation (1).

In Ref. 3 it is proved that a sufficient condition for the existence and uniqueness of a solution of equation (1) for all $F \in \mathfrak{F}^n$ and $B \in E^n$ is that A satisfy a weak row-sum dominance condition:

$$a_{ii} \geqq \sum_{j \neq i} | a_{ij} |, \qquad i = 1, \cdots, n.*$$

Other information concerning the location and the computation of the solution is also given in Ref. 3.

The class of matrices satisfying the condition of our theorem (which is defined in Section III and denoted by P_0) includes all positive semidefinite matrices as well as all matrices which satisfy any one of several

* Appendix A contains a simpler proof of a similar result and a proof of a new related result. These results specify convergent algorithms for obtaining the solution.

dominance conditions. Many other matrices are included in P_0; and since the condition of our theorem is both a necessary and a sufficient one, we are assured that P_0 is the largest class of matrices A for which equation (1) has a unique solution for all $F \varepsilon \mathfrak{F}^n$ and all $B \varepsilon E^n$.

II. NONLINEAR NETWORKS

Equation (1) is often encountered in the study of nonlinear electrical networks. In the case of networks containing only resistors (that is, linear resistors with nonnegative resistance), dependent and independent sources, and two-terminal nonlinear resistors that are described by functions in $\mathfrak{F}^1$ (diodes, for example), this is rather obvious.[3] Even for networks which contain more general nonlinear devices, however, equation (1) can often provide a convenient characterization. For example, D. A. Calahan shows in his recent book that the transistor network of Fig. 1 may be described by the equation

$$\begin{bmatrix} I_{es}(e^{qV_r/kT} - 1) \\ I_{cs}(e^{qV_f/kT} - 1) \end{bmatrix} + \begin{bmatrix} 0.0225 & 0.309 \\ -0.168 & 0.494 \end{bmatrix} \begin{bmatrix} V_r \\ V_f \end{bmatrix} = \begin{bmatrix} 0.00177\,V_{cc} \\ -0.188\,V_{cc} \end{bmatrix}$$

if the Ebers–Moll model is used to represent the transistor. (See pp. 13ff of Ref. 4.) In this equation I_{es}, I_{cs}, q, k, T, and V_{cc} all represent fixed real parameters. It is quite trivial to apply the theory of this paper (in particular, Corollary 3 of Section IV) to Calahan's example and prove that this equation has a solution, the solution is unique, and the solution depends continuously on V_{cc}. We also show how bounds on the solution can be obtained.

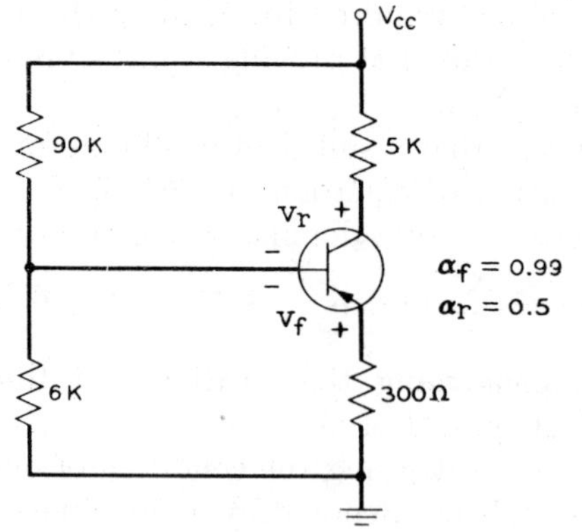

Fig. 1 — Biased transistor-stage.

More generally, it is frequently the case that networks which contain transistors, as well as the previously mentioned linear and nonlinear elements, may be described by the equation

$$TF(x) + Ax = B. \tag{2}$$

In this case, x is a vector whose components are the voltages across the nonlinear resistors and the transistor base-emitter and base-collector voltages. The $n \times n$ matrix A is the y-parameter matrix of the linear n-port network which is obtained by removing all nonlinear resistors and transistors and setting the value of each independent source to zero. The function $TF(x)$ describes the behavior of the nonlinear resistors and the transistors. It happens that the matrix T is nonsingular; therefore equation (2) can be put into the form of equation (1).

Networks which contains inductors and capacitors as well as the memoryless elements already mentioned are of course described by differential equations. Even the study of such networks, however, can often lead to the consideration of equations of the same type as equation (1). One usually finds the solution of such an equation is necessary, for example, when computing the solution of the differential equations by using some implicit numerical integration formula.

The problem of determining the equilibrium states of the above-mentioned dynamic networks is one in which the consideration of equations of type (1) often arises in perhaps a more direct manner. In this regard, if it happens that equation (1) has a unique solution, then the network cannot possibly be bistable.

When the determination of equilibrium states of a transistor network leads first to the consideration of equation (2), then as a rather direct application of our existence and uniqueness theorem it follows that if the matrix A satisfies a weak column-sum dominance condition,

$$a_{ii} \geqq \sum_{j \neq i} |a_{ji}|, \qquad i = 1, \cdots, n,$$

then $T^{-1}A \in P_0$ and hence the network has exactly one equilibrium state. This result and related results which are proved in Section IV have the following interesting corollary: One cannot synthesize a bistable network which consists of resistors, inductors, capacitors, diodes, independent voltage and current sources, and one (Ebers–Moll modeled) transistor—or even an arbitrary number of (Ebers–Moll modeled) transistors with a common base connection.

The authors feel that in many respects the main contributions of this paper are in the techniques used to prove the results. For this reason, we have not chosen to summarize all of the results at the outset and relegate proofs to later sections. But rather, the results and the proofs will appear in the order in which they will best illustrate the techniques developed.

III. MATRICES OF CLASSES P AND P_0

The following notation will be used throughout the remainder of the paper: The origin in E^n will be denoted by θ. If D is a diagonal matrix then $D > 0$ ($D \geqq 0$) means that each element of D on the main diagonal is positive (nonnegative).

In Ref. 5 and Ref. 6 M. Fiedler and V. Pták define the classes of matrices denoted by P and P_0. They in fact prove that the following properties of a square matrix A are equivalent:

(*i*) All principal minors of A are positive.

(*ii*) For each vector $x \neq \theta$ there exists an index k such that $x_k y_k > 0$ where $y = Ax$.

(*iii*) For each vector $x \neq \theta$ there exists a diagonal matrix $D_x > 0$ such that the scalar product $\langle Ax, D_x x\rangle > 0$.

(*iv*) For each vector $x \neq \theta$ there exists a diagonal matrix $H_x \geqq 0$ such that $\langle Ax, H_x x\rangle > 0$.

(*v*) Every real eigenvalue of A, as well as of each principal submatrix of A, is positive.

The class of all matrices satisfying one of the above conditions is denoted by P. Fiedler and Pták prove that the following properties of a square matrix A are also equivalent:

(*i*) All principal minors of A are nonnegative.

(*ii*) For each vector $x \neq \theta$ there exists an index k such that $x_k \neq 0$ and $x_k y_k \geqq 0$ where $y = Ax$.

(*iii*) For each vector $x \neq \theta$ there exists a diagonal matrix $D_x \geqq 0$ such that $\langle x, D_x x\rangle > 0$ and $\langle Ax, D_x x\rangle \geqq 0$.

(*iv*) Every real eigenvalue of A, as well as of each principal submatrix of A, is nonnegative.

The class of all matrices satisfying one of the above conditions is denoted by P_0.

The following theorems follow directly from the above definitions.

Theorem 1. If $A \in P_0$ then for every diagonal matrix $\Delta \geqq 0$ ($\Delta > 0$), $\Delta + A \in P_0$ ($\Delta + A \in P$).

Proof: Let $x \neq \theta$. Then, since $A \in P_0$, there exists an index k such that $x_k \neq 0$ and $x_k(Ax)_k \geqq 0$. Thus, $x_k(\Delta x + Ax)_k \geqq 0$ (>0). □

In particular, Theorem 1 implies that if $A \varepsilon P_0$ and $\Delta \geqq 0$ ($\Delta > 0$) then det $(\Delta + A) \geqq 0$ (>0).

Theorem 2. If $A \varepsilon P$ *then* $A^{-1} \varepsilon P$.

Proof: Suppose $A \varepsilon P$. Let $x \neq \theta$ be given and let $y = A^{-1}x$. $y \neq \theta$ since A^{-1} is nonsingular. Thus, there exists a diagonal matrix $D > 0$ such that $\langle Ay, Dy\rangle > 0$, which implies $\langle x, DA^{-1}x\rangle > 0$, or $\langle Dx, A^{-1}x\rangle > 0$, or $\langle A^{-1}x, Dx\rangle > 0$. That is, for every $x \neq \theta$ there exists $D > 0$ such that $\langle A^{-1}x, Dx\rangle > 0$. Hence $A^{-1} \varepsilon P$. □

Because of the similarity of the definitions of the classes of matrices P and P_0, one might conjecture that this proposition is also true: *If* $A \varepsilon P_0$, *and det* $A \neq 0$, *then* $A^{-1} \varepsilon P_0$. This conjecture is in fact true. Interestingly enough, however, its proof is not obtained as one might at first suspect, by simply modifying the proof of Theorem 2. Moreover, the proof of this conjecture does not even seem to follow directly from any of the above definitions of P_0. Rather, upon making the trivial observation that for every diagonal matrix $D > 0$, det $(A^{-1} + D) =$ det (A^{-1}) $\cdot$det $(D^{-1} + A)\cdot$det (D), the conjecture is easily seen to follow from the fact that det $(D + A) \neq 0$ for every diagonal $D > 0$ if and only if $A \varepsilon P_0$. This fact is a direct corollary to the proof of Theorem 3.

IV. EXISTENCE AND UNIQUENESS THEOREM

The following theorem is the principal result of this paper.

Theorem 3. If A is an $n \times n$ *matrix then there exists a unique solution of equation* (1) *for each* $F \varepsilon \mathfrak{F}^n$ *and for each* $B \varepsilon E^n$ *if and only if* $A \varepsilon P_0$.

Proof: (*if*) Let $A \varepsilon P_0$, $F \varepsilon \mathfrak{F}^n$, and $B \varepsilon E^n$. The solution of equation (1) is then unique (if it exists) since if x and y are both solutions then, using the strict monotonicity property of F, there exists a diagonal matrix $D > 0$ such that $F(x) - F(y) = D(x - y)$. But $[D + A](x - y) = \theta$ and, by Theorem 1, $D + A$ is nonsingular. This means that $x = y$.

We prove the existence of a solution of equation (1) by induction. For $k = 1, \cdots, n$, let

$$F_k(x) = \begin{bmatrix} f_1(x_1) \\ \vdots \\ f_k(x_k) \end{bmatrix}, \qquad A_k = \begin{bmatrix} a_{11} & \cdots & a_{1k} \\ \cdot & \cdot & \cdot \\ a_{k1} & \cdots & a_{kk} \end{bmatrix}, \qquad B_k = \begin{bmatrix} b_1 \\ \vdots \\ b_k \end{bmatrix}.$$

Clearly, $A_k \varepsilon P_0$, $F_k \varepsilon \mathfrak{F}^k$, and $B_k \varepsilon E^k$. Also, it is clear that there exists a unique solution of $F_1(x) + A_1x = B_1$ for each $F_1 \varepsilon \mathfrak{F}^1$ and for each $B_1 \varepsilon E^1$, and that this solution is a continuous function of b_1.

Assume that there exists a unique solution of $F_k(x)+A_kx=B_k$ for each $F_k \varepsilon \mathfrak{F}^k$, $B_k \varepsilon E^k$, and that this solution depends continuously on any scalar parameter η upon which B_k depends continuously. Let the matrices $A_{k,k+1}$ and $A_{k+1,k}$ be defined by

$$A_{k,k+1} = \begin{bmatrix} a_{1,k+1} \\ \vdots \\ a_{k,k+1} \end{bmatrix},$$

$$A_{k+1,k} = [a_{k+1,1} \cdots a_{k+1,k}].$$

Then, for every real number x_{k+1}, the equation

$$F_k(x) + A_kx + A_{k,k+1}x_{k+1} = B_k \tag{3}$$

has a (unique) solution which is a continuous function of x_{k+1} and of η. Let the components of this solution be denoted by $x_i = m_i(x_{k+1}, \eta)$, for $i = 1, \cdots, k$, and define the vector $M_k(x_{k+1}, \eta)$ by $M_k = (m_1, \cdots, m_k)^t$.

We now prove that the function

$$\varphi(x_{k+1}, \eta) \equiv A_{k+1,k}M_k(x_{k+1}, \eta) + a_{k+1,k+1}x_{k+1} - b_{k+1}(\eta)$$

is monotone increasing in x_{k+1}: Let $x^1_{k+1}, x^2_{k+1} \varepsilon E^1$ with $x^1_{k+1} < x^2_{k+1}$. Then, if M^1 (M^2) denotes the solution of equation (3) when $x_{k+1} = x^1_{k+1}$ (x^2_{k+1}), we have

$$F_k(M^2) - F_k(M^1) + A_k(M^2 - M^1) + A_{k,k+1}(x^2_{k+1} - x^1_{k+1}) = \theta.$$

Because of the strict monotonicity of the function F_k, however, there exists a $k \times k$ diagonal matrix $\Delta > 0$ such that

$$F_k(M^2) - F_k(M^1) = \Delta(M^2 - M^1).$$

Hence,

$$M^2 - M^1 = -[\Delta + A_k]^{-1}A_{k,k+1}(x^2_{k+1} - x^1_{k+1}).$$

Thus,

$$\varphi(x^2_{k+1}) - \varphi(x^1_{k+1}) = \{a_{k+1,k+1} - A_{k+1,k}[\Delta + A_k]^{-1}A_{k,k+1}\}(x^2_{k+1} - x^1_{k+1}).$$

But then, from the easily verified relation

$$a_{k+1,k+1} - A_{k+1,k}[\Delta + A_k]^{-1}A_{k,k+1} = \frac{\det\left(\begin{bmatrix} \Delta & \begin{matrix} 0 \\ \vdots \end{matrix} \\ 0 \cdots & 0 \end{bmatrix} + A_{k+1}\right)}{\det(\Delta + A_k)}$$

and since

$$\det(\Delta + A_k) > 0, \qquad \det\left\{\begin{bmatrix} \Delta & \begin{matrix}0\\ \vdots\end{matrix} \\ 0 \cdots & 0 \end{bmatrix} + A_{k+1}\right\} \geqq 0 \text{ (Theorem 1)},$$

and $x_{k+1}^2 - x_{k+1}^1 > 0$, it follows that $\varphi(x_{k+1}^2) \geqq \varphi(x_{k+1}^1)$.

Now since φ is monotone increasing and, obviously, continuous in x_{k+1}, it follows that the left side of the equation

$$f_{k+1}(x_{k+1}) + \varphi(x_{k+1}) = 0 \tag{4}$$

is a strictly monotone increasing function mapping E^1 onto E^1, and hence equation (4) has a unique solution. If x_{k+1}^0 denotes this solution then

$$x^0 \equiv \begin{bmatrix} m_1(x_{k+1}^0) \\ \vdots \\ m_k(x_{k+1}^0) \\ x_{k+1}^0 \end{bmatrix}$$

is the (unique) solution of

$$F_{k+1}(x) + A_{k+1}x = B_{k+1}.$$

We must now prove that this solution is a continuous function of any scalar parameter η upon which B_{k+1} depends continuously. It suffices to prove that x_{k+1} depends continuously on η (see equation (3)). This may be done as follows:

Let x_{k+1}^0 be the solution of equation (4) corresponding to $\eta = \eta^0$. That is, let

$$f_{k+1}(x_{k+1}^0) + \varphi(x_{k+1}^0, \eta^0) = 0,$$

and let $\epsilon > 0$ be given. Since f_{k+1} is a strictly monotone increasing mapping of E^1 onto E^1, so is f_{k+1}^{-1}, and hence f_{k+1}^{-1} is continuous. Hence, there exists $\delta' > 0$ such that if $| f_{k+1}(x_{k+1}^0) - f_{k+1}(x_{k+1}) | < \delta'$ then $| x_{k+1}^0 - x_{k+1} | < \epsilon$. Since φ is a continuous function of η, there exists $\delta > 0$ such that $| \eta^0 - \eta | < \delta$ implies $| \varphi(x_{k+1}^0, \eta^0) - \varphi(x_{k+1}^0, \eta) | < \delta'$. If $| \eta^0 - \eta | < \delta$, and

$$f_{k+1}(x_{k+1}) + \varphi(x_{k+1}, \eta) = 0,$$

then,

$$f_{k+1}(x^0_{k+1}) - f_{k+1}(x_{k+1}) + \varphi(x^0_{k+1}, \eta) - \varphi(x_{k+1}, \eta)$$
$$= -[\varphi(x^0_{k+1}, \eta^0) - \varphi(x^0_{k+1}, \eta)].$$

But since both f_{k+1} and φ are monotone increasing in x_{k+1},

$$(x^0_{k+1} - x_{k+1})[f_{k+1}(x^0_{k+1}) - f_{k+1}(x_{k+1})] \geqq 0,$$

and

$$(x^0_{k+1} - x_{k+1})[\varphi(x^0_{k+1}, \eta) - \varphi(x_{k+1}, \eta)] \geqq 0.$$

Therefore,

$$| (x^0_{k+1} - x_{k+1})[f_{k+1}(x^0_{k+1}) - f_{k+1}(x_{k+1})] |$$
$$\leqq | (x^0_{k+1} - x_{k+1})[\varphi(x^0_{k+1}, \eta^0) - \varphi(x^0_{k+1}, \eta)] |.$$

Now, if $x^0_{k+1} = x_{k+1}$ then of course $| x^0_{k+1} - x_{k+1} | < \epsilon$. Otherwise,

$$| f_{k+1}(x^0_{k+1}) - f_{k+1}(x_{k+1}) | \leqq | \varphi(x^0_{k+1}, \eta^0) - \varphi(x^0_{k+1}, \eta) |.$$

But then,

$$| f_{k+1}(x^0_{k+1}) - f_{k+1}(x_{k+1}) | < \delta',$$

and hence $| x^0_{k+1} - x_{k+1} | < \epsilon$. Thus, x_{k+1} is a continuous function of η.

(*only if*) Suppose $A \notin P_0$. If $\det A < 0$ then for sufficiently small $\zeta > 0$, $\det (\zeta I + A) < 0$. For sufficiently large ζ, however,

$$\det (\zeta I + A) = \zeta^n \cdot \det \left(I + \frac{1}{\zeta} A\right) > 0.$$

Thus, since $\det (\zeta I + A)$ is a continuous function of ζ, there is some value of $\zeta > 0$ such that $\det (\zeta I + A) = 0$. For this value of ζ let $F(x) = \zeta I x$. Clearly, for this choice of $F \in \mathfrak{F}^n$, equation (1) cannot have a unique solution.

If $\det A \geqq 0$, but A has a negative principal minor, we can still find a diagonal matrix $\Delta > 0$ such that $\det (\Delta + A) = 0$; however, in this case Δ will not, in general, be simply the identity matrix multiplied by a positive constant ζ.

For some positive integer $k < n$ let A have a $k \times k$ principal minor which is negative and let

$$\Delta^{(1)} = \text{diag} [\delta_1, \cdots, \delta_n].$$

Since the determinant of $\Delta + A$ is not altered if any two rows and then the corresponding pair of columns are interchanged we may, without loss of generality, assume that the matrix A is partitioned as

$$A = \begin{bmatrix} A_1 & A_2 \\ A_3 & A_4 \end{bmatrix},$$

where A_1 is a $k \times k$ matrix with $\det A_1 < 0$. Let $\xi > 0$ be chosen so small that $\det(\xi I + A_1) < 0$, and let $\delta_1 = \cdots = \delta_k = \xi$. Now, if $\delta_{k+1} = \cdots = \delta_n = \zeta > 0$, then

$$\det(\Delta^{(1)} + A) = \det \begin{bmatrix} \xi I + A_1 & A_2 \\ A_3 & \zeta I + A_4 \end{bmatrix}$$

$$= \zeta^{n-k} \cdot \det \begin{bmatrix} \xi I + A_1 & A_2 \\ \frac{1}{\zeta} A_3 & I + \frac{1}{\zeta} A_4 \end{bmatrix}.$$

Thus, for $\zeta > 0$ chosen to be sufficiently large, $\det(\Delta^{(1)} + A) < 0$. ($\det(\Delta^{(1)} + A) \to \zeta^{n-k} \cdot \det(\xi I + A_1) < 0$ as $\zeta \to \infty$.) Now, if for $\eta > 0$, $\Delta^{(2)} = \eta I$, then it is clear that for η chosen sufficiently large, $\det[\Delta^{(2)} + A] = \eta^n \cdot \det(I + (1/\eta)A) > 0$. Thus, if

$$\Delta(\epsilon) = \epsilon \Delta^{(1)} + (1 - \epsilon)\Delta^{(2)},$$

it is clear, since $\det[\Delta(0) + A] > 0$ and $\det[\Delta(1) + A] < 0$ and since $\det[\Delta(\epsilon) + A]$ is a continuous function on $0 \leqq \epsilon \leqq 1$, that there is a value of $\epsilon > 0$ ($0 < \epsilon < 1$) such that $\det[\Delta(\epsilon) + A] = 0$. For this value of ϵ, $\Delta(\epsilon) > 0$ is the required diagonal matrix. □

Notice that our proof shows that if $F \in \mathfrak{F}^n$ and $A \in P_0$, then the solution of equation (1) depends continuously on any scalar parameter upon which B depends continuously. The arguments of Section V show, under these assumptions on F and A, that the operator $(F + A)^{-1}$ is in fact a continuous map of E^n into itself.

In the proof of Theorem 3 we see that the uniqueness of the solution follows simply from the hypotheses that each f_i is strictly monotone increasing and that $A \in P_0$. The additional hypotheses that each f_i is continuous and maps E^1 *onto* E^1 are not necessary (continuity of each f_i is not explicitly hypothesized, but follows from the "monotonicity" and "onto" hypotheses). Hence, we have:

Corollary 1. If, for $i = 1, \cdots, n$, S_i is a subset of E^1, and if $S = S_1 \times \cdots \times S_n$, and if $F(x) = (f_1(x_1), \cdots, f_n(x_n))^t$, where each f_i maps E^1 into E^1 and is strictly monotone increasing on S_i, then if $A \in P_0$ and $B \in E^n$, there exists at most one solution of equation (1) *in S.*

We now prove another interesting corollary of Theorem 3. We first define some additional notation.

For each positive integer n let $\mathcal{S}^n$ denote the collection of all subsets of E^n defined by: $S \in \mathcal{S}^n$ if and only if $S = S^1 \times \cdots \times S^n$ where, for $i = 1, \cdots, n$, $S^i \subset E^1$ and S^i has the same cardinality as E^1. For each $S \subset \mathcal{S}^n$ we define the collection $\mathfrak{F}^n(S)$ of functions mapping S onto E^n by: $F \in \mathfrak{F}^n(S)$ if and only if there exist, for $i = 1, \cdots, n$, strictly monotone increasing functions f_i mapping S^i onto E^1 such that for each $x \in S^n$, $F(x) = (f_1(x_1), \cdots, f_n(x_n))^t$.

Corollary 2. If A is an $n \times n$ matrix and the collection $\mathfrak{F}^n(S)$ is non-empty then there exists a unique solution of the equation

$$F_1(x) + AF_2(x) = B \tag{5}$$

for each $F_1 \in \mathfrak{F}^n(S)$, $F_2 \in \mathfrak{F}^n(S)$, and each $B \in E^n$ if and only if $A \in P_0$.

Proof: Since $F_2 \in \mathfrak{F}^n(S)$, $F_2^{-1} : E^n \to S$ exists and $F_1 \circ F_2^{-1} \in \mathfrak{F}^n$. Thus, there exists a unique solution of equation (5) if and only if there exists a unique solution of

$$F_1(F_2^{-1}(y)) + Ay = B. \quad \square$$

As special cases of Corollary 2 we have: there exists a unique solution of each of the equations

$$F_1(x) + AF_2(x) = B,$$

and

$$x + AF(x) = B,$$

for each F_1, F_2, $F \in \mathfrak{F}^n$ and each $B \in E^n$ if and only if $A \in P_0$.

In Theorem 3 (and Corollary 2) the hypothesis that each of the functions f_i is an onto mapping is quite necessary in order to guarantee the *existence* of a solution for each $A \in P_0$. In the following example all of the hypotheses of Theorem 3 except this one are satisfied:

$$e^{x_1} + x_1 - x_2 = 1$$

$$e^{x_2} - x_1 + x_2 = -2.$$

It is of course impossible for these equations to have a solution since, by adding both sides, we find that the solution would have to satisfy

$$e^{x_1} + e^{x_2} = -1,$$

which is absurd.

Even though the functions f_i are not "onto," it is still possible to specify sufficient conditions for the existence of a unique solution of equation (5) [and equation (1)] by strengthening the hypothesis on the matrix A—namely, by requiring that $A \in P$. This is the essence of Corollary 3. We first require additional notation.

With $\mathcal{S}^n$ defined as above, we define, for each $S \in \mathcal{S}^n$, the collection of functions $\mathcal{F}_0^n(S)$ mapping S into E^n by: $F \in \mathcal{F}_0^n(S)$ if and only if there exist, for $i = 1, \cdots, n$, monotone increasing functions f_i mapping S^i onto a connected set in E^1 such that, for each $x \in S$, $F(x) = (f_1(x_1), \cdots, f_n(x_n))^t$. When $S = E^n$ we denote $\mathcal{F}_0^n(S)$ by $\mathcal{F}_0^n$.

Corollary 3. If A is an $n \times n$ matrix then there exists a unique solution of equation (5) *for each $F_1 \in \mathcal{F}_0^n(S)$, $F_2 \in \mathcal{F}^n(S)$, or $F_1 \in \mathcal{F}^n(S)$, $F_2 \in \mathcal{F}_0^n(S)$, and for each $B \in E^n$, if $A \in P$.*

Proof: If $F_2 \in \mathcal{F}^n(S)$, $F_2^{-1} : E^n \rightarrow S$ exists and $F_1 \circ F_2^{-1} \in \mathcal{F}_0^n$. Thus, in this case, there exists a unique solution of equation (5) if there exists a unique solution of

$$F_1(F_2^{-1}(y)) + Ay = B. \tag{6}$$

Now, since $A \in P$, it follows from the fact that the determinant of a matrix is a continuous function of each of its elements, that there is a matrix $A^* \in P \subset P_0$ and an $\epsilon > 0$, such that $A = \epsilon I + A^*$. Hence, equation (6) is equivalent to

$$F(y) + A^*y = B, \tag{7}$$

where we have defined

$$F(y) \equiv F_1(F_2^{-1}(y)) + \epsilon Iy.$$

But, since $F_1 \circ F_2^{-1} \in \mathcal{F}_0^n$ and $\epsilon I \in \mathcal{F}^n$, it follows that $F \in \mathcal{F}^n$. Therefore, since $A^* \in P_0$, equation (7) and hence equation (6) and hence equation (5) have unique solutions.

The case when $F_1 \in \mathcal{F}^n(S)$ and $F_2 \in \mathcal{F}_0^n(S)$ can be reduced to the case just considered by making the simple observations that, in this case, equation (5) has a unique solution if

$$A^{-1}F_1(x) + F_2(x) = A^{-1}B$$

has a unique solution, and $A \in P$ implies $A^{-1} \in P$ (Theorem 2). □

In Corollary 3 a *sufficient* condition is given for the existence of a unique solution to say equation (1) when the functions f_i which specify F are not necessarily mappings onto E^1. That the condition ($A \in P$) is not *necessary* is easily demonstrated by the counterexample: Let $F \in \mathcal{F}_0^2$ and

$B \, \varepsilon \, E^2$; then the equations

$$f_1(x_1) - x_2 = b_1 , \text{ and } f_2(x_2) + x_1 = b_2$$

have a unique solution in spite of the fact that the matrix

$$A \equiv \begin{bmatrix} 0 & -1 \\ 1 & 0 \end{bmatrix} \not\varepsilon P.$$

This is true because the function $f_2(f_1(x_1) - b_1)$ is obviously a continuous monotone increasing function of x_1 , and hence the left side of the equation

$$f_2(f_1(x_1) - b_1) + x_1 = b_2 \tag{8}$$

is a strictly monotone increasing mapping of E^1 onto E^1. Thus equation (8) has a unique solution.

V. BOUNDED SOLUTIONS AND RELATED PROBLEMS

For many systems whose behavior is described by an equation having the form of equation (1), the vector B may be regarded as the system's input and the vector x may be regarded as the system's response, or output. Thus, if a sequence $B^1, B^2, B^3, \cdots$ of input vectors for the system is given, the corresponding sequence $x^1, x^2, x^3, \cdots$ of output vectors is specified by equation (1). An important property that such systems might have is that of producing a bounded sequence of output vectors for each bounded sequence of input vectors; that is, the property that whenever an input sequence $B^1, B^2, B^3, \cdots$ is contained in some bounded region of E^n, then the corresponding output sequence $x^1, x^2, x^3, \cdots$ (exists and) also is contained in some bounded region of E^n. By considering matrices A which are not members of P_0 , it is easy to demonstrate that all equations having the form of equation (1) do not have this property. For example, if $f(x) \equiv x + e^x$ ($f \, \varepsilon \, \mathfrak{F}^1$), then the sequence of solutions of the equation $f(x) + (-1)x = b$ is unbounded, even though the sequence $b = 1, \frac{1}{2}, \frac{1}{3}, \cdots$ of inputs is bounded. The fact that one *must* resort to matrices A which are not in P_0 , and the fact that by choosing *any* $A \not\varepsilon P_0$, an example of the above kind can be constructed by an appropriate choice of $F \, \varepsilon \, \mathfrak{F}^n$, follows from our next theorem

Theorem 4. If A is an $n \times n$ matrix then $A \, \varepsilon \, P_0$ if and only if for each $F \, \varepsilon \, \mathfrak{F}^n$ and each unbounded sequence of points $x^1, x^2, x^3, \cdots$ in E^n, the corresponding sequence $B^1, B^2, B^3, \cdots$ ($B^k = F(x^k) + Ax^k$, $k = 1, 2, 3, \cdots$) is unbounded.

Proof: (*if*) If $A \notin P_0$ then, as shown in the "only if" part of the proof of Theorem 3, there exists a diagonal matrix $D > 0$ such that $D + A$ is singular. Hence, there exists some point $p \in E^n$, $p \neq \theta$, such that $Dp + Ap = \theta$. Let p_j, the j-th component of p, be nonzero. Let the diagonal elements of the matrix D be denoted by $d_1, \cdots, d_n$ and let the mapping $F \in \mathfrak{F}^n$ be defined by

$$f_i(x_i) = \begin{cases} d_i x_i, & \text{for } i \neq j, \\ d_i x_i + e^{x_i}, & \text{for } i = j. \end{cases}$$

If $p_j < 0$ let $\epsilon = 1$, if $p_j > 0$ let $\epsilon = -1$. Consider the unbounded sequence $x^1, x^2, x^3, \cdots$ defined by $x^k = k \cdot \epsilon \cdot p$, for $k = 1, 2, 3, \cdots$. The members of the corresponding sequence $B^1, B^2, B^3, \cdots$ are $B^k = (0, \cdots, 0, e^{k\epsilon p_j}, 0, \cdots, 0)^t$, $k = 1, 2, 3, \cdots$, where the j-th element of each B^k is nonzero. Since for $k = 1, 2, 3, \cdots$, $k \epsilon p_j < 0$, the sequence $B^1, B^2, B^3, \cdots$ is bounded.

(*only if*) Our proof of the "only if" part of Theorem 4 consists of proving Theorem 5 which is referred to later for another purpose. □

Theorem 5. Let $F \equiv (f_1(\cdot), \cdots, f_n(\cdot))^t \in \mathfrak{F}^n$, $A \in P_0$, *and, for* $i = 1, \cdots, n$, $\alpha_i \leqq \beta_i$ *be given. There exist, for* $i = 1, \cdots, n$, *real numbers* $\gamma_i \leqq \delta_i$ *such that for any* $B \equiv (b_1, \cdots, b_n)^t \in E^n$ *with* $\alpha_i \leqq b_i \leqq \beta_i$ *for* $i = 1, \cdots, n$, *if* x *satisfies equation* (1) *then* $\gamma_i \leqq x_i \leqq \delta_i$ *for* $i = 1, \cdots, n$.

Proof of Theorem 5: We first prove a useful lemma.

Lemma 1. Let f *be a strictly monotone increasing mapping of* E^1 *onto itself. Let* x, b, α, β *be real numbers such that* $xf(x) \leqq xb$ *with* $\alpha \leqq b \leqq \beta$. *Then* $\gamma \leqq x \leqq \delta$, *where* $\gamma = \min \{f^{-1}(\alpha), 0\}$ *and* $\delta = \max \{f^{-1}(\beta), 0\}$.

Proof: Let $\alpha \leqq b \leqq \beta$ and define $\gamma = \min \{f^{-1}(\alpha), 0\}$ and $\delta = \max \{f^{-1}(\beta), 0\}$. Let x satisfy $xf(x) \leqq xb$. Then $x(f(x) - b) \leqq 0$. Clearly, $\gamma \leqq 0 \leqq \delta$ and hence if $x = 0$ then $\gamma \leqq x \leqq \delta$. If $x > 0$ then $f(x) \leqq b \leqq \beta$ which implies $x \leqq f^{-1}(\beta) \leqq \delta$ and hence $\gamma \leqq 0 < x \leqq \delta$. If $x < 0$, then $f(x) \geqq b \geqq \alpha$ which implies $x \geqq f^{-1}(\alpha) \geqq \gamma$ and hence $\gamma \leqq x < 0 \leqq \delta$. □

(*Proof of Theorem 5*) Since $A \in P_0$ there exists $k_1 \in \{1, \cdots, n\}$ such that $x_{k_1}(Ax)_{k_1} \geqq 0$ and hence,

$$x_{k_1} b_{k_1} = x_{k_1} f_{k_1}(x_{k_1}) + x_{k_1}(Ax)_{k_1} \geqq x_{k_1} f_{k_1}(x_{k_1}).$$

Thus, by Lemma 1, there exist $\gamma_{k_1}^{(1)} = \gamma_{k_1}^{(1)}(f_{k_1}, \alpha_{k_1})$ and $\delta_{k_1}^{(1)} = \delta_{k_1}^{(1)}(f_{k_1}, \beta_{k_1})$

such that $\gamma_{k_1}^{(1)} \leqq x_{k_1} \leqq \delta_{k_1}^{(1)}$. Now if F_{n-1} denotes the mapping of E^{n-1} onto E^{n-1} defined by

$$F_{n-1} \equiv (f_1(\cdot), \cdots, f_{k_1-1}(\cdot), f_{k_1+1}(\cdot), \cdots, f_n(\cdot))^t,$$

if A_{n-1} denotes the $(n-1) \times (n-1)$ matrix obtained from A by deleting the k_1-st row and column (note that $A_{n-1} \in P_0$), if

$$a_{n-1} = (a_{1.k_1}, \cdots, a_{k_1-1.k_1}, a_{k_1+1.k_1}, \cdots, a_{n.k_1})^t,$$

and if

$$B_{n-1} = (b_1, \cdots, b_{k_1-1}, b_{k_1+1}, \cdots, b_n)^t,$$

then

$$F_{n-1}(x) + A_{n-1}x = B_{n-1} - a_{n-1}x_{k_1}.^*$$

Since $A_{n-1} \in P_0$, there is a $k_2 \in \{1, \cdots, k_1-1, k_1+1, \cdots, n\}$ such that $x_{k_2}(A_{n-1}x)_{k_2} \geqq 0$ and hence, as before,

$$x_{k_2}(b_{k_2} - a_{k_2.k_1}x_{k_1}) \geqq x_{k_2}f_{k_2}(x_{k_2}).$$

But, if $a_{k_2.k_1} \leqq 0$, then

$$\alpha_{k_2} - a_{k_2.k_1}\gamma_{k_1}^{(1)} \leqq b_{k_2} - a_{k_2.k_1}x_{k_1} \leqq \beta_{k_2} - a_{k_2.k_1}\delta_{k_1}^{(1)},$$

and if $a_{k_2.k_1} > 0$, then

$$\alpha_{k_2} - a_{k_2.k_1}\delta_{k_1}^{(1)} \leqq b_{k_2} - a_{k_2.k_1}x_{k_1} \leqq \beta_{k_2} - a_{k_2.k_1}\gamma_{k_1}^{(1)}.$$

Therefore, by Lemma 1, there is a $\gamma_{k_2}^{(1)} = \gamma_{k_2}^{(1)}(f_{k_2}, \alpha_{k_2} - a_{k_2.k_1}\gamma_{k_1}^{(1)})$ and $\delta_{k_2}^{(1)} = \delta_{k_2}^{(1)}(f_{k_2}, \beta_{k_2} - a_{k_2.k_1}\delta_{k_1}^{(1)})$ such that $\gamma_{k_2}^{(1)} \leqq x_{k_2} \leqq \delta_{k_2}^{(1)}$ if $a_{k_2.k_1} \leqq 0$, and similarly for $a_{k_2.k_1} > 0$.

The above process may be repeated successively until the n pairs of real numbers $\gamma_{k_i}^{(1)}, \delta_{k_i}^{(1)}, (i = 1, \cdots, n)$ have been obtained. Thus, for any given B with $\alpha_i \leqq b_i \leqq \beta_i$ for $i = 1, \cdots, n$, the components of the solution x of equation (1) will be bounded by these pairs of numbers, provided it is known at each step which coordinate k_i to choose. The appropriate coordinate choice, however, will in general depend on the particular solution x which is associated with the given B. For different input vectors B the appropriate choice will in general be different. Therefore, in order to obtain bounds on x which are valid for all B with $\alpha_i \leqq b_i \leqq \beta_i$ $(i = 1, \cdots, n)$ we must consider each of the $n!$ permutations of the coordinates $\{1, \cdots, n\}$ and, for each one, generate the set of bounds $\{\gamma_{k_i}^{(\nu)}, \delta_{k_i}^{(\nu)}: i = 1, \cdots, n\}$ for $\nu = 1, \cdots, n!$. We then define $\gamma_i =$

* In this equation x is understood to be $(x_1, \cdots, x_{k_1-1}, x_{k_1+1}, \cdots, x_n)^t$.

$\min \{\gamma_i^{(\nu)}: \nu = 1, \cdots, n!\}$ and $\delta_i = \max \{\delta_i^{(\nu)}: \nu = 1, \cdots, n!\}$ for $i = 1, \cdots, n$. Then, for each B with $\alpha_i \leqq b_i \leqq \beta_i$ for $i = 1, \cdots, n$, we have that $\gamma_i \leqq x_i \leqq \delta_i$ for $i = 1, \cdots, n$, since at least one of the sets of bounds $\{\gamma_{k_i}^{(\nu)}, \delta_{k_i}^{(\nu)}: i = 1, \cdots, n\}$ must always apply. □

If the matrix A of Theorem 5 satisfies a stronger condition than $A \varepsilon P_0$ (that is, if A satisfies a weak row-sum dominance condition),

$$a_{ii} \geqq \sum_{j \neq i} | a_{ij} |, \qquad \text{for} \quad i = 1, \cdots, n,$$

it is possible to use a method that requires much less computational effort than that of Theorem 5 to compute the vectors γ and δ whose components bound the corresponding components of the solution of equation (1). This method of computing the bounds, a straightforward generalization of an idea presented in Ref. 3, is explained in Appendix B.

From Theorems 3 and 4 we now have the result: *Every bounded input sequence* $B^1, B^2, B^3, \cdots$ *is mapped by equation* (1) *into a bounded output sequence* $x^1, x^2, x^3, \cdots$, *for each* $F \varepsilon \mathfrak{F}^n$, *if and only if* $A \varepsilon P_0$.

In the proof of Theorem 5, the number of real numbers $\gamma_{k_i}^{(\nu)}, \delta_{k_i}^{(\nu)}$ which must be computed, in order to determine bounds for x, is $2n \times (n!)$. At the expense of obtaining poorer bounds it is easy to reduce this number to $2n^2$. Suppose we compute, at the first step, the $2n$ numbers $\gamma_1^{(1)}, \delta_1^{(1)}, \cdots, \gamma_n^{(1)}, \delta_n^{(1)}$ and set $\lambda_1 = \min \{\gamma_1^{(1)}, \cdots, \gamma_n^{(1)}\}$, $\mu_1 = \max \{\delta_1^{(1)}, \cdots, \delta_n^{(1)}\}$. Then, for each B with $\alpha_i \leqq b_i \leqq \beta_i$ for $i = 1, \cdots, n$, one of the components of the corresponding x will be bounded by λ_1 (from below) and μ_1 (from above). We next compute the $2n$ numbers $\gamma_i^{(2)} = \gamma_i^{(2)}(f_i, \alpha_i - p_i^{(1)})$, $\delta_i^{(2)} = \delta_i^{(2)}(f_i, \beta_i - q_i^{(1)})$, where $p_i^{(1)} = \max \{a_{ij}\lambda_1, a_{ij}\mu_1 : j \neq i\}$, $q_i^{(1)} = \min \{a_{ij}\lambda_1, a_{ij}\mu_1 : j \neq i\}$, and denote the smallest $\gamma_i^{(2)}$ by λ_2 and the largest $\delta_i^{(2)}$ by μ_2. Then we have bounds which apply for two of the components of the x which corresponds to any B with $\alpha_i \leqq b_i \leqq \beta_i$ for $i = 1, \cdots, n$. By computing $\gamma_i^{(3)} = \gamma_i^{(3)}(f_i, \alpha_i - p_i^{(1)} - p_i^{(2)})$, $\delta_i^{(3)} = \delta_i^{(3)}(f_i, \beta_i - q_i^{(1)} - q_i^{(2)})$, etc., the above process may be continued to obtain the numbers $\lambda_1, \cdots, \lambda_n, \mu_1, \cdots, \mu_n$. Each component of the x corresponding to any B with $\alpha_i \leqq b_i \leqq \beta_i$ for $i = 1, \cdots, n$ will be bounded by $\lambda = \min \{\lambda_1, \cdots, \lambda_n\}$ (from below) and $\mu = \max \{\mu_1, \cdots, \mu_n\}$ (from above).

A matter that is closely related to the proofs of the above theorems on the boundedness of solutions of equation (1) is that of proving: *For each* $F \varepsilon \mathfrak{F}^n$ *and each* $A \varepsilon P_0$ *the solution* x *of equation* (1) *is a continuous function of the vector* B. It is obvious that it will suffice to prove that for each $F \varepsilon \mathfrak{F}^n$ with $F(\theta) = \theta$, and for each $A \varepsilon P_0$, the solution x of equation (1) is continuous in B at $B = \theta$. We then note that if f

satisfies the hypotheses of Lemma 1 and, in addition, if $f(0) = 0$ then, due to the continuity of f^{-1}, for every $\epsilon > 0$ there exists $\zeta > 0$ such that if α, β in Lemma 1 satisfy $-\zeta < \alpha \leqq b \leqq \beta < \zeta$ then γ, δ in Lemma 1 satisfy $-\epsilon < \gamma \leqq x \leqq \delta < \epsilon$. This observation may be used to incorporate a simple "ϵ-δ argument" into the steps of the previous paragraph to show that when $F(\theta) = \theta$ then for arbitrary $\epsilon > 0$, one can determine $\zeta > 0$ such that $\| B \| < \zeta$ implies $\| x \| < \epsilon$.

At this point we return to the matter of the existence and uniqueness of solutions of equation (1). We state first a theorem of R. S. Palais (Ref. 7—see also the Appendix of Ref. 8) which shows the connection between the concepts of existence and uniqueness of solutions and the boundedness of solutions.

Palais' Theorem. Let $f_1, \cdots, f_n$ be n continuously differentiable real valued functions of n real variables. Necessary and sufficient conditions that the mapping $f : E^n \rightarrow E^n$ defined by $f(x) = (f_1(x), \cdots, f_n(x))^t$ be a diffeomorphism of E^n onto itself are:

(*i*) $\det [\partial f_i / \partial x_j]$ *never vanishes.*
(*ii*) $\lim_{\|x\| \to \infty} \| f(x) \| = \infty$.

Palais' Theorem may be used to prove a result which is almost equivalent to our Theorem 3, that is:

Theorem 6. If A is an $n \times n$ matrix then there exists a unique solution of equation (1) *for each $F \equiv (f_1(x_1), \cdots, f_n(x_n))^t$ with continuously differentiable, strictly monotone increasing functions f_i which map E^1 onto itself, and whose slopes are everywhere positive, and for each $B \in E^n$, if and only if $A \in P_0$.*

A proof of Theorem 6 which is independent of our Theorem 3 is easy to construct: For all $A \in P_0$, the rather trivial Theorem 1 guarantees that condition (*i*) of Palais' Theorem is satisfied, and Theorem 5 guarantees that condition (*ii*) is satisfied. If $A \notin P_0$ then a choice of F such as is specified in the "*if*" part of the proof of Theorem 4 provides a case in which condition (*ii*) of Palais' Theorem is violated.

VI. SUFFICIENT CONDITIONS FOR $A \in P_0$ OR P

For a given matrix A, it is not in general an easy task to determine whether or not A satisfies any one of the four equivalent conditions of Fiedler and Pták which are given in Section III and which serve to define the class of matrices P_0 (or the conditions which define P). This is particularly true when the order of A is large. For this reason, we now give several conditions which are sufficient to insure that a matrix A is in P_0 or P (and which are not so difficult to verify).

Suppose it were known that every eigenvalue of A as well as every eigenvalue of each principal submatrix of A had a nonnegative (positive) real part. Then this would guarantee that $A \in P_0$ (P). This is the main idea involved in the following theorem.

Theorem 7. If any one of the following inequalities is satisfied by the elements a_{ij} of the matrix A, for all $i = 1, \cdots, n$, then $A \in P_0$.

(*i*) $$a_{ii} \geqq \Big(\sum_{j \neq i} | a_{ij} |\Big)^{\alpha}\Big(\sum_{k \neq i} | a_{ki} |\Big)^{1-\alpha}, \qquad 0 \leqq \alpha \leqq 1;$$

(*ii*) $$a_{ii} \geqq \alpha_i^{1/q}\Big(\sum_{j \neq i} | a_{ij} |^{p}\Big)^{1/p}, \qquad p \geqq 1, \qquad p^{-1} + q^{-1} = 1,$$

α_i positive numbers satisfying $\sum_{i=1}^{n} (1 + \alpha_i)^{-1} \leqq 1$;

(*iii*) $a_{ii} \geqq \alpha \max_{j \neq i} | a_{ij} |$, α positive satisfying

$$\sum_{i=1}^{n} \Big\{\sum_{j \neq i} | a_{ij} | \big(\max_{j \neq i} | a_{ij} |\big)^{-1}\Big\} \leqq \alpha(1 + \alpha), \qquad (0/0 = 0).$$

If any one of the above inequalities with $\geqq$ replaced by $>$ is satisfied for $i = 1, \cdots, n$, then $A \in P$.

Proof: If the right-hand side of any of the above inequalities is denoted by the nonnegative number r_i then it is well known that all of the eigenvalues of the matrix A are contained in the union $\cup\{C_i : i = 1, \cdots, n\}$ of the disks $C_i = \{z: | z - a_{ii} | \leqq r_i\}$.[9] But the condition $a_{ii} \geqq (>)\ r_i$ guarantees that if $z \in C_i$ then $\mathrm{Re}(z) \geqq (>)\ 0$. Thus, each of the eigenvalues of the matrix A has a nonnegative (positive) real part. The same is true of each eigenvalue of every principal submatrix of A, for if one of the above inequalities is satisfied by the elements of A it is also satisfied by the elements of any principal submatrix. □

VII. COMPUTATION OF THE SOLUTION

At present, the authors know of no single computational algorithm which is guaranteed to yield the solution of equation (1) for all $F \in \mathfrak{F}^n$, $A \in P_0$, $B \in E^n$. However, there are several ways that the solution may be computed for large classes of such equations.

If, for example, the matrix A satisfies either a weak row-sum or weak column-sum dominance condition (inequality (*i*) of Theorem 7 with either $\alpha = 1$ or $\alpha = 0$) and if $F \in \mathfrak{F}^n$ with, roughly speaking, the slopes of each f_i bounded from below by some positive constant, then it can be shown (see Appendix A) that an algorithm for computing the solution can be obtained by the use of Banach's contraction-mapping fixed point theorem.

If the matrix A is positive semidefinite then, as mentioned in Section I, the existence of a unique solution of equation (1) for all $F \in \mathfrak{F}^n$ follows from the earlier work of Sandberg and Minty. If, in addition, there exists for $i = 1, \cdots, n$, positive constants α_i and β_i such that

$$\alpha_i \leqq \frac{f_i(u) - f_i(v)}{u - v} \leqq \beta_i$$

for all $u \neq v$, then Sandberg's iteration scheme (also resulting from an application of the contraction-mapping fixed point theorem) can be used to compute the solution.[1] In this regard, if the techniques of Section V are first used to obtain bounds on the location of the solution then one could modify equation (1) by changing the nature of the functions f_i outside the domain in which the solution is known to lie (but still keeping the f_i strictly monotone increasing from E^1 onto E^1) and obtain a new equation which has the same solution as the original equation. By doing this, the functions f_i in the new equation might be made to satisfy the above inequalities in cases where this was impossible for the original f_i. Also, even if these inequalities could be satisfied for the original equation, larger values of α_i and smaller values of β_i might be used for the modified equation. This can result in a more rapidly converging iteration process (see Section VII of Ref. 3). Similarly, the bounds can be used to improve the performance of other iteration schemes.

In case $A \in P_0$ is not positive semidefinite, it might be that there exist diagonal matrices $\Delta_1, \Delta_2 > 0$ such that $\Delta_1 A \Delta_2$ is positive semidefinite. If such matrices can be found, then Sandberg's iteration scheme could be used to compute the solution of the equation

$$\Delta_1 F(\Delta_2 x) + \Delta_1 A \Delta_2 x = \Delta_1 B,$$

from which the solution of equation (1) may be obtained directly. Unfortunately, it is not the case that such $\Delta_1, \Delta_2 > 0$ exist for all $A \in P_0$. For example, it is quite easily verified that for

$$A = \begin{bmatrix} 1 & 0 \\ 1 & 0 \end{bmatrix},$$

even though $A \varepsilon P_0$, the matrix $\Delta_1 A \Delta_2$ is not positive semidefinite for any choice of $\Delta_1, \Delta_2 > 0$.

It is easily verified, however, that appropriate $\Delta_1, \Delta_2 > 0$ can be found for all 2×2 matrices $A \varepsilon P_0$ except those for which

(*i*) $a_{11}a_{22} = 0$,

and

(*ii*) $a_{12}a_{21} = 0$,

and

(*iii*) either $a_{12} \neq 0$ or $a_{21} \neq 0$.

In particular, for all nonsingular 2×2 matrices $A \varepsilon P_0$, appropriate Δ_1, Δ_2 ($\Delta_2 = I$) can be found. Thus, Sandberg's iteration scheme could be used, for example, to compute the solution of the example problem of Section II which was taken from Calahan's book.

VIII. APPLICATION TO EQUATIONS FOR TRANSISTOR NETWORKS

In this section some of the above theory is applied to the equations which describe the behavior of electrical networks containing transistors. By the word transistor we refer to the three-terminal device whose equivalent circuit is shown in Fig. 2.* Considering the transistor as a nonlinear two-port network, the following equations which express the port currents in terms of the port voltages follow immediately from inspection of Fig. 2:

$$\begin{bmatrix} i_1 \\ i_2 \end{bmatrix} = \begin{bmatrix} 1 & -\alpha_{12} \\ -\alpha_{21} & 1 \end{bmatrix} \begin{bmatrix} f_1(v_1) \\ f_2(v_2) \end{bmatrix}.$$

We assume, as is the case for the usual large-signal model of a physical transistor, that $0 < \alpha_{12} < 1$, $0 < \alpha_{21} < 1$, and that both of the functions f_1 and f_2 are continuous and strictly monotone increasing. The character of the functions f_1 and f_2 which describe the transistor's nonlinear conductances will depend on whether the transistor is designated as NPN or PNP. We shall, however, have no occasion to distinguish between these two cases in what is to follow.

Suppose an electrical network is synthesized by connecting together,

* In some respects this equivalent circuit is an *ideal* model of a transistor. Nevertheless, since this model is often used in the design and the computer analysis of transistor networks, consideration of it is important. The presence of series resistance at the base, emitter, and collector terminals of a transistor will be considered by the authors in another paper.

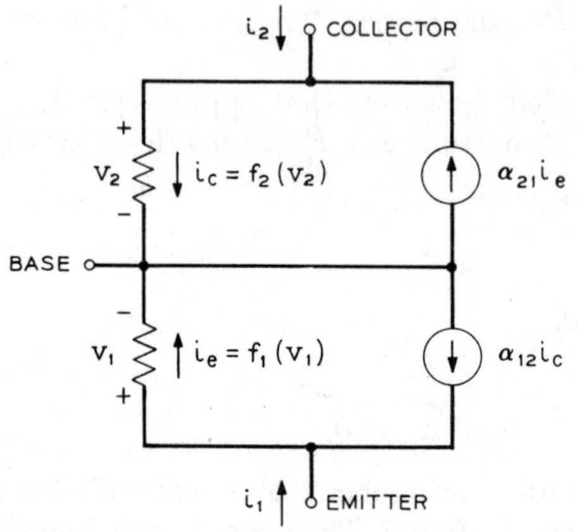

Fig. 2 — The equivalent circuit of a transistor.

in an arbitrary manner, any (finite) number of transistors, resistors (that is, linear resistors with nonnegative resistance), voltage sources, current sources, and nonlinear resistors which are described by strictly monotone increasing conductance functions (and which we shall henceforth refer to as "diodes"). Suppose the network contains n transistors and d diodes. For $k = 1, \cdots, n$, let x_{2k-1}, x_{2k}, y_{2k-1}, and y_{2k} denote the voltage and current variables v_1, v_2, i_1, and i_2, respectively, for the k-th transistor. For $k = 1, \cdots, d$, let x_{2n+k} and y_{2n+k} denote the voltage across, and the current through, the kth diode. Let these variables be related by $y_{2n+k} = f_{2n+k}(x_{2n+k})$. Then, if $x = (x_1, \cdots, x_{2n+d})^t$ and $y = (y_1, \cdots, y_{2n+d})^t$, we have

$$y = TF(x), \tag{9}$$

where $T = \text{diag}(T_1, T_2)$, with T_1 a block diagonal matrix with n 2×2 diagonal blocks of the form

$$\begin{bmatrix} 1 & -\alpha_{12}^{(k)} \\ -\alpha_{21}^{(k)} & 1 \end{bmatrix},$$

and T_2 a $d \times d$ identity matrix. The nonlinear function F has the form $F(x) \equiv (f_1(x_1), \cdots, f_{2n+d}(x_{2n+d}))^t$.

Consider now the $(2n+d)$-port network of resistors and independent sources which is formed from the original network by removing the transistors and diodes. If the y-parameter matrix G of this $(2n+d)$-port exists then we have the additional equation relating the vectors

x and y:

$$y = -Gx + u, \tag{10}$$

where u is some vector of constants which is, in general, nonzero since sources are present in the $(2n+d)$-port.

Combining equations (9) and (10) we obtain

$$TF(x) + Gx = u. \tag{11}$$

Now T is a nonsingular matrix and hence, if equation (11) is multiplied by T^{-1}, we obtain an equation having the form of equation (1). If the matrix $T^{-1}G \in P_0$ then, by Corollary 1, there exists at most one set of transistor and diode voltages satisfying equation (11). Moreover, if each of the nonlinear functions describing the transistors and diodes in our network maps E^1 *onto* E^1, or if $T^{-1}G \in P$, then Theorem 3, or Corollary 3, guarantees the *existence* of a unique solution of equation (11).

We have been careful to distinguish between the case when our theory guarantees only the uniqueness of a solution and the case when it guarantees both the solution's existence and its uniqueness for the following reason: In the analysis of transistor networks the nonlinear functions which are used to describe diodes or to describe the nonlinear conductances in the equivalent circuit of a transistor are often taken to be of the form

$$f(x) = I_0(e^{\lambda x} - 1),$$

where I_0 and λ are constants. The range of such a function is not the entire real line. Presumably, therefore, one can construct transistor networks having the property that if functions of the above type are used in a transistor's equivalent circuit then the network admits no solution. We now give a simple example of such a network. We wish to emphasize, though, that even for these networks whose equations may sometimes have no solution, our theory still guarantees that if $T^{-1}G \in P_0$ and if a solution of equation (11) exists, then it is unique.

Consider the network of Fig. 3. For this network, equation (11) becomes

$$\begin{bmatrix} 1 & -\alpha_{12} \\ -\alpha_{21} & 1 \end{bmatrix} \begin{bmatrix} f_1(v_1) \\ f_2(v_2) \end{bmatrix} + \begin{bmatrix} g & -g \\ -g & g \end{bmatrix} \begin{bmatrix} v_1 \\ v_2 \end{bmatrix} = \begin{bmatrix} I_a \\ I_b \end{bmatrix}.$$

Suppose $\alpha_{12} = 0.5$, $\alpha_{21} = 0.9$, and $g = 5.5$ mhos. Then, the above equation is equivalent to

$$\begin{bmatrix} f_1(v_1) \\ f_2(v_2) \end{bmatrix} + \begin{bmatrix} 5 & -5 \\ -1 & 1 \end{bmatrix} \begin{bmatrix} v_1 \\ v_2 \end{bmatrix} = \frac{1}{11} \begin{bmatrix} 20 & 10 \\ 18 & 20 \end{bmatrix} \begin{bmatrix} I_a \\ I_b \end{bmatrix}.$$

Hence, v_1 and v_2 must satisfy

$$f_1(v_1) + 5f_2(v_2) = 10(I_a + I_b).$$

If we now assume that the transistor's nonlinear conductances are described by the functions

$$f_1(v_1) = -I_e(e^{-\lambda_e v_1} - 1),$$
$$f_2(v_2) = -I_c(e^{-\lambda_c v_2} - 1),$$

where the parameters I_e, I_c, λ_e, and λ_c are each positive, then for all v_1, v_2 we have

$$f_1(v_1) + 5f_2(v_2) < I_e + 5I_c .$$

Hence, if the values of the independent current sources of Fig. 3 are chosen such that

$$I_a + I_b \geqq \tfrac{1}{10}I_e + \tfrac{1}{2}I_c ,$$

then the equation for this network has no solution.

Let us now consider the problem of determining whether or not, for a given network, the matrices T and G in equation (11) satisfy the condition $T^{-1}G \in P_0$ (or $T^{-1}G \in P$). (The existence of many transistor bistable circuits assures us that this condition is not always satisfied.)

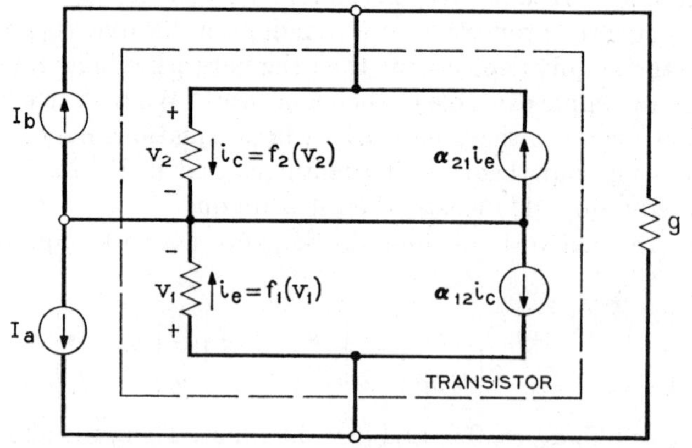

Fig. 3 — A transistor network whose equations may have no solution.

There is a large class of networks for which this condition is satisfied, and for which a simple inspection of the G matrix suffices to identify a member of the class.

Since the matrix T satisfies a strong column-sum dominance condition, that is, since

$$t_{ii} > \sum_{j \neq i} | t_{ji} | \qquad \text{for} \quad i = 1, \cdots, 2n + d,$$

the following theorem guarantees that if the matrix G also satisfies a strong column-sum dominance condition, then $T^{-1}G \in P$, and that if the matrix G satisfies a weak column-sum dominance condition,

$$g_{ii} \geqq \sum_{j \neq i} | g_{ji} |,$$

then $T^{-1}G \in P_0$ and, hence, the above conclusions concerning the existence and the uniqueness of a solution follow.

Theorem 8. If the square matrix A satisfies a strong column-sum dominance condition and if the square matrix B satisfies a weak (strong) column-sum dominance condition, then $A^{-1}B \in P_0$ (P).

Proof: Suppose $A^{-1}B \notin P_0$. Then, by the main result of the "only if" part of the proof of Theorem 3, there exists some diagonal matrix $D > 0$ such that $\det (D + A^{-1}B) = 0$. But $\det (D + A^{-1}B) = \det (A^{-1}) \cdot \det (AD + B)$, and $\det (A^{-1}) \neq 0$. Likewise, $\det (AD + B) \neq 0$ since $AD + B$ satisfies a strong column-sum dominance condition. Hence, $A^{-1}B \in P_0$.

With B strongly column-sum dominant, let $\delta > 0$ be such that $B - \delta A$ also possesses the strong dominance property. Suppose that $A^{-1}B - \delta I \notin P_0$. Then, as above, there is a $D > 0$ such that $A^{-1}B - \delta I + D = A^{-1}[B - \delta A + AD]$ is singular, which is a contradiction. Therefore $A^{-1}B - \delta I \in P_0$, and, by Theorem 1, $A^{-1}B \in P$. □

IX. COMMON-BASE TRANSISTOR NETWORKS

We now consider a special class of the networks which are comprised of transistors, resistors, diodes, and independent sources. We consider the class of all such networks for which there is a single node (called ground) to which the base terminal of each transistor is connected. Let us first consider a subclass of this class of networks; that is, let us temporarily assume that no diodes are present. For all networks in this subclass it is easily verified that when the G matrix for equation (11) exists, then it satisfies the above weak column-sum dominance condition and hence, by Theorem 8, $T^{-1}G \varepsilon P_0$. This fact is made evident if we consider the network of resistors which is described by G (that is, the linear multiport to which the transistors are connected, with all sources removed) and first simplify this network by using the star-mesh transformation to remove all internal nodes. Of course for many networks of this sublcass G is strongly column-sum dominant, in which case $T^{-1}G \varepsilon P$.

It is clear that the networks for which the G matrix fails to exist are exactly those networks in which either one or more of the collector or emitter terminals are connected, through the resistor network, directly to ground (that is, through a branch having infinite conductance), or else two (or more) of the transistors' collector or emitter terminals are connected directly together (through a branch of the resistor network having infinite conductance). These direct connections can exist in the resistor network either because of corresponding short-circuits in the original linear multiport, or because of corresponding connections involving branches which contain only ideal voltage sources.

If one assumes that each transistor in the network has a nonzero series resistance associated with both its emitter and its collector terminals (this assumption certainly being consistent with physical reality) then one need not be concerned about the possibility of the nonexistence of the G matrix since the situations mentioned in the previous paragraph cannot occur. We now show, however, that one need not rely upon this assumption in order to prove the uniqueness of the solution of the equations which describe the networks that we are considering.

We have observed that the matrix G will not exist if and only if the linear multiport has fewer independent port voltages than it has ports. In this case we modify the nonlinear multiport in such a manner that we can break some of the connections to the linear multiport so that it then possesses a G matrix and hence can be described by an equation having the form of equation (10). The modifications to the nonlinear multiport which are called for are obviously the addition of voltage sources between certain nodes, the values of these sources being the same as those of the voltage sources connecting the corresponding nodes in the linear multiport. This simple concept is illustrated in Fig. 4. Here, the network of Fig. 4a, containing a linear 6-port, has been replaced by the "equivalent" network of Fig. 4b containing a linear 3-port. Although the G matrix of the 6-port does not

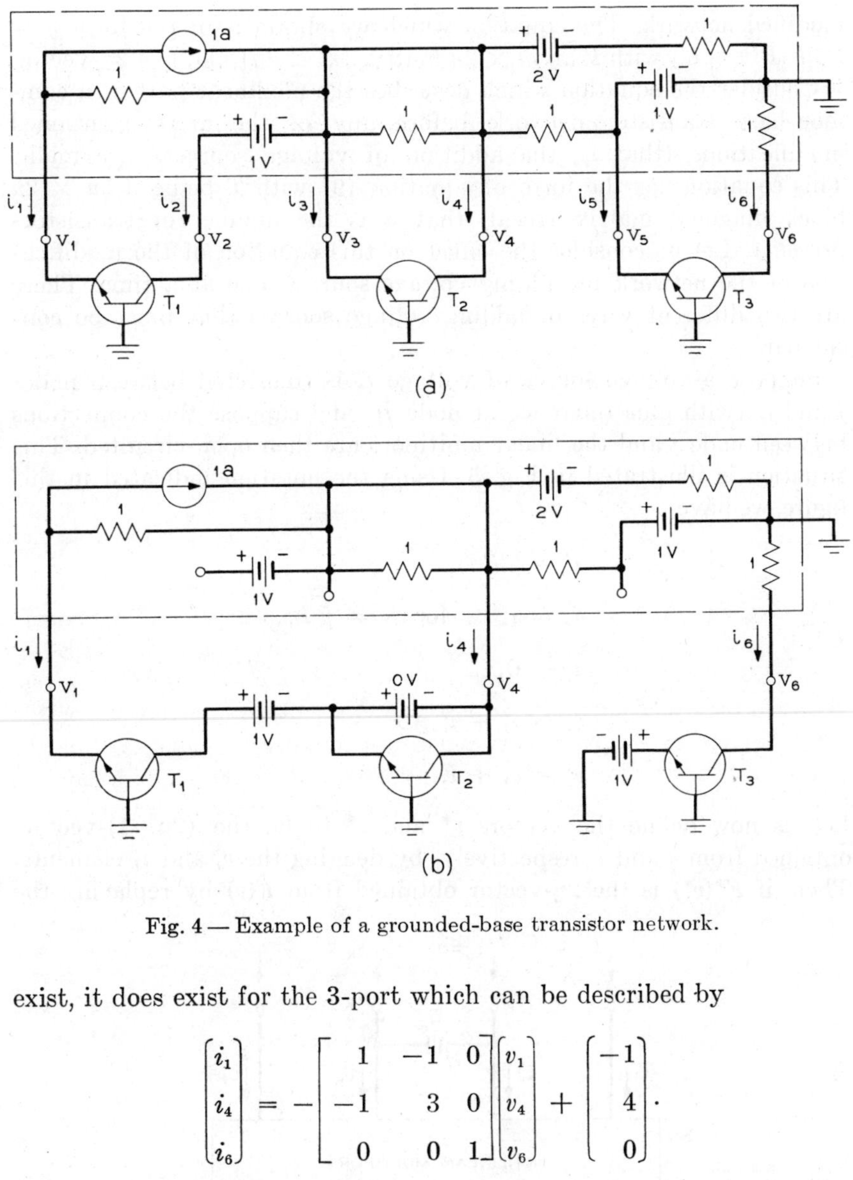

Fig. 4 — Example of a grounded-base transistor network.

exist, it does exist for the 3-port which can be described by

$$\begin{bmatrix} i_1 \\ i_4 \\ i_6 \end{bmatrix} = -\begin{bmatrix} 1 & -1 & 0 \\ -1 & 3 & 0 \\ 0 & 0 & 1 \end{bmatrix}\begin{bmatrix} v_1 \\ v_4 \\ v_6 \end{bmatrix} + \begin{bmatrix} -1 \\ 4 \\ 0 \end{bmatrix}.$$

We have shown that the above artifice allows an equation having the form of equation (10) to always be written to describe the linear multiport contained in our network. We now show that an equation like equation (9) can be written to describe the nonlinear part of our modified network. The equation which we obtain is of the form $y = PTF(P^t x + C)$ with P an $m \times 2n$ matrix $(m < 2n)$ and C a $2n$-vector.

Consider the equation which describes the nonlinear part of a common-base transistor network before any of the above-mentioned modifications (that is, the addition of voltage sources) are made. This equation has the form of equation (9) with T being a $2n \times 2n$ block diagonal matrix (recall that n is the number of transistors present). Let us consider the effect on this equation of the modification of the network by adding voltage sources, one at a time. There are two different ways of adding voltage sources that must be considered.

Suppose a voltage source of voltage E is connected between nodes j and k (with plus reference at node j), and suppose the connections between node j and the linear multiport are then open-circuited. This situation is illustrated in Fig. 5. Using the notation indicated in this figure, we have

$$i' = TF(v),$$

$$i_\nu = i'_\nu \qquad \text{for} \quad \nu \neq j, k,$$

$$i_j = 0,$$

$$i_k = i'_j + i'_k\,,$$

$$v_j = v_k + E.$$

Let us now define the vectors v^* and i^* to be the $(2n-1)$-vectors obtained from v and i, respectively, by deleting the v_j and i_j elements. Then, if $F^*(v^*)$ is the $2n$-vector obtained from $F(v)$ by replacing the

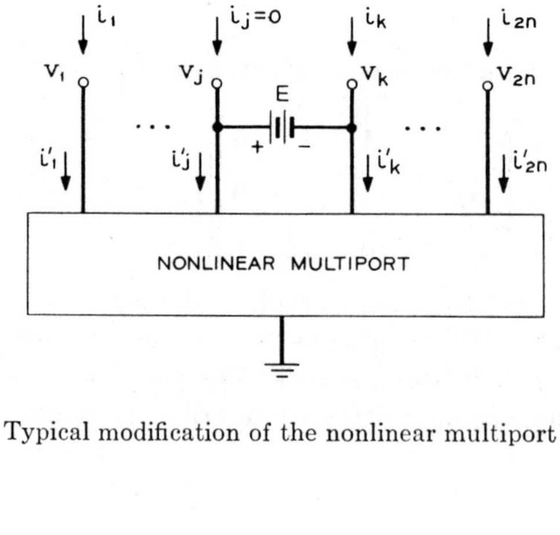

Fig. 5 — Typical modification of the nonlinear multiport network.

argument v_j by $v_k + E$, we then have that

$$i^* = T^*F^*(v^*), \tag{12}$$

where the $(2n-1) \times 2n$ matrix T^* is obtained from T by adding the j-th row to the k-th row and then deleting the j-th row. Observe that $T^*F^*(v^*)$ can be written as $QTF(Q^t v^* + R)$ in which the j-th element of the $2n$-vector R is E, all other elements of R are zero, and Q is obtained from the identity matrix of order $2n$ by adding the j-th row to the k-th row and then deleting the j-th row.

In case a voltage source of voltage E is connected between node j and ground (with the plus reference at node j) and all connections between node j and the linear multiport are open-circuited, then we can again form equation (12) from equation (9) by simply replacing v_j by E wherever it appears in the argument of F, to form $F^*(v^*)$, and deleting the j-th row of the matrix T, to form T^*. In this case $T^*F^*(v^*)$ can be written as $QTF(Q^t v^* + R)$ in which R is as defined earlier, but in this case Q is obtained from the identity matrix of order $2n$ by simply deleting the j-th row.

The above processes can be applied repeatedly to account for the addition of an arbitrary number of voltage sources to the nonlinear multiport. The resulting equation which describes the multiport will have the form

$$\begin{aligned} y &= Q_p \cdots Q_2 Q_1 T F(Q_1^t Q_2^t \cdots Q_p^t x + C) \\ &\equiv \tilde{T}\tilde{F}(x) \end{aligned}$$

with C some constant $2n$-vector and each of the Q_i obtained from the identity matrix of the appropriate order in one of the two ways described above.

Consider equation (9) in which T is a square matrix. Due to the strict monotonicity of each component function of F, the mapping $TF(x)$ has the following property: If p, q are arbitrary $2n$-vectors then there is a diagonal matrix $D > 0$ such that

$$TF(p) - TF(q) = TD(p - q), \tag{13}$$

and furthermore, the matrix TD is strongly column-sum dominant (since T is strongly column-sum dominant). We now wish to show that a similar fact is true in the more general case.

With m the number of rows of Q_p, let p and q denote arbitrary m-vectors. Then since there is a diagonal $D > 0$ such that

$$F(Q_1^t Q_2^t \cdots Q_p^t p + C) - F(Q_1^t Q_2^t \cdots Q_p^t q + C) = DQ_1^t Q_2^t \cdots Q_p^t (p - q),$$

we have

$$\tilde{T}\tilde{F}(p) - \tilde{T}\tilde{F}(q) = Q_p \cdots Q_2 Q_1 T D Q_1^t Q_2^t \cdots Q_p^t (p - q).$$

The fact that $Q_p \cdots Q_2 Q_1 T D Q_1^t Q_2^t \cdots Q_p^t$ is strongly column-sum dominant follows from the very easily verified proposition that the product $Q_k M Q_k^t$ ($k = 1, 2, \cdots, p$) possesses that property whenever M does.

Therefore if x^1 and x^2 denote two solutions of the "generalized equation (11)," then $[\tilde{T}\tilde{D} + G](x^1 - x^2) = \theta$ in which

$$\tilde{T} = Q_p \cdots Q_2 Q_1 T \qquad \text{and} \qquad \tilde{D} = D Q_1^t Q_2^t \cdots Q_p^t .$$

But $\tilde{T}\tilde{D}$, and hence $\tilde{T}\tilde{D} + G$, is strongly column-sum dominant and hence, nonsingular. This implies that $x^1 = x^2$.

We have now shown that in any network constructed from resistors, independent sources, and transistors having a common-base connection, the transistors' base-emitter and base-collector voltages are unique. It is a trivial matter to show that the same result applies when diodes are also allowed to be present in the network.

Suppose the result was not true for some network containing at least one diode. Then there would be two different sets of voltages and currents which satisfy Kirchoff's laws. Thus for each diode in the network there would be two (not necessarily distinct) pairs of points $(v_d^{(1)}, i_d^{(1)})$, $(v_d^{(2)}, i_d^{(2)})$ at which the diode is biased, corresponding to each solution. Letting f denote the strictly monotone increasing function which characterizes the diode we have $i_d^{(1)} = f(v_d^{(1)})$ and $i_d^{(2)} = f(v_d^{(2)})$. But then, suppose the diode is replaced by the series combination of a resistor r and a voltage source E whose values are chosen so that the line $i_d = (1/r)v_d - E/r$ passes through the points $(v_d^{(1)}, i_d^{(1)})$ and $(v_d^{(2)}, i_d^{(2)})$. (Due to the strict monotonicity of f, this can certainly be done with some positive choice of r.) Performing the above type of substitution for each diode in the network, we obtain a new network of the type already considered. This new network would possess two different sets of transistor base-emitter and base-collector voltages (the same as before). This contradicts our previous result, and hence the previous result must apply, even when diodes are present in the network.

To determine the equilibrium solutions of the differential equations which describe a network containing inductors and capacitors as well as the elements mentioned above, one must determine the solutions of a dc equation for a network of the above class. Thus, in summary, what we have shown is: *One cannot synthesize a bistable network which consists of resistors, inductors, capacitors, diodes, independent voltage and current sources, and an arbitrary number of (Fig. 2) transistors having a common base connection (or, in particular, only one transistor).*

X. ACKNOWLEDGMENT

The authors would like to acknowledge the helpful conversations with their colleague H. C. So.

APPENDIX A

Algorithms for Computing Solutions of Equation (1)

In this appendix two algorithms for computing the solution of equation (1) are presented. It is proved that one of the algorithms will always converge to the solution of equation (1) if the matrix A satisfies either a weak row-sum or column-sum dominance condition (inequality (i) of Theorem 7 with either $\alpha = 1$ or $\alpha = 0$) and if, roughly speaking, the slopes of each f_i are bounded from below by some positive constant. In each case the proof of convergence relies upon Banach's contraction-mapping fixed point theorem, and therefore also represents an independent proof of the existence and uniqueness of a solution of equation (1) for the conditions stated above.

The following notation will be used: For fixed $F \in \mathfrak{F}^n$, $B \in E^n$, let $f(x) \equiv F(x) - B$; also, if A is a given $n \times n$ matrix with elements a_{ij}, we define the diagonal matrix D by $D = \text{diag}\,[a_{11}, a_{22}, \cdots, a_{nn}]$, and let $\Delta = A - D$.

Theorem A. If the $n \times n$ matrix A satisfies

$$a_{ii} \geqq \sum_{j \neq i} | a_{ij} |, \qquad \text{for } i = 1, \cdots, n,$$

and if $F \in \mathfrak{F}^n$, $B \in E^n$, and if there exists some $\epsilon > 0$ such that for each $\alpha, \beta \in E^1$, $\epsilon \mid \alpha - \beta \mid \leqq \mid f_i(\alpha) - f_i(\beta) \mid$ for $i = 1, \cdots, n$, then equation (1) possesses a unique solution, and if x^0 is an arbitrary point in E^n, the sequence $x^0, x^1, x^2, \cdots$ defined by

$$x^{k+1} = (f + D)^{-1}(-\Delta)x^k$$

converges to the solution.

Proof: Equation (1) may be rewritten as

$$f(x) + Dx + \Delta x = \theta.$$

Hence, if the operation $T: E^n \rightarrow E^n$ is defined by $T = (f + D)^{-1}(-\Delta)$, then the solution of the equation $x = Tx$ is identical to the solution of equation (1). We now prove that the sequence $x^0, x^1, x^2, \cdots$ converges to this solution by proving that T is a contraction.

Let x and y be arbitrary points in E^n and let $g = Tx$, $h = Ty$. Then, $f(g) + Dg = -\Delta x$ and $f(h) + Dh = -\Delta y$. Thus, for $i = 1, \cdots, n$,

$$f_i(g_i) - b_i + a_{ii}g_i = -(\Delta x)_i ,$$

and

$$f_i(h_i) - b_i + a_{ii}h_i = -(\Delta y)_i .$$

Subtracting, we obtain

$$f_i(g_i) - f_i(h_i) + a_{ii}(g_i - h_i) = (\Delta y)_i - (\Delta x)_i .$$

Since f_i is strictly monotone increasing, we have

$$| f_i(g_i) - f_i(h_i) | + a_{ii} | g_i - h_i | = | (\Delta x)_i - (\Delta y)_i |,$$

and hence, since $\epsilon + a_{ii} > 0$,

$$| g_i - h_i | \leqq \frac{1}{a_{ii} + \epsilon} | (\Delta x)_i - (\Delta y)_i |.$$

Now,

$$\begin{aligned} | (\Delta x)_i - (\Delta y)_i | &= | \sum_{j \neq i} a_{ij}(x_j - y_j) | \\ &\leqq \sum_{j \neq i} (| a_{ij} | \cdot | x_j - y_j |) \\ &\leqq (\sum_{j \neq i} | a_{ij} |) \cdot \max_j | x_j - y_j |. \end{aligned}$$

Thus, defining the metric ρ on E^n by $\rho(x, y) = \max_i | x_i - y_i |$, we have, for $i = 1, \cdots, n$,

$$| g_i - h_i | \leqq \frac{1}{a_{ii} + \epsilon} (\sum_{j \neq i} | a_{ij} |) \cdot \rho(x, y).$$

But, since $0 \leqq \sum_{j \neq i} | a_{ij} | < a_{ii} + \epsilon$, there exists K, $0 \leqq K < 1$, such that $| g_i - h_i | \leqq K \cdot \rho(x, y)$ for $i = 1, \cdots, n$, and in particular, $\rho(Tx, Ty) = \max_i | g_i - h_i | \leqq K \cdot \rho(x, y)$. Hence T is a contraction. □

Theorem B. If the $n \times n$ *matrix* A *satisfies*

$$a_{ii} \geqq \sum_{j \neq i} | a_{ji} |, \qquad \text{for} \quad i = 1, \cdots, n,$$

and if $F \in \mathfrak{F}^n$, $B \in E^n$, *and if there exists some* $\epsilon > 0$ *such that for each*

$\alpha, \beta \in E^1$, $\epsilon \mid \alpha - \beta \mid \leqq \mid f_i(\alpha) - f_i(\beta) \mid$ for $i = 1, \cdots, n$, then equation (1) possesses a unique solution, and if z^0 is an arbitrary point in E^n, the sequence $z^0, z^1, z^2, \cdots$ defined by

$$z^{k+1} = -\Delta(f + D)^{-1}z^k$$

converges to some point z^* and the solution of equation (1) *is given by*

$$x^* = (f + D)^{-1}z^*.$$

Proof: As in Theorem A, the solution of equation (1) is also the solution of $x = (f + D)^{-1}(-\Delta)x$. For each $x \in E^n$, let $z = (f + D)x$ and hence $x = (f + D)^{-1}z$. Thus, x^* is the solution of equation (1) if $x^* = (f + D)^{-1}z^*$, where z^* is the solution of $z = -\Delta(f + D)^{-1}z$. The theorem is thus proved if it is proved that the operator $T \equiv -\Delta(f + D)^{-1}$ is a contraction.

Let P denote the operator $(f + D)^{-1}$, and let x and y be arbitrary points in E^n. Then, proceeding as in the proof of Theorem A, we obtain

$$\mid (Px)_j - (Py)_j \mid \leqq \frac{1}{a_{jj} + \epsilon} \mid x_j - y_j \mid, \qquad \text{for} \quad j = 1, \cdots, n.$$

Thus, if $g = Tx$ and $h = Ty$, then for $i = 1, \cdots, n$,

$$g_i = -\sum_{j \neq i} a_{ij}(Px)_j \qquad \text{and} \qquad h_i = -\sum_{j \neq i} a_{ij}(Py)_j .$$

Hence

$$\begin{aligned} \mid g_i - h_i \mid &= \mid \sum_{j \neq i} a_{ij}((Px)_j - (Py)_j) \mid \\ &\leqq \sum_{j \neq i} (\mid a_{ij} \mid \cdot \mid (Px)_j - (Py)_j \mid) \\ &\leqq \sum_{j \neq i} \left(\mid a_{ij} \mid \cdot \frac{1}{a_{jj} + \epsilon} \cdot \mid x_j - y_j \mid \right). \end{aligned}$$

Therefore,

$$\begin{aligned} \sum_{i=1}^{n} \mid g_i - h_i \mid &\leqq \sum_{i=1}^{n} \sum_{j \neq i} \frac{\mid a_{ij} \mid}{a_{jj} + \epsilon} \mid x_j - y_j \mid \\ &= \sum_{j=1}^{n} \left(\sum_{i \neq j} \frac{\mid a_{ij} \mid}{a_{jj} + \epsilon} \right) \mid x_j - y_j \mid. \end{aligned}$$

But, there exists K, $0 \leqq K < 1$, such that, for $j = 1, \cdots, n$,

$$\sum_{i \neq j} \frac{\mid a_{ij} \mid}{a_{jj} + \epsilon} \leqq K,$$

and hence

$$\sum_{i=1}^{n} | g_i - h_i | \leqq K \sum_{j=1}^{n} | x_j - y_j |.$$

Defining the metric ρ on E^n by

$$\rho(x, y) = \sum_{i=1}^{n} | x_i - y_i |,$$

we therefore have

$$\rho(Tx, Ty) = \sum_{i=1}^{n} | g_i - h_i | \leqq K \cdot \rho(x, y),$$

and hence T is a contraction. □

APPENDIX B

Determination of Bounds on the Solution of Equation (1)

In this appendix we present a method for determining bounds on the solution of equation (1) when $F \varepsilon \mathfrak{F}^n$, A is weakly row-sum dominant, and (for given $\alpha \equiv (\alpha_1, \cdots, \alpha_n)^t$, $\beta \equiv (\beta_1, \cdots, \beta_n)^t \varepsilon E^n$) $B \equiv (b_1, \cdots, b_n)^t$ satisfies $\alpha_i \leqq b_i \leqq \beta_i$ for $i = 1, \cdots, n$. The solution bounds are, in general, easier to compute than those of Theorem 5. The method presented here is a generalization of an idea presented in Ref. 3.

The computation of the solution bounds proceeds in two steps. First, one solves each of the equations

$$F(x) = \alpha \tag{14a}$$

and

$$F(x) = \beta. \tag{14b}$$

Denoting the solutions of equations (14a) and (14b) by $\mu \equiv (\mu_1, \cdots, \mu_n)^t$ and $\nu \equiv (\nu_1, \cdots, \nu_n)^t$, respectively, and defining

$$\lambda = \max\{| \mu_1 |, \cdots, | \mu_n |, | \nu_1 |, \cdots, | \nu_n |\},$$

and

$$B' = (\sum_{j \neq 1} | a_{1j} |, \cdots, \sum_{j \neq n} | a_{nj} |)^t,$$

one then solves each of the equations

$$F(x) + \text{diag}\,[a_{11}, \cdots, a_{nn}]x = \alpha - \lambda B', \tag{15a}$$

$$F(x) + \text{diag}\,[a_{11}, \cdots, a_{nn}]x = \beta + \lambda B'. \quad (15b)$$

Denoting the solutions of equations (15a) and (15b) by $\gamma \equiv (\gamma_1, \cdots, \gamma_n)^t$ and $\delta \equiv (\delta_1, \cdots, \delta_n)^t$, respectively, one has $\gamma_i \leqq x_i^0 \leqq \delta_i$ for $i = 1, \cdots, n$, where x^0 is the solution of equation (1) that corresponds to any B satisfying $\alpha_i \leqq b_i \leqq \beta_i$ for $i = 1, \cdots, n$.

To prove that the components of the vectors γ and δ, determined by the above procedure, are indeed bounds for the corresponding components of the solution x^0 involves no more than a word-for-word repetition of the proof of Theorem 2 of Ref. 3, with several quite obvious modifications. We omit the details.

REFERENCES

1. Sandberg, I. W., "On the Properties of Some Systems that Distort Signals-I," B.S.T.J., *42*, No. 5 (September 1963), pp. 2033–2046.
2. Minty, G. J., "Two Theorems on Nonlinear Functional Equations in Hilbert Space," Bull. Amer. Math. Soc., *69*, No. 5 (September 1963), pp. 691–692.
3. Willson, Jr., A. N., "On the Solutions of Equations for Nonlinear Resistive Networks," B.S.T.J., *47*, No. 8 (October 1968), pp. 1755–1773.
4. Calahan, D. A., *Computer-Aided Network Design* (Preliminary Ed.), New York: McGraw-Hill, 1968.
5. Fiedler, M. and Pták, V., "On Matrices with Non-Positive Off-Diagonal Elements and Positive Principal Minors," Czech. Math. J., *12*, No. 3 (1962), pp. 382–400.
6. Fiedler, M. and Pták, V., "Some Generalizations of Positive Definiteness and Monotonicity," Numer. Math., *9*, No. 2 (1966), pp. 163–172.
7. Palais, R. S., "Natural Operations on Differential Forms," Trans. Amer. Math. Soc., *92*, No. 1 (1959), pp. 125–141.
8. Holzmann, C. A. and Liu, R., "On the Dynamical Equations of Nonlinear Networks with n-Coupled Elements," Proc. Third Ann. Allerton Conf. on Circuit and System Theory., (U. of Illinois, 1965), pp. 536–545.
9. Marcus, M., *Basic Theorems in Matrix Theory,* Washington, D. C., National Bureau of Standards Applied Mathematics Series, *57* (1960).

Some Network-Theoretic Properties of Nonlinear DC Transistor Networks

By I. W. SANDBERG and A. N. WILLSON, JR.

(Manuscript received September 9, 1968)

This paper extends, in several directions, some of the results of earlier work concerned with the existence and uniqueness of solutions of the dc equations of nonlinear transistor networks. In particular, here we develop techniques which enable us to deal directly with a more complicated transistor model.

I. INTRODUCTION

Several results are presented in Ref. 1 concerning the equation

$$F(x) + Ax = B \tag{1}$$

(with $F(\cdot)$ a "diagonal" nonlinear mapping of real Euclidean n-space E^n into itself, and A a real $n \times n$ matrix) which plays a central role in the dc analysis of transistor networks. In particular, a necessary and sufficient condition on A is given such that the equation possesses a unique solution x for each real n-vector B and each strictly monotone increasing $F(\cdot)$ that maps E^n onto itself. Several circuit-theoretic implications of the results are also described in Ref. 1; for example, it is shown that the short-circuit admittance matrix of the linear portion of the dc model of an interesting class of switching circuits must violate a certain dominance condition.

In Ref. 1 the word *transistor* was used to refer to the three-terminal device whose dc equivalent circuit is shown in Fig. 1(a). Although this equivalent circuit is frequently used in the design and computer analysis of transistor networks it is, from a physical standpoint, somewhat incomplete. A more exact dc model of a physical transistor is that of Fig. 1(b) in which the presence of series resistance in each of the transistor's leads has been accounted for.

In this paper we report on several extensions of the previous results. The motivation for much of this work was to enable the model of Fig. 1(b) to be taken into account. In addition, we present here

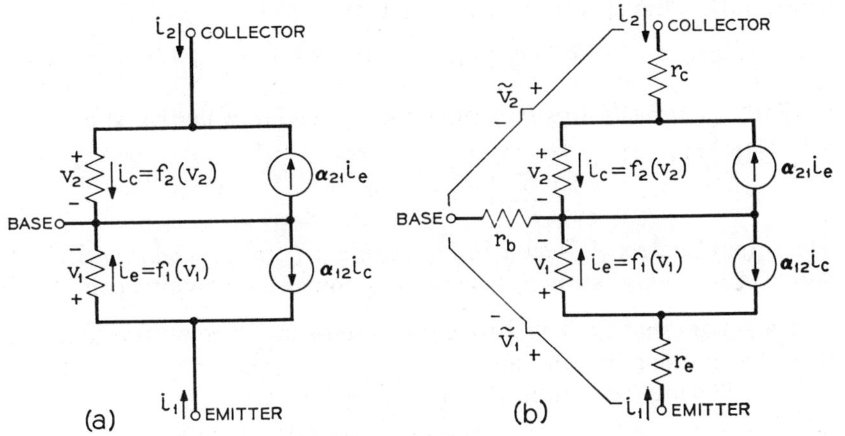

Fig. 1 — DC transistor models.

further material concerning cases in which (in accordance with standard assumptions) the nonlinear functions of Fig. 1(b) do not map E^1 *onto* itself. Finally, we prove a considerably stronger result than that of Ref. 1, to the effect that a certain class of networks cannot be bistable.

We now summarize some of the material of Ref. 1 that will be needed in the sequel:

For each positive integer n, we let $\mathfrak{F}^n$ denote that collection of mappings of the real n-dimensional Euclidean space E^n onto itself defined by: $F \in \mathfrak{F}^n$ if and only if there exist, for $i = 1, \cdots, n$, strictly monotone increasing functions f_i mapping E^1 onto E^1 such that,† for each $x \equiv (x_1, \cdots, x_n)^t \in E^n$, $F(x) \equiv (f_1(x_1), \cdots, f_n(x_n))^t$.

The origin in E^n will be denoted by θ. Throughout this article we consider only matrices whose elements are real. If D is a diagonal matrix then $D > 0$ ($D \geqq 0$) means that each element on the main diagonal of D is positive (nonnegative).

The classes of matrices P and P_0 have been defined by M. Fiedler and V. Pták in Refs. 2 and 3. They prove that these classes can be defined by any one of several equivalent properties. We shall need only the following characterization of the classes P and P_0 : A square matrix A is a member of the class P (P_0) if and only if all principal minors of A are positive (nonnegative). In the appendix it is proved that $A \in P_0$ if and only if det $[A + D] \neq 0$ for every diagonal matrix $D > 0$.

† If M is an arbitrary matrix, then the transpose of M is denoted in this article by M^t.

The following theorem is proved in Ref. 1:

Theorem 1: *If A is an $n \times n$ matrix then there exists a unique solution of (1) for each $F \varepsilon \mathfrak{F}^n$ and each $B \varepsilon E^n$ if and only if $A \varepsilon P_0$.*

We say that an $n \times n$ matrix A is *strongly (weakly) row-sum dominant* if and only if the elements a_{ij} of A satisfy

$$a_{ii} > (\geqq) \sum_{j \neq i} | a_{ij} |, \quad \text{for} \quad i = 1, \cdots, n.$$

Similarly, a *strongly (weakly) column-sum dominant* matrix is one that satisfies

$$a_{ii} > (\geqq) \sum_{j \neq i} | a_{ji} |, \quad \text{for} \quad i = 1, \cdots, n.$$

The square matrix A is said to be *dominant (strongly dominant)* if and only if A is weakly (strongly) row-sum dominant and symmetric.

If a square matrix A is strongly column-sum or row-sum dominant then A is nonsingular, in fact $A \varepsilon P$.

The following theorem is also proved in Ref. 1:

Theorem 2: *If the square matrix A satisfies a strong column-sum dominance condition and if the square matrix B satisfies a weak (strong) column-sum dominance condition, then $A^{-1}B \varepsilon P_0$ (P).*

An analogous theorem involving row-sum dominant matrices is also true, and can be proved with trivial modifications of the proof of Theorem 2 given in Ref. 1.

II. FURTHER RESULTS CONCERNING THE EXISTENCE AND UNIQUENESS OF SOLUTIONS

The proof of Theorem 1 given in Ref. 1 exploits the fact that the straight line described by the equation $y = -ax + b$ has exactly one intersection with the graph of each strictly monotone increasing function $f(x)$ which maps E^1 onto E^1 if and only if $a \geqq 0$.

It happens that a useful result that is slightly more general than that of Theorem 1 can be proved easily if use is made of a proposition that is similar to, but stronger than, the elementary fact mentioned in the preceding paragraph. That proposition is stated below.

Definition: For all α, β with $-\infty \leqq \alpha < \beta \leqq \infty$, let $I(\alpha, \beta)$ denote the interval $I(\alpha, \beta) = \{x : \alpha < x < \beta\}$.

The following proposition is quite easily verified:

Proposition: For $-\infty \leqq \alpha < \beta \leqq \infty$, the straight line described by the equation $y = -ax + b$ has exactly one intersection with the graph of each strictly monotone increasing function $f(x)$ which maps $I(\alpha, \beta)$ onto E^1 if and only if $a \geqq 0$.

Definition: For each positive integer n and each pair of n-vectors α, β whose components α_i, β_i lie in the extended real number system, with $\alpha < \beta$ (that is, with $-\infty \leqq \alpha_i < \beta_i \leqq \infty$ for $i = 1, \cdots, n$) let $\mathfrak{F}^n(\alpha, \beta; E^n)$ denote that collection of mappings of $I(\alpha_1, \beta_1) \times \cdots \times I(\alpha_n, \beta_n)$ onto E^n defined by: $F \in \mathfrak{F}^n(\alpha, \beta; E^n)$ if and only if there exist, for $i = 1, \cdots, n$, strictly monotone increasing functions f_i mapping (α_i, β_i) onto E^1 such that for each $x \equiv (x_1, \cdots, x_n)^t \in I(\alpha_1, \beta_1) \times \cdots \times I(\alpha_n, \beta_n)$,

$$F(x) \equiv (f_1(x_1), \cdots, f_n(x_n))^t.$$

Let the collection of strictly monotone increasing mappings of E^n onto $I(\alpha_1,\beta_1) \times \cdots \times I(\alpha_n,\beta_n)$ be similarly defined, and denoted by $\mathfrak{F}^n(E^n; \alpha, \beta)$. Note that $F \in \mathfrak{F}^n(\alpha, \beta; E^n)$ if and only if F^{-1} exists and $F^{-1} \in \mathfrak{F}^n(E^n; \alpha, \beta)$. Also, in case $I(\alpha_1,\beta_1) \times \cdots \times I(\alpha_n,\beta_n) = E^n$, then $\mathfrak{F}^n(\alpha, \beta; E^n) = \mathfrak{F}^n(E^n; \alpha, \beta) = \mathfrak{F}^n$.

Using the above proposition it is now easy to prove:

Theorem 3: For the n-vectors $\alpha < \beta$ whose components lie in the extended real number system, if A is an $n \times n$ matrix then there exists a unique solution of (1) *for each $F \in \mathfrak{F}^n(\alpha, \beta; E^n)$ and each $B \in E^n$ if and only if $A \in P_0$.*

Proof: (*if*) The proof of this part of the theorem is identical to the proof (given in Ref. 1) of the corresponding part of Theorem 1 with the exception that appropriate use is made of the above proposition. Since the necessary modifications are quite obvious we omit the details.

(*only if*) Suppose $A \notin P_0$. Then there exists a diagonal matrix $D \equiv \text{diag}\,[d_1, \cdots, d_n] > 0$ such that $\det\,[A + D] = 0$. Let x^0 be an arbitrary point in $I(\alpha_1, \beta_1) \times \cdots \times I(\alpha_n, \beta_n)$ and let y^0 be an arbitrary point in E^n. Let

$$B = y^0 + Ax^0.$$

Let $\delta > 0$ be chosen such that

$$\alpha_i < x_i^0 - \delta < x_i^0 + \delta < \beta_i, \qquad \text{for } i = 1, \cdots, n,$$

and choose $F \equiv (f_1(\cdot), \cdots, f_n(\cdot))^t$ in $\mathfrak{F}^n(\alpha, \beta; E^n)$ such that for $i = 1, \cdots, n$, and for $x_i^0 - \delta < x_i < x_i^0 + \delta$,

$$f_i(x_i) = y_i^0 + d_i(x_i - x_i^0).$$

Thus, $F(x^0) = y^0$ and hence, x^0 is a solution of (1) for this choice of F.

Since det $[A + D] = 0$, there exists some n-vector $x^* \neq \theta$ having the property that

$$Ax^* + Dx^* = \theta.$$

Thus, for each real number ϵ,

$$y^0 + D\epsilon x^* + A(x^0 + \epsilon x^*) = B.$$

In particular, if $\epsilon \neq 0$ is chosen such that $|\epsilon|$ is sufficiently small, then $|\epsilon x_i^*| < \delta$ for $i = 1, \cdots, n$. Hence, for such ϵ, if $x = x^0 + \epsilon x^*$, $F(x) = y^0 + D\epsilon x^*$ and therefore $x \neq x^0$ is also a solution of (1). □

An important special case of Corollary 3 of Ref. 1 is:

Corollary 1: For the n-vectors $\alpha < \beta$ *whose components lie in the extended real number system, if* A *is an* $n \times n$ *matrix then there exists a unique solution of* (1) *for each* $F \in \mathfrak{F}^n(E^n; \alpha, \beta)$ *and each* $B \in E^n$ *if* $A \in P$.

Theorem 3 may be used to prove a sharper (and, from the viewpoint of transistor networks, a more useful) result than Corollary 1. We have:

Theorem 4: For the n-vectors $\alpha < \beta$ *whose components lie in the extended real number system (in the real number system), if* A *is an* $n \times n$ *matrix then there exists a unique solution of* (1) *for each* $F \in \mathfrak{F}^n(E^n; \alpha, \beta)$ *and each* $B \in E^n$ *if (and only if)* $A \in P_0$ *and det* $A \neq 0$.

Proof: (*if*) As pointed out in Ref. 1, $A \in P_0$ and det $A \neq 0$ imply that $A^{-1} \in P_0$. Also, F^{-1} exists and $F^{-1} \in \mathfrak{F}^n(\alpha, \beta; E^n)$. Now x satisfies (1) if and only if y satisfies

$$F^{-1}(y) + A^{-1}y = A^{-1}B, \tag{2}$$

where $y = F(x)$. But, according to Theorem 3, there exists a unique y which satisfies (2).

(*only if*) We assume here that the components of α and β are real. Suppose $A \notin P_0$. Then, in a manner similar to that used in the proof of the "only if" part of Theorem 3, we can choose a mapping $F \in \mathfrak{F}^n(E^n; \alpha, \beta)$ and a point $B \in E^n$, such that the solution of (1) is not unique.

If, on the other hand, det $A = 0$, then there exists $x^* \neq \theta$ such that $A^t x^* = \theta$. Assume that (1) has a solution x for each $B \in E^n$. Then, since $\langle x^*, Ax \rangle = 0$ for all x, we have

$$\langle x^*, F(x) \rangle = \langle x^*, B \rangle,$$

for each $B \in E^n$ (and the corresponding x). It is clear, since the components of α and β are finite, that there exists some constant M such that

$$|\langle x^*, F(x)\rangle| \leqq M$$

for all $x \in E^n$. But B can certainly be chosen such that $\langle x^*, B\rangle > M$. This contradiction completes the proof of the theorem. □

The following theorem provides an alternative method of characterizing the class of matrices that are in P_0 and are nonsingular (compare with the theorem of the appendix).

Theorem 5: If A is a real square matrix then $A \in P_0$ and det $A \neq 0$ if and only if det $[A + D] \neq 0$ for every diagonal matrix $D \geqq 0$.

Proof: (*if*) It is clear, by the theorem of the appendix, that $A \in P_0$, since det $[A + D] \neq 0$ for all diagonal $D > 0$. Moreover, det $A \neq 0$, by hypothesis.

(*only if*) It is shown in Ref. 1 that, for each $A \in P_0$ and each diagonal $D \geqq 0$, $A + D \in P_0$. It suffices, therefore, to show that if $D_i = \text{diag}\,[0, \cdots, 0, d_i, 0, \cdots, 0]$ with $d_i \geqq 0$, and $A \in P_0$ with det $A > 0$, then det $[A + D_i] > 0$. Letting A_i denote the principal submatrix obtained from A by deleting the ith row and the ith column, we have

$$\det\,[A + D_i] = \det A + d_i \det A_i\,.$$

But det $A > 0$ and d_i det $A_i \geqq 0$. □

III. APPLICATION TO EQUATIONS FOR TRANSISTOR NETWORKS

In the analysis of a transistor network one could account for the presence of series lead resistance, while using the model of Fig. 1(a) to represent the transistor, by including appropriate additional resistors in the rest of the network. Indeed, there is at least one good reason for doing this. When treated in this manner, the presence of nonzero series resistance in the base, collector, and emitter leads of each transistor ensures that the y-parameter matrix exists for the circuit to which the transistors are connected—and hence ensures that the transistor network can be described by an equation having the form of (1). On the other hand, there are also good reasons for representing the transistor, for analysis purposes, by the model of Fig. 1(b). Using this model it will be shown, for example, that it is often possible to determine that there is a unique solution of the equation describing a given transistor network *regardless* of the (nonnegative) values of the transistors' series lead resistances. Since these resistances are usually parasitic and unavoidable in nature it is significant that one might be able to show that their introduction in, say, a certain monostable circuit will not cause the circuit to become bistable.

Using the model of Fig. 1(b) it is quite easy to see that the port variables for the transistor, when considered as a nonlinear two-port network, obey the following relationship

$$\begin{bmatrix} i_1 \\ i_2 \end{bmatrix} = \begin{bmatrix} 1 & -\alpha_{12} \\ -\alpha_{21} & 1 \end{bmatrix} \begin{bmatrix} f_1(v_1) \\ f_2(v_2) \end{bmatrix}$$

where

$$\begin{bmatrix} v_1 \\ v_2 \end{bmatrix} = \begin{bmatrix} \tilde{v}_1 \\ \tilde{v}_2 \end{bmatrix} - \begin{bmatrix} r_e + r_b & r_b \\ r_b & r_c + r_b \end{bmatrix} \begin{bmatrix} i_1 \\ i_2 \end{bmatrix}.$$

As in Ref. 1 we assume that $0 < \alpha_{12} < 1$, $0 < \alpha_{21} < 1$, and that both of the functions f_1 and f_2 are strictly monotone increasing mappings of E^1 into E^1.

Suppose an electrical network is synthesized containing transistors, resistors (that is, linear resistors having nonnegative resistance), independent voltage and current sources, and nonlinear resistors which are described by strictly monotone increasing conductance functions (and which shall henceforth be called "diodes"). Suppose the network contains n transistors and d diodes $(n + d > 0)$. For $k = 1, \cdots, n$ let x_{2k-1}, x_{2k}, $\tilde{x}_{2k-1}$, $\tilde{x}_{2k}$, y_{2k-1}, and y_{2k} denote the voltage and current variables v_1, v_2, $\tilde{v}_1$, $\tilde{v}_2$, i_1, and i_2, respectively, for the kth transistor. For $k = 1, \cdots, d$, let x_{2n+k} and y_{2n+k} denote the voltage across, and the current through, the kth diode; also (for $k=1, \cdots, d$) let $\tilde{x}_{2n+k} = x_{2n+k}$. Let these variables be related by $y_{2n+k} = f_{2n+k}(x_{2n+k})$. Then, if $x = (x_1, \cdots, x_{2n+d})^t$, $\tilde{x} = (\tilde{x}_1, \cdots, \tilde{x}_{2n+d})^t$, and $y = (y_1, \cdots, y_{2n+d})^t$, we have

$$y = TF(x), \qquad x = \tilde{x} - Ry, \tag{3}$$

where $T = \text{diag}[T_1, T_2]$, with T_1 a block diagonal matrix with n 2×2 diagonal blocks of the form

$$\begin{bmatrix} 1 & -\alpha_{12}^{(k)} \\ -\alpha_{21}^{(k)} & 1 \end{bmatrix}, \tag{4}$$

and T_2 the $d \times d$ identity matrix. Also, $R = \text{diag}[R_1, R_2]$, with R_1 a block diagonal matrix with n 2×2 diagonal blocks of the form

$$\begin{bmatrix} r_e^{(k)} + r_b^{(k)} & r_b^{(k)} \\ r_b^{(k)} & r_e^{(k)} + r_b^{(k)} \end{bmatrix}, \tag{5}$$

and R_2 the $d \times d$ null matrix.

Consider now the $(2n + d)$-port network of resistors and independent sources which is formed from the original network by removing the transistors and diodes. If the y-parameter matrix G of this $(2n + d)$-port exists then we have the additional equation relating the vectors $\tilde{x}$ and y:

$$y = -G\tilde{x} + \tilde{u} \tag{6}$$

where $\tilde{u}$ is some vector of constants that is, in general, nonzero since sources are present in the $(2n + d)$-port.

The vectors $\tilde{x}$ and y can easily be eliminated from (3) and (6), resulting in the equation

$$TF(x) + [I + GR]^{-1}Gx = u, \tag{7}$$

where we have defined the vector u by

$$u = [I + GR]^{-1}\tilde{u}.$$

According to Theorem 6, below, the matrix $[I + GR]$ must be nonsingular.

In case the matrix R contains all zeros (that is, in case all series lead resistors are omitted from the transistors) (7) reduces immediately to the equation which was studied in Ref. 1. Even when R does not contain all zeros, however, the results of Ref. 1 can be applied directly to (7). By applying Theorem 2 we have: *If the matrix* $[I + GR]^{-1} G$ *is dominant*† *then there is at most one solution of* (7). *If, furthermore,* F *maps* E^n *onto* E^n, *or if* $[I + GR]^{-1}G$ *is strongly dominant, then there exists a unique solution of* (7).

Making use of Theorem 4, we also have the stronger result: *There exists a unique solution of* (7) *if* $[I + GR]^{-1}G$ *is dominant and* G *is nonsingular.*

Although it is not, in general, true that the inverse of a strongly column-sum (row-sum) dominant matrix is strongly row-sum (column-sum) dominant, the statement is true when the order of the matrix is less than three. This elementary observation turns out to be quite useful in the proof of Theorem 6, which yields results that focus attention on the properties of G, concerning the existence and uniqueness of a solution of (7).

† For symmetric matrices the properties (*i*) weak column-sum dominance, and (*ii*) dominance, are identical. Since it is easily verified that for symmetric G and R, $[I + GR]^{-1}G$ is also symmetric, we simply specifiy that $[I + GR]^{-1}G$ be dominant.

Theorem 6: Let A (B) be the direct sum of n 2×2 and d 1×1 strongly column-sum (weakly row-sum) dominant matrices. Let B be symmetric and let C be a square matrix of order $2n + d$. Then:

(*i*) $\det [I + CB] \neq 0$, *provided that C is positive semidefinite,*
(*ii*) $A^{-1}[I + CB]^{-1}C \in P_0$, *provided that C is dominant,*
(*iii*) $A^{-1}[I + CB]^{-1}C \in P$, *provided that C is strongly dominant.*

Proof: (*i*) Here C is positive semidefinite. Let $B^{\frac{1}{2}}$ be the symmetric nonnegative square root of B, so that $I + CB = I + CB^{\frac{1}{2}}B^{\frac{1}{2}}$. Since (see Appendix A of Ref. 4) $\det [I + CB^{\frac{1}{2}}B^{\frac{1}{2}}] = \det [I + B^{\frac{1}{2}}CB^{\frac{1}{2}}]$, and since $I + B^{\frac{1}{2}}CB^{\frac{1}{2}}$ is positive definite, we have $\det [I + CB] > 0$.

(*ii*) Here C is dominant (which, as is well known, implies that C is positive semidefinite and hence, by (*i*), implies that $[I + CB]^{-1}$ exists). Suppose $A^{-1}[I + CB]^{-1}C \notin P_0$. Then, by the theorem of the appendix, there exists a diagonal matrix $D > 0$ such that $A^{-1}[I + CB]^{-1}C + D$ is singular. But

$$A^{-1}[I + CB]^{-1}C + D = A^{-1}[I + CB]^{-1}[C(D^{-1}A^{-1} + B) + I]AD,$$

which means that $C(D^{-1}A^{-1} + B) + I$ must be singular. Since A is a direct sum of 1×1 and 2×2 strongly column-sum dominant matrices, it follows that A^{-1} is a direct sum of 1×1 and 2×2 strongly row-sum dominant matrices. Thus, $D^{-1}A^{-1}$ and hence $D^{-1}A^{-1} + B$ is strongly row-sum dominant. Therefore, $(D^{-1}A^{-1} + B)$ is nonsingular, and $(D^{-1}A^{-1} + B)^{-1}$ is strongly column-sum dominant. But,

$$C(D^{-1}A^{-1} + B) + I = [C + (D^{-1}A^{-1} + B)^{-1}](D^{-1}A^{-1} + B)$$

in which the right-hand side is nonsingular since $C + (D^{-1}A^{-1} + B)^{-1}$ is strongly column-sum dominant, which is a contradiction.

(*iii*) Here C is strongly dominant. Since $C(I + BC) = (I + CB)C$, we have $\det(I + BC) > 0$ and

$$(I + CB)^{-1}C = C(I + BC)^{-1}.$$

Suppose that there is no constant $\delta > 0$ such that $A^{-1}C(I + BC)^{-1} - \delta I \in P_0$. Then, for each $\delta > 0$ there is a diagonal matrix $D > 0$ such that $A^{-1}C(I + BC)^{-1} - \delta I + D$ is singular. But,

$$\begin{aligned} A^{-1}C(I + BC)^{-1} - \delta I + D \\ &= A^{-1}[C - \delta A(I + BC) + AD(I + BC)](I + BC)^{-1} \\ &= D\{I + BC + D^{-1}A^{-1}[C - \delta A(I + BC)]\}(I + BC)^{-1} \\ &= \{D + [DB + A^{-1} - \delta(C^{-1} + B)]C\}(I + BC)^{-1}, \end{aligned}$$

which leads to the conclusion that for each $\delta > 0$ there is a $D > 0$ such that $D + [DB + A^{-1} - \delta(C^{-1} + B)]C$ is singular. We now establish a contradiction:

For all $x \in E^n$, let $\| x \| = \max_i | x_i |$. If $x, y \in E^n$ such that $\| x \| = 1$ and

$$[DB + A^{-1}]y = x$$

then it is easy to show that

$$\| y \| \leqq \max_k \frac{1}{\alpha_{kk} - \sum_{i \neq k} | \alpha_{ki} |}$$

in which the α_{kj} are the elements of A^{-1}. Thus, the norm of $[DB + A^{-1}]^{-1}$ can be bounded from above uniformly in $D > 0$. Therefore,

$$D + [DB + A^{-1} - \delta(C^{-1} + B)]C = (DB + A^{-1})\{(DB + A^{-1})^{-1}D + [I - \delta(DB + A^{-1})^{-1}(C^{-1} + B)]C\}$$

in which $\delta > 0$ can be chosen so small that $[I - \delta(DB+A^{-1})^{-1}(C^{-1}+B)]C$ is strongly column-sum dominant for all $D > 0$. Since $(DB + A^{-1})^{-1}D$ is also column-sum dominant, we have a contradiction. It follows that for some $\delta > 0$, $A^{-1}(I + CB)^{-1}C - \delta I \in P_0$ and hence, by Theorem 1 of Ref. 1, $A^{-1}(I + CB)^{-1}C \in P$. □

The matrices T, R, and G of (7) satisfy the hypotheses on A, B, and C, respectively, of Theorem 6 if it happens that G is dominant (strongly dominant for (*iii*)). Thus, we have the result: *If the y-parameter matrix G is dominant then there is at most one solution of* (7). *If, furthermore, F maps E^n onto E^n, or if G is strongly dominant, then there exists a unique solution of* (7).

Making use of Theorem 4 and since $\det C \neq 0$ implies $\det [A^{-1} (I + CB)^{-1}C] \neq 0$, we also have: *There exists a unique solution of* (7) *if G is dominant and nonsingular.*

These results show that if the solution of the equation

$$TF(x) + Gx = \tilde{u}, \tag{8}$$

describing a given transistor network (with the transistors represented by the model of Fig. 1(a) is shown to (exist and) be unique by showing that the y-parameter matrix G is dominant (and $\det G \neq 0$, or that F maps E^n onto E^n), then any other network obtained from the original by adding arbitrary (nonnegative) resistances in series with any of the transistor leads will be described by (7) and, furthermore, the solution of (7) will also (exist and) be unique. Thus, the addition of series lead resistance does not affect the existence and uniqueness of the solution, provided G is dominant.

We now prove another result concerning the relationship between the existence and uniqueness of solutions of the two equations (7) and (8). We prove that, roughly speaking, whenever (8) has a unique solution for all transistors and diodes then so does (7). More precisely, let us define, for a given transistor network, the class of matrices $\mathfrak{T}$:

Definition: Let (8) describe the given network for some choice of transistor parameters α_{12}, α_{21}, for each transistor. Let $\mathfrak{T}$ then denote that class of matrices T obtained by considering all possible combinations of values of α_{12}, α_{21} $(0 < \alpha_{12} < 1, 0 < \alpha_{21} < 1)$ for each transistor. We then have:

Theorem 7: If (8) *has a unique solution for each* $T \in \mathfrak{T}$, *and each* $F \in \mathfrak{F}^n(E^n; \alpha, \beta)$ *for all* $\alpha < \beta$ *whose components lie in the extended real number system then, for each* R, *so does* (7).

Proof: The hypotheses imply (using Theorem 4) that $T^{-1}G \in P_0$ and $\det [T^{-1}G] \neq 0$ for each $T \in \mathfrak{T}$. Thus, G^{-1} exists. Letting

$$H \equiv [I + GR]^{-1}G,$$

H^{-1} exists and,

$$H^{-1} = G^{-1} + R.$$

As pointed out in Ref. 1, since $\det [T^{-1}G] \neq 0$, $T^{-1}G \in P_0$ for every $T \in \mathfrak{T}$ implies that $G^{-1}T \in P_0$ for every $T \in \mathfrak{T}$. Hence

$$\det [G^{-1}T + D] > 0, \qquad \text{for all } T \in \mathfrak{T} \text{ and all } D > 0.$$

But then,

$$\det [G^{-1} + DT^{-1}] > 0, \qquad \text{for all } T \in \mathfrak{T} \text{ and all } D > 0.$$

Now, due to the special structure of the matrix R (that is, block diagonal with dominant blocks that are "compatible" with T^{-1}) it is clear that, for any such R, any diagonal $D > 0$, and any $T \in \mathfrak{T}$, there exists a diagonal $\Delta > 0$ and some $M \in \mathfrak{T}$, such that $R + DT^{-1} = \Delta M^{-1}$. Hence, it is clear that

$$\det [G^{-1} + R + DT^{-1}] > 0, \qquad \text{for all } T \in \mathfrak{T} \text{ and all } D > 0.$$

It easily follows that $H^{-1}T \in P_0$ and hence $T^{-1}H \in P_0$ for all $T \in \mathfrak{T}$. Applying Theorem 4, we thus have that there exists a unique solution of (7) for each $T \in \mathfrak{J}$, and each $F \in \mathfrak{F}^n(E^n; \alpha, \beta)$ for all $\alpha < \beta$ whose components lie in the extended real number system. □

It is not difficult to show that there exist transistor networks for which $[I + GR]^{-1}G$ is dominant while G is not, and also networks for which G is dominant while $[I + GR]^{-1}G$ is not. For the first case, pick any network for which G is not dominant and $\det G \neq 0$. If the values of the series lead resistors in each transistor lead are then allowed to become large, since

$$[I + GR]^{-1}G = [I + R^{-1}G^{-1}]^{-1}R^{-1},$$

and since each element of R^{-1} approaches zero as the lead resistor values approach infinity, we see that $[I + GR]^{-1}G \to R^{-1}$. But R^{-1} is strongly dominant and hence there certainly exist sufficiently large values for the lead resistors such that $[I + GR]^{-1}G$ is dominant. The network of Fig. 2 is an example of the other case. For this network,

$$G = \begin{bmatrix} 1 & 0 & -1 & 0 \\ 0 & 1 & 0 & -1 \\ -1 & 0 & 1 & 0 \\ 0 & -1 & 0 & 1 \end{bmatrix}, \quad R = \begin{bmatrix} 9 & 9 & 0 & 0 \\ 9 & 9 & 0 & 0 \\ 0 & 0 & 9 & 9 \\ 0 & 0 & 9 & 9 \end{bmatrix},$$

while

$$[I + GR]^{-1}G = \frac{1}{37}\begin{bmatrix} 19 & -18 & -19 & 18 \\ -18 & 19 & 18 & -19 \\ -19 & 18 & 19 & -18 \\ 18 & -19 & -18 & 19 \end{bmatrix}.$$

IV. A SPECIAL CLASS OF TRANSISTOR NETWORKS

Transistor networks in which the base terminal of each transistor is connected to a common node are considered in Ref. 1 using the model of Fig. 1(a) to represent the transistor. It is shown there that there is at most one pair of base-collector and base-emitter voltages for each transistor in such a network—even in the cases in which the network is not described by an equation having the form of (1).

In this section we show that the class of common-base transistor networks is but a subset of a considerably more extensive special class of transistor networks for which the same statement is true. We show that there is at most one pair of base-collector and base-emitter volt-

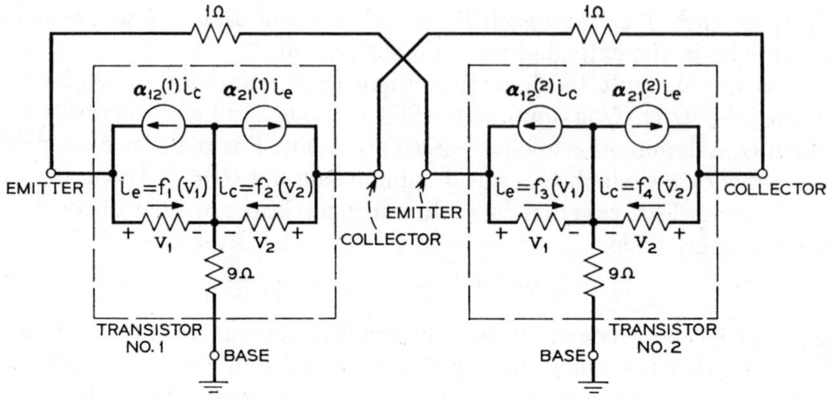

Fig. 2 — A two-transistor network.

ages for each transistor in any dc network which has the structure shown in Fig. 3. The box at the top of Fig. 3 represents, assuming that there are n transistors, any $(2n + 1)$-terminal network consisting of independent voltage and current sources, resistors (that is, linear resistors having nonnegative resistance), and diodes (that is, nonlinear resistors which are described by strictly monotone increasing conductance functions). Each of the n boxes at the bottom of Fig. 3 represents an

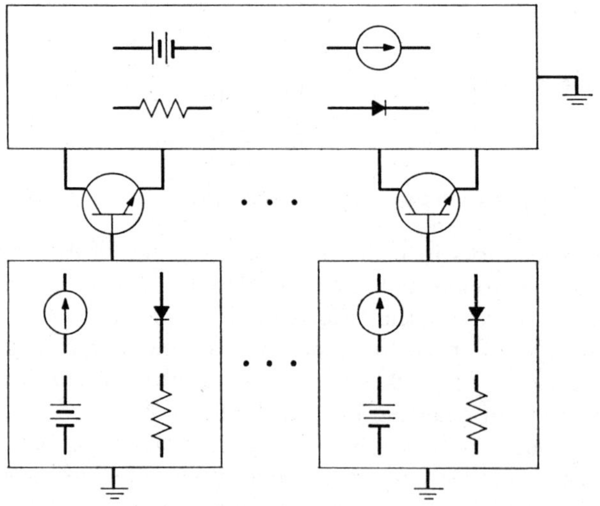

Fig. 3 — A special class of transistor networks.

arbitrary 2-terminal network consisting of independent sources, resistors, and diodes. Each of the transistors in Fig. 3 is represented by the model of Fig. 1(b), in which the value of each of the resistors r_b , r_c , r_e may be any nonnegative number. In this regard, we note here that it suffices in what follows to show, for each transistor, the uniqueness of the voltages v_1 and v_2 (in Fig. 1(b)) since, clearly, the voltages $\tilde{v}_1$ and $\tilde{v}_2$ are then uniquely determined.

As in Ref. 1 we assume, temporarily, that no diodes are present in the network. This assumption allows each of the n boxes at the bottom of Fig. 3 to be replaced by either a current source or else a Thévenin's equivalent circuit in which the value of the Thévenin's resistor is not infinite. Let us temporarily ignore the possibility that any of these boxes is equivalent to a current source. Following the technique presented in Section IX of Ref. 1, we may then consider the network of Fig. 4 instead of that of Fig. 3. In Fig. 4 we have explicitly shown the base, emitter, and collector resistors of each transistor, and we consider the Thévenin's resistor of each base circuit to be lumped in with the corresponding base resistor. The m-vectors v^* and i^* ($m \leqq 2n$) and the $2n$-vectors v' and i' are related by the four equations:

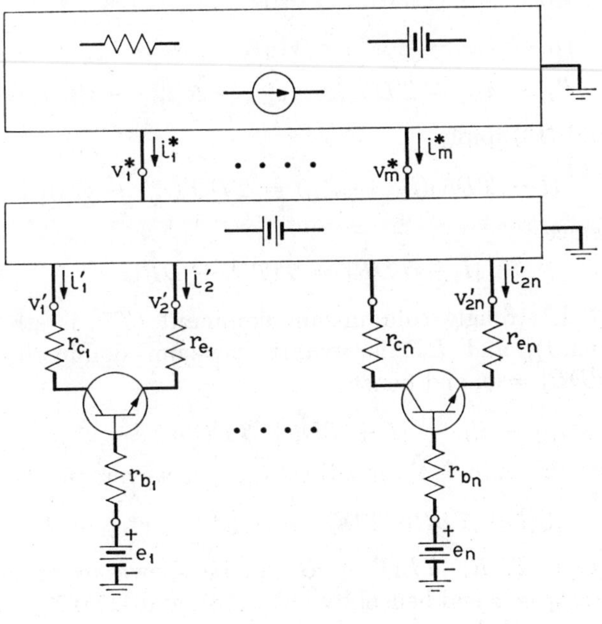

Fig. 4 — Network derived from that of Fig. 3.

$$i^* = -Gv^* + b, \tag{9}$$

$$i^* = Qi', \tag{10}$$

$$v' = Q^t v^* + c, \tag{11}$$

$$i' = TF(v' - e - Ri'), \tag{12}$$

in which b, c, and e are vectors whose elements are constants, G is a dominant matrix, Q is an $m \times 2n$ matrix having the property that whenever the $2n \times 2n$ matrix M is strongly column-sum dominant then so is the $m \times m$ matrix QMQ^t, T and R are $2n \times 2n$ block diagonal matrices having 2×2 diagonal blocks of the form (4) and (5), respectively.

We now show that the vectors v^*, i^*, v', and i' which satisfy (9) through (12) are unique (if they exist). Let $\{v^*_{(1)}, i^*_{(1)}, v'_{(1)}, i'_{(1)}\}$ and $\{v^*_{(2)}, i^*_{(2)}, v'_{(2)}, i'_{(2)}\}$ denote two sets of vectors, each of which satisfies (9) through (12). Subtracting corresponding equations, and observing the strictly monotone character of F, we see that there exists a diagonal matrix $D > 0$ such that:

$$i^*_{(1)} - i^*_{(2)} = -G(v^*_{(1)} - v^*_{(2)}), \tag{13}$$

$$i^*_{(1)} - i^*_{(2)} = Q(i'_{(1)} - i'_{(2)}), \tag{14}$$

$$v'_{(1)} - v'_{(2)} = Q^t(v^*_{(1)} - v^*_{(2)}), \tag{15}$$

$$i'_{(1)} - i'_{(2)} = TD(v'_{(1)} - v'_{(2)} - R(i'_{(1)} - i'_{(2)})). \tag{16}$$

But (15) and (16) imply

$$[I + TDR](i'_{(1)} - i'_{(2)}) = TDQ^t(v^*_{(1)} - v^*_{(2)}).$$

However, since

$$[I + TDR] = T[T^{-1} + DR],$$

in which T is strongly column-sum dominant (T^{-1} is strongly row-sum dominant), and DR is weakly row-sum dominant, we have $\det [I + TDR] \neq 0$, and hence,

$$i'_{(1)} - i'_{(2)} = [I + TDR]^{-1}TDQ^t(v^*_{(1)} - v^*_{(2)}). \tag{17}$$

Substituting this into (14) and then (13), however, yields:

$$\{Q[I + TDR]^{-1}TDQ^t + G\}(v^*_{(1)} - v^*_{(2)}) = \theta.$$

Now if $Q[I + TDR]^{-1}TDQ^t + G$ can be shown to be nonsingular then $v^*_{(1)} - v^*_{(2)} = \theta$ and hence, by (13), (15), and (17): $i^*_{(1)} - i^*_{(2)} = \theta$, $v'_{(1)} - v'_{(2)} = \theta$, and $i'_{(1)} - i'_{(2)} = \theta$, which, together, show that the vectors which satisfy (9) through (12) are unique. Since G is dominant it suffices to show that $[I+TDR]^{-1}TD$ (and hence $Q[I+TDR]^{-1}TDQ^t$) is strongly column-sum dominant. But

$$[I + TDR]^{-1}TD = [D^{-1}T^{-1} + R]^{-1},$$

which is the inverse of the direct sum of 2×2 strongly row-sum dominant matrices and is, therefore, strongly column-sum dominant.

Let us now consider the case in which diodes are present in the box at the top of Fig. 3. In this case, arguing as in Section IX of Ref. 1, if the set of base-emitter and base-collector voltages for Fig. 3 was not unique, we could replace all of the diodes by an appropriate series combination of a voltage source and a (nonnegative) resistor and thus synthesize a network of the type just considered, for which the set of base-emitter and base-collector voltages is not unique. This is a contradiction, and hence establishes that the set of base-emitter and base-collector voltages for the network of Fig. 3 is unique even when diodes are present in the top box.

A somewhat similar argument may now be used to show the uniqueness of the voltage across each of the diodes in the box at the top of Fig. 3. Assume that there exist two sets of branch voltages and currents, S_1 and S_2, which satisfy Kirchoff's and Ohm's laws for the network of Fig. 3. Since we have just proved the uniqueness of the base-emitter and base-collector voltages of each transistor, the elements of S_1 and S_2 which correspond to any such voltage must be identical. Thus, if each transistor is replaced by, say, an appropriate pair of voltage sources, the sets S_1 and S_2 still satisfy Kirchoff's and Ohm's laws for the modified network. Let us now choose (arbitrarily) any diode in the network and, as in the previous argument, replace all other diodes by a series combination of a voltage source and a (nonnegative) resistor, thus obtaining a new network, containing only one diode, for which the sets S_1 and S_2 still satisfy Kirchoff's and Ohm's laws. Suppose this remaining diode is characterized by the equation $i = f(v)$. The (now linear) network to which this diode is connected contains only independent sources and nonnegative resistors, and hence is characterized by one of the equations: $-i = gv + I_0$, $v = V_0$, where $g \geqq 0$, I_0, and V_0 are constants. Due to the strictly monotone increasing character of f, however, the graph of either of the above equations can intersect the graph of f in at most one point. Thus, the elements of S_1 and S_2 that specify the voltage across this diode must be equal. We can therefore conclude that the corresponding elements of S_1 and S_2 which specify the voltage across any diode are equal. That is, the diode voltages are unique for all diodes in the box at the top of Fig. 3.

We now consider the case in which some box at the bottom of Fig. 3 is equivalent to a current source. Let I_b denote the value of this current source (with reference direction chosen to be *out* of the base of the associated transistor). In this case, using the notation of Fig. 1(b), the variables v_1, i_1, v_2, and i_2, for the associated transistor, are constrained by the relationships:

$$i_1 = \frac{(1 - \alpha_{12}\alpha_{21})f_1(v_1) - \alpha_{12}I_b}{(1 - \alpha_{12})},$$

$$i_2 = \frac{(1 - \alpha_{12}\alpha_{21})f_2(v_2) - \alpha_{21}I_b}{(1 - \alpha_{21})}. \tag{18}$$

Thus, this transistor can be replaced by a pair of diodes (each in series with one of the resistors r_e, r_c) whose nonlinear conductance functions are specified by (18). We may now consider these diodes, these resistors, and the current source, all to be components of the box at the *top* of Fig. 3. We have thus shown, in summary, that when one (or more) of the boxes at the bottom of Fig. 3 is equivalent to a current source, the base-emitter and base-collector voltages of each transistor are still unique, since the network is then equivalent to a network of a type already considered.†

By use of the same type of argument that was applied to the case in which diodes are present in the box at the top of Fig. 3, the above results may, finally, be shown to be valid when diodes are present in the boxes at the bottom of Fig. 3.

The above results show the validity of the following statement concerning bistable networks: *One cannot synthesize a bistable network which consists of resistors, inductors, capacitors, diodes, independent voltage and current sources, and an arbitrary number of (Fig. 1b) transistors, and which has the structure of Fig. 3 when all capacitors are open-circuited and all inductors are short-circuited.*

APPENDIX

In this appendix we give the proof of a theorem which is used here and which is implied in Ref. 1 but is not stated explicitly there.

Theorem: If A is a real square matrix then $A \in P_0$ if and only if $\det [A + D] \neq 0$ for every diagonal matrix $D > 0$.

† Here, of course, we use the proposition, proved above, that the voltage across each diode in the box at the top of Fig. 3 is unique.

Proof: (*if*) Suppose $A \notin P_0$. If $\det A < 0$ then for sufficiently small $\zeta > 0$, $\det [\zeta I + A] < 0$. For sufficiently large ζ, however,

$$\det [\zeta I + A] = \zeta^n \cdot \det \left[I + \frac{1}{\zeta} A \right] > 0.$$

Thus, since $\det [\zeta I + A]$ is a continuous function of ζ, there exists some value of $\zeta > 0$ such that $\det [\zeta I + A] = 0$. For this value of ζ let $D = \zeta I$.

If $\det A \geqq 0$ but, for some positive integer $k < n$, A has a $k \times k$ principal minor which is negative we may, without loss of generality, assume that A is partitioned as

$$A = \begin{bmatrix} A_1 & A_2 \\ A_3 & A_4 \end{bmatrix},$$

where A_1 is a $k \times k$ matrix with $\det A_1 < 0$. This is so because $\det [D + A]$ is not altered if any two rows and then the corresponding pair of columns are interchanged. Let $D^{(1)} = \operatorname{diag}[d_1, \cdots, d_n]$ with $d_1 = \cdots = d_k = \xi$, where $\xi > 0$ is chosen so small that $\det[\xi I + A_1] < 0$. Then, with $d_{k+1} = \cdots = d_n = \zeta > 0$, we have

$$\det [D^{(1)} + A] = \det \begin{bmatrix} \xi I + A_1 & A_2 \\ A_3 & \zeta I + A_4 \end{bmatrix}$$

$$= \zeta^{n-k} \cdot \det \begin{bmatrix} \xi I + A_1 & A_2 \\ \frac{1}{\zeta} A_3 & I + \frac{1}{\zeta} A_4 \end{bmatrix}.$$

Thus, for $\zeta > 0$ chosen to be sufficiently large, $\det[D^{(1)} + A] < 0$. Now, if $D^{(2)} = \eta I$, for $\eta > 0$, then it is clear that for η chosen sufficiently large,

$$\det [D^{(2)} + A] = \eta^n \cdot \det \left[I + \frac{1}{\eta} A \right] > 0.$$

Thus, if

$$D(\epsilon) = \epsilon D^{(1)} + (1 - \epsilon) D^{(2)},$$

it is clear that there exists a value of ϵ, $0 < \epsilon < 1$, such that $\det [D(\epsilon) + A] = 0$.

(*only if*) By Theorem 1 of Ref. 1, since $A \in P_0$ and $D > 0$, $[D + A] \in P$. Thus, $\det [D + A] \neq 0$. □

REFERENCES

1. Sandberg, I. W. and Willson, A. N., Jr., "Some Theorems on Properties of DC Equations of Nonlinear Networks," B.S.T.J., *48,* No. 1 (January 1969), pp. 1–34.
2. Fiedler, M. and Pták, V., "On Matrices with Non-Positive Off-Diagonal Elements and Positive Principal Minors," Czech. Math. J., *12,* No. 3 (1962), pp. 382–400.
3. Fiedler, M. and Pták, V., "Some Generalizations of Positive Definiteness and Monotonicity," Numer. Math., *9,* No. 2 (1966), pp. 163–172.
4. Sandberg, I. W., "On the Theory of Linear Multi-Loop Feedback Systems," B.S.T.J., *42,* No. 2 (March 1963), pp. 355–382.

New Theorems on the Equations of Nonlinear DC Transistor Networks

By ALAN N. WILLSON, JR.

(Manuscript received March 26, 1970)

It has long been recognized that equations describing dc transistor networks do not necessarily have unique solutions. The Eccles-Jordan (flip-flop) circuit is an excellent example of one for which the dc equations may have more than one solution.

Only recently, however, has a comprehensive theory concerning matters such as the existence and uniqueness of solutions of the dc equations of general transistor networks begun to take shape. This paper represents another contribution to the evolution of that theory.

A key concept in the development of the recent theory is the concept of a "P_0 matrix." We give a generalization of that concept, showing that one can specify properties possessed by certain pairs of square matrices, analogous to the properties possessed by a single P_0 matrix. Pairs of matrices possessing these properties are called $\mathcal{W}_0$ pairs. Use is made of this $\mathcal{W}_0$ pair concept to prove results which are more general than some of the existing ones. We provide an extension of much of the existing theory in such a manner that a broader class of dc transistor networks may be considered. In particular, the new results provide one with the ability to answer certain questions concerning the existence, uniqueness, boundedness, and so on, of solutions of the equations for any *network which is comprised of transistors, diodes, resistors, and independent sources.*

I. INTRODUCTION

Suppose a network is constructed by connecting in an arbitrary manner any number of transistors, diodes, resistors, and independent voltage and current sources. Without loss of generality, we may consider the network to have the canonical form shown in Fig. 1; that is, we may consider the network to be a multiport containing resistors and inde-

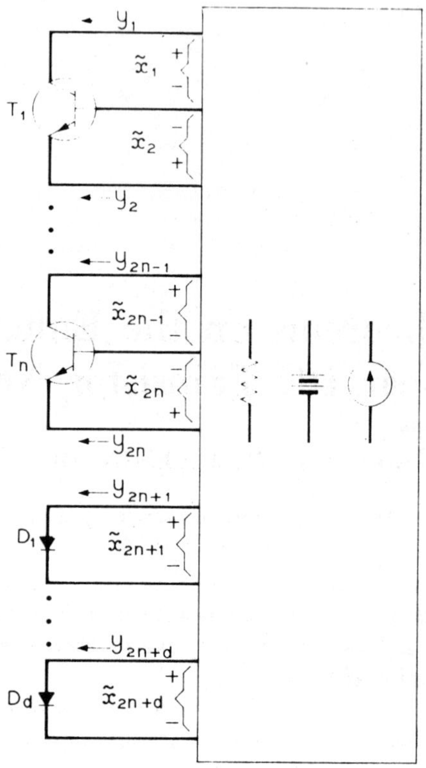

Fig. 1—Canonical form of a transistor network.

pendent sources, with transistors and diodes connected to the ports.*

There are some fundamental questions that one should then, hopefully, be able to answer. For example: Do the equations that describe this dc network have a unique solution? With the exception of certain uniqueness results for a special (but none the less important) class of transistor networks, all of the previous explicit results in Refs. 1, 2, and 3, which have shown methods for obtaining answers to such questions, have been concerned only with the class of transistor networks for which, after setting the value of each independent source to zero, there exists a short-circuit admittance matrix (a G matrix) to characterize the linear

* It will become apparent that the theory can also accommodate many other structures which are of the Fig. 1 type except that the multiport contains additional linear elements (such as controlled sources). We do not stress this point though, since in the present context such elements seem somewhat unnatural.

multiport of Fig. 1. It is the primary purpose of this paper to show how that restriction can be removed. We shall show in fact that almost all of the previous results are but special cases of results that follow from a more general theory in which the assumption of the existence of a G matrix for the linear multiport is unnecessary.*

Section II concerns methods for characterizing a general multiport containing resistors and independent sources. In Section III, we consider the model for a transistor. An equation for dc transistor networks is then developed in Section IV and, after explaining some notation in Section V, we develop the $\mathcal{W}_0$ pair concept in Section VI. Sections VII, VIII and IX show how the $\mathcal{W}_0$ pair concept provides a generalization of the existing results concerning dc transistor networks. Finally, we consider an example network in Section X.

II. LINEAR MULTIPORT CHARACTERIZATION

A multiport having n ports (an n-port) is *characterized* by determining every combination of the $2n$ port voltages and currents that the network admits (see Ref. 4). We discuss here two methods of characterizing multiports that contain resistors and independent sources. The first method makes use of the familiar concept of a hybrid matrix. The second method uses a pair of matrices in a manner that was apparently first suggested—for multiports containing no independent sources—by V. Belevitch.[5]

2.1 *The Hybrid Formalism*

When the value of each independent source is set to zero, for a multiport containing only resistors and independent sources, the multiport becomes, of course, a *resistive* multiport. H. C. So has proved (as a special case of a theorem in Ref. 6) that *any resistive multiport has a hybrid matrix description*. That is, for any resistive n-port, it is always possible to label the port voltage and current variables in such a way that there

* Pragmatists might argue that in any "physical" network, there will always be enough "stray" resistance present which, if taken into account, will guarantee the existence of, say, a G matrix. It seems to this writer, however, that by taking such a point of view, one does not obtain an entirely satisfactory understanding of matters (even *practical* matters). To know that fundamental results do not *depend* (if, in fact, they don't) upon such fortunate occurrences as these (and for many transistor networks this is the case) seems to be the more satisfactory situation. Furthermore, it should be noted that in the analysis of a physical network, to obtain a tractable problem, it often behooves one to neglect the presence of unimportant elements. Thus, it is not necessarily true that such stray resistors will always be present in the model of the network which the analyst desires to consider.

exists an integer m, $0 \leqq m \leqq n$, a pair of n-vectors*

$$x = (i_1, \cdots, i_m, v_{m+1}, \cdots, v_n)^T,$$
$$y = (v_1, \cdots, v_m, i_{m+1}, \cdots, i_n)^T,$$

and a real $n \times n$ matrix H, the hybrid matrix, such that the network admits the port variables v_k, i_k as the voltage and current, respectively, at the kth port, for $k = 1, \cdots, n$, if and only if the vectors x and y satisfy

$$y = Hx. \tag{1}$$

Thus, a resistive multiport may always be characterized by a hybrid matrix.

When independent sources whose values are nonzero are present in an otherwise resistive multiport, a hybrid matrix will not generally suffice to characterize the multiport. Clearly the vectors $x = y = (0, 0, \cdots, 0)^T$ which satisfy equation (1) for any matrix H do not always specify an admissible combination of port variables when independent sources are present. One might hope, however, that a characterization of the type

$$y = Hx + c, \tag{2}$$

where c is some constant vector (whose elements are real numbers), might always be possible. Indeed, we are about to show that this is the case. There is one problem, however, that was not present in the consideration of resistive n-ports that must first be dealt with: there are ways to interconnect independent sources and resistors such that the resulting structure doesn't make sense. That is, the independent sources might impose self-contradictory constraints on the network. We rule out such possibilities by agreeing that, when we refer to "a multiport containing resistors and independent sources," we always assume that the multiport possesses the following property:

Assumption: The linear graph that is formed by associating an edge with each resistor, each independent source, and each port, has no cut-sets containing only current source edges for which the values of the current sources cause a violation of Kirchhoff's current law. Similarly, no circuits of voltage source edges for which the values of the voltage sources cause a violation of Kirchhoff's voltage law are present.

This assumption in no way restricts the generality of our work. We

* We use the superscript T to denote the transpose of a vector or a matrix. Thus, the vectors x and y above are both column vectors.

are simply ruling out multiports, like the 2-port of Fig. 2, for which the set of admissible port voltage and current combinations is empty.

We have worded the Assumption so that the presence of, say, a series connection of two 1-ampere current sources in an otherwise resistive multiport does not cause the multiport to be inadmissible. We have done this because no violation of Kirchhoff's laws results from such interconnections of resistors and sources; the network is perfectly legitimate. One should be aware, however, that if "superfluous" sources are present in a network, it will follow that one cannot uniquely determine the value of each branch voltage and current in the network. That is, even though one might be able to uniquely determine the value of the voltage across the *pair* of 1-ampere sources, there is no way to determine the value of the voltage across each individual source. Aside from such ambiguities, it follows (see below and the proof of Theorem 1 in Ref. 6) that the value of all branch voltages and currents can be uniquely determined for a multiport satisfying the Assumption, whenever the values of the "independent" port variables are known.

Theorem: *Any multiport containing resistors and independent sources can be characterized by equation* (2), *where* H *is a hybrid matrix characterization of the corresponding resistive multiport that is obtained by setting all independent source values to zero, and* c *is a vector of real numbers.*

A proof of this theorem can be constructed by incorporating a few simple observations and minor modifications into the arguments used by So in Ref. 6. We therefore simply sketch the main ideas: First, if the linear graph mentioned in the Assumption contains any current source cut-sets, then it must be the case (because of that Assumption) that these sources have values such that Kirchhoff's current law is satisfied. That being the case, the port behavior of the multiport will clearly be unaltered if a sufficient number of current sources are removed (by coalescing appropriate nodes) to eliminate such cut-sets. A similar observation applies to voltage source circuits. Therefore without any loss of generality, we may consider the linear graph to have no current source cut-sets and no voltage source circuits. Next, by Lemmas 1 and 2 of Ref. 6, it then follows that there exists a tree* for the linear graph for which all voltage source edges are branches and all current source edges are links. At each port, one of the two port variables is then designated as "independent," the choice depending upon whether the edge corresponding to that port is a branch or a link. The existence of the

* In case the linear graph is not *connected* each reference to the word *tree* should, of course, be changed to *forest*.

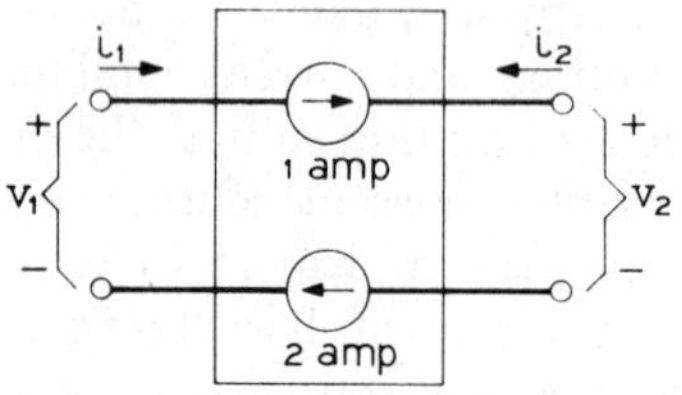

Fig. 2—An inadmissible n-port.

hybrid matrix H and the vector c for the characterization (2) then follows in the same manner as the existence of a hybrid matrix for a resistive multiport follows from So's arguments.

2.2 *Belevitch's Formalism*

For some multiports, it might be that (after setting all independent source values to zero) a hybrid matrix exists such that the vectors x and y in equation (1) satisfy $x = v \equiv (v_1, \cdots, v_n)^T$ and $y = i \equiv (i_1, \cdots, i_n)^T$. In this case the hybrid matrix is given the special name, *admittance matrix*. Similarly, if it happens that H exists such that $x = i$ and $y = v$, then H is called the *impedance matrix*. For many resistive multiports, neither an impedance matrix nor an admittance matrix exists. It is still possible, however, to characterize any n-port for which a hybrid matrix exists in terms of the vectors v and i. Obviously, x and y satisfy equation (1) if and only if v and i satisfy

$$[I_l \,\vdots\, -H_r]v = [H_l \,\vdots\, -I_r]i, \tag{3}$$

where the $n \times m$ matrix H_l and the $n \times (n - m)$ matrix H_r are defined by $H = [H_l \,\vdots\, H_r]$, and similarly $[I_l \,\vdots\, I_r]$ is the $n \times n$ identity matrix.

The characterization (3), being equivalent to equation (1), is perfectly adequate for any resistive n-port. It is, however, but a special case of a more general characterization due to Belevitch, namely:

$$Pv = Qi, \tag{4}$$

where P and Q are $n \times n$ real matrices. Belevitch's characterization can be used for quite a broad class of networks, including some rather pathological ones which require dependent sources, or gyrators and negative resistors to realize, and for which no hybrid characterization exists. For example, the one-port called a *norator*, for which the set of admissible port voltage and current combinations is the set of all pairs of real numbers, may be characterized by $[0]v = [0]i$. We should note, however, that if one allows the aforementioned elements to be present in an n-port, then even equation (4) cannot always provide a characterization. The *nullator*, for example, a one-port whose only admissible combination of port voltage and current variables is the pair (0, 0), is such an n-port.

When an n-port contains independent sources it can often be characterized by the equation

$$Pv = Qi + c, \tag{5}$$

where P and Q are real $n \times n$ matrices, and c is a constant vector. Clearly, any n-port containing only resistors and independent sources has such a characterization. It is this class of n-ports which is our primary concern in the study of transistor networks. We note, however, that equation (5) is adequate for characterizing a much broader class of n-ports.

III. NONLINEAR TRANSISTOR CHARACTERIZATION

In Fig. 3, a commonly used large signal dc transistor model is displayed. It is easily verified that the voltage and current variables defined in that figure obey the following relationships:

$$\begin{bmatrix} i_1 \\ i_2 \end{bmatrix} = \begin{bmatrix} 1 & -\alpha_r \\ -\alpha_f & 1 \end{bmatrix} \begin{bmatrix} f_1(v_1) \\ f_2(v_2) \end{bmatrix}, \tag{6}$$

$$\begin{bmatrix} v_1 \\ v_2 \end{bmatrix} = \begin{bmatrix} \tilde{v}_1 \\ \tilde{v}_2 \end{bmatrix} - \begin{bmatrix} r_e + r_b & r_b \\ r_b & r_c + r_b \end{bmatrix} \begin{bmatrix} i_1 \\ i_2 \end{bmatrix}. \tag{7}$$

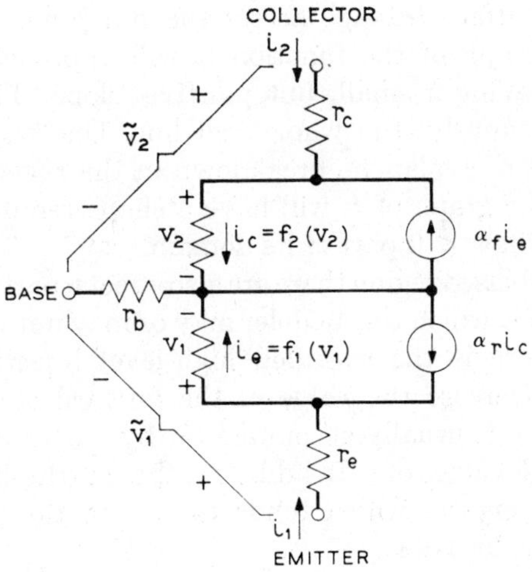

Fig. 3—Large signal dc transistor model.

Each of the parameters α_f and α_r may assume any value in the open interval (0, 1). The parameters r_b, r_c, and r_e, which account for lead resistances, are sometimes omitted by device modelers (their presence is sometimes accounted for by including appropriate additional resistors in the network to which the transistor model is connected). To accommodate these various points of view we specify only, therefore, that the values of the parameters r_b, r_c, and r_e be nonnegative. Thus any or all of them may be zero.

Depending upon whether the transistor being modeled is a pnp or an npn, the graph of each of the functions f_1 and f_2 has one of the general shapes shown in Fig. 4 (at least for values of $|v|$ that are "not too large"). Often these functions are described by an equation of the form

$$f_k(v) = m_k[\exp(n_k v) - 1], \qquad (k = 1, 2), \tag{8}$$

where m_k and n_k are appropriately chosen constants, both being positive for a pnp transistor, and both negative for an npn. On the other hand, for example, a piecewise-linear representation is sometimes specified for f_1 and f_2.

The nature of the functions f_1 and f_2 for large values of $|v|$ depends upon which assumptions the modeler is willing to make, and which effects he is interested in considering. For large negative (in the pnp case) values of v, for example, the graph of f_k approaches—according to equation (8)—the horizontal asymptote $i = -m_k$. Thus, if the modeler chooses to use equation (8) to describe f_k for all values of v, the range of f_k will not be the entire real line. If, on the other hand, the effect of ohmic surface leakage across the p-n junction is included in the model, the graph of the function f_k will approach asymptotically a straight line having a small, but positive, slope. The range of such a function is, obviously, the whole real line. One might also wish to include the effect of avalanche breakdown in the reverse-biased region. If this is done, the graph of f_k will have a shape reminiscent of that of a Zener diode in the $v < 0$ part of its domain.

In the forward-biased region there are also effects, particularly apparent for large values of v, which the modeler may or may not wish to recognize. For example, there is the so-called high-level injection phenomenon which tends to decrease the value of the forward current and which, using equation (8), is usually accounted for by a decrease in the magnitude of n_k for large values of v. In addition, there is the effect of the ohmic resistance of the crystal which tends to reduce the value of forward current for large values of v.

From the point of view of the device modeler, the question of whether or not to include some of the effects mentioned above is often a minor issue. For many networks the behavior will be essentially the same whether or not, say, surface leakage is accounted for in the transistor model. From the point of view of the network analyst, however, the situation is somewhat different. For example, the matter of whether or not the functions f_k map the real line *onto* the real line can, in some cases, make the difference between whether or not there exists a solution of the network's equations. Similarly, other results that have been obtained recently (presented later, beginning in Section VII) also seem to depend upon the graphs of the functions f_k having certain special properties.

It seems safe to say that no matter which "special effects" are included (or omitted) in the description of the transistor, the functions f_k may at least be considered to be strictly monotone increasing mappings of the real line into itself. For the purpose of formulating the equations for transistor networks, this is the only hypothesis that we shall make. When additional hypotheses regarding the nature of these functions are needed (to obtain certain results concerning properties of these equations) those hypotheses will be mentioned explicitly. In each case it will be clear that the additional hypotheses are, in some appropriate sense, rather weak.

Similar remarks can be made for the diodes that are shown in Fig. 1, which might also be present in transistor networks. Thus, we assume that each diode is described by an equation of the type $i = f(v)$ where, at this point, we only assume that the function f is a strictly monotone increasing mapping of the real line into itself.

IV. EQUATIONS FOR TRANSISTOR NETWORKS

Suppose we are given a dc network consisting of transistors, diodes, resistors, and independent voltage and current sources, connected together in an arbitrary manner. Let there be n transistors and d diodes. Clearly, there is no loss of generality if we consider the network to be of the type shown in Fig. 1. Using the results of Section III, we may describe the nonlinear devices in the network by the equations

$$y = TF(x), \qquad x = \tilde{x} - Ry, \tag{9}$$

where $T = \operatorname{diag}[T_1, T_2]$, with T_1 a block diagonal matrix with n 2×2 diagonal blocks of the form

$$\begin{bmatrix} 1 & -\alpha_r^{(k)} \\ -\alpha_f^{(k)} & 1 \end{bmatrix}, \quad \text{for } k = 1, \cdots, n, \tag{10}$$

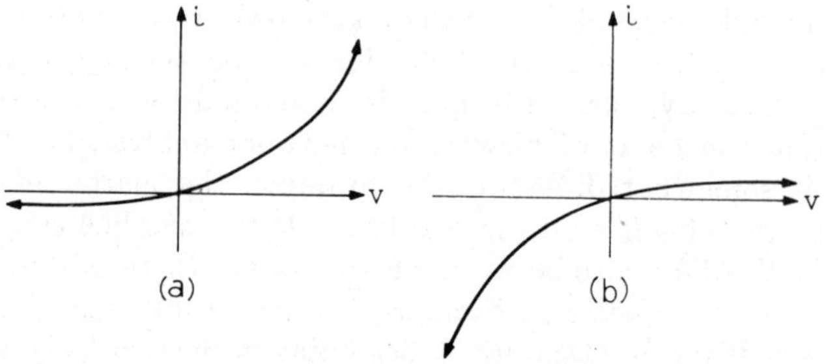

Fig. 4—General shape of the functions f_1 and f_2; (a) pnp transistor, (b) npn transistor.

and T_2 the $d \times d$ identity matrix. Also, $R = \text{diag}\,[R_1\,,\,R_2]$, with R_1 a block diagonal matrix with n 2×2 diagonal blocks of the form

$$\begin{bmatrix} r_e^{(k)} + r_b^{(k)} & r_b^{(k)} \\ r_b^{(k)} & r_c^{(k)} + r_b^{(k)} \end{bmatrix}, \qquad k = 1, \cdots, n, \tag{11}$$

and R_2 the $d \times d$ matrix whose elements are all zeros. The function F has the form $F(x) \equiv [f_1(x_1), \cdots, f_{2n+d}(x_{2n+d})]^T$, where each of the f_k is a strictly monotone increasing mapping of the real line into itself.

Using the results of Section II, the effect of the linear multiport in Fig. 1 is to constrain the vectors of port variables, $\tilde{x}$ and y, to obey the relationship

$$P\tilde{x} = -Qy + c, \tag{12}$$

where P and Q are $(2n + d) \times (2n + d)$ real matrices and c is a real $(2n + d)$-vector. The minus sign appears in equation (12) as a consequence of having chosen the reference direction for the port currents (the elements of the vector y) to be opposite to that which is usually assumed.

By using equations (9), we may easily eliminate the variables $\tilde{x}$ and y from equation (12), resulting in the equation

$$(PR + Q)TF(x) + Px = c. \tag{13}$$

The central problem in determining the values of all branch voltages and currents in a dc transistor network is the determination of a solution of equation (13). The rest is relatively straightforward, for if x is a (unique) solution of equation (13), then the (unique) vectors $\tilde{x}$ and y, such that equations (9) and (12) are satisfied, may immediately be computed from equations (9).

Since the matrix T is nonsingular, it follows that whenever either $(PR + Q)$ or P is nonsingular, equation (13) can be transformed into, respectively, one of the equations

$$F(x) + Ax = b, \tag{14}$$

$$AF(x) + x = b. \tag{15}$$

The first of these equations has been studied rather extensively (see Refs. 1–3 and 7) and for most of the results obtained there, it can be shown that parallel results are possible for equation (15). Both of these equations, however, are but special cases of the equation

$$AF(x) + Bx = c, \tag{16}$$

which accommodates equation (13) directly. It is, therefore, this equation to which we shall now direct our attention. It will be shown that most of the results which have been obtained to date for equation (14) have rather natural (though not obvious) extensions to equation (16). It is important that such extensions be possible because one is often forced to deal with equations like (16) in the analysis of transistor networks. Clearly, this is the case whenever both of the matrices $(PR + Q)$ and P of equation (13) are singular—and this can easily happen (for example, if the matrix R contains all zeros, then it will happen whenever there exists no admittance matrix nor impedance matrix for the linear multiport of Fig. 1).

V. NOTATION

The following notation shall be used throughout the remainder of the paper: For each positive integer n we denote by E^n the n-dimensional Euclidean space, the elements of which are ordered n-tuples of real numbers, which we consider to be column vectors. The origin in E^n is denoted by θ. If $x = (x_1, \cdots, x_n)^T$ and $y = (y_1, \cdots, y_n)^T$ are elements of E^n we denote their inner product by $\langle x, y \rangle = \sum_{k=1}^{n} x_k y_k$. The norm of each $x \in E^n$ is denoted by $||x|| = \langle x, x \rangle^{\frac{1}{2}}$.

If A is an $n \times n$ matrix, then for $k = 1, \cdots, n$, A_k denotes the kth column of A. A principal submatrix of a square matrix A is any square submatrix of A whose main diagonal is contained in the main diagonal of A. A principal minor of A is the determinant of any principal submatrix of A. If D is a diagonal matrix, then $D > 0$ means that each element of the main diagonal is a positive number; similarly, $D \geqq 0$ denotes that each element of the main diagonal is nonnegative. We denote the $n \times n$ identity matrix by either I_n or, when the dimension is unimportant or is clear from the context, simply by I. The direct sum of two matrices A, B is denoted by $A \oplus B$. A square matrix of real numbers A is said to be strongly row-sum dominant if its elements a_{ij} satisfy $a_{ii} > \sum_{j \neq i} |a_{ij}|$ for $i = 1, \cdots, n$.

If f is a real valued function defined on E^1 then f is said to be monotone increasing if for all $x < y$ it follows that $f(x) \leqq f(y)$. We say that f is strictly monotone increasing if $f(x) < f(y)$ for all $x < y$. For each positive integer n, we denote by $\mathfrak{F}^n$ that collection of mappings of E^n onto itself defined by: $F \in \mathfrak{F}^n$ if and only if there exist, for $i = 1, \cdots, n$, strictly monotone increasing functions f_i mapping E^1 onto E^1 such that for each $x = (x_1, \cdots, x_n)^T \in E^n$, $F(x) = [f_1(x_1), \cdots, f_n(x_n)]^T$.

VI. PAIRS OF MATRICES OF TYPE $\mathcal{W}_0$

Many of the recent results referred to above, concerning equation (14), have relied heavily upon certain properties that a matrix is known to possess whenever it is a member of a class of matrices that has been given the name P_0. In a similar way the results that follow rely upon useful properties that are possessed by certain *pairs of matrices.* We shall define a class, the elements of which are these pairs of matrices, and give it the name $\mathcal{W}_0$.

The class of matrices called P_0 was defined by M. Fiedler and V. Pták.[8] They proved that the following properties of a square matrix of real numbers, A, are equivalent:

(*i*) All principal minors of A are nonnegative.
(*ii*) For each vector $x \neq \theta$ there exists an index k such that $x_k \neq 0$ and $x_k(Ax)_k \geqq 0$.
(*iii*) For each vector $x \neq \theta$ there exists a diagonal matrix $D_x \geqq 0$ such that $\langle x, D_x x\rangle > 0$ and $\langle Ax, D_x x\rangle \geqq 0$.
(*iv*) Every real eigenvalue of A, as well as of each principal submatrix of A, is nonnegative.

Sandberg and Willson proved that another property can be added to this list of equivalent properties,[2,3] namely:

(*v*) $\det (D + A) \neq 0$ for every diagonal matrix $D > 0$.

The class of all matrices possessing one (and hence all) of the above properties is called P_0.

We shall now state a theorem which provides a useful generalization of the concept of the class of P_0 matrices.

Definition: For each pair of $n \times n$ matrices (A, B) we shall denote by $\mathcal{C}(A, B)$ the collection of all the $n \times n$ matrices that can be constructed by juxtaposing columns taken from either A or B while maintaining the original relative ordering of the columns. Thus, $M \in \mathcal{C}(A, B)$ if and only if for each $k = 1, \cdots, n$, either $M_k = A_k$ or $M_k = B_k$.

Obviously $\mathcal{C}(A, B)$ contains 2^n matrices (for certain pairs (A, B)—namely for those having $A_k = B_k$ for one or more values of k—it can happen that two or more matrices in $\mathcal{C}(A, B)$ are identical).

Definition: The pair of $n \times n$ matrices (M, N) is said to be a complementary pair taken from $\mathcal{C}(A, B)$ if and only if both M and N are members of $\mathcal{C}(A, B)$ and for each $k = 1, \cdots, n$, either $M_k = A_k$ and $N_k = B_k$, or else $M_k = B_k$ and $N_k = A_k$.

It is obvious that (A, B) is a complementary pair taken from $\mathcal{C}(A, B)$. It is also clear that $\mathcal{C}(A, B) = \mathcal{C}(B, A)$ and, moreover, that if (M, N) is any complementary pair taken from $\mathcal{C}(A, B)$, then $\mathcal{C}(M, N) = \mathcal{C}(A, B)$. Furthermore, for each $M \in \mathcal{C}(A, B)$ there exists $N \in \mathcal{C}(A, B)$ such that (M, N) is a complementary pair.

Theorem 1: The following properties of a pair of $n \times n$ matrices of real numbers (A, B) are equivalent:

(*i*) *det $(AD + B) \neq 0$ for every diagonal matrix $D > 0$.*

(*ii*) *There exists a matrix $M \in \mathcal{C}(A, B)$ such that det $M \neq 0$ and such that det $M \cdot$ det $N \geqq 0$ for all $N \in \mathcal{C}(A, B)$.*

(*iii*) *For each vector $x \neq \theta$ there exists an index k such that either $(A^T x)_k \neq 0$ or $(B^T x)_k \neq 0$, and such that $(A^T x)_k (B^T x)_k \geqq 0$.*

(*iv*) *For each vector $x \neq \theta$ there exists a diagonal matrix $D_x \geqq 0$ such that either $\langle A^T x, D_x A^T x \rangle > 0$ or $\langle B^T x, D_x B^T x \rangle > 0$ (that is, such that $\langle A^T x, D_x A^T x \rangle + \langle B^T x, D_x B^T x \rangle > 0$), and such that $\langle A^T x, D_x B^T x \rangle \geqq 0$.*

(*v*) *For each complementary pair of matrices (M, N) taken from $\mathcal{C}(A, B)$, each real value of λ that satisfies det$(M - \lambda N) = 0$ is nonnegative.*

(*vi*) *There exists a complementary pair of matrices (M, N) taken from $\mathcal{C}(A, B)$ such that $M^{-1}N \in P_0$.*

(*vii*) *There exists a matrix $M \in \mathcal{C}(A, B)$ such that det $M \neq 0$; and, for any complementary pair of matrices (M, N) taken from $\mathcal{C}(A, B)$ with det $M \neq 0$, $M^{-1}N \in P_0$.*

In this paper, we do not make use of properties (*iii*), (*iv*), or (*v*) of Theorem 1. The proof that the remaining four properties are equivalent is given in the Appendix. A complete proof of Theorem 1 is given elsewhere.[9]

Definition: The class of all pairs of matrices which possess one (and hence all) of the properties listed in Theorem 1 is called $\mathcal{W}_0$.

To see that properties (*i*) and (*ii*) of Theorem 1 are in fact generalizations of the previously mentioned properties (*v*) and (*i*), respectively, that define P_0 is a simple matter. It happens that for any $n \times n$ matrix B the pair $(I_n, B) \in \mathcal{W}_0$ if and only if $B \in P_0$. (This follows from property (*vii*) of Theorem 1.) With our attention restricted to pairs of matrices of the type (I_n, B), it is clear that property (*i*) of Theorem 1 is equivalent to property (*v*) which determines those matrices B that that are in P_0. Concerning property (*ii*) of Theorem 1, an arbitrary matrix $N \in \mathcal{C}(I_n, B)$ is either the matrix I_n or else, a matrix formed from B by replacing some of the columns of B by the corresponding columns of I_n. Consequently, $\det N = \det B_N$ where B_N is the principal submatrix of B that is formed by removing from B the columns that are not present in N and then removing the corresponding rows. Hence, since $\det I_n \neq 0$, we may take I_n to be the matrix M in property (*ii*) of Theorem 1, and observe that this property then becomes: $\det B_N \geqq 0$ for all $N \in \mathcal{C}(I_n, B)$. It is now clear that this property is equivalent to the property (*i*) that defines the class of P_0 matrices. (Note that there are exactly $2^n - 1$ principal minors for each $n \times n$ matrix, and that the set $\mathcal{C}(I_n, B) \setminus \{I_n\}$ contains exactly $2^n - 1$ members.)

VII. THEOREMS ON EXISTENCE AND UNIQUENESS

7.1 *First Existence and Uniqueness Theorem*

The following theorem, which is proved in Ref. 2, provides a necessary and sufficient condition for the existence of a unique solution of equation (14) for all F that are strictly monotone increasing "diagonal" mappings of E^n *onto* E^n and for all $b \in E^n$.

Theorem 2: *If A is an $n \times n$ matrix of real numbers, then there exists a unique solution of equation* (14) *for each $F \in \mathfrak{F}^n$ and for each $b \in E^n$ if and only if $A \in P_0$.*

Using this theorem along with the results of Section VI we can prove the following (more general) theorem.

Theorem 3: *If A and B are $n \times n$ matrices of real numbers, then there exists a unique solution of equation* (16) *for each $F \in \mathfrak{F}^n$ and each $c \in E^n$ if and only if $(A, B) \in \mathcal{W}_0$.*

Proof: (*if*) Let $(A, B) \in \mathcal{W}_0$. Then, by Theorem 1, there exists a complementary pair (M, N) taken from $\mathcal{C}(A, B)$ such that $M^{-1}N \in P_0$. For each $F \equiv [f_1(\cdot), \cdots, f_n(\cdot)]^T \in \mathfrak{F}^n$ let $G \equiv [g_1(\cdot), \cdots, g_n(\cdot)]^T$ denote the mapping (also in $\mathfrak{F}^n$) defined by

$$g_k(\cdot) = \begin{cases} f_k(\cdot) & \text{if} \quad M_k = A_k , \\ f_k^{-1}(\cdot) & \text{if} \quad M_k \neq A_k , \end{cases} \quad \text{for} \quad k = 1, \cdots, n.$$

Clearly, the vectors x and y satisfy

$$AF(x) + Bx = MG(y) + Ny$$

if they satisfy the relation

$$y_k = \begin{cases} x_k & \text{if} \quad M_k = A_k , \\ f_k(x_k) & \text{if} \quad M_k \neq A_k , \end{cases} \quad \text{for} \quad k = 1, \cdots, n, \tag{17}$$

and since this relation defines a homeomorphism of E^n onto itself, it follows that there exists a unique solution of equation (16) for each $c \in E^n$ if there exists a unique solution of the equation

$$MG(y) + Ny = c \tag{18}$$

for each $c \in E^n$. But, that this is so follows immediately from Theorem 2 and from the fact that $M^{-1}N \in P_0$.

(*only if*) Suppose $(A, B) \notin \mathcal{W}_0$. Then, by Theorem 1, there exists a diagonal matrix $D > 0$ such that $\det(AD + B) = 0$. Choosing $F(x) \equiv Dx$, we have $F \in \mathfrak{F}^n$, while equation (16) does not have, with this choice of F, a unique solution for all $c \in E^n$. □

There are corollaries to Theorem 2, given in Ref. 2, that also may be generalized in a similar manner. For example, the following result is a generalization of an important special case of Corollary 1 of Ref. 2; it shows that the condition $(A, B) \in \mathcal{W}_0$ is still sufficient to insure the uniqueness of a solution of equation (16) (if a solution exists) even when the mapping F is not *onto*.

Theorem 4: *If* $F(x) \equiv [f_1(x_1), \cdots, f_n(x_n)]^T$, *where each* f_k *is a strictly monotone increasing mapping of* E^1 *into* E^1, *and if* $(A, B) \in \mathcal{W}_0$, *then there exists at most one solution of equation* (16) *for each* $c \in E^n$.

Proof: Suppose that, for some $c \in E^n$, x^1 and x^2 are solutions of equation (16) with $x^1 - x^2 \neq \theta$. Then, $A[F(x^1) - F(x^2)] + B(x^1 - x^2) = \theta$. But then, since F is a strictly monotone increasing "diagonal" mapping, there exists a diagonal matrix $D > 0$ such that $F(x^1) - F(x^2) = D(x^1 - x^2)$, and hence $(AD + B)(x^1 - x^2) = \theta$. Since $x^1 - x^2 \neq \theta$ it follows that $\det(AD + B) = 0$, which implies that $(A, B) \notin \mathcal{W}_0$. □

7.2 *A Nonuniqueness Theorem*

From the proof of the "only if" part of Theorem 2 (given in Ref. 2) it follows that whenever $A \notin P_0$, there exists a mapping $F \in \mathfrak{F}^n$ and a vector $b \in E^n$ such that equation (14) has more than one solution. On the other hand, even if $A \notin P_0$, if the mapping $F \in \mathfrak{F}^n$ is "fixed," then it is easy to see that the nonuniqueness of solutions of equation (14) need not necessarily follow for any $b \in E^n$[take $F(x) \equiv x$ and $Ax \equiv -2x$, for example]. I. W. Sandberg has shown,[10] however, that if one assumes that the "fixed" mapping F has another special property, rather than assuming that $F \in \mathfrak{F}^n$, then the nonuniqueness of solutions of equation (14) follows, for some $b \in E^n$, whenever $A \notin P_0$. Moreover, he has shown that under these hypotheses and for any $\delta > 0$, there exists some $b \in E^n$ such that equation (14) has two solutions, x and y, which satisfy $||x - y|| = \delta$. The special property that F is assumed to have is given in the following definition (in words, the property is: that it be possible to draw a straight line having any given positive slope, and any given length, between some pair of points on the graph of each of the functions f_k).

Definition: For each positive integer n we denote by $\mathcal{E}^n$ that collection of mappings of E^n into itself defined by: $F \in \mathcal{E}^n$ if and only if there exist, for $k = 1, \cdots, n$, continuous functions f_k mapping E^1 into E^1 such that for each $x \in E^n$, $F(x) = [f_1(x_1), \cdots, f_n(x_n)]^T$, with each of the f_k satisfying, for all $\beta > 0$,

$$\inf \{f_k(\alpha + \beta) - f_k(\alpha): -\infty < \alpha < \infty\} = 0,$$

$$\sup \{f_k(\alpha + \beta) - f_k(\alpha): -\infty < \alpha < \infty\} = \infty.$$

By using Theorem 1 it is possible to prove the following generalization of Sandberg's result:

Theorem 5: Let $F \in \mathcal{E}^n$, let $(A, B) \notin \mathcal{W}_0$ be a pair of real $n \times n$ matrices, and let δ be a positive constant. Then, for some $c \in E^n$ there exist solutions of equation (16), *x and y, satisfying $||x - y|| = \delta$.*

Proof: Since $(A, B) \notin \mathcal{W}_0$ there exists a diagonal matrix $D = \mathrm{diag}(d_1, \cdots, d_n) > 0$, such that $\det(AD + B) = 0$. Therefore, there exists $x^* \in E^n$, with $||x^*|| = \delta$, such that $(AD + B)x^* = \theta$. Since $F \in \mathcal{E}^n$ there exists $x \in E^n$ such that

$$f_k(x_k) - f_k(x_k - x_k^*) = x_k^* d_k, \qquad \text{for} \qquad k = 1, \cdots, n.$$

Let $c = AF(x) + Bx$, and let $y = x - x^*$. Then

$$\begin{aligned} A[F(x) - F(y)] + B(x - y) &= A[F(x) - F(x - x^*)] + Bx^* \\ &= (AD + B)x^* = \theta. \quad \square \end{aligned}$$

For a mapping F to be a member of $\mathcal{E}^n$, it is not necessary that $F \in \mathfrak{F}^n$. It follows from the above definition of $\mathcal{E}^n$ that $F \in \mathcal{E}^n$ implies that each of the functions f_k is a monotone increasing function from E^1 onto some interval in E^1 whose length is infinite; the f_k need not, however, be *strictly* monotone increasing, nor *onto* E^1. For those $F \in \mathcal{E}^n$ for which each of the functions f_k is strictly monotone increasing, we have the following corollary to the two preceding theorems.

Corollary: *Let* $F(x) \equiv [f_1(x_1), \cdots, f_n(x_n)]^T \in \mathcal{E}^n$ *and let each of the functions* f_k *be strictly monotone increasing. Then there exists at most one solution of equation* (16) *for each* $c \in E^n$ *if and only if* $(A, B) \in \mathcal{W}_0$.

VIII. RESULTS ON CONTINUITY AND BOUNDEDNESS

For many systems whose behavior is described by an equation having the form (16), the vector c may be regarded as the system's input and the vector x may be regarded as the system's response or output. Those properties that one might expect well-behaved systems to possess are likely to include continuity and boundedness. Thus, one might expect (*i*) "small" changes to result in the value of the system's output when "small" changes are made in the value of the system's input, and (*ii*) a bounded sequence of input vectors to yield a bounded sequence of outputs. We now show that such properties are indeed possessed by the type of system that is the main concern of this paper.

8.1 *Continuity*

When the $n \times n$ matrix A is a member of the class P_0 and the mapping $F \in \mathfrak{F}^n$, it follows that the solution x of equation (14) is a continuous function of the (input) vector b.[2] Using this fact, it is easy to prove the following theorem.

Theorem 6: *For each* $F \in \mathfrak{F}^n$ *and each pair of* $n \times n$ *matrices* $(A, B) \in \mathcal{W}_0$ *the solution* x *of equation* (16) *is a continuous function of the vector* c.

Proof: Proceeding as in the "if" part of the proof of Theorem 3, we see that the theorem follows immediately from the facts that equation (17) is a homeomorphism and that the aforementioned result guarantees that y, the solution of equation (18), is a continuous function of c. □

8.2 *Boundedness*

In Ref. 2 a theorem (Theorem 5) is proved which shows that, when $F \in \mathfrak{F}^n$ and $A \in P_0$, bounds can be obtained for the solution of equation (14) whenever bounds for $b \in E^n$ are given. The proof of a more general theorem concerning equation (16) can be constructed quite easily by using that theorem, and by using the same technique that was used in the proof of the preceding theorem, along with the trivial observations:

(*i*) For any nonsingular $n \times n$ matrix of real numbers, M, and any real numbers $\alpha_i \leqq \beta_i$, $i = 1, \cdots, n$, there exist real numbers, $\alpha'_i \leqq \beta'_i$, $i = 1, \cdots, n$, such that when each of the components c_i of the vector c satisfies $\alpha_i \leqq c_i \leqq \beta_i$, it follows that $\alpha'_i \leqq (M^{-1}c)_i \leqq \beta'_i$, for $i = 1, \cdots, n$.

(*ii*) For any given real numbers $\gamma_i \leqq \delta_i$, $i = 1, \cdots, n$, there exist for the homeomorphism (17), real numbers $\gamma'_i \leqq \delta'_i$, $i = 1, \cdots, n$, such that whenever x, y satisfy equation (17) with $\gamma_i \leqq y_i \leqq \delta_i$, for $i = 1, \cdots, n$, it follows that $\gamma'_i \leqq x_i \leqq \delta'_i$, for $i = 1, \cdots, n$.

The more general theorem, whose quite obvious proof is omitted, is the following:

Theorem 7: *Let $F \in \mathfrak{F}^n$, let $(A, B) \in \mathfrak{W}_0$ be a pair of $n \times n$ matrices, and, for $i = 1, \cdots, n$, let $\alpha_i \leqq \beta_i$ be given. There exist, for $i = 1, \cdots, n$, real numbers $\gamma_i \leqq \delta_i$ such that for any $c = (c_1, \cdots, c_n)^T \in E^n$ with $\alpha_i \leqq c_i \leqq \beta_i$ for $i = 1, \cdots, n$, if x satisfies equation* (16), *then $\gamma_i \leqq x_i \leqq \delta_i$ for $i = 1, \cdots, n$.*

According to Theorem 7, $(A, B) \in \mathfrak{W}_0$ is a sufficient condition for a bounded sequence of vectors c to yield a bounded sequence of solution vectors of equation (16), for all $F \in \mathfrak{F}^n$. The following theorem shows that $(A, B) \in \mathfrak{W}_0$ is also a necessary condition.

Theorem 8: *If (A, B) is a pair of real $n \times n$ matrices, then $(A, B) \in \mathfrak{W}_0$ if and only if for each $F \in \mathfrak{F}^n$ and each unbounded sequence of points $x^1, x^2, x^3, \cdots$ in E^n, the corresponding sequence $c^1, c^2, c^3, \cdots$ $[c^k = AF(x^k) + Bx^k$, $k = 1, 2, 3, \cdots]$ is unbounded.*

This theorem, which is a generalization of Theorem 4 of Ref. 2, can be proved in a manner which is a quite obvious generalization of the proof, given there, of that theorem. Thus, an appeal to Theorem 7 proves the "only if" part, and the "if" part is proved by assuming that $(A, B) \notin \mathfrak{W}_0$ and then choosing the same kind of mapping $F \in \mathfrak{F}^n$ as was chosen in Ref. 2, for which an unbounded sequence of vectors x^k yields a bounded sequence of vectors c^k.

IX. COMPUTATION OF THE SOLUTION

A. Gersho[7] has shown that whenever $F \in \mathfrak{F}^n \cap C^1$ (that is, whenever each of the functions f_k is a continuously differentiable strictly monotone increasing mapping of the real line onto itself), it is possible to compute the solution of equation (14), for any $A \in P_0$ and any $b \in E^n$, by making use of a gradient descent algorithm due to A. A. Goldstein.[11] The following theorem extends this result to the class of equations of the type (16).

Theorem 9: *Let M be an arbitrary positive definite symmetric matrix, and let $Q : E^n \to E^1$ be defined by*

$$Q(x) = [AF(x) + Bx - c]^T M[AF(x) + Bx - c],$$

where $F \in \mathfrak{F}^n \cap C^1$, $(A, B) \in \mathcal{W}_0$, and $c \in E^n$. For each $x \in E^n$ and each $\gamma \geqq 0$ let

$$g(x, \gamma) = \begin{cases} \dfrac{Q(x) - Q[x - \gamma \nabla Q(x)]}{\gamma \, \| \nabla Q(x) \|^2}, & \gamma > 0; \\ 1, & \gamma = 0; \end{cases}$$

where $\nabla Q(x)$ denotes the gradient of Q at the point x. Then, if δ is any real number satisfying $0 < \delta \leqq \frac{1}{2}$, and if x^0 is an arbitrary point in E^n, the sequence $\{x^k : k = 0, 1, 2, \cdots\}$ converges to the solution of equation (16), *where (for $k = 0, 1, 2, \cdots$) the x^k satisfy*

$$x^{k+1} = x^k - \gamma^k \nabla Q(x^k),$$

each γ^k being any real number that satisfies $\delta \leqq g(x^k, \gamma^k) \leqq 1 - \delta$ if $g(x^k, 1) < \delta$, or $\gamma^k = 1$ if $g(x^k, 1) \geqq \delta$.

Proof: This proof uses generalizations of some of the ideas in Ref. 7 and relies ultimately upon the Goldstein algorithm.[11]

We first remark that the sequence $\{x^k\}$ is well-defined: It is easy to show (see the first part of the proof of Theorem 1, p. 31, Ref. 11) that for each $x \in E^n$, $g(x, \cdot)$ is a continuous function on $[0, \infty)$. This being the case, it is clear that if $g(x^k, 1) < \delta$, then for each ξ in the interval $[\delta, 1)$—and, in particular, for each ξ in the interval $[\delta, 1 - \delta]$—there is some γ^k in the interval $(0, 1)$ such that $g(x^k, \gamma^k) = \xi$.

Let $S = \{x \in E^n : Q(x) \leqq Q(x^0)\}$. Using the fact that M is a positive definite symmetric matrix, and using the fact that $F \in \mathfrak{F}^n$, $(A, B) \in \mathcal{W}_0$ implies that $\| AF(x) + Bx \| \to \infty$ if and only if $\| x \| \to \infty$ (Theorem 8) we have that the set $S \subset E^n$ is bounded. By continuity of Q, S is closed. Thus, S is compact and, therefore, the gradient ∇Q (which is continuous on E^n, since $F \in C^1$) is uniformly continuous on S, and ∇Q is bounded on S. Also, Q is bounded below on S. [Indeed, we have $Q \geqq 0$ on E^n and by the existence and uniqueness theorem, Theorem 3, there exists exactly one point x^* ($x^* \in S$) at which $Q(x^*) = 0$.]

It is easily verified that, for each $x \in E^n$,

$$\nabla Q(x) = 2(AD_x + B)^T M[AF(x) + Bx - c],$$

where, for $k = 1, \cdots, n$, the kth diagonal element of the diagonal matrix $D_x > 0$ has the value of the derivative of the function f_k, evaluated at the point x_k. Since $(A, B) \in \mathcal{W}_0$ implies that $\det (AD_x + B) \neq 0$, and since det $M \neq 0$, it follows that $\nabla Q(x) = \theta$ if and only if x is the solution of equation (16).

In view of the above, it follows directly from Goldstein's theorem that the sequence $\{x^k\}$ converges to the solution of equation (16). □

Other methods of computing the solution of equation (16), in certain cases, also exist. If one performs a transformation of the type (17) on the independent variable x (in theory this can always be done) then the solution of equation (16) can easily be computed by first computing the solution of an equation of the type $G(y) + M^{-1}Ny = M^{-1}c$, where $G \in \mathfrak{F}^n$ and $M^{-1}N \in P_0$. Methods of computing the solution of certain equations of this type may be found in Refs. 1–3.

X. EXAMPLE

With the aid of the modern computing facilities that are commonly available today, it is clearly a rather routine matter to obtain an equation of the type (16) for any given transistor network. Moreover, it is not unfeasible, even for networks of moderately large size (say, up to 4 or 5 transistors), to consider the straightforward evaluation of the 2^n determinants specified in property (2) of Theorem 1, and thereby resolve the issue of whether or not the matrices involved in the equation are a $\mathcal{W}_0$ pair. Due regard would of course have to be paid to the matter of performing sufficiently accurate computations.

On the other hand, even without the aid of a computer, it should often be possible to use a little ingenuity and a few devices* to reduce the computations involved in the application of the above theory to many specific problems to a point where they will just about fit onto the back of an envelope. Consider, for example, the following analysis of a three-transistor network:

For the network of Fig. 5, the voltage and current variables defined there must satisfy the following equations:

$$\begin{bmatrix} i_1 \\ i_2 \\ i_3 \\ i_4 \\ i_5 \\ i_6 \end{bmatrix} = \begin{bmatrix} T^{(1)} & & \\ & T^{(2)} & \\ & & T^{(3)} \end{bmatrix} \begin{bmatrix} f_1(v_1) \\ f_2(v_2) \\ f_3(v_3) \\ f_4(v_4) \\ f_5(v_5) \\ f_6(v_6) \end{bmatrix}, \tag{19}$$

* According to R. Bellman: "a device is a trick that works at least twice." [12]

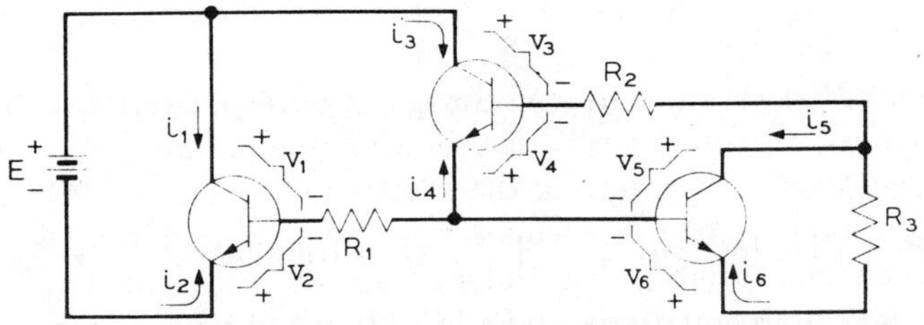

Fig. 5—Example of a three transistor network.

$$\begin{bmatrix} v_1 \\ v_3 \\ v_6 \\ -i_2 \\ -i_4 \\ -i_5 \end{bmatrix} = \left[\begin{array}{ccc|ccc} 0 & 0 & 0 & 1 & 0 & 0 \\ 0 & R_1 & 0 & 1 & 1 & 0 \\ 0 & 0 & R_3 & 0 & 0 & 1 \\ \hline -1 & -1 & 0 & 0 & 0 & 0 \\ 0 & -1 & 0 & 0 & G_2 & G_2 \\ 0 & 0 & -1 & 0 & G_2 & G_2 \end{array}\right] \begin{bmatrix} -i_1 \\ -i_3 \\ -i_6 \\ v_2 \\ v_4 \\ v_5 \end{bmatrix} + \begin{bmatrix} E \\ E \\ 0 \\ 0 \\ 0 \\ 0 \end{bmatrix}, \tag{20}$$

where (we are using the transistor model of Fig. 3, with $r_b = r_c = r_e = 0$) each of the 2×2 matrices $T^{(k)}$, $k = 1, 2, 3$, is of the form (10). A hybrid characterization has been used for the linear part of the network. As indicated in equation (3), this hybrid characterization can easily be converted into a characterization of the Belevitch type. Thus, denoting the 3×3 blocks of the hybrid matrix in equation (20) by H_{11}, H_{12}, H_{21}, H_{22}, in the usual manner, one obtains

$$\begin{bmatrix} I & -H_{12} \\ 0 & -H_{22} \end{bmatrix} v = -\begin{bmatrix} H_{11} & 0 \\ H_{21} & -I \end{bmatrix} i + c, \tag{21}$$

where $v = (v_1, v_3, v_6, v_2, v_4, v_5)^T$ and i is similarly defined. We could now simply reorder the columns of each matrix in equation (21) in such a way that the resulting equation would have the same form, except that the subscripts on the components of the vectors v and i would occur in the natural order (1, 2, 3, 4, 5, 6) and then use that equation, along with equation (19), to produce an equation of the type (16) for our network. In this example, though, it's probably easier to reorder the rows and columns of the matrix T (recall, $T = T^{(1)} \oplus T^{(2)} \oplus T^{(3)}$) to obtain from equation (19) an equation that is compatible with equation (21). Thus,

$$i = \begin{bmatrix} I & -P \\ -Q & I \end{bmatrix} F(v), \tag{22}$$

where

$$F(v) \equiv [f_1(v_1), f_3(v_3), f_6(v_6), f_2(v_2), f_4(v_4), f_5(v_5)]^T,$$

and

$$P = \operatorname{diag}[\alpha_r^{(1)}, \alpha_r^{(2)}, \alpha_f^{(3)}], \qquad Q = \operatorname{diag}[\alpha_f^{(1)}, \alpha_f^{(2)}, \alpha_r^{(3)}].$$

Eliminating i from equations (21) and (22), we obtain

$$\begin{bmatrix} H_{11} & 0 \\ H_{21} & -I \end{bmatrix}\begin{bmatrix} I & -P \\ -Q & I \end{bmatrix} F(v) + \begin{bmatrix} I & -H_{12} \\ 0 & -H_{22} \end{bmatrix} v = c. \tag{23}$$

Note that since $\det H_{11} = \det H_{22} = 0$, it is impossible to put this equation into either of the forms (14) or (15). Clearly this would be the same situation no matter which ordering of subscripts was chosen for the components of v. The cause of the difficulty is simply the fact that neither an impedance matrix nor an admittance matrix exists for the linear part of our network.

Let us determine whether or not the pair of matrices

$$\left(\begin{bmatrix} H_{11} & 0 \\ H_{21} & -I \end{bmatrix}\begin{bmatrix} I & -P \\ -Q & I \end{bmatrix}, \begin{bmatrix} I & -H_{12} \\ 0 & -H_{22} \end{bmatrix}\right)$$

is a $\mathcal{W}_0$ pair. We shall try to verify property (1) of Theorem 1. Let $\delta_1, \cdots, \delta_6$ denote arbitrary positive real numbers, and let $\Delta_I = \operatorname{diag}(\delta_1, \delta_2, \delta_3)$, $\Delta_{II} = \operatorname{diag}(\delta_4, \delta_5, \delta_6)$. We wish to show that

$$\det \left\{\begin{bmatrix} H_{11} & 0 \\ H_{21} & -I \end{bmatrix}\begin{bmatrix} I & -P \\ -Q & I \end{bmatrix}\begin{bmatrix} \Delta_I^{-1} & 0 \\ 0 & \Delta_{II} \end{bmatrix} + \begin{bmatrix} I & -H_{12} \\ 0 & -H_{22} \end{bmatrix}\right\} \neq 0.$$

By multiplying the above matrix on the left by the (nonsingular) matrix $\operatorname{diag}(I_3, -I_3)$ and then multiplying on the right by $\operatorname{diag}(\Delta_I, I_3)$, we obtain the equivalent statement:

$$\det \left[\begin{array}{c|c} H_{11} + \Delta_I & -H_{12} - H_{11}P\,\Delta_{II} \\ \hline -H_{21} - Q & H_{22} + (I + H_{21}P)\,\Delta_{II} \end{array}\right] \neq 0.$$

The 3×3 submatrix in the upper left corner is nonsingular and diagonal. The 3×3 submatrix in the lower left corner can be diagonalized by performing a single elementary row operation on the matrix; namely, by subtracting $1/(\delta_2 + R_1)$ times the second row from the fourth row. Having done this, our problem reduces to one of showing that

$$\det \left[\begin{array}{ccc:ccc} \delta_1 & 0 & 0 & -1 & 0 & 0 \\ 0 & \delta_2 + R_1 & 0 & -1 & -(1 + \alpha_r^{(2)} R_1\, \delta_5) & 0 \\ 0 & 0 & \delta_3 + R_3 & 0 & 0 & -(1 + \alpha_f^{(3)} R_3\, \delta_6) \\ \hdashline 1 - \alpha_f^{(1)} & 0 & 0 & (1 - \alpha_r^{(1)})\, \delta_4 + \dfrac{1}{\delta_2 + R_1} & \dfrac{1 - \alpha_r^{(2)}\, \delta_2\, \delta_5}{\delta_2 + R_1} & 0 \\ 0 & 1 - \alpha_f^{(2)} & 0 & 0 & G_2 + (1 - \alpha_r^{(2)})\, \delta_5 & G_2 \\ 0 & 0 & 1 - \alpha_r^{(3)} & 0 & G_2 & G_2 + (1 - \alpha_f^{(3)})\, \delta_6 \end{array}\right] \neq 0.$$

It is easy to verify that whenever $\det A_{11} \neq 0$, then

$$\det \begin{bmatrix} A_{11} & A_{12} \\ A_{21} & A_{22} \end{bmatrix} \neq 0$$

if and only if $\det (A_{22} - A_{21} A_{11}^{-1} A_{12}) \neq 0$. In our case both A_{11} and A_{21} are diagonal and hence we can immediately reduce our problem to:

$$\det \begin{bmatrix} (1-\alpha_r^{(1)})\delta_4 + \dfrac{1}{\delta_2+R_1} + \dfrac{1-\alpha_f^{(1)}}{\delta_1} & \dfrac{1-\alpha_r^{(2)}\delta_2\delta_5}{\delta_2+R_1} & 0 \\ \dfrac{1-\alpha_f^{(2)}}{\delta_2+R_1} & G_2 + (1-\alpha_r^{(2)})\delta_5 + \dfrac{1-\alpha_f^{(2)}}{\delta_2+R_1}(1+\alpha_r^{(2)}R_1\delta_5) & G_2 \\ 0 & G_2 & G_2 + (1-\alpha_f^{(3)})\delta_6 + \dfrac{1-\alpha_r^{(3)}}{\delta_3+R_3}(1+\alpha_f^{(3)}R_3\delta_6) \end{bmatrix} \neq 0.$$

It is obvious that the above determinant is always positive. First, note that every term in the matrix is nonnegative except, possibly, the (1, 2) term, which may be either positive or negative (or zero). In the event that the (1, 2) term is positive (or zero), we have $1/(\delta_2 + R_1) \geqq (1 - \alpha_r^{(2)} \delta_2 \delta_5)/(\delta_2 + R_1)$, and hence we observe that the matrix is strongly row-sum dominant. This implies that its determinant is positive.

In the event that the (1, 2) term is negative, we do not necessarily have dominance; however, considering an expansion of the determinant along its first row we see that, because of the assumption that the (1, 2) term is negative, the value of the determinant is computed as the sum of two *positive* terms.

We have thus shown that, no matter which (positive) values are assigned to R_1, R_2, R_3, or which values the transistor's current gains assume $[0 < \alpha_f^{(k)} < 1, 0 < \alpha_r^{(k)} < 1]$, the pair of 6×6 matrices that appear in equation (23) is a $\mathcal{W}_0$ pair. Thus, all of the results concerning a solution's existence, uniqueness, continuity, boundedness, and so on, hold for this equation.

XI. ACKNOWLEDGMENT

The author has benefited from discussions with his colleagues I. W. Sandberg and H. C. So.

APPENDIX

Proof of Part of Theorem 1

In this appendix we prove the equivalence of properties (*i*), (*ii*), (*vi*), and (*vii*) of Theorem 1, which define the class of pairs of matrices $\mathcal{W}_0$. We omit the proof of the equivalence of the three remaining properties, since those properties are not referred to in this paper. A complete proof of Theorem 1 is given elsewhere.[9] We begin by proving a useful lemma:

Lemma 1: *For each positive integer n the polynomial*

$$(c_0)\, d_1\, d_2 \cdots d_n + (c_1)\, d_1\, d_2 \cdots d_{n-1} + \cdots + (c_n)\, d_2 \cdots d_n$$
$$+ (c_{n+1})\, d_1\, d_2 \cdots d_{n-2} + \cdots + (c_{n(n+1)/2})\, d_3 \cdots d_n + \cdots + (c_{2^n-1})$$

in the n variables $d_1, d_2, \cdots, d_n$ *is nonzero for all positive values of the variables if and only if at least one of the coefficients* $c_0, \cdots, c_{2^n-1}$ *is nonzero, and all nonzero coefficients have the same sign.*

Proof: (By induction) For $n = 1$ the statement is obviously true. Let N be a positive integer. Then any polynomial of the above type in $N + 1$ variables, $(c_0)d_1 \cdots d_{N+1} + \cdots + (c_{2^{N+1}-1})$, can be written as $P(d_1, \cdots, d_N)\cdot d_{N+1} + Q(d_1, \cdots, d_N)$ where P and Q are both polynomials of the above type in N variables. Then, assuming that the statement is true for $n = N$, $P + Q \neq 0$ and $P\cdot Q \geqq 0$ for all positive values of the variables $d_1, \cdots, d_N$ if and only if at least one of the coefficients $c_0, \cdots, c_{2^{N+1}-1}$ is nonzero and all nonzero coefficients have the same sign. But, we know that $P\cdot d_{N+1} + Q \neq 0$ for all $d_{N+1} > 0$ if and only if $P + Q \neq 0$ and $P\cdot Q \geqq 0$. □

A.1 *Property* (*i*) *is Equivalent to* (*ii*)

Let $D = \text{diag}\,(d_1, \cdots, d_n)$. By expanding $\det\,(AD + B)$ along the first column we have

$$\det\,(AD + B) = d_1\cdot\det P + \det Q,$$

where the first columns of P and Q satisfy $P_1 = A_1$, $Q_1 = B_1$, and for $k = 2, \cdots, n$, $P_k = Q_k = (AD + B)_k$. Both P and Q are independent of d_1. We now expand $\det P$ and $\det Q$ along their second columns, resulting in

$$\det P = d_2\cdot\det R + \det S,$$
$$\det Q = d_2\cdot\det U + \det V,$$

and hence,

$$\det (AD + B) = d_1 d_2 \cdot \det R + d_1 \cdot \det S + d_2 \cdot \det U + \det V,$$

where

$$\begin{aligned} R_1 &= A_1 , & R_2 &= A_2 , \\ S_1 &= A_1 , & S_2 &= B_2 , \\ U_1 &= B_1 , & U_2 &= A_2 , \\ V_1 &= B_1 , & V_2 &= B_2 , \end{aligned}$$

and for $k = 3, \cdots, n$,

$$R_k = S_k = U_k = V_k = (AD + B)_k .$$

Proceeding in this manner until all columns of $(AD + B)$ have been encountered, we obtain an expansion of $\det (AD + B)$ as a polynomial in the variables $\{d_1, d_2, \cdots, d_n\}$ whose coefficients are the determinants of the matrices in $\mathcal{C}(A, B)$. By using Lemma 1 it thus follows that (*i*) and (*ii*) are equivalent.

A.2 Property (vi) Follows from (i) and (ii)

According to (*ii*) there exists a complementary pair of matrices (M, N) taken from $\mathcal{C}(A, B)$ such that $\det M \neq 0$. Let $D = \text{diag}\,(d_1, \cdots, d_n) > 0$, then $\det (M^{-1}N + D) \neq 0$ if and only if $\det (MD + N) \neq 0$. But, using property (*i*), $\det (MD + N) = \det (A\hat{D} + B) \cdot \det \tilde{D} \neq 0$, where the matrices $\hat{D} = \text{diag}\,(\hat{d}_1, \cdots, \hat{d}_n) > 0$ and $\tilde{D} = \text{diag}\,(\tilde{d}_1, \cdots, \tilde{d}_n) > 0$ are defined by $\hat{d}_k = d_k$ and $\tilde{d}_k = 1$ if $M_k = A_k$, and $\hat{d}_k = 1/d_k$, $\tilde{d}_k = d_k$ otherwise (for $k = 1, \cdots, n$). Thus, $M^{-1}N \in P_0$.

A.3 Property (i) Follows from (vi)

Using the notation above, it is clear that for each diagonal matrix $D > 0$, $\det (AD + B) = \det (M\hat{D} + N) \cdot \det \tilde{D}$. Thus, if $M^{-1}N \in P_0$ it follows that $\det (AD + B) \neq 0$.

A.4 Property (vii) is Equivalent to (vi)

Clearly property (*vi*) follows from property (*vii*). Thus, we need only prove that (*vi*) implies (*vii*). Let (M, N) and (P, Q) both be complementary pairs taken from $\mathcal{C}(A, B)$ with $M^{-1}N \in P_0$ and $\det P \neq 0$. For any $D = \text{diag}\,(d_1, \cdots, d_n) > 0$, $\det (P^{-1}Q + D) \neq 0$ if and only if $\det (PD + Q) \neq 0$. But $\det (PD + Q) = \det (M\hat{D} + N) \cdot \det \tilde{D} \neq 0$, where the matrices $\hat{D} = \text{diag}\,(\hat{d}_1, \cdots, \hat{d}_n) > 0$ and $\tilde{D} = \text{diag}\,(\tilde{d}_1, \cdots, \tilde{d}_n) > 0$ are defined by $\hat{d}_k = d_k$ and $\tilde{d}_k = 1$ if $P_k = M_k$, and $\hat{d}_k = 1/d_k$, $\tilde{d}_k = d_k$ otherwise (for $k = 1, \cdots, n$). Thus, $P^{-1}Q \in P_0$. □

REFERENCES

1. Willson, A. N., Jr., "On the Solutions of Equations for Nonlinear Resistive Networks," B.S.T.J., *47*, No. 8 (October 1968), pp. 1755–1773.
2. Sandberg, I. W., and Willson, A. N., Jr., "Some Theorems on Properties of DC Equations of Nonlinear Networks," B.S.T.J., *48*, No. 1 (January 1969), pp. 1–34.
3. Sandberg, I. W., and Willson, A. N., Jr., "Some Network-Theoretic Properties of Nonlinear DC Transistor Networks," B.S.T.J., *48*, No. 5 (May–June 1969), pp. 1293–1311.
4. Kuh, E. S., and Rohrer, R. A., *Theory of Linear Active Networks*, San Francisco: Holden-Day, Inc., 1967, p. 2.
5. Belevitch, V., "Four-Dimensional Transformations of 4-Pole Matrices with Applications to the Synthesis of Reactance 4-Poles," IRE Trans. Circuit Theory, *CT-3*, No. 2 (June 1956), pp. 105–111.
6. So, H. C., "On the Hybrid Description of a Linear n-Port Resulting from the Extraction of Arbitrarily Specified Elements," IEEE Trans. Circuit Theory, *CT-12*, No. 3 (September 1965), pp. 381–387.
7. Gersho, A., "Solving Nonlinear Network Equations Using Optimization Techniques," B.S.T.J., *48*, No. 9 (November 1969), pp. 3135–3138.
8. Fiedler, M., and Pták, V., "Some Generalizations of Positive Definiteness and Monotonicity," Numer. Math., *9*, No. 2 (1966), pp. 163–172.
9. Willson, A. N., Jr., "A Useful Generalization of the P_0 Matrix Concept," to be published.
10. Sandberg, I. W., "Theorems on the Analysis of Nonlinear Transistor Networks," B.S.T.J., *49*, No. 1 (January 1970), pp. 95–114.
11. Goldstein, A. A., *Constructive Real Analysis*, New York: Harper and Row, 1967, p. 31.
12. Bellman, R., *Stability Theory of Differential Equations*, New York: Dover, 1969, p. ix.

Theorems on the Analysis of Nonlinear Transistor Networks*

By I. W. SANDBERG

(Manuscript received August 19, 1969)

This paper reports on further results concerning nonlinear equations of the form $F(x) + Ax = B$, in which $F(\cdot)$ is a "diagonal nonlinear mapping" of real Euclidean n-space E^n into itself, A is a real $n \times n$ matrix, and B is an element of E^n. Such equations play a central role in the dc analysis of transistor networks, the computation of the transient response of transistor networks, and the numerical solution of certain nonlinear partial-differential equations.

Here a nonuniqueness result, which focuses attention on a simple special property of transistor-type nonlinearities, is proved; this result shows that under certain conditions the equation $F(x) + Ax = B$ has at least two solutions for some $B \in E^n$. The result proves that some earlier conditions for the existence of a unique solution cannot be improved by taking into account more information concerning the nonlinearities, and therefore makes more clear that the set of matrices denoted in earlier work by P_0 plays a very basic role in the theory of nonlinear transistor networks. In addition, some material concerned with the convergence of algorithms for computing the solution of the equation $F(x) + Ax = B$ is presented, and some theorems are proved which provide more of a theoretical basis for the efficient computation of the transient response of transistor networks. In particular, the following proposition is proved.

If the dc equations of a certain general type of transistor network possess at most one solution for all $B \in E^n$ for "the original set of α's as well as for an arbitrary set of not-larger α's", then the nonlinear equations encountered at each time step in the use of certain implicit numerical integration algorithms possess a unique solution for all values of the step size, and hence then for all step-size values it is possible to carry out the algorithms.

* The material of this paper was presented at the Advanced Study Institute on Network Theory (sponsored by the N.A.T.O.; Knokke, Belgium; September 1-12, 1969).

I. INTRODUCTION AND DISCUSSION OF RESULTS

References 1 and 2 present some results concerning the equation

$$F(x) + Ax = B, \tag{1}$$

in which, with n an arbitrary positive integer, A is a real $n \times n$ matrix, B is an element of real Euclidean n-space E^n, and $F(\cdot)$ is a mapping of E^n into E^n defined by the condition that* for all $x = (x_1, x_2, \cdots, x_n)^{tr} \in E^n$,

$$F(x) = [f_1(x_1), f_2(x_2), \cdots, f_n(x_n)]^{tr} \tag{2}$$

with each $f_i(\cdot)$ a strictly monotone increasing mapping of E^1 into itself. Equation (1) plays a central role in the dc analysis of transistor networks,** the transient analysis of transistor networks (see Section 1.4), and the numerical solution of certain nonlinear partial differential equations.

In Ref. 1 it is proved that there exists a unique solution x of equation (1) for each strictly monotone increasing mapping $F(\cdot)$ of E^n onto E^n (that is, for each set of strictly monotone increasing mappings $f_i(\cdot)$ of E^1 onto itself) and each $B \in E^n$ if and only if A is a member of the set P_0 of real $n \times n$ matrices with all principal minors nonnegative. It is also proved in Ref. 1 that equation (1) possesses a unique solution x for each continuous monotone nondecreasing mapping $F(\cdot)$ of E^n into E^n (that is, for each set of continuous monotone nondecreasing mappings of E^1 into E^1) and each $B \in E^n$ if A belongs to the set P of all real $n \times n$ matrices with all principal minors positive†. A direct modification of the existence proof given in Ref. 1, as indicated in Ref. 2, shows that equation (1) possesses a unique solution for each strictly monotone increasing mapping $F(\cdot)$ of E^n onto $(\alpha_1, \beta_1) \times (\alpha_2, \beta_2) \times \cdots \times (\alpha_n, \beta_n)$ with each α_i and β_i elements of the extended real line‡ (real line) such that $\alpha_i < \beta_i$ and each $B \in E^n$ if (and only if) $A \in P_0$ and $\det A \neq 0$. Some network theoretic implications of these and related results are discussed in Refs. 1 and 2, where the matter of determining whether or not $A \in P_0$ or $A \in P$ is considered in some detail.

---

* Throughout the paper the superscript tr denotes transpose.

** See Ref. 1 for a derivation of the equation within the context of the transistor dc-analysis problem.

† There are some interesting applications of this result in the study of numerical methods for solving certain nonlinear partial-differential equations, in which A has nonpositive off-diagonal terms and is irreducibly diagonally dominant.[3]

‡ The numbers α_i and β_i are members of the *extended real line* if $-\infty \leqq \alpha_i \leqq \infty$ and $-\infty \leqq \beta_i \leqq \infty$.

This paper presents a proof of a nonuniqueness result. The proof focuses attention on a simple special property of transistor-type nonlinearities. The result shows that under certain conditions equation (1) has at least two solutions for some $B \in E^n$. In addition, the paper presents some material concerned with the convergence of algorithms for computing the solution of equation (1) and proves some theorems which provide more of a theoretical basis for the efficient computation of the transient response of transistor networks. The remaining portion of Section I is concerned with a detailed discussion of the results and their significance.

1.1 *An Application of the Nonuniqueness Theorem*

The standard Ebers–Moll transistor model, which is widely used, gives rise to functions $f_i(\cdot)$ which, while continuous and strictly monotone increasing, are mappings of E^1 onto open semi-infinite intervals. For such $f_i(\cdot)$, the results stated above assert that the equation (1) possesses at most one solution x for each $B \in E^n$ if $A \in P_0$; and if $A \in P_0$ and det $A \neq 0$, then equation (1) possesses a solution for each $B \in E^n$. Since, as indicated in Ref. 1, $A = T^{-1}G$ with T a nonsingular matrix which takes into account the forward and reverse transistor α's, and G is the short circuit conductance matrix of the linear portion of the network, the condition that det A not vanish is equivalent to the rather weak assumption that the linear portion of the network possess an open-circuit resistance matrix.

It is natural to ask whether the use of more-detailed information concerning the nonlinearities of the transistor model would enable us to make assertions concerning the existence of a unique solution of equation (1) for all $B \in E^n$ under weaker assumptions on A. In particular, can the condition that A belong to P_0 be relaxed? The first result proved in this paper, Theorem 1 of Section II, shows that if the $f_i(\cdot)$ are exponential nonlinearities of the type associated with the Ebers–Moll model, then the condition that A belong to P_0 cannot be replaced by a weaker condition. More explicitly, in Section II a set $\mathfrak{F}_0^n$ of mappings of E^n into E^n is defined, and $\mathfrak{F}_0^n$ contains all of the mappings $F(\cdot)$ that correspond to Ebers–Moll type $f_i(\cdot)$'s. It is proved there that if $A \notin P_0$, then for *any* $F(\cdot) \in \mathfrak{F}_0^n$, there is a $B \in E^n$ such that equation (1) possesses at least two solutions. In fact, it is proved that if $A \notin P_0$ and if δ is an arbitrary positive number, then for any $F(\cdot) \in \mathfrak{F}_0^n$, there is a $B \in E^n$ such that equation (1) possesses two solutions such that the distance in E^n between the two solutions is δ.

Thus Theorem 1 together with the earlier results mentioned above concerning existence of solutions show that the set of matrices P_0 plays a quite fundamental role in the theory of nonlinear transistor networks.

1.2 *An Algorithm for Computing the Solution of Equation (1)*

Several results which assert that $A \in P_0$ under certain conditions on the transistor α's and the short-circuit conductance matrix of the linear portion of the network are proved in Refs. 1 and 2. In particular, Ref. 1 proves that $A \in P$, and hence that $A \in P_0$, if $A = P^{-1}Q$ with P and Q real $n \times n$ matrices such that for all $j = 1, 2, \cdots, n$

$$p_{jj} > \sum_{i \neq j} | p_{ij} | \quad \text{and} \quad q_{jj} > \sum_{i \neq j} | q_{ij} |.^*$$

Theorem 2 of Section II shows that a relatively simple and entirely constructive algorithm can be used to generate a sequence $x^{(0)}, x^{(1)}, \cdots$ of elements of E^n that converges to the unique solution of (1) if $A = P^{-1}Q$ with P and Q as defined above and each $f_i(\cdot)$ is a continuous (but not necessarily differentiable) monotone nondecreasing mapping of E^1 into E^1.**

1.3 *Palais' Theorem, Existence of Solutions of Equation (1), and Algorithms for Computing the Solution of Equation (1)*

Reference 1 gives two existence proofs concerning equation (1). One proof, the more basic of the two, is based on first principles and employs an inductive argument in which, with k an arbitrary positive integer less than n, the existence proposition is assumed to be true with n replaced by k and it is proved that then the proposition is true with n replaced by $(k + 1)$. The second proof uses a theorem of R. S. Palais and requires that the $f_i(\cdot)$ be continuously differentiable throughout E^1. More explicitly, Palais' theorem† asserts that if $R(\cdot)$ is a continuously differentiable mapping of E^n into itself with values $R(q)$ for $q \in E^n$, then $R(\cdot)$ is a diffeomorphism‡ of E^n onto itself if and only if

(*i*) det $J_q \neq 0$ for all $q \in E^n$, in which J_q is the Jacobian matrix of $R(\cdot)$ with respect to q, and
(*ii*) $\| R(q) \| \to \infty$ as $\| q \| \to \infty$.††

* It is proved also that $A \in P_0$ if $A = P^{-1}Q$ with $p_{jj} > \sum_{i \neq j} | p_{ij} |$ and $q_{jj} \geqq \sum_{i \neq j} | q_{ij} |$ for all j.

** A related result given in Ref. 4 is not directly applicable here because of assumptions made in Ref. 4 concerning the existence and boundedness of a certain Jacobian matrix.

† See Ref. 5 and the appendix of Ref. 6.

‡ A diffeomorphism of E^n onto itself is a continuously differentiable mapping of E^n into E^n which possesses a continuously differentiable inverse.

†† Here $\| \cdot \|$ denotes any norm on E^n.

And the second proof of Ref. 1 shows that, with

$$R(q) = F(q) + Aq$$

for all $q \in E^n$, the two conditions (*i*) and (*ii*) are met when $A \in P_0$ and each $f_i(\cdot)$ is a continuously differentiable strictly-monotone-increasing function which maps E^1 onto E^1 and whose slope is positive throughout E^1.*

There are some problems which arise in connection with, for example, the numerical solution of certain nonlinear partial-differential equations** in which one encounters an equation of the form (1) with $A \in P_0$ and det $A \neq 0$, but with functions $f_i(\cdot)$ which, while continuously differentiable, are monotone nondecreasing (rather than strictly monotone increasing) mappings of E^1 into E^1. We can prove that even in such cases equation (1) possesses a unique solution for each $B \in E^n$ as follows. Here the Jacobian matrix of $F(q)+Aq$ exists and is of the form $D(q)+A$ in which $D(q)$ is a diagonal matrix with nonnegative diagonal elements. Since $A \in P_0$ and det $A \neq 0$, we have[2] det $[D(q) + A] \neq 0$ for all $q \in E^n$. An immediate application of Theorem 3 of Section II shows that $\| F(q) + Aq \| \to \infty$ as $\| q \| \to \infty$.† Therefore, by Palais' theorem, $F(x) + Ax = B$ possesses a unique solution for each B.

Theorem 3 is of use not only in connection with the proof given in the preceding paragraph; it also plays a key role in showing that there is an algorithm which generates a sequence of elements of E^n $x^{(0)}, x^{(1)}, \cdots$ that converges to the unique solution of $F(x) + Ax = B$ whenever each $f_i(\cdot)$ is twice continuously differentiable on E^1 and the conditions on A and $F(\cdot)$ of the preceding paragraph are satisfied.‡

More generally, if $R(\cdot)$ is any twice-differentiable mapping of E^n into itself such that conditions (*i*) and (*ii*) of Palais' theorem are satisfied, then, with $R^{-1}(\cdot)$ the continuously-differentiable inverse of $R(\cdot)$, $x = R^{-1}(\theta)$ satisfies $R(x) = \theta$ in which θ is the zero element of E^n, and there are steepest decent as well as Newton-type algorithms each of

---

* The reasons that two proofs were presented in Ref. 1, with the second proof a proof of a somewhat weaker result, are that the arguments needed for the application of Palais' theorem had already been developed in Ref. 1 and used for other purposes there, and it was felt desirable to indicate an alternative approach to essentially the same problem.

** The writer is indebted to J. McKenna and E. Wasserstrom for bringing this fact to his attention.

† More explicitly, Theorem 3 shows that there is a vector $C \in E^n$ such that $\| F(q) + Aq + C \| \to \infty$ as $\| q \| \to \infty$, which is equivalent to the statement concerning $\| F(q) + Aq \|$ made above.

‡ The differentiability assumption here is introduced as a matter of convenience, and is certainly satisfied when the $f_i(\cdot)$ are Ebers–Moll exponential-type nonlinearities.

which generates a sequence in E^n that converges to x. To show this, let[7] $f(y) = \|R(y)\|^2$ for all $y \in E^n$ in which $\|\cdot\|$ denotes the usual Euclidean norm (that is, the square-root of the sum of squares). Since condition (*i*) of Palais' theorem is satisfied, the gradient ∇f of $f(\cdot)$ satisfies $(\nabla f)(y) \neq \theta$ unless $f(y) = 0$,* and since condition (*ii*) of Palais' theorem is satisfied the set $S = \{y \in E^n : f(y) \leqq f(x^{(0)})\}$ is bounded for any $x^{(0)} \in E^n$. Therefore we may appeal to, for example, the theorem of page 43 of Ref. 7 according to which for any $x^{(0)} \in E^n$, for any member of a certain class of mappings $\varphi(\cdot)$ of S into E^n, and for suitably chosen constants γ_0, γ_1, $\cdots$, the sequence $x^{(0)}, x^{(1)}, \cdots$ defined by

$$x^{(k+1)} = x^{(k)} + \gamma_k \varphi(x^{(k)}) \quad \text{for all} \quad k \geqq 0$$

belongs to S and is such that $\|R(x^{(k)})\| \to 0$ as $k \to \infty$. However, since $R^{-1}(\cdot)$ exists and is continuous,† it follows from

$$x^{(k)} = R^{-1}[R(x^{(k)})] \quad \text{for all} \quad k \geqq 0$$

and the fact that $R(x^{(k)}) \to \theta$ as $k \to \infty$, that $\lim_{k\to\infty} x^{(k)}$ exists and

$$\lim_{k\to\infty} x^{(k)} = R^{-1}(\theta),$$

which means that $x = \lim_{k\to\infty} x^{(k)}$.‡

1.4 *Transient Response of Transistor-Diode Networks and Implicit Numerical-Integration Formulas*

At this point we briefly consider some aspects of the manner in which the previous material bears on the important problem of providing more of a theoretical basis for numerically integrating the ordinary differential equations which govern the transient response of nonlinear transistor networks. Although we consider explicitly only networks containing transistors, diodes, and resistors, the material to be presented can be extended to take into account other types of elements as well. In addition, we shall focus attention on the use of linear multipoint integration formulas of closed (that is, of implicit) type, since such

* Here we have used the fact that $(\nabla f)(y) = 2J_y^{tr}R(y)$ for all $y \in E^n$.[7]

† By Palais' theorem $R(\cdot)$ is a diffeomorphism of E^n onto itself.

‡ The material of the second part of Section 1.3 was motivated by previous recent work of the writer's colleague A. Gersho who made the observation that the convergence of an algorithm for the solution of equation (1) could be shown by combining results of Ref. 1 with the approaches described by Goldstein.[7] (See the November 1969 B.S.T.J. Brief by A. Gersho.)

formulas are of considerable use in connection with the typically "stiff systems" of differential equations encountered.

A very large class of networks containing resistors, transistors, and diodes modeled in a standard manner is governed by the equation[8]

$$\frac{du}{dt} + TF[C^{-1}(u)] + (I + GR)^{-1}GC^{-1}(u) = B(t), \qquad t \geqq 0 \qquad (3)$$

where, assuming that there are q diodes and p transistors,

(*i*) $T = I_q \oplus T_1 \oplus T_2 \oplus \cdots \oplus T_p$, the direct sum of the identity matrix of order q and p 2×2 matrices T_k in which

$$T_k = \begin{Bmatrix} 1 & -\alpha_r^{(k)} \\ -\alpha_f^{(k)} & 1 \end{Bmatrix}$$

with $0 < \alpha_r^{(k)} < 1$ and $0 < \alpha_f^{(k)} < 1$ for $k = 1, 2, \cdots, p$.

(*ii*) $R = R_0 \oplus R_1 \oplus R_2 \oplus \cdots \oplus R_p$, the direct sum of a diagonal matrix $R_0 = \operatorname{diag}(r_1, r_2, \cdots, r_q)$ with $r_k \geqq 0$ for $k = 1, 2, \cdots, q$ and p 2×2 matrices R_k in which for all $k = 1, 2, \cdots, p$

$$R_k = \begin{Bmatrix} r_e^{(k)} + r_b^{(k)} & r_b^{(k)} \\ r_b^{(k)} & r_c^{(k)} + r_b^{(k)} \end{Bmatrix}$$

with $r_e^{(k)} \geqq 0$, $r_b^{(k)} \geqq 0$, and $r_c^{(k)} \geqq 0$. (The matrix R takes into account the presence of bulk resistance in series with the diodes and the emitter, base, and collector leads of the transistors.)

(*iii*) G is the short-circuit conductance matrix associated with the resistors of the network. (It does not take into account the bulk resistances of the semiconductor devices.)

(*iv*) $F(\cdot)$ is a mapping of $E^{(2p+q)}$ into $E^{(2p+q)}$ defined by the condition that

$$F(x) = [f_1(x_1), f_2(x_2), \cdots, f_{2p+q}(x_{2p+q})]^{tr}$$

for all $x \in E^{(2p+q)}$ with each $f_i(\cdot)$ a continuously differentiable strictly-monotone increasing mapping of E^1 into E^1.

(*v*) $C^{-1}(\cdot)$ is the inverse of the mapping $C(\cdot)$, of $E^{(2p+q)}$ into itself, defined by

$$C(x) = \operatorname{diag}(c_1, c_2, \cdots, c_{2p+q})x + \operatorname{diag}(\tau_1, \tau_2, \cdots, \tau_{2p+q})F(x)$$

for all $x \in E^{(2p+q)}$ with each c_i and each τ_i a positive constant.

(*vi*) $B(t)$ is a $(2p + q)$-vector which takes into account the voltage and current generators present in the network, and (*vii*) u is related to v the vector of junction voltages of the semiconductor devices through $C(v) = u$ for all $v \in E^{(2p+q)}$.

Equation (3) is equivalent to*

$$\dot{u} + f(u, t) = \theta_{(2p+q)}, \qquad t \geqq 0 \tag{4}$$

in which of course

$$f(u, t) = TF[C^{-1}(u)] + (I + GR)^{-1}GC^{-1}(u) - B(t) \tag{5}$$

and $\theta_{(2p+q)}$ is the zero vector of order $(2p + q)$.

It is well known that certain specializations of the general multipoint formula[9,10]

$$y_{n+1} = \sum_{k=0}^{r} a_k y_{n-k} + h \sum_{k=-1}^{r} b_k \tilde{y}_{n-k} \tag{6}$$

in which

$$\tilde{y}_{n-k} = -f[y_{n-k}, (n - k)h] \tag{7}$$

can be used as a basis for computing the solution of equation (4). Here h, a positive number, is the step size, the a_k and the b_k are real numbers, and of course y_n is the approximation to $u(nh)$ for $n \geqq 1$.

In the literature dealing with formulas of the type (6) in connection with systems of equations of the type (4), information concerning the location of the eigenvalues of the Jacobian matrix J_u of $f(u, t)$ with respect to u plays an important role in determining whether or not a given formula will be (in some suitable sense) stable. In particular, an assumption often made is that all of the eigenvalues of J_u lie in the strict right-half plane for all $t \geqq 0$ and all u. For $f(u, t)$ given by equation (5), we have

$$J_u = T \operatorname{diag} \left\{ \frac{f_i'[g_i(u_i)]}{c_i + \tau_i f_i'[g_i(u_i)]} \right\} + (I + GR)^{-1}G \operatorname{diag} \left\{ \frac{1}{c_i + \tau_i f_i'[g_i(u_i)]} \right\} \tag{8}$$

in which for $j = 1, 2, \cdots, (2p + q)$ $g_i(u_i)$ is the jth component of $C^{-1}(u)$ Thus here J_u is a matrix of the form

$$TD_1 + (I + GR)^{-1}GD_2 \tag{9}$$

where D_1 and D_2 are diagonal matrices with positive diagonal elements.

* Ref. 8 shows that if $B(\cdot)$ is a continuous mapping of $[0, \infty)$ into $E^{(2p+q)}$, then for any initial condition $u^{(0)} \in E^{(2p+q)}$ there exists a unique continuous $(2p + q)$- vector-valued function $u(\cdot)$ such that $u(0) = u^{(0)}$ and (3) is satisfied for all $t > 0$.

A simple result concerning equation (9), Theorem 4 of Section II, asserts that if there exists a diagonal matrix D with positive diagonal elements such that*

(*i*) DT is strongly column-sum dominant, and
(*ii*) $D(I + GR)^{-1}G$ is weakly column-sum dominant,

then for all diagonal matrices D_1 and D_2 with positive diagonal elements, all eigenvalues of (9) lie in the strict right-half plane. This condition on T, G, and R is often satisfied.†

The subclass of numerical integration formulas (6) defined by the condition that $b_{-1} > 0$ are of considerable use[11,12,13] in applications involving the typically "stiff systems" of differential equations encountered in the analysis of nonlinear transistor networks. With $b_{-1} > 0$, y_{n+1} is defined *implicitly* through

$$y_{n+1} + hb_{-1}f(y_{n+1}\,,\,(n+1)h) = \sum_{k=0}^{r} a_k y_{n-k} + h\sum_{k=0}^{r} b_k \tilde{y}_{n-k}$$

in which the right side depends on y_{n-k} only for $k \in \{0, 1, 2, \cdots, r\}$, and for $f(u, t)$ given by equation (5), we have

$$y_{n+1} + hb_{-1}\{TF[C^{-1}(y_{n+1})] + (I + GR)^{-1}GC^{-1}(y_{n+1})\} = q_n \quad (10)$$

in which

$$q_n = \sum_{k=0}^{r} a_k y_{n-k} + h\sum_{k=0}^{r} b_k \tilde{y}_{n-k} + hb_{-1}B[(n+1)h].$$

Obviously, the numerical integration formula (10) makes sense only if there exists for each n a $y_{n+1} \in E^{(2p+q)}$ such that equation (10) is satisfied.

Let $x_{n+1} = C^{-1}(y_{n+1})$ for each n. Then equation (10) possesses a unique solution y_{n+1} if and only if there exists a unique $x_{n+1} \in E^{(2p+q)}$ such that

$$C(x_{n+1}) + hb_{-1}[TF(x_{n+1}) + (I + GR)^{-1}Gx_{n+1}] = q_n\,. \quad (11)$$

Since $C(x_{n+1}) = cx_{n+1} + \tau F(x_{n+1})$, in which

$$c = \text{diag}\,(c_1\,, c_2\,, \cdots, c_{2p+q})$$

and

$$\tau = \text{diag}\,(\tau_1\,, \tau_2\,, \cdots, \tau_{2p+q}),$$

* The terms "strongly-column-sum dominant" and "weakly-column-sum dominant" are reasonably standard. However they are defined in Section II.
† See Ref. 8 for examples.

equation (11) is equivalent to

$$[\tau + hb_{-1}T]F(x_{n+1}) + [c + hb_{-1}(I + GR)^{-1}G]x_{n+1} = q_n . \quad (12)$$

The matrices τ and c are both diagonal with positive diagonal elements. Thus it is clear that for all positive h

$$\det [\tau + hb_{-1}T] \neq 0$$

and

$$\det [c + hb_{-1}(I + GR)^{-1}G] \neq 0.*$$

For all sufficiently small positive h

$$[\tau + hb_{-1}T]^{-1}[c + hb_{-1}(I + GR)^{-1}G] \in P_0 .^{\dagger}$$

Consequently†† for all sufficiently small $h > 0$, equation (12) possesses a unique solution for each q_n .‡ However, our interest in equation (12) is primarily in connection with "large-h" algorithms.

Suppose that $\det G \neq 0$ and that $T^{-1}G \in P_0$ for all possible combinations of α_r and α_f ($0 < \alpha_r < 1, 0 < \alpha_f < 1$) for each transistor (see Ref. 1 for examples). Then, according to Theorem 6 of Section II, for any particular T and R

$$[\tau + hb_{-1}T]^{-1}[c + hb_{-1}(I + GR)^{-1}G] \in P_0$$

for all $h > 0$, and hence equation (10) possesses a unique solution y_{n+1} for all positive values of h.

An important and general proposition concerning (10) is as follows. Suppose that

$$T^{-1}[(I + GR)^{-1}G] \in P_0 \quad (13)$$

and that condition (13) is satisfied whenever $\alpha_r^{(k)}$ and $\alpha_f^{(k)}$ are replaced with positive constants $\delta_r^{(k)}$ and $\delta_f^{(k)}$, respectively, such that $\delta_r^{(k)} \leqq \alpha_r^{(k)}$ and $\delta_f^{(k)} \leqq \alpha_f^{(k)}$ for $k = 1, 2, \cdots, p$. In other words, assuming that $F(\cdot)$ is as defined in this section and that $F(\cdot) \in \mathfrak{F}_0^{(2p+q)}$ (see Definition 1 of Section 2.1), suppose that the dc equation

$$F(x) + T^{-1}[(I + GR)^{-1}G]x = B$$

possesses at most one solution x for each $B \in E^{(2p+q)}$ for "the original set of α's as well as for an arbitrary set of *not-larger* α's." Then an

* Here we have used the fact that $(I + GR)^{-1}G$ is positive semidefinite.
† See Section 1.2.
†† See Section 1.3.
‡ Alternatively, this conclusion could have been obtained by applying the contraction-mapping fixed-point principle to (10), in view of the fact that each of the elements of J_u is bounded on $u \in E^{(p+q)}$ and $t \in [0, \infty)$.

immediate application of Theorem 5 of Section II shows that

$$[\tau + hb_{-1}T]^{-1}[c + hb_{-1}(I + GR)^{-1}G] \in P_0$$

for all $h > 0$, and hence that equation (10) possesses a unique solution y_{n+1} for all $h > 0$ and all $q_n \in E^{(2p+q)}$.

II. THEOREMS, PROOFS, AND SOME DISCUSSION

Throughout this section,

(i) n is an arbitrary positive integer,

(ii) P_0 denotes the set of all real $n \times n$ matrices M such that all principal minors of M are nonnegative,

(iii) real Euclidean n-space is denoted by E^n, and θ is the zero element of E^n,

(iv) v^{tr} denotes the transpose of the row vector $v = (v_1, v_2, \cdots, v_n)$,

(v) $\| v \|$ denotes $(\sum_{i=1}^{n} v_i^2)^{1/2}$ for all $v \in E^n$,

(vi) if D is a real diagonal matrix, then $D > 0$ $(D \geqq 0)$ means that the diagonal elements of D are positive (nonnegative),

(vii) I_q denotes the identity matrix of order q, and I denotes the identity matrix of order determined by the context in which the symbol is used, and

(viii) we shall say that a real $n \times n$ matrix M is strongly (weakly) column-sum dominant if and only if for $j = 1, 2, \cdots, n$

$$m_{jj} > (\geqq) \sum_{i \neq j} | m_{ij} |.$$

2.1 *Definition 1*

For each positive integer n, let $\mathfrak{F}_0^n$ denote that collection of mappings of E^n into itself defined by: $F \in \mathfrak{F}_0^n$ if and only if there exist for $j = 1, 2, \cdots, n$, continuous functions $f_j(\cdot)$ mapping E^1 into E^1 such that for each $x = (x_1, x_2, \cdots, x_n)^{tr} \in E^n$, $F(x) = [f_1(x_1), f_2(x_2), \cdots, f_n(x_n)]^{tr}$, and

(i) $$\inf_{\alpha \in (-\infty, \infty)} [f_j(\alpha + \beta) - f_j(\alpha)] = 0$$

(ii) $$\sup_{\alpha \in (-\infty, \infty)} [f_j(\alpha + \beta) - f_j(\alpha)] = +\infty$$

for all $\beta > 0$ and all $j = 1, 2, \cdots, n$.

2.2 *Theorem 1*

Let $F \in \mathfrak{F}_0^n$, let A be a real $n \times n$ matrix such that $A \notin P_0$, and let δ be a positive constant. Then there exist $B \in E^n$, $x \in E^n$, and $y \in E^n$

such that

(i) $$F(x) + Ax = B,$$

(ii) $$F(y) + Ay = B,$$

and

(iii) $$\| x - y \| = \delta.$$

2.3 Proof of Theorem 1

Since $A \notin P_0$, there exists[2] a real diagonal matrix $D > 0$ such that $\det(D + A) = 0$. Thus there exists a $x^* \in E^n$ such that $\| x^* \| = \delta$ and $(D + A)x^* = \theta$.

Since $F \in \mathfrak{F}_0^n$, there exists a $x \in E^n$ such that

$$f_j(x_j) - f_j(x_j - x_j^*) = x_j^* d_j$$

for all $j = 1, 2, \cdots, n$ in which d_j is the jth diagonal element of D. Let

$$B = F(x) + Ax,$$

and let $y = x - x^*$. Then $A(x - y) = Ax^* = -Dx^*$, and

$$F(x) - F(y) + A(x - y) = \theta. \quad \square$$

2.4 Remarks Concerning Theorem 1

If, as in the case of standard transistor models,

$$f_j(x_j) = e^{\lambda_j x_j} - 1$$

or

$$f_j(x_j) = 1 - e^{-\lambda_j x_j}$$

with $\lambda_j > 0$, we have, respectively,

$$f_j(\alpha + \beta) - f_j(\alpha) = e^{\lambda_j \alpha}(e^{\lambda_j \beta} - 1)$$

or

$$f_j(\alpha + \beta) - f_j(\alpha) = e^{-\lambda_j \alpha}(1 - e^{-\lambda_j \beta})$$

and it is clear that for either type of function conditions (i) and (ii) of Definition 1 are satisfied.

In Ref. 1 it is proved that if $F(\cdot) \in \mathfrak{M}$ the set of all $F(\cdot)$ of the form (2) with each $f_i(\cdot)$ a strictly monotone increasing mapping of E^1 into E^1, and if $A \in P_0$, then equation (1) possesses at most one solution.

Thus, using Theorem 1, we see that for each $F(\cdot) \in \mathfrak{M} \cap \mathfrak{F}_0^n$ there exists at most one solution of $F(x) + Ax = B$ for each $B \in E^n$ if and only if $A \in P_0$. Similarly, with $\mathfrak{M}_0$ the set of all $F(\cdot)$ of the form (2) with each $f_i(\cdot)$ a strictly monotone increasing mapping of E^1 onto (α_i, β_i) with each α_i and β_i such that $-\infty \leqq \alpha_i < \beta_i \leqq \infty$, if $F(\cdot) \in \mathfrak{M}_0 \cap \mathfrak{F}_0^n$ and $\det A \neq 0$, then there exists a unique solution x of $F(x) + Ax = B$ for each $B \in E^n$ if and only if $A \in P_0$. (The "if" part of this statement is proved in Ref. 2.) A parallel development can be carried out for equations of the form $AF(x) + x = B$ with A a real $n \times n$ matrix, $F(\cdot) \in \mathfrak{M}_0 \cap \mathfrak{F}_0^n$, and $B \in E^n$. More explicitly, we can prove that if $F(\cdot) \in \mathfrak{M}_0 \cap \mathfrak{F}_0^n$, then there exists a unique solution x of $AF(x) + x = B$ for each $B \in E^n$ if and only if $A \in P_0$.

There may be a temptation to conjecture that whenever $F(\cdot) \in \mathfrak{M} \cap \mathfrak{F}_0^n$ and $A \notin P_0$ then the equation $F(x) + Ax = B$ does not possess a solution for some $B \in E^n$. The conjecture is false. In fact, with $n = 2$, $f_1(x_1) = e^{x_1}$, $f_2(x_2) = e^{x_2}$, and

$$A = \begin{bmatrix} 0 & 1 \\ 1 & 0 \end{bmatrix},$$

we have a situation in which (it is easy to show that) there exists a solution for all $B \in E^2$. Of course here for some choices of B the solution is not unique.

2.5 *Theorem 2*

Let P and Q denote real $n \times n$ matrices such that

$$p_{jj} > \sum_{i \neq j} |p_{ij}| \quad \text{and} \quad q_{jj} > \sum_{i \neq j} |q_{ij}|$$

for all $j = 1, 2, \cdots, n$. For $j = 1, 2, \cdots, n$ let $f_j(\cdot)$ denote a continuous monotone nondecreasing (but not necessarily differentiable) mapping of E^1 into itself, and let $F(x) = [f_1(x_1), f_2(x_2), \cdots, f_n(x_n)]^{tr}$ for all $x \in E^n$. Then for each $R \in E^n$, there exists a unique $x \in E^n$ such that

$$PF(x) + Qx = R,$$

and, for any $y_0 \in E^n$, x is the limit of the sequence $x^{(0)}, x^{(1)}, \cdots$ defined by

$$y^{(n)} = D_P F(x^{(n)}) + D_Q x^{(n)}$$

$$y^{(n+1)} + (P - D_P)F(x^{(n)}) + (Q - D_Q)x^{(n)} = R$$

for $n \geqq 0$, in which D_P and D_Q are diagonal matrices whose diagonal elements coincide with those of P and Q, respectively.

2.6 *Proof of Theorem 2*

Since the continuous mapping $[D_P F(\cdot) + D_Q]$ of E^n into E^n possesses an inverse $[D_P F(\cdot) + D_Q]^{-1}$, the equation

$$PF(x) + Qx = R$$

possesses a unique solution x if and only if $y = D_P F(x) + D_Q x$ is the unique solution of

$$y + \tilde{P}F[(D_P F(\cdot) + D_Q)^{-1}y] + \tilde{Q}[(D_P F(\cdot) + D_Q)^{-1}y] = R$$

in which $\tilde{P} = (P - D_P)$ and $\tilde{Q} = (Q - D_Q)$.

Therefore, by Banach's contraction-mapping fixed-point theorem, it suffices to show that with the metric $\rho(y, z) = \sum_{i=1}^{n} | y_i - z_i |$, the operator H defined by

$$H(y) = \tilde{P}F[(D_P F(\cdot) + D_Q)^{-1}y] + \tilde{Q}[(D_P F(\cdot) + D_Q)^{-1}y]$$

for all $y \in E^n$, is a contraction mapping of E^n into itself. We show this as follows. Let $y \in E^n$ and $z \in E^n$. Using the fact that

$$\alpha = d_{Qj}[(d_{Pj}f_j(\cdot) + d_{Qj})^{-1}\alpha] + d_{Pj}f_j[(d_{Pj}f_j(\cdot) + d_{Qj})^{-1}\alpha]$$

for all real α and all $j = 1, 2, \cdots, n$, in which d_{Pj} and d_{Qj} is the jth diagonal element of D_P and D_Q, respectively, it is a simple matter to verify that for all j:

$$f_j[(d_{Pj}f_j(\cdot) + d_{Qj})^{-1}y_j] - f_j[(d_{Pj}f_j(\cdot) + d_{Qj})^{-1}z_j]$$

$$= \frac{r_j}{d_{Qj} + d_{Pj}r_j}(y_j - z_j),$$

and

$$(d_{Pj}f_j(\cdot) + d_{Qj})^{-1}y_j - (d_{Pj}f_j(\cdot) + d_{Qj})^{-1}z_j = \frac{1}{d_{Qj} + d_{Pj}r_j}(y_j - z_j)$$

in which $r_j = 1$ if $y_j = z_j$, and, if $y_j \neq z_j$,

$$r_j \triangleq \frac{f_j[(d_{Pj}f_j(\cdot) + d_{Qj})^{-1}y_j] - f_j[(d_{Pj}f_j(\cdot) + d_{Qj})^{-1}z_j]}{(d_{Pj}f_j(\cdot) + d_{Qj})^{-1}y_j - (d_{Pj}f_j(\cdot) + d_{Qj})^{-1}z_j}.$$

Thus

$$H(y) - H(z)$$

$$= \tilde{P}\operatorname{diag}\left\{\frac{r_j}{d_{Qj} + d_{Pj}r_j}\right\}(y - z) + \tilde{Q}\operatorname{diag}\left\{\frac{1}{d_{Qj} + d_{Pj}r_j}\right\}(y - z)$$

in which $r_j \geq 0$. Therefore

$$\rho(H(y), H(z)) \leq \max_j \left(\frac{\sigma_{Qj} + \sigma_{Pj} r_j}{d_{Qj} + d_{Pj} r_j}\right) \rho(y, z)$$

in which $\sigma_{Qj} = \sum_{i \neq j} |q_{ij}|$ and $\sigma_{Pj} = \sum_{i \neq j} |p_{ij}|$. Since $\sigma_{Qj} < d_{Qj}$ and $\sigma_{Pj} < d_{Pj}$ for all j, there exists a positive constant $\beta < 1$ such that

$$\max_j \left(\frac{\sigma_{Qj} + \sigma_{Pj} r_j}{d_{Qj} + d_{Pj} r_j}\right) \leq \beta$$

for all $r_j \geq 0$. □

2.7 *Theorem 3*

If $A \in P_0$ and $\det A \neq 0$, if for each $j = 1, 2, \cdots, n: f_j(\cdot)$ is a continuous mapping of E^1 into itself such that

$$f_j(x_j) = 0 \quad \text{for all } x_j$$

or

$$f_j(x_j) > 0 \quad \text{for all } x_j > c$$

and

$$f_j(x_j) < 0 \quad \text{for all } x_j < -c$$

for some $c \geq 0$, then, with $F(x) = [f_1(x_1), f_2(x_2), \cdots, f_n(x_n)]^{tr}$ for all $x \in E^n$,

$$\|F(x) + Ax\| \to \infty \quad \text{as} \quad \|x\| \to \infty.$$

2.8 *Proof of Theorem 3*

We note that

$$\|F(x) + Ax\| \to \infty \quad \text{as} \quad \|x\| \to \infty$$

if and only if

$$\|A^{-1} F(x) + x\| \to \infty \quad \text{as} \quad \|x\| \to \infty.$$

With $M = A^{-1}$, let

$$MF(x) + x = q. \tag{14}$$

Since $A \in P_0$, we have $M \in {}_t P_0$.[1] Since $M \in P_0$, we have[1] for any $y \in E^n$ and $y \neq \theta$

$$y_k (My)_k \geq 0$$

for some index k such that $y_k \neq 0$.

Suppose that $F(x) \neq \theta$. Then there exists an index k_1 such that

$$f_{k_1}(x_{k_1})[MF(x)]_{k_1} \geqq 0$$

with $f_{k_1}(x_{k_1}) \neq 0$. Thus, using (14),

$$f_{k_1}(x_{k_1})[MF(x)]_{k_1} + f_{k_1}(x_{k_1})x_{k_1} = f_{k_1}(x_{k_1})q_{k_1}$$

and

$$f_{k_1}(x_{k_1})x_{k_1} \leqq f_{k_1}(x_{k_1})q_{k_1} .$$

Either $x_{k_1} \in [-c, c]$ or not. If not, then $f_{k_1}(x_{k_1})x_{k_1} > 0$ and $| x_{k_1} | \leqq | q_{k_1} |$. Therefore for some index k_1, $| x_{k_1} | \leqq \delta_1 \triangleq \max (c, | q_{k_1} |)$, whether or not $F(x) = \theta$.

Let $M^{(k_1)}$ denote the matrix obtained from M by deleting the k_1 row and column, and let $M_{(k_1)}$ denote the k_1 column of M with the k_1 entry removed. Similarly, let $x_{(k_1)}$, $q_{(k_1)}$ and $F_{(k_1)}(x_{(k_1)})$ denote the $(n - 1)$-vectors obtained from x, q, and $F(x)$, respectively, by removing the k_1 entry. Then

$$M^{(k_1)}F_{(k_1)}(x_{(k_1)}) + x_{(k_1)} = q_{(k_1)} - M_{(k_1)}f_{k_1}(x_{k_1}).$$

Since $M^{(k_1)} \in P_0$, we can repeat the argument given above. Thus there exists an index k_2, different from k_1, such that

$$| x_{k_2} | \leqq \delta_2 \triangleq \max (c, | q_{(k_1,k_2)} |)$$

in which

$$| q_{(k_1,k_2)} | = \max_{|x_{k_1}| \leqq \delta_1} | [q_{(k_1)} - M^{(k_1)}f_{k_1}(x_{k_1})]_{l_2} |$$

and l_2 is the index of the component of $x_{(k_1)}$ that corresponds to the k_2 component of x. By continuing in this manner we can determine positive constants $\delta_1, \delta_2, \cdots, \delta_n$ depending only on q, F, M, and c such that, with $\delta = \max_j \{\delta_j\}$,

$$| x_j | \leqq \delta \quad \text{for all} \quad j = 1, 2, \cdots, n$$

and each δ_i depends on q such that for any positive constant α, there exists a constant $\beta_i(\alpha)$ with the property that $\delta_i \leqq \beta_i(\alpha)$ provided that $\| q \| \leqq \alpha$. Therefore for any $\alpha > 0$ there is a $\beta(\alpha)$ such that $\| x \| \leqq \beta(\alpha)$ whenever $\| q \| \leqq \alpha$, which implies that $\| q \| \to \infty$ as $\| x \| \to \infty$. □

2.9 *Theorem 4*

Let P and Q denote real $n \times n$ matrices with P strongly column-sum dominant. Suppose that there exists a real diagonal matrix $D > 0$ such *that DP is strongly column-sum dominant and DQ is weakly column-sum dominant. Then for all real diagonal matrices $D_1 > 0$ and $D_2 > 0$, all eigenvalues of $(PD_1 + QD_2)$ lie in the strict (that is, open) right-half plane.*

2.10 *Proof of Theorem 4*

Since the strict right-half plane contains all of the eigenvalues of P, there exists choices of $D_1 > 0$ and $D_2 > 0$ such that every eigenvalue of $(PD_1 + QD_2)$ lies in the strict right-half plane. Thus it suffices to show that $(PD_1 + QD_2)$ does not possess an eigenvalue on the boundary of the complex plane for all $D_1 > 0$ and all $D_2 > 0$. In other words, it suffices to show that (with $i = (-1)^{\frac{1}{2}}$)

$$PD_1 + QD_2 + i\omega I \tag{15}$$

is nonsingular for all $D_1 > 0$, all $D_2 > 0$, and all real constants ω.

Suppose that (15) is singular for some ω and some $D_1 > 0$ and some $D_2 > 0$. Then $(DPD_1 + DQD_2 + i\omega D)$ is singular. But DPD_1 is strongly column-sum dominant and DQD_2 is weakly column-sum dominant. Thus $M = (DPD_1 + DQD_2)$ is strongly column-sum dominant, and, since

$$| m_{jj} + i\omega d_j | > \sum_{i \neq j} | m_{ij} |$$

for all j, in which d_j is the jth diagonal element of D, it follows that det $(M + i\omega D) \neq 0$, which is a contradiction. □

2.11 *Definition 2*

With q and p nonnegative integers such that $(p + q) > 0$, let $\mathfrak{I}$ denote the set of all matrices M such that $M = I_q \oplus M_1 \oplus M_2 \oplus \cdots \oplus M_p$ with

$$M_k = \begin{bmatrix} 1 & -\alpha_r^{(k)} \\ -\alpha_f^{(k)} & 1 \end{bmatrix}$$

and

$$0 < \alpha_r^{(k)} < 1$$
$$0 < \alpha_f^{(k)} < 1$$

for all $k = 1, 2, \cdots, p$.*

* As suggested, if $q = 0$, then $M = M_1 \oplus M_2 \oplus \cdots \oplus M_p$, while if $p = 0$, then $M = I_q$.

2.12 *Definition 3*

With q and p nonnegative integers such that $(p + q) > 0$, let $\mathfrak{I}(\alpha)$ denote the set of all matrices M such that $M = I_q \oplus M_1 \oplus M_2 \oplus \cdots \oplus M_p$ with

$$M_k = \begin{bmatrix} 1 & -\delta_r^{(k)} \\ -\delta_f^{(k)} & 1 \end{bmatrix}$$

and

$$0 < \delta_r^{(k)} \leqq \alpha_r^{(k)}$$

$$0 < \delta_f^{(k)} \leqq \alpha_f^{(k)}$$

for all $k = 1, 2, \cdots, p$.*

2.13 *Theorem 5*

Let $T \in \mathfrak{I}$, let H be a real matrix of order $(2p + q)$, and suppose that $M^{-1}H \in P_0$ for all $M \in \mathfrak{I}(\alpha)$. Then

$$(T + D_1)^{-1}(H + D_2) \in P_0$$

for all diagonal matrices $D_1 \geqq 0$ and $D_2 \geqq 0$.

2.14 *Proof of Theorem 5*

Suppose that for some $D_1 \geqq 0$ and $D_2 \geqq 0$

$$(T + D_1)^{-1}(H + D_2) \notin P_0 .$$

Then there exists[2] a diagonal matrix $D > 0$ such that

$$(T + D_1)^{-1}(H + D_2) + D$$

is singular. It follows that

$$H + \Delta + TD$$

is singular, in which $\Delta = D_2 + D_1 D$. Since

$$\Delta + TD = M(\Delta + D)$$

in which $M \in \mathfrak{I}(\alpha)$, it follows that

$$H + M(\Delta + D)$$

is singular, and therefore that

* As suggested, if $q = 0$, then $M = M_1 \oplus M_2 \oplus \cdots \oplus M_p$, while if $p = 0$, then $M = I_q$.

$$M^{-1}H + (\Delta + D)$$

is singular, which is a contradiction since[1] $M^{-1} H \in P_0$ and $(\Delta + D)$ is a diagonal matrix with positive diagonal elements. □

2.15 *Theorem 6*

Let $M^{-1}G \in P_0$ for all $M \in \mathfrak{J}$, and let $\det G \neq 0$. Let R be as defined in Section 1.4. Then for any $T \in \mathfrak{J}$

$$(T + D_1)^{-1}[(I + GR)^{-1}G + D_2] \in P_0$$

for all diagonal matrices $D_1 \geqq 0$ and $D_2 \geqq 0$.

2.16 *Proof of Theorem 6*

Since $\det G \neq 0$ and $M^{-1}G \in P_0$ for all $M \in \mathfrak{J}$, it follows (see the proof of Theorem 7 of Ref. 2) that

$$M^{-1}(I + GR)^{-1}G \in P_0$$

for all $M \in \mathfrak{J}$.

Suppose that for some $T \in \mathfrak{J}$ and some $D_1 \geqq 0$ and $D_2 \geqq 0$

$$(T + D_1)^{-1}[(I + GR)^{-1}G + D_2] \notin P_0 .$$

Then, following the proof of Theorem 5, we would have

$$\det \{M^{-1}(I + GR)^{-1}G + (\Delta + D)\} = 0$$

for some $M \in \mathfrak{J}$ and some diagonal matrix $(\Delta + D)$ with positive diagonal elements, which is a contradiction. □

III. ACKNOWLEDGMENT

The writer is indebted to A. N. Willson, Jr. for carefully reading the draft.

REFERENCES

1. Sandberg, I. W., and Willson, A. N., Jr., "Some Theorems on Properties of DC Equations of Nonlinear Networks," B.S.T.J., *48*, No. 1 (January 1969), pp. 1–34.
2. Sandberg, I. W., and Willson, A. N., Jr., "Some Network-Theoretic Properties of Non-Linear DC Transistor Networks," B.S.T.J., *48*, No. 5 (May–June 1969), pp. 1293–1312.
3. Varga, R. S., *Matrix Iterative Analysis,* Englewood Cliffs, New Jersey: Prentice-Hall, 1962, p. 23.
4. Stern, T. E., *Theory of Nonlinear Networks and Systems,* Reading, Mass.: Addison-Wesley, 1965, pp. 42–43.

5. Palais, R. S., "Natural Operations on Differential Forms," Trans. Amer. Math. Soc., *92,* No. 1 (1959), pp. 125–141.
6. Holzmann, C. A., and Liu, R., "On the Dynamical Equations of Nonlinear Networks with n-Coupled Elements," Proc. Third Ann. Allerton Conf. on Circuit and System Theory, U. of Illinois, 1965, pp. 536–545.
7. Goldstein, A. A., *Constructive Real Analysis,* New York: Harper & Row, 1967, pp. 41–45.
8. Sandberg, I. W., "Some Theorems on the Dynamic Response of Nonlinear Transistor Networks," B.S.T.J., *48,* No. 1 (January 1969), pp. 35–54.
9. Hamming, R. W., *Numerical Methods for Scientists and Engineers,* New York: McGraw-Hill, 1962.
10. Ralston, A. A., *A First Course in Numerical Analysis,* New York: McGraw-Hill, 1965.
11. Hachtel, G. D., and Rohrer, R. A., "Techniques for the Optimal Design and Synthesis of Switching Circuits," Proc. of the IEEE, *55,* No. 11 (November 1967), pp. 1864–1876.
12. Sandberg, I. W., and Shichman, H., "Numerical Integration of Systems of Stiff Nonlinear Differential Equations," B.S.T.J., *47,* No. 4 (April 1968), pp. 511–527.
13. Calahan, D. A., "Efficient Numerical Analysis of Non-Linear Circuits," Proc. Sixth Ann. Allerton Conf. on Circuit and System Theory, U. of Illinois, 1968, pp. 321–331.

Conditions for the Existence of a Global Inverse of Semiconductor-Device Nonlinear-Network Operators

IRWIN W. SANDBERG, MEMBER, IEEE

Abstract—Necessary and sufficient conditions are presented for the global invertibility of a certain general operator that arises in the analysis of realistically modeled nonlinear networks containing transistors and diodes. It is often possible to determine by inspection whether or not a certain key condition is satisfied; whenever the conditions are satisfied there are convergent algorithms for computing the solution of the associated network problem.

I. Introduction and Brief Summary

This paper is essentially a sequel to [1], which is concerned with operators of the form $[F(\cdot)+A]$ in which A is a not necessarily nonsingular real $n\times n$ matrix and $F(\cdot)$ is a "diagonal" strictly monotone-increasing mapping of the set of all real n vectors E^n onto an open subset of E^n. In [1] necessary and sufficient conditions are given for the existence of the global inverse of a large class of operators of the form $[F(\cdot)+A]$. Operators of the type $[F(\cdot)+A]$ frequently arise in the analysis of nonlinear networks and are encountered in other areas as well. In particular, for A the short-circuit conductance matrix of a resistance network, and $F(x)$ the transpose of $(f_1(x_1), f_2(x_2), \cdots, f_n(x_n))$ for all $x\in E^n$ in which the $f_j(\cdot)$ are the usual exponential diode functions, [1] presents a complete solution to the problem of determining whether or not $[F(\cdot)+A]$ possesses a global inverse on E^n.[1]

Although it is not assumed in [1] that A is a symmetric nonnegative-definite matrix, there is an important case that is not covered by the results given there. More explicitly, with the exception of some relatively special situations, the results of [1] were not shown to be applicable to the operator $[TF(\cdot)+G]$, in which T and G are real matrices, which arises (see [2]) in the analysis of networks containing resistors and Ebers–Moll-modeled transistors and diodes. Here, using an approach different from that of [1], we present, as Theorem 2 (Section II), a complete solution to the problem of determining whether or not $[TF(\cdot)+G]$ possesses a global inverse on E^n, assuming merely that the transistor models are passive.

It can be shown that there are convergent algorithms for computing the solution x of $TF(x)+Gx=b$ for any $b\in E^n$ when $[TF(\cdot)+G]$ possesses a global inverse on E^n. Indeed this can easily be done by using arguments similar to those of [3, Sec. 2.3].

Manuscript received December 22, 1970; revised April 12, 1971.

The author is with Bell Telephone Laboratories, Inc., Murray Hill, N. J. 07974.

[1] See [1] for further motivation and for a discussion of the relation between the results given there and in earlier related work.

II. Theorems 1 and 2

A. Notation and Definitions

We use the following notation throughout Section II. With n an arbitrary positive integer, E^n denotes the set of all real n vectors, and θ denotes the zero vector of E^n. If $x\in E^n$, then x_j denotes the jth component of x, $\langle x, y\rangle$ denotes $\sum_{j=1}^n x_j y_j$ for all $x\in E^n$ and $y\in E^n$, and $\|x\| = \langle x, x\rangle^{1/2}$ for all $x\in E^n$.

The symbol G denotes an arbitrary real symmetric nonnegative-definite matrix of order n, $\mathfrak{N}(G)\triangleq\{x: x\in E^n, Gx=\theta\}$ and $\mathfrak{R}(G)\triangleq\{x: x=Gu \text{ for some } u\in E^n\}$. Clearly, $\mathfrak{N}(G)$ is the orthogonal complement of $\mathfrak{R}(G)$ with respect to E^n.

The following theorem plays a central role in the Proof of Theorem 2.

B. Theorem 1

Let $\mathfrak{N}(G)\neq\{\theta\}$, let $\mathfrak{R}(G)\neq\{\theta\}$, and let $N(\cdot)$ be a continuous mapping of E^n into E^n such that $[N(\cdot)+G]$ is a local homeomorphism in E^n. Suppose that 1) for some positive constant c, $\langle x, N(x)\rangle\geq 0$ for all $\|x\|\geq c$ and 2) for each pair of positive constants k_1 and ρ, there exists a positive constant k_2 such that $\|N(w_y+y)\|\geq k_1\|y\|^{1/2}$ whenever $\|y\|\geq k_2$, $y\in\mathfrak{N}(G)$, $w_y\in\mathfrak{R}(G)$, and $\|w_y\|\leq\rho\|y\|^{1/2}$. Then $[N(\cdot)+G]$ is a homeomorphism of E^n onto E^n.

C. Proof of Theorem 1

We shall use a theorem of Palais (see [4], [5, Appendix]), according to which if $R(\cdot)$ is a local homeomorphism in E^n, then $R(\cdot)$ is a homeomorphism of E^n onto E^n if and only if $\|R(x)\|\to\infty$ as $\|x\|\to\infty$. It suffices to show that if $N(x)+Gx=b$ with $\|b\|\leq\beta$, then there is a positive constant k depending on only β, $N(\cdot)$, and G, such that $\|x\|\leq k$.

Let $N(x)+Gx=b$ with $\|b\|\leq\beta$, and let $b=Gu+\varphi$ with $u\in\mathfrak{R}(G)$ and $\varphi\in\mathfrak{N}(G)$. Then $N(x)+G(x-u)=\varphi$, and with $z=x-u$, $N(z+u)+Gz=\varphi$. Let $z=w+y$ with $w\in\mathfrak{R}(G)$ and $y\in\mathfrak{N}(G)$. Thus

$$N(w+y+u)+Gw=\varphi \tag{1}$$

and hence

$$\langle x, N(x)\rangle+\langle w+u, Gw\rangle=\langle y, \varphi\rangle.$$

Reprinted from *IEEE Transactions on Circuit Theory*, vol. CT-19, pp. 34–36, January 1972.

Therefore, by hypothesis 1) either $\|x\|<c$ or $\|x\|\geq c$ and

$$\langle w+u, Gw\rangle = \langle w+u, G(w+u)\rangle - \langle w+u, Gu\rangle \leq \|\varphi\|\cdot\|y\|. \quad (2)$$

Suppose that $\|x\|\geq c$ and that (2) is satisfied. Since $(w+u)\in\mathcal{R}(G)$, we have $\lambda\|w+u\|^2\leq\langle w+u, G(w+u)\rangle$ in which λ is the smallest nonzero eigenvalue of G. Also, $|\langle w+u, Gu\rangle|\leq\|Gu\|\cdot\|w+u\|$. Thus

$$\lambda\|w+u\|^2 - \|Gu\|\cdot\|w+u\| \leq \|\varphi\|\cdot\|y\|.$$

We note that

$$\tfrac{1}{2}\lambda\|w+u\|^2 \leq \lambda\|w+u\|^2 - \|Gu\|\cdot\|w+u\|$$

for $\|w+u\|\geq 2(\lambda)^{-1}\|Gu\|$. Therefore at least one of the following two inequalities is met:

$$\|w+u\| \leq 2(\lambda)^{-1}\beta \quad (3)$$

or

$$\|w+u\|^2 \leq 2(\lambda)^{-1}\beta\|y\|. \quad (4)$$

We observe that if (4) is not satisfied, then (3) is satisfied, $\|y\|<2(\lambda)^{-1}\beta$, and hence $\|x\|\leq 2^{3/2}(\lambda)^{-1}\beta$. On the other hand, if (4) is met, then since

$$\|N(w+y+u)\| \leq \|G(w+u)\| + \|Gu\| + \|\varphi\| \leq \|G(w+u)\| + 2\beta$$

[see (1)], we have

$$\|N(w+y+u)\| \leq (2\lambda^{-1}\beta)^{1/2}\|G\|\cdot\|y\|^{1/2} + 2\beta. \quad (5)$$

By hypothesis 2) there exists a positive constant k_2 that depends on only β, $N(\cdot)$, and G such that (5) is violated for all $\|y\|\geq k_2$ [recall that $y\in\mathfrak{N}(G)$ and that $(w+u)\in\mathcal{R}(G)$ satisfies (4)].

To summarize, our hypotheses and $N(x)+Gx=b$ with $\|b\|\leq\beta$ imply that at least one of the following statements is true.

1) $\|x\|<c$;
2) $\|x\|\geq c$ and $\|x\|\leq 2^{3/2}(\lambda)^{-1}\beta$;
3) $\|x\|\geq c$, $\|y\|\leq k_2$, and $\|w+u\|^2\leq 2(\lambda)^{-1}\beta k_2$.

D. Some Further Definitions

Definition

Let T denote a real nonsingular matrix of order n, and let $F(\cdot)$ denote a mapping of E^n into E^n with the property that $F(x)$ is equal to the transpose of the row vector $(f_1(x_1), f_2(x_2), \cdots, f_n(x_n))$ for all $x\in E^n$, in which for each j either

$$f_j(x_j) = c_j[\exp(b_jx_j) - 1]$$

or

$$f_j(x_j) = c_j[1 - \exp(-b_jx_j)]$$

for all $x_j\in E^1$ where c_j and b_j are positive constants.

Definition

The set of all real $n\times n$ matrices A, such that all principal minors of A are nonnegative, is denoted by P_0.

Definition

The set $\{x: x\in E^n, \|F(\alpha x)\| \text{ is bounded as } \alpha\to\infty\}$ is denoted by $\mathcal{B}(F)$.

E. Theorem 2

Suppose that $\langle x, TF(x)\rangle\geq 0$ for all $x\in E^n$. Then the mapping $[TF(\cdot)+G]$ of E^n into E^n is a homeomorphism of E^n onto E^n if and only if

1) $T^{-1}G\in P_0$;
2) $\mathcal{B}(F)\cap\mathfrak{N}(G)=\{\theta\}$.

If $T^{-1}G\notin P_0$, then there are at least two solutions of $TF(x)+Gx=b$ for some $b\in E^n$; and if $T^{-1}G\in P_0$ but $\mathcal{B}(F)\cap\mathfrak{N}(G)\neq\{\theta\}$, then for some $b\in E^n$ there is no solution of $TF(x)+Gx=b$.

F. Proof of Theorem 2

If $\mathcal{B}(F)\cap\mathfrak{N}(G)\notin\{\theta\}$, then there exists $v\in\mathcal{B}(F)\cap\mathfrak{N}(G)$ such that $v\neq\theta$ and $\|TF(\alpha v)+\alpha Gv\|$ is bounded as $\alpha\to\infty$. Then, by the result of Palais, referred to in the Proof of Theorem 1, $[TF(\cdot)+G]$ is not a homeomorphism of E^n onto E^n.

In [6] it is proved that if $T^{-1}G\notin P_0$, then there are at least two solutions of $TF(x)+Gx=b$ for some b. If $T^{-1}G\in P_0$ but $\mathcal{B}(F)\cap\mathfrak{N}(G)\neq\{\theta\}$, then by essentially the same argument used in the last part of the proof of [1, lemma 2], it follows that there is no solution of $TF(x)+Gx=b$ for some $b\in E^n$.

We now prove that $[TF(\cdot)+G]$ is a homeomorphism of E^n onto E^n when $T^{-1}G\in P_0$ and $\mathcal{B}(F)\cap\mathfrak{N}(G)=\{\theta\}$.

Suppose that $T^{-1}G\in P_0$ and that $\mathcal{B}(F)\cap\mathfrak{N}(G)=\{\theta\}$. Since $T^{-1}G\in P_0$, it follows [2] that $[TF(\cdot)+G]$ is a local homeomorphism in E^n. It has been proved that $[TF(\cdot)+G]$ is a homeomorphism of E^n onto E^n if $\det G>0$ [7]. We now assume that $\det G=0$. By Theorem 1, it suffices to show (since T is nonsingular) that for each pair of positive constants k_1 and ρ, there exists a positive constant k_2 such that $\|F(w_y+y)\|\geq k_1\|y\|^{1/2}$ whenever $\|y\|\geq k_2$, $y\in\mathfrak{N}(G)$, $w_y\in\mathcal{R}(G)$, and $\|w_y\|\leq\rho\|y\|^{1/2}$.

We have $\mathcal{B}(F)\cap\mathfrak{N}(G)=\{\theta\}$. Let G be of nullity r, and let $\{\varphi^{(j)}\}_{j=1}^r$ be an orthonormal set of elements of E^n which span $\mathfrak{N}(G)$. Thus any $\varphi\in\mathfrak{N}(G)$ can be written in the form

$$\sum_{j=1}^{r} a_j\varphi^{(j)}$$

in which the a_j are real constants. Suppose that $\varphi\in\mathfrak{N}(G)$ and that $\varphi\neq\theta$. Then $\|F(\alpha\varphi)\|\to\infty$ as $\alpha\to\infty$, for if that were not so, we would have $\varphi\in\mathcal{B}(F)$, and hence, since $\mathcal{B}(F)\cap\mathfrak{N}(G)=\{\theta\}$, $\varphi\notin\mathfrak{N}(G)$, a contradiction. For each $j=1, 2, \cdots, n$, let $s_j=1$ or -1, depending on whether $|f_j(x_j)|\to\infty$ as $x_j\to\infty$ or $x_j\to-\infty$, respectively. We observe therefore that if $\varphi\in\mathfrak{N}(G)$ and $\varphi\neq\theta$, then there exists at least one value of the index j such that $\varphi_j\neq 0$ and $\operatorname{sgn}(\varphi_j)=s_j$.

Let $\varphi=\sum_{j=1}^{r}a_j\varphi^{(j)}$ be an arbitrary element of $\mathfrak{N}(G)$ such that $\sum_{j=1}^{r}a_j^2=1$, and let S denote the set of all indices j such that $\varphi_j\neq 0$ and $\operatorname{sgn}(\varphi_j)=s_j$. The number

$$\sum_{j\in S}|\varphi_j|$$

is positive and depends continuously on the r-vector $(a_1, a_2, \cdots, a_r)$. Thus there is a positive number k such that

$$\inf_{\sum_{j=1}^{r} a_j^2 = 1} \sum_{j \in S} |\varphi_j| = k$$

and hence for at least one j, $|\varphi_j| \geq kn^{-1}$ and sgn $(\varphi_j) = s_j$. More generally, if $y \in \mathfrak{N}(G)$ and $y \neq \theta$, then for at least one value of the index j, $|y_j| \geq c\|y\|$ and sgn $(y_j) = s_j$, in which $c = kn^{-1}$.

For each $\beta \in [0, \infty)$, let $f(\beta) = \min\{\min\{f_j(\beta): s_j = 1\}, \min\{-f_j(-\beta): s_j = -1\}\}$, with the understanding that a minimum over the empty set is ∞. Thus $f(\cdot)$ is monotone nondecreasing, $\beta^{-1/2}f(\beta) \rightarrow \infty$ as $\beta \rightarrow \infty$, and

$$\|F(x)\| \geq f(|x_j|)$$

for all j such that $x_j \neq 0$ and sgn $(x_j) = s_j$.

Let $\rho > 0$ be given, let $\|w_y\| \leq \rho\|y\|^{1/2}$ with $y \in \mathfrak{N}(G)$, and let $\|y\|^{1/2} \geq 2\rho c^{-1}$. Then for any j, $|(w_y)_j| \leq \|w_y\| \leq \frac{1}{2}c\|y\|$, and for at least one j, $|(w_y + y)_j| \geq \frac{1}{2}c\|y\|$ and sgn $((w_y + y)_j) = s_j$. Then

$$\|F(w_y + y)\| \geq f(\tfrac{1}{2}c\|y\|)$$

and $\|y\|^{-1/2}f(\frac{1}{2}c\|y\|) \rightarrow \infty$ as $\|y\| \rightarrow \infty$.

G. Remarks

In the main part of the Proof of Theorem 2 we used the hypothesis that the $f_j(\cdot)$ are exponential functions in order to define a monotone-nondecreasing function $f(\cdot)$ such that $\beta^{-1/2}f(\beta) \rightarrow \infty$ as $\beta \rightarrow \infty$. It is clear that much weaker assumptions concerning the $f_j(\cdot)$ would have sufficed for that purpose.

The condition that $\mathfrak{R}(F) \cap \mathfrak{N}(G) = \{\theta\}$ can easily be shown to be equivalent to the condition that the set $Q \triangleq \{x: x \in E^n, x \neq \theta, Gx = \theta$, and $x_j s_j \geq 0$ for all $j\}$ is empty, in which the s_j are defined in Section II-F.

Since (within the context of the network problem of principal interest here) G is the short-circuit conductance matrix of a resistor network, and since $\{x: Gx = \theta\}$ is simply the set of voltage vectors that produce zero current at all ports, it is frequently possible to determine by inspection of the network whether or not Q is empty. Reference [1] contains a simple example which illustrates this point.

References

[1] I. W. Sandberg, "Necessary and sufficient conditions for the global invertibility of certain nonlinear operators that arise in the analysis of networks," *IEEE Trans. Circuit Theory*, vol. CT-18, Mar. 1971, pp. 260–263.

[2] I. W. Sandberg and A. N. Willson, Jr., "Some theorems on properties of dc equations of nonlinear networks," *Bell Syst. Tech. J.*, vol. 48, Jan. 1969, pp. 1–34.

[3] I. W. Sandberg, "Theorems on the computation of the transient response of nonlinear networks containing transistors and diodes," *Bell Syst. Tech. J.*, vol. 49, Oct. 1970, pp. 1739–1776.

[4] R. S. Palais, "Natural operations on differential forms," *Trans. Amer. Math. Soc.*, vol. 92, no. 1, 1959, pp. 125–141.

[5] C. A. Holzmann and R. Liu, "On the dynamical equations of nonlinear networks with *n*-coupled elements," in *Proc. 3rd Annu. Allerton Conf. Circuit and System Theory*, 1965, pp. 536–545.

[6] I. W. Sandberg, "Theorems on the analysis of nonlinear transistor networks," *Bell Syst. Tech. J.*, vol. 49, Jan. 1970, pp. 95–114.

[7] I. W. Sandberg and A. N. Willson, Jr., "Some network-theoretic properties of nonlinear dc transistor networks," *Bell Syst. Tech. J.*, vol. 48, May–June 1969, pp. 1293–1312.

Existence of Solutions for the Equations of Transistor–Resistor–Voltage Source Networks

IRWIN W. SANDBERG, MEMBER, IEEE, AND ALAN N. WILLSON, JR., MEMBER, IEEE

Abstract—We consider the dc equations of nonlinear networks containing resistors, independent voltage sources (independent current sources are excluded), and certain types of nonlinear devices (such as Ebers–Moll-modeled transistors and diodes) that possess a certain property closely related to passivity. It is proved for the first time that the equations always possess at least one solution. This result complements some of the writers' previous work in which attention was not focused on networks containing only independent sources of the voltage type. In fact it was shown by simple examples given earlier that there exist transistor networks (containing ideal independent current sources) for which the network equations have no solution.

Here we also complete a study of conditions under which it is possible to carry out certain implicit numerical integration algorithms for the computation of the transient response of an important class of nonlinear networks containing transistors and diodes. We in fact prove that the assumption of passivity for the transistors and diodes implies that it is always possible to carry out the algorithms (in the sense that for any value of the step size there is always at least one solution of a certain key set of nonlinear equations).

Manuscript received March 10, 1971; revised June 11, 1971. This paper was presented at the 1970 IEEE International Symposium on Circuit Theory, Atlanta, Ga.

The authors are with Bell Telephone Laboratories, Inc., Murray Hill, N. J. 07974.

I. Introduction

THIS PAPER is concerned primarily with networks containing only transistors, diodes, resistors, and independent sources. We shall, however, focus attention on the more general network of Fig. 1 in order to emphasize that our results apply also to a much larger class of networks. It is, of course, clear that any network containing only transistors, diodes, resistors, and independent sources can be viewed as a network of the type shown in Fig. 1, with the nonlinear n-port containing the transistors and diodes, and the linear n-port containing the resistors and independent sources.

The results of this paper differ considerably from those of the writers' earlier related papers (see, for example, [1], [2]). In particular, here the emphasis is on 1) special properties of networks containing only independent sources of the voltage type, and 2) the implications of certain passivity, and passivity-like conditions.

The following notation is used throughout the paper. For each positive integer n we denote by E^n the n-dimensional

Reprinted from *IEEE Transactions on Circuit Theory*, vol. CT-18, pp. 619–625, November 1971.

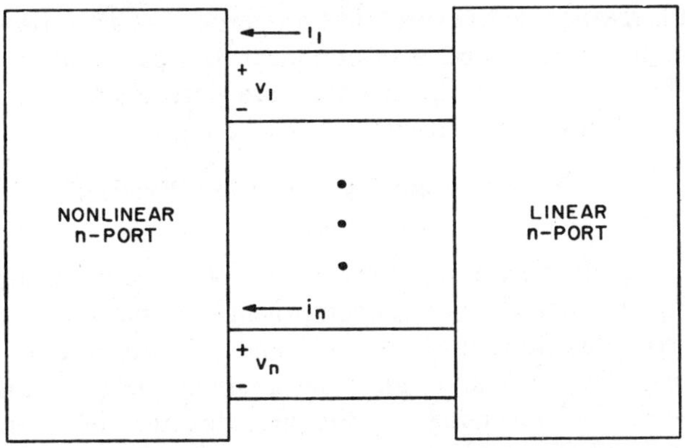

Fig. 1. Network constructed by interconnecting two n-ports.

Euclidean space, the elements of which are ordered n-tuples of real numbers which we consider to be column vectors. The origin in E^n is denoted by θ. If $x \in E^n$, then for $k=1, \cdots, n$, x_k denotes the kth component of x. For all $x, y \in E^n$ the inner product $\sum_{k=1}^{n} x_k y_k$ is denoted by $\langle x, y \rangle$. The norm of each $x \in E^n$ is denoted by $\|x\| = \langle x, x \rangle^{1/2}$. For each subset $S \subset E^n$ we denote the closure of S by $\bar{S}$; also ∂S denotes the boundary of S. All matrices considered here contain only real elements. We use the superscript T to denote the transpose of a vector or a matrix. An $n \times n$ matrix A is said to be positive definite (positive semidefinite) if $\langle x, Ax \rangle > 0$ (≥ 0) for all $x \neq \theta$ in E^n. Note that, unlike the conventions adopted by some writers, we do not require that positive definite and positive semidefinite matrices necessarily be symmetric.

II. The Linear n-Port

Let $v=(v_1, \cdots, v_n)^T$ and $i=(i_1, \cdots, i_n)^T$ be real n-vectors whose components represent the values of the port voltages and currents, respectively, of a given n-port. We say that the n-port has a *hybrid characterization* if there exists a real $n \times n$ matrix H, a real n-vector c, and two disjoint subsets $\tilde{N}$, $\hat{N}$ of the set of indices $N=\{1, \cdots, n\}$ satisfying $\tilde{N} \cup \hat{N} = N$ such that the n-port admits the port variables v, i if and only if

$$y = Hx + c \qquad (1)$$

with x, y, v, i related by

$$\begin{aligned} y_k &= v_k \text{ and } x_k = i_k, && \text{if } k \in \tilde{N} \\ y_k &= i_k \text{ and } x_k = v_k, && \text{if } k \in \hat{N}. \end{aligned} \qquad (2)$$

We say that a pair of $n \times n$ matrices (P, Q) is a *passive pair* if it follows from $Pv = Qi$ that $\langle v, i \rangle \geq 0$.

The following lemma (which is used in the proof of our main result) shows the relationship between certain hybrid characterizations (1), (2) and characterizations of the form

$$Pv = Qi + c. \qquad (3)$$

Lemma 1

An n-port has a characterization of the form $\hat{P}v = \hat{Q}i + \hat{c}$ with $(\hat{P}, \hat{Q})$ a passive pair only if the n-port has both a hybrid characterization (1), (2) with H a positive semidefinite matrix, as well as a characterization of the form (3) with (P, Q) a passive pair, and with the same vector c appearing in (1) and (3).

The proof of Lemma 1 is given in the Appendix.

For n-ports containing only resistors (i.e., resistors of nonnegative resistance) and independent voltage and current sources, a detailed discussion of the relationship between (1) and (3) is given in [4]. In particular, it is shown there that for any such n-port there exist characterizations of the type (1) and (3) with the same vector c appearing in both (1) and (3). It is clear that for such n-ports the matrix H in (1) is positive semidefinite, and that the pair (P, Q) in (3) is a passive pair. In order to make further progress, we need the following result which plays a key role in the physical interpretation of Theorem 2, our main theorem.

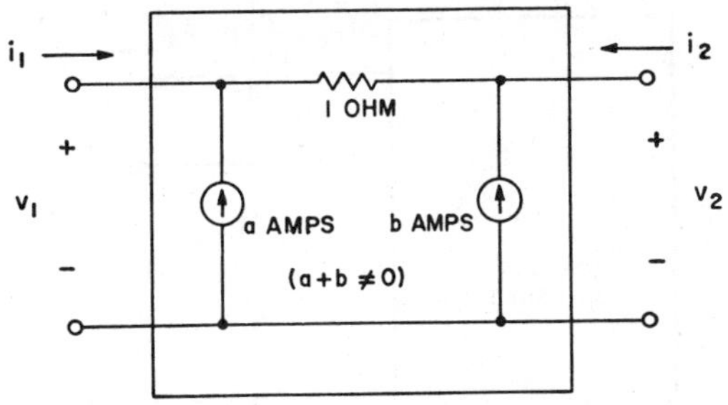

Fig. 2. Highly idealized 2-port containing independent current sources.

Theorem 1

If (3) is a characterization of an n-port containing only resistors and independent voltage sources, then the vector c is contained in the range of the matrix operator P.

A proof of Theorem 1 is given in the Appendix. As a consequence of Theorem 1 it is also evident that c is contained in the range of P for any n-port characterized by (3) and containing independent current sources as well as independent voltage sources and resistors, provided that the n-port's topology is such that it would be possible to eliminate the presence of the current sources by appropriate application of Thévenin's theorem. Indeed, the range of P need not contain c only for highly idealized n-ports such as, for example, the 2-port shown in Fig. 2.[1]

In view of Theorem 1, attention is restricted throughout the remainder of the paper to those linear n-ports which have a characterization of the form

$$P(v - c') = Qi \qquad (4)$$

where (P, Q) is a passive pair of real $n \times n$ matrices and c' is a real n-vector. It follows that we may consider all such n-ports to also possess a hybrid characterization of the

[1] While the fact that c is contained in the range of P for essentially all networks of current practical interest has important implications concerning the existence of solutions, consideration of only the case in which c and P are related in this way does not enable one to prove the global inversion theorems of [1]–[6] which, for example, enable one to prove that certain iteration techniques can be used to compute a solution to the associated network problem. In other words, the study reported on here by no means displaces the results of [1]–[6] for practical networks; it simply complements those results.

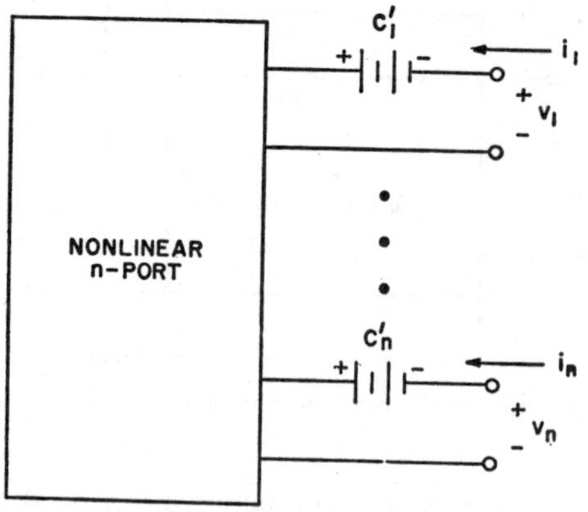

Fig. 3. Physical interpretation of property 2.

form

$$y = Hx + Pc' \tag{5}$$

with the vectors x, y, v, i related as described above.

III. The Nonlinear n-Port

We consider the class of all nonlinear n-ports that can be characterized in terms of their port voltage and current vectors v, i by

$$i = \mathfrak{N}(v) \tag{6}$$

where $\mathfrak{N}$ is a nonlinear mapping possessing the following four properties.

Property 1: $\mathfrak{N}$ is continuous from E^n into E^n.

Property 2: For each vector $c' \in E^n$ there exists a real number $R>0$ such that $\langle v, \mathfrak{N}(v+c')\rangle \geq 0$ for all $v \in E^n$ satisfying $\|v\| \geq R$.

Property 3: If $\tilde{N} \cap \hat{N} = \varnothing$ and $\tilde{N} \cup \hat{N} = N = \{1, \cdots, n\}$, then it follows from $v_k = v_k'$ for all $k \in \hat{N}$, and $\mathfrak{N}_k(v) = \mathfrak{N}_k(v')$ for all $k \in \tilde{N}$, that $v_k = v_k'$ for all $k \in \tilde{N}$.

Property 4: If $\tilde{N} \cap \hat{N} = \varnothing$ and $\tilde{N} \cup \hat{N} = N = \{1, \cdots, n\}$ and if, for each $k \in \tilde{N}$, the real numbers c_k' are given, then there exists $v \in E^n$ such that $v_k = c_k'$ for each $k \in \hat{N}$ and $\mathfrak{N}_k(v) = 0$ for each $k \in \tilde{N}$.

The physical interpretation of property 1, continuity, is clear. Property 2 specifies that if voltage sources of arbitrary fixed values c_k' are connected in series with each port to form a new n-port (as shown in Fig. 3), then that n-port is "eventually passive." That is, for all port voltage vectors v with $\|v\|$ sufficiently large, the direction of net power flow is *into* the n-port.

Properties 3 and 4 specify, in particular, that the mapping $\mathfrak{N}$ is one-to-one, and that $\mathfrak{N}$ contains the origin θ in its range. Moreover, they specify that the m-ports formed by attaching voltage sources of arbitrary values to an arbitrary collection of $n-m$-ports, and then considering only the remaining m-ports as being extefnally accessible, have those same properties. Thus for all such m-ports, no two different excitations (sets of port voltages) produce the same response (set of port currents), and there always exists some excitation for which the response is zero.

It is reasonable to expect that properties 1–4 are possessed by many realistic models of nonlinear devices. It will in fact be shown in Section V that they are possessed by passive Ebers–Moll-modeled transistors and diodes.

IV. The Network Equation and Existence of a Solution

An equation governing the behavior of the network shown in Fig. 1, when the n-ports comprising that network are of the type described in Sections II and III, can be obtained at once from (4) and (6). After making the substitution $w = v - c'$ and reversing the reference direction of the nonlinear n-port's port-current vector, we obtain

$$Q\mathfrak{N}(w + c') + Pw = \theta. \tag{7}$$

In particular, an equation of this type can be written for any network containing only transistors, diodes, resistors, and independent voltage sources.

The main result of this paper, Theorem 2 (below), asserts that the equations of networks of the type already described (in particular, the equations of transistor–resistor–voltage source networks) always possess at least one solution.

Theorem 2

If (P, Q) is a passive pair of real $n \times n$ matrices and if $\mathfrak{N}: E^n \to E^n$ possesses properties 1–4, then for each real n-vector c' there exists (at least one) $w \in E^n$ such that (7) is satisfied.

Proof: The following lemma, which is proved in [7],[2] will be needed.

Lemma 2

Let C be an open bounded set in E^n and assume that $F: \bar{C} \subset E^n \to E^n$ is continuous and satisfies $\langle x - x^0, F(x)\rangle \geq 0$ for some $x^0 \in C$ and all $x \in \partial C$. Then $F(x) = \theta$ has a solution in $\bar{C}$.

Let H be the positive semidefinite (hybrid) matrix for which (4) and (5) are satisfied by all vectors x, y, v, i satisfying (2). Let the mappings $g, h: E^n \to E^n$ be defined by

$$g(w) = [g_1(w), \cdots, g_n(w)]^T$$
$$h(w) = [h_1(w), \cdots, h_n(w)]^T$$

where for each $w \in E^n$

$$g_k(w) = -\mathfrak{N}_k(w + c') \quad \text{and} \quad h_k(w) = w_k, \qquad \text{if } k \in \tilde{N}$$
$$g_k(w) = w_k \quad \text{and} \quad h_k(w) = -\mathfrak{N}_k(w + c'), \qquad \text{if } k \in \hat{N}.$$

Since $\mathfrak{N}$ possesses properties 3 and 4 it is clear that the mappings g and h are one-to-one and that there is some $w^0 \in E^n$ such that $g(w^0) = \theta$. Let $B \subset E^n$ be an open ball, centered at the origin, of radius sufficiently large that $\langle w, \mathfrak{N}(w + c')\rangle \geq 0$ for each $w \in \partial B$, and such that B contains w^0. Since $\bar{B}$ is compact and $g_{\bar{B}}$ and $h_{\bar{B}}$, the restrictions of g and h, respectively, to $\bar{B}$ are continuous and one-to-one, it

[2] The proof of Lemma 2 given in [7] involves a straightforward application of a fixed-point theorem due to Leray and Schauder.

follows[3] that $\overline{B}$ is homeomorphic to each of the sets

$$\overline{B}_g = \{x : x = g_{\overline{B}}(w),\ w \in \overline{B}\}$$
$$\overline{B}_h = \{x : x = h_{\overline{B}}(w),\ w \in \overline{B}\}.$$

Thus for each $x \in \partial \overline{B}_g$ there corresponds exactly one point $g_{\overline{B}}^{-1}(x) \in \partial B$ and exactly one point $h_{\overline{B}}(g_{\overline{B}}^{-1}(x)) \in \partial \overline{B}_h$. Also B_g, the interior of $\overline{B}_g$, contains the origin θ.

We consider the continuous mapping $Hx - h_{\overline{B}}(g_{\overline{B}}^{-1}(x))$ of $\overline{B}_g$ into E^n. On $\partial \overline{B}_g$,

$$\langle x, Hx - h_{\overline{B}}(g_{\overline{B}}^{-1}(x))\rangle = \langle x, Hx\rangle - \langle x, h_{\overline{B}}(g_{\overline{B}}^{-1}(x))\rangle$$
$$= \langle x, Hx\rangle - \langle g_{\overline{B}}(w), h_{\overline{B}}(w)\rangle$$

where $w = g_{\overline{B}}^{-1}(x)$ satisfies $w \in \partial B$. Thus for each $x \in \partial \overline{B}_g$ there exists $w \in \partial B$ such that

$$\langle x, Hx - h_{\overline{B}}(g_{\overline{B}}^{-1}(x))\rangle = \langle x, Hx\rangle + \langle w, \mathfrak{N}(w + c')\rangle.$$

Since, however, H is positive semidefinite and $\mathfrak{N}$ possesses property 2 on ∂B, it follows that

$$\langle x, Hx - h_{\overline{B}}(g_{\overline{B}}^{-1}(x))\rangle \geqq 0$$

on $\partial \overline{B}_g$. Therefore, by Lemma 2, there exists $x \in \overline{B}_g$ such that

$$h_{\overline{B}}(g_{\overline{B}}^{-1}(x)) = Hx.$$

For this x, if $w = g_{\overline{B}}^{-1}(x)$, $w \in \overline{B}$ satisfies $h(w) = Hg(w)$ and hence, due to the relationship expressed in (2), (4), and (5), w satisfies (7).

Our proof of Theorem 2 shows that for any particular equation having the form (7), the hypotheses concerning the mapping $\mathfrak{N}$ are stronger than they need be. (Nevertheless, they are satisfied by many nonlinear n-ports of practical importance.) More precisely, properties 3 and 4 may be weakened in that they need hold only for the particular choice of $\tilde{N}$ and $\hat{N}$ associated with the hybrid characterization (1), (2) of the linear n-port.

A special case of (7), is the equation

$$\mathfrak{N}(w + c') + Pw = \theta \qquad (8)$$

in which P is a positive semidefinite matrix. For this equation the allowable weakening of properties 3 and 4 causes them to vanish altogether. If it is known, furthermore, that det $P \neq 0$, then (8) has the following equivalent formulation:

$$\mathfrak{N}(v) + Pv = c \qquad (9)$$

in which c denotes an arbitrary vector in E^n.

We therefore have the following theorem.

Theorem 3

If P is any nonsingular positive semidefinite $n \times n$ matrix of real numbers, and if $\mathfrak{N}: E^n \rightarrow E^n$ possesses properties 1 and 2, then for each real n-vector c there exists (at least one) $v \in E^n$ such that (9) is satisfied.

[3] This follows from the easily proved lemma: if f is a continuous one-to-one mapping of a compact set $X \subset E^n$ onto a set $Y \subset E^n$, then f^{-1} is a continuous mapping of Y onto X.

The following theorem is closely related to Theorem 3. It shows that the same conclusion holds, with weaker hypotheses concerning the nonlinear mapping $\mathfrak{N}$, provided the hypotheses concerning the linear operator P are suitably strengthened.

Theorem 4

If P is any positive definite $n \times n$ matrix of real numbers, and if $\mathfrak{N}: E^n \rightarrow E^n$ is continuous and satisfies $\langle v, \mathfrak{N}(v)\rangle \geqq 0$ for all $v \in E^n$ with $\|v\|$ sufficiently large, then for each real n-vector c there exists (at least one) $v \in E^n$ such that (9) is satisfied.

Proof: Let $\rho = \inf \{\langle v, Pv\rangle / \|v\|^2 : v \in E^n,\ v \neq \theta\}$. Since P is positive definite, $\rho > 0$. Let C be any open bounded set in E^n containing the subset $\{v : \|v\| \leqq \|c\|/\rho\}$, and for which $v \in \partial C$ implies $\langle v, \mathfrak{N}(v)\rangle \geqq 0$. Then, for each $v \in \partial C$, $\langle v, \mathfrak{N}(v) + Pv - c\rangle \geqq \langle v, Pv - c\rangle \geqq \rho\|v\|^2 - \|v\| \cdot \|c\| > 0$. Thus, using Lemma 2, there exists a solution of (9) in $\overline{C}$.

From either Theorem 3 or Theorem 4 it follows that the equation

$$[\tau + hb_{-1}T]F(x) + [c + hb_{-1}(I + GR)^{-1}G]x = q \qquad (10)$$

which is equation (12) of [8], has a solution for all $h > 0$, $b_{-1} > 0$, and all $q \in E^n$ (see [8] for explanation of the notation). This result completes a study of conditions under which it is possible to carry out certain implicit numerical integration algorithms for the computation of the transient response of an important class of nonlinear networks containing transistors and diodes. The result shows that the assumption of passivity for the transistors and diodes implies that it is always possible to carry out the algorithms, in the sense that for any value of the step size h there is always at least one solution of (10).

V. Transistor–Resistor–Voltage Source Networks

In order to show that the main result of this paper, Theorem 2, applies to the equations of all networks containing transistors, diodes, resistors, and voltage sources, we now show that the mapping (6) characterizing the nonlinear n-port for such networks possesses properties 1–4. We consider any such network (containing, say, t transistors and d diodes) in which a passive Ebers–Moll model is used to represent the transistors and diodes. For such networks it follows [1], [2], [4] that the nonlinear n-port can be characterized by an equation of the type (6) in which

$$\mathfrak{N}(v) \equiv TF(v). \qquad (11)$$

In (11), T is a matrix of the form $T = T_1 \oplus \cdots \oplus T_t \oplus I_d$; that is, T is the direct sum of I_d, the identity matrix of order d, and the 2×2 matrices T_k, $k = 1, \cdots, t$, each of which is of the form

$$T_k = \begin{bmatrix} 1 & -\alpha_r^{(k)} \\ -\alpha_f^{(k)} & 1 \end{bmatrix}$$

with $0 < \alpha_r^{(k)} < 1$ and $0 < \alpha_f^{(k)} < 1$. The nonlinear mapping F in (11) is a "diagonal" nonlinear mapping of E^n into E^n defined in terms of the functions $f_k: E^1 \to E^1$ by the relation $F(v) = [f_1(v_1), \cdots, f_{d+2t}(v_{d+2t})]^T$, for each $v \in E^{d+2t}$. The functions f_k are each of the form

$$f_k(v_k) = m_k[\exp(n_k v_k) - 1] \tag{12}$$

where, for $k=1, \cdots, d+2t$, $m_k n_k > 0$. For $k=1, \cdots, t$, it follows [9] from the physical requirement that each transistor be passive[4] that

$$\alpha_r^{(k)} \leqq \frac{m_{2k-1}}{m_{2k}} \leqq \frac{1}{\alpha_f^{(k)}}$$

$$\alpha_r^{(k)} \leqq \frac{n_{2k-1}}{n_{2k}} \leqq \frac{1}{\alpha_f^{(k)}}$$

It is clear that the mapping $\mathfrak{N}$ of (11) possesses the continuity property 1. Furthermore, due to the special (decoupled) structure of the mapping, it is easy to show that it also possesses properties 3 and 4. It suffices to consider the typical mapping

$$F(x) \equiv \begin{bmatrix} 1 & -\alpha_r \\ -\alpha_f & 1 \end{bmatrix} \begin{pmatrix} f_1(x_1) \\ f_2(x_2) \end{pmatrix}$$

with $0<\alpha_r<1$, $0<\alpha_f<1$, and for $k=1, 2$, the functions f_k satisfying (12) with

$$\alpha_r \leqq \frac{m_1}{m_2} \leqq \frac{1}{\alpha_f}. \tag{13}$$

To verify property 3, we observe that if $F(x)=F(y)$, it follows that $x=y$; also if $x_1=y_1$ and $-\alpha_f f_1(x_1)+f_2(x_2) = -\alpha_f f_1(y_1) + f_2(y_2)$, then $x_2=y_2$. To verify property 4, since $F(\theta)=\theta$, it suffices to show that for each real number x_1 there exists x_2 such that $-\alpha_f f_1(x_1)+f_2(x_2)=0$. But, using (12) and (13), this follows from the fact that $m_1/m_2 \leqq 1/\alpha_f$.

It remains only to show that the mapping $\mathfrak{N}$ of (11) possesses property 2. In order to do this, we first prove the following lemma.

Lemma 3

For $i=1, \cdots, k$ let n_i denote a positive integer and let F_i denote a continuous mapping of E^{n_i} into E^{n_i} with the property that

$$\lim_{\|x\| \to \infty} \langle x, F_i(x) \rangle = \infty.$$

Let $m_0=0$ and for $i=1, \cdots, k$ let $m_i = \sum_{j=1}^{i} n_j$. Let $N=m_k$, and let the (continuous) mapping F of E^N into E^N be defined by

$$F(x) = [F_1^T(x^1), \cdots, F_k^T(x^k)]^T$$

where, for $i=1, \cdots, k$,

$$x^i = (x_{1+m_{i-1}}, \cdots, x_{m_i})^T.$$

Then

$$\lim_{\|x\| \to \infty} \langle x, F(x) \rangle = \infty.$$

Proof: Let $R>0$ be given. It is clear that for $i=1, \cdots, k$ there exists $L_i \leqq 0$ such that $\langle x, F_i(x) \rangle \geqq L_i$ for all $x \in E^{n_i}$. Let $L = \sum_{i=1}^{k} L_i$. There exists for $i=1, \cdots, k$, $\delta_i > 0$, such that $\|x\| > \delta_i/\sqrt{k}$ implies that $\langle x, F_i(x) \rangle > R - L$. Let $\delta = \max\{\delta_i : i=1, \cdots, k\}$. Then, if $x \in E^N$ and $\|x\| > \delta$, there exists $l \in \{1, \cdots, k\}$ such that $\|x^l\| > \delta/\sqrt{k}$. But then $\langle x^l, F_l(x^l) \rangle > R - L$; and hence

$$\langle x, F(x) \rangle = \langle x^l, F_l(x^l) \rangle + \sum_{\substack{i=1 \\ i \neq l}}^{k} \langle x^i, F_i(x^i) \rangle > R - L$$

$$+ \sum_{\substack{i=1 \\ i \neq l}}^{k} L_i = R - L_l \geqq R.$$

That is, $\|x\| > \delta$ implies that $\langle x, F(x) \rangle > R$.

In view of Lemma 3, and since (obviously) for each $c=(c_1, \cdots, c_d)^T \in E^d$, the mapping $I_d F_d(\cdot) \equiv [f_{2t+1}(\cdot), \cdots, f_{2t+d}(\cdot)]^T : E^d \to E^d$ satisfies $\langle x, I_d F_d(x+c) \rangle \to \infty$ as $\|x\| \to \infty$, the following theorem shows that the mapping $\mathfrak{N}$ of (11) possesses property 2 whenever the four "passivity inequalities" are satisfied with strict inequality.[5]

Theorem 5

Let the real numbers b_1, b_2 be given. Let $0<\alpha_1<1$, $0<\alpha_2<1$ and, for $k=1, 2$, let the functions $f_k: E^1 \to E^1$ be defined by (12), where $m_k n_k > 0$, $\alpha_1 < m_1/m_2 < 1/\alpha_2$, and $\alpha_1 < n_1/n_2 < 1/\alpha_2$. Then

$$\lim_{\|v\| \to \infty} \psi(v) = \infty$$

where $\psi: E^2 \to E^1$ is defined by

$$\psi(v) = (v_1, v_2) \begin{bmatrix} 1 & -\alpha_1 \\ -\alpha_2 & 1 \end{bmatrix} \begin{pmatrix} f_1(v_1 + b_1) \\ f_2(v_2 + b_2) \end{pmatrix}. \tag{14}$$

We in fact prove a result more general than Theorem 5. The functions f_1, f_2 defined by (12) are special cases of functions f_1, f_2 defined by

$$f_k(v_k) = m_k g(n_k v_k), \qquad k = 1, 2 \tag{15}$$

where, for $k=1, 2$, $m_k n_k > 0$, and where g is a continuous monotone nondecreasing function satisfying $g(0)=0$, and such that there exist real numbers $\epsilon>0$, $\delta>0$ such that $|g(x)| > \epsilon$ for all $|x| > \delta$.

If, in (14), the functions f_1, f_2 are defined by (15), and if b_1, b_2 are arbitrary real numbers, then

$$\lim_{\|v\| \to \infty} \psi(v) = \infty \quad \text{if and only if} \quad \lim_{\|y\| \to \infty} \phi(y) = \infty,$$

[4] The transistor is certainly a passive circuit component; it always absorbs, rather than delivers, electrical energy. This, of course, in no way contradicts the common engineering practice of referring to transistors as *active* elements when considering only their *small-signal* behavior.

[5] However, it can be shown that the conclusion of Theorem 5 remains unchanged if the inequalities $\alpha_1 < m_1/m_2 < 1/\alpha_2$ and $\alpha_1 < n_1/n_2 < 1/\alpha_2$ are replaced with $\alpha_1 \leqq m_1/m_2 \leqq 1/\alpha_2$ and $\alpha_1 \leqq n_1/n_2 \leqq 1/\alpha_2$. In this connection, observe that since $\alpha_1 < 1$ and $1/\alpha_2 > 1$, at least two of the inequalities must be met with strict inequality.

where [with

$$y_k = n_k(v_k + b_k), \qquad k = 1, 2$$

and $c_1 = -b_1 n_1$, $c_2 = -b_2 n_2$] we define

$$\phi(y) = (y_1 + c_1, y_2 + c_2) A \begin{pmatrix} g(y_1) \\ g(y_2) \end{pmatrix} \tag{16}$$

in which

$$A = \begin{bmatrix} m_1/n_1 & -\alpha_1 m_2/n_1 \\ -\alpha_2 m_1/n_2 & m_2/n_2 \end{bmatrix}.$$

If the inequalities $\alpha_1 < m_1/m_2 < 1/\alpha_2$, $\alpha_1 < n_1/n_2 < 1/\alpha_2$ of Theorem 5 are also met, the elements a_{ij} of the matrix A satisfy

$$\begin{aligned} a_{11} > -a_{12} > 0 \qquad a_{11} > -a_{21} > 0 \\ a_{22} > -a_{12} > 0 \qquad a_{22} > -a_{21} > 0. \end{aligned} \tag{17}$$

We now prove the following generalization of Theorem 5.

Theorem 6

Let the real numbers c_1, c_2 be given. Let the elements a_{ij} of the real 2×2 matrix A satisfy (17). Let g be a continuous monotone nondecreasing function with the property that $g(0) = 0$, and such that there exist real numbers $\epsilon > 0$, $\delta > 0$ such that $|g(x)| > \epsilon$ for all $|x| > \delta$. Then

$$\lim_{\|y\| \to \infty} \phi(y) = \infty$$

where $\phi: E^2 \to E^1$ is defined by (16).

Proof: We show that $\phi(y) \to \infty$ as $\|y\| \to \infty$ in each of the three closed regions shown in Fig. 4. Due to the symmetric nature of the mapping ϕ (with respect to the subscripts 1, 2) it is clear that the behavior is similar in the half-plane not explicitly considered.

In region I: $y = (y_1, \rho y_1)^T$, where $y_1 \geqq 0$ and $0 \leqq \rho \leqq 1$. Thus

$$\begin{aligned} \phi(y) = {} & y_1[a_{11} g(y_1) + a_{12} g(\rho y_1) + \rho a_{21} g(y_1) + \rho a_{22} g(\rho y_1)] \\ & + (c_1 a_{11} + c_2 a_{21}) g(y_1) + (c_2 a_{22} + c_1 a_{12}) g(\rho y_1). \end{aligned}$$

Let

$$K = |c_1 a_{11} + c_2 a_{21}| + |c_2 a_{22} + c_1 a_{12}|.$$

Then

$$\phi(y) \geqq y_1[(a_{11} + \rho a_{21}) g(y_1) + (\rho a_{22} + a_{12}) g(\rho y_1)] - K g(y_1).$$

Obviously, either $(\rho a_{22} + a_{12}) \geqq 0$ or else $(\rho a_{22} + a_{12}) < 0$. If $(\rho a_{22} + a_{12}) \geqq 0$,

$$\begin{aligned} \phi(y) & \geqq [(a_{11} + \rho a_{21}) y_1 - K] g(y_1) \\ & \geqq [(a_{11} + a_{21}) y_1 - K] g(y_1). \end{aligned}$$

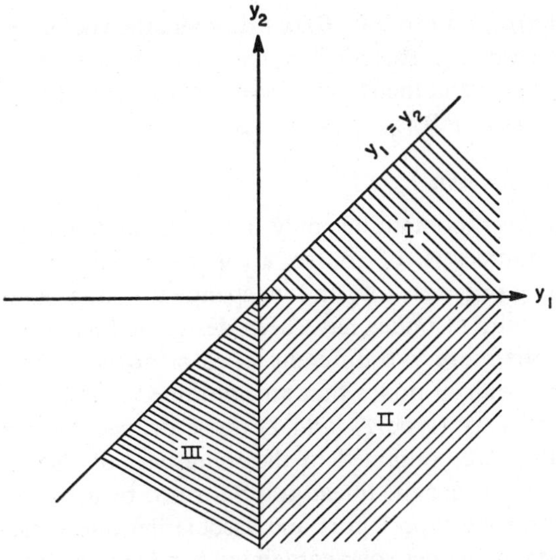

Fig. 4. Regions of y_1–y_2 plane considered in the proof of Theorem 6.

On the other hand, if $(\rho a_{22} + a_{12}) < 0$, then, since $g(\rho y_1) \leqq g(y_1)$, we have

$$\begin{aligned} \phi(y) & \geqq y_1[(a_{11} + \rho a_{21}) g(y_1) + (\rho a_{22} + a_{12}) g(y_1)] - K g(y_1) \\ & \geqq [(a_{11} + a_{12}) y_1 + \rho(a_{22} + a_{21}) y_1 - K] g(y_1) \\ & \geqq [(a_{11} + a_{12}) y_1 - K] g(y_1). \end{aligned}$$

It follows that $\phi(y) \to \infty$ as $y_1 \to \infty$. Thus in region I, $\|y\| \to \infty$ implies $y_1 \to \infty$ which implies $\phi(y) \to \infty$.

In region II: $y_1 \geqq 0$ and $y_2 \leqq 0$. Thus

$$\begin{aligned} \phi(y) = {} & a_{11} y_1 g(y_1) + a_{22} y_2 g(y_2) + a_{12} y_1 g(y_2) + a_{21} y_2 g(y_1) \\ & + (c_1 a_{11} + c_2 a_{21}) g(y_1) + (c_2 a_{22} + c_1 a_{12}) g(y_2) \\ \geqq {} & [a_{11} y_1 + (c_1 a_{11} + c_2 a_{21})] g(y_1) \\ & + [a_{22} y_2 + (c_2 a_{22} + c_1 a_{12})] g(y_2). \end{aligned}$$

Thus it is clear that, in region II, $\phi(y) \to \infty$ as $\|y\| \to \infty$.

In region III: $y = (y_1, \rho y_1)^T$, where $y_1 \leqq 0$ and $0 \leqq \rho \leqq 1$. By making a few obvious changes in the argument used for region I, we find that, in region III, $\|y\| \to \infty$ implies $y_2 \to -\infty$ which implies $\phi(y) \to \infty$.

Appendix

Proof of Lemma 1

Observe that if $(\hat{P}, \hat{Q})$ is a passive pair, then $(\hat{P}, \hat{Q})$ is a $\mathcal{W}_0$ pair (see [3]); hence, there exists a complementary pair $(A, B) \in \mathcal{C}(\hat{P}, \hat{Q})$ such that $\det A \neq 0$. (See [4] for an explanation of these concepts.) Let the subset $\tilde{N}$ of the set of indices $N = \{1, \cdots, n\}$ be defined by $k \in \tilde{N}$ if and only if the kth column of the matrix A equals the kth column of the matrix $\hat{P}$. Let $\hat{N} = N \backslash \tilde{N}$. For $k = 1, \cdots, n$ let $d_k = 1$ if $k \in \tilde{N}$, and $d_k = -1$ if $k \in \hat{N}$. Let $D = \operatorname{diag}(d_1, \cdots, d_n)$. It is clear that the characterization $Pv = Qi + c$, where $P = (AD)^{-1}\hat{P}$, $Q = (AD)^{-1}\hat{Q}$, and $c = (AD)^{-1}\hat{c}$, is equivalent to the given characterization $\hat{P}v = \hat{Q}i + \hat{c}$; furthermore, it is equivalent to the characterization $y = DA^{-1}BDx + c$, where the vectors x, y, v, i are related by (2). It is a trivial matter to show that (P, Q) is a passive pair and then (since $Pv = Qi$ is equivalent to $y = Hx$) that $H = DA^{-1}BD$ is a positive semidefinite matrix.

Proof of Theorem 1

An *n-port* containing only resistors and independent voltage sources is, of course, a (possibly disjoint) *network* containing only resistors and independent voltage sources in which n pairs of nodes have been designated as "ports." In general there are two different types of ports: 1) those defined by a pair of nodes that are contained in *different* parts of the network (type 1), and 2) those defined by a pair of nodes that are contained in the *same* part of the network (type 2). Consider the linear graph formed by associating an edge with each type-1 port and associating one vertex with each part of the network containing a node associated with a type-1 port. Choose an arbitrary tree for this linear graph. (In case the linear graph is not connected, choose a *forest*.) Let an arbitrary value of port voltage be assigned to each port associated with an edge of the tree (forest). Let all other ports (including all type-2 ports) remain open-circuited. It is clear that the port current of *each* port then has the value zero and that the port voltage of each port is uniquely determined. Thus we determine a pair of vectors v, i, with $i = \theta$, and such that (3) is satisfied. For this vector v, $Pv = c$.

References

[1] I. W. Sandberg and A. N. Willson, Jr., "Some theorems on properties of dc equations of nonlinear networks," *Bell Syst. Tech. J.*, vol. 48, Jan. 1969, pp. 1–34.

[2] ——, "Some network-theoretic properties of nonlinear dc transistor networks," *Bell Syst. Tech. J.*, vol. 48, May–June 1969, pp. 1293–1311.

[3] ——, "Existence and uniqueness of solutions for the equations of nonlinear dc networks," *SIAM J. Appl. Math.*, to be published.

[4] A. N. Willson, Jr., "New theorems on the equations of nonlinear dc transistor networks," *Bell Syst. Tech. J.*, vol. 49, Oct. 1970, pp. 1713–1738.

[5] I. W. Sandberg, "Necessary and sufficient conditions for the global invertibility of certain nonlinear operators that arise in the analysis of networks," *IEEE Trans. Circuit Theory*, vol. CT-18. Mar. 1971, pp. 260–263.

[6] ——, "Conditions for the existence of a global inverse of semiconductor-device nonlinear-network operators," to be published in *IEEE Trans. Circuit Theory*, vol. CT-19, Jan. 1972.

[7] J. M. Ortega and W. C. Rheinboldt, *Iterative Solution of Nonlinear Equations in Several Variables.* New York: Academic Press, 1970.

[8] I. W. Sandberg, "Theorems on the analysis of nonlinear transistor networks," *Bell Syst. Tech. J.*, vol. 49, Jan. 1970, pp. 95–114.

[9] B. Gopinath and D. Mitra, "When are transistors passive?" to be published.

A Charge-Control Transistor Model for Network Analysis Programs

Abstract—**A representation is given of an Ebers-Moll model with charge control that employs only standard circuit elements. Storage of carriers associated with current flow is represented with the aid of two current-controlled voltage generators.**

The most widely used large-signal transistor model is that of Ebers and Moll.[1] With the original model, transient analysis is complicated by the fact that in the forward and reverse current generators α_f and α_r are functions of frequency. Incorporation of charge control[2],[3] has overcome this difficulty.

This letter contains a representation of the charge-control Ebers-Moll model that makes it useful for network analysis programs. This representation does not require storances or other nonstandard circuit elements. In the idealized model of Fig. 1, the currents i_f and i_r through the emitter and collector diodes each control a current and voltage generator.

Consider diode voltages and currents to be related by

$$i_f = i_{fo}[\exp(qV_{eb}/kT) - 1] \quad (1)$$

$$i_r = i_{ro}[\exp(qV_{cb}/kT) - 1]. \quad (2)$$

Then the model is characterized by the following parameters:

$$\begin{matrix} i_{fo} & i_{ro} \\ C_e & C_c \\ \alpha_f & \alpha_r \\ \tau_f & \tau_r \end{matrix}$$

with the constraint

$$\alpha_f i_{fo} = \alpha_r i_{ro}. \quad (3)$$

The parameters are frequency-independent and are, for the idealized model, constants. In refined models they may be slowly-varying functions of currents and voltages.

The charge Q_f stored on capacitor C_e is

$$Q_f = V_{eb}C_e + i_f \tau_f. \quad (4)$$

A corresponding relation obtains for the charge on C_c. The first term on the right-hand side of (4) represents charge stored capacitively at the edge of the emitter-base transition region, while the second term represents minority carriers stored predominantly in the base and collector depletion regions.

The physical content of the model is identical to that of a model given by Koehler[3] and to that of an analog computer representation of Balaban and Logan.[4] The present model is attractive for digital network analysis programs because it uses only standard circuit elements. Two state variables are required. Because it is essentially a single-pole model, it is reliable only for frequencies below f_t. This is a minor limitation in the analysis of switching circuits.

So far, a highly idealized case has been treated. A more realistic model is given in Fig. 2. Here diodes 1 and 2 represent the active region, i.e., the region under the emitter. Diodes 1 and 2 are "ideal" diodes, having a diode coefficient n of unity in the exponent $qV/(nkT)$. The corresponding current gains α_f and α_r are unity, and the saturation currents of diodes 1 and 2 are equal. This is in keeping with the fact that typically recombination in the base contributes negligibly to the alpha-defect current and that I_c versus V_{eb} and I_e versus V_{cb} obey closely the ideal diode law.[5] Diode 3 represents nonideal base current, with $1<n<2$. Diode 4 represents the portion of the collector-base junction that is not under the emitter. The resistances depend on the geometry. For example, in a top-collector-contact planar transistor the collector current flows mostly through r_{c3}, i.e., r_{c1} is large. Base-width modulation[6] may be accounted for in terms of collector-voltage dependence of i_{fo}, and, if required, through a collector-voltage-controlled voltage source, not shown, in the base leg. As with most models, fine points can be represented at the expense of an increasing number of parameters.

Manuscript received February 2, 1968.

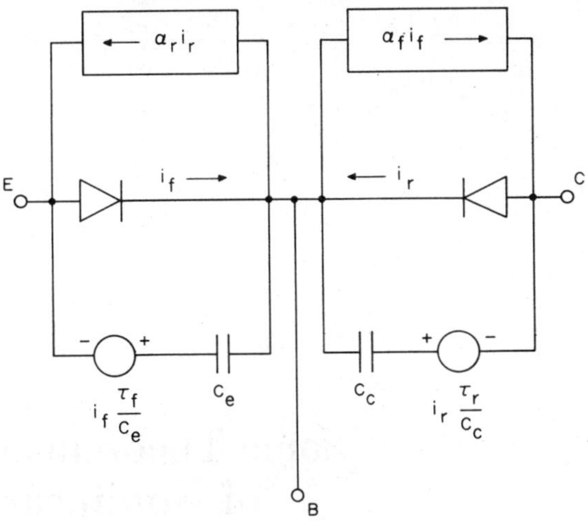

Fig. 1. Idealized transistor model.

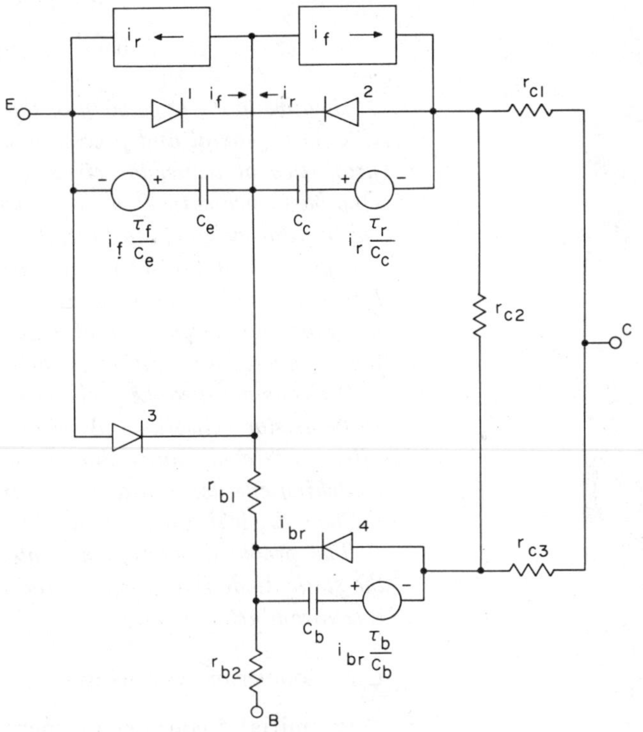

Fig. 2. Realistic transistor model.

Conclusion

By the addition of two current-controlled voltage generators, the basic Ebers-Moll model can be modified so as to contain frequency-independent conventional network elements. The modified form shown in Fig. 1 is convenient for transient analysis.

H. K. Gummel
Bell Telephone Laboratories, Inc.
Murray Hill, N. J. 07974

References

[1] J. J. Ebers and J. L. Moll, "Large-signal behavior of junction transistors," *Proc. IRE*, vol. 42, pp. 1761–1772, December 1954.

[2] R. Beaufoy and J. J. Sparkes, "The junction transistor as a charge-controlled device," *ATE J.* (London), vol. 13, pp. 310–324, October 1957.

[3] D. Koehler, "The charge-control concept in the form of equivalent circuits, representing a link between the classic large signal diode and transistor models," *Bell Sys. Tech. J.*, vol. 46, pp. 523–576, March 1967.

[4] P. Balaban and J. Logan, "Analog computer simulation of semiconductor circuits," *1968 Spring Joint Computer Conf., AFIPS Proc.*, vol. 32.

[5] H. K. Gummel, "Measurement of the number of impurities in the base layer of a transistor," *Proc. IRE (Correspondence)*, vol. 49, p. 834, April 1961.

[6] J. M. Early, "Effects of space-charge layer widening in junction transistors," *Proc. IRE*, vol. 40, pp. 1401–1406, November 1952.

Reprinted from *Proceedings of the IEEE*, vol. 56, p. 751, April 1968.

Some Theorems on the Dynamic Response of Nonlinear Transistor Networks

By I. W. SANDBERG

(Manuscript received July 16, 1968)

Relative to the huge body of theory of linear time-invariant systems, very little of a general and precise nature is known about the network-theoretic properties of transistor circuits operating under large-signal conditions. One basic property P which a transitor network might have is that if the input approaches a constant, then the output approaches a constant which is independent of the initial conditions. In this paper we prove a stability theorem concerning a nonlinear differential equation that governs the behavior of a large class of networks. A corollary of this theorem asserts that if a certain condition is satisfied, then property P holds.

We consider also the problem of estimating the rate of decay of transients in transistor networks and we prove theorems which allow us to make some often quite conservative, but definite, statements concerning limitations on switching speeds. A practical example considered shows that in some cases the bounds, which are frequently very easy to evaluate, can be quite useful.

The proofs depend in an interesting way on the relationship between the static diode characteristic and the nonlinear capacitance associated with a semiconductor junction.

I. INTRODUCTION AND DERIVATION OF THE DIFFERENTIAL EQUATION

We initially consider the network of Fig. 1, which contains transistors, linear resistors, voltage sources, and current sources. Each transistor is represented by a model of the type shown in Fig. 2 (see Gummel[1] and Koehler[2]) which takes into account nonlinear dc properties as well as the presence of nonlinear junction capacitances. Associated with this model are six parameters: α_f, α_r, τ_e, τ_c, c_e, and c_c (all positive constants; $\alpha_f < 1$, $\alpha_r < 1$) and two nonlinear functions $f_e(\cdot)$ and $f_c(\cdot)$.

Concerning $f_e(\cdot)$ and $f_c(\cdot)$, for our purposes it is necessary to assume only that

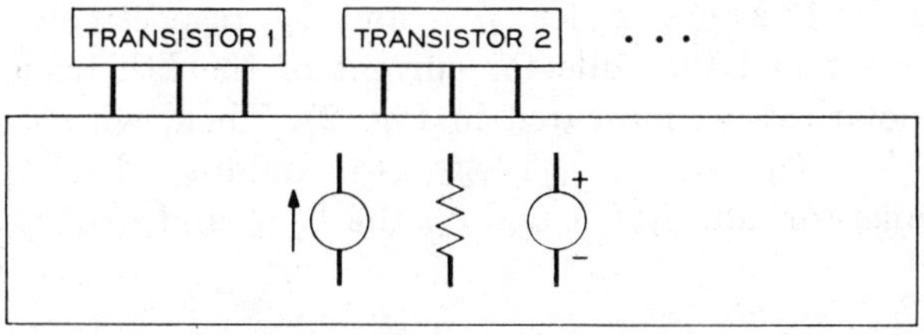

Fig. 1 — General network containing transistors, sources, and resistors.

Assumption 1: For each transistor: $f_e(\cdot)$ and $f_c(\cdot)$ are strictly-monotone increasing mappings of the real interval $(-\infty, \infty)$ into itself; $f_e(0) = f_c(0) = 0$, and $f_e(\cdot)$ and $f_c(\cdot)$ are continuously differentiable on $(-\infty, \infty)$.

The functions $f_e(\cdot)$ and $f_c(\cdot)$ of Gummel's model[1] are of simple exponential type and satisfy Assumption 1.

From Fig. 2:

$$i_e = \frac{d}{dt}\,[c_e v_e + \tau_e f_e(v_e)] + f_e(v_e) - \alpha_r f_c(v_c),$$

$$i_c = \frac{d}{dt}\,[c_c v_c + \tau_c f_c(v_c)] - \alpha_f f_e(v_e) + f_c(v_c).$$

Suppose that the network of Fig. 1 contains p transistors; for $k = 1, 2, \cdots, p$, let v_{2k-1} and v_{2k}, respectively, denote the emitter to base voltage and the collector to base voltage of the kth transistor. Simi-

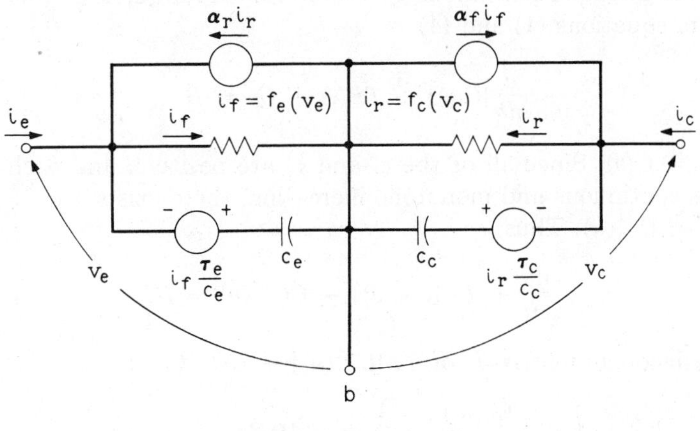

Fig. 2 — Transistor model.

larly, for $k = 1, 2, \cdots, p$, let i_{2k-1} and i_{2k}, respectively, denote the emitter current and the collector current of the kth transistor (with reference polarities as indicated in Fig. 2). Then, with $v = (v_1, v_2, \cdots, v_{2p})^{tr}$, $i = (i_1, i_2, \cdots, i_{2p})^{tr}$, $f_{2k-1}(\cdot)$ and c_{2k-1} the $f_e(\cdot)$ and c_e of the kth transistor, and $f_{2k}(\cdot)$ and c_{2k} the $f_c(\cdot)$ and c_c of the kth transistor,

$$i = \frac{d}{dt}[C(v)] + TF(v) \tag{1}$$

where, for $j = 1, 2, \ldots, 2p$,

$$[C(v)]_j = c_j v_j + \tau_j f_j(v_j) \tag{2}$$

$$[F(v)]_j = f_j(v_j), \tag{3}$$

and $T = T_1 \oplus T_2 \oplus \cdots \oplus T_p$, the direct sum of p 2×2 matrices T_k in which

$$T_k = \begin{bmatrix} 1 & -\alpha_r^{(k)} \\ -\alpha_f^{(k)} & 1 \end{bmatrix}$$

for $k = 1, 2, \cdots, p$.

We assume that the linear resistive portion of the structure of Fig. 1 introduces the constraint

$$i = -Gv + B \tag{4}$$

in which G is a conductance matrix and B is an element of the set $\mathcal{B}$ of all real bounded continuous $2p$-vector-valued functions of t on $[0, \infty)$.

From equations (1) and (4)

$$\frac{d}{dt}[C(v)] + TF(v) + Gv = B. \tag{5}$$

Let $u = C(v)$. Since all of the c_j and τ_j are positive, and each of the $f_j(\cdot)$ is continuous and monotone increasing, there exists a $C^{-1}(\cdot)$ such that $v = C^{-1}(u)$. Thus,

$$\frac{du}{dt} + TF[C^{-1}(u)] + GC^{-1}(u) = B. \tag{6}$$

The Jacobian matrix J_u of $TF[C^{-1}(u)] + GC^{-1}(u)$ is

$$T \operatorname{diag}\left\{\frac{f_i'[g_i(u_i)]}{c_i + \tau_i f_i'[g_i(u_i)]}\right\} + G \operatorname{diag}\left\{\frac{1}{c_i + \tau_i f_i'[g_i(u_i)]}\right\}$$

in which for all $j = 1, 2, \cdots, 2p$

$$g_j(u_j) = [C^{-1}(u)]_j$$

with each of the $g_j(\cdot)$ continuously differentiable.

Since J_u is continuously dependent on u, and $\| J_u \|$ ($\| \cdot \|$ any norm) is bounded from above uniformly in u, it follows that there exists a constant L such that

$$\| TF[C^{-1}(u_a)] + GC^{-1}(u_a) - TF[C^{-1}(u_b)] - GC^{-1}(u_b) \| \leqq L \| u_a - u_b \| \quad (7)$$

for all u_a and u_b belonging to real Euclidean $2p$-space E^{2p}. In particular, we have

$$\| TF[C^{-1}(u)] + GC^{-1}(u) - B \| \leqq L \| u \| + \|B\| \quad (8)$$

for all $t \geqq 0$ and all $u \in E^{2p}$. Therefore (see, for example, Nemytskii and Stepanov[3]), for any initial condition $u_0 \in E^{2p}$, there exists a unique continuous $2p$-vector-valued function $u(\cdot)$ such that $u(0) = u_0$ and equation (6) is satisfied for all $t > 0$. In other words, under the assumptions we have introduced, it makes sense to study the properties of the solution of the equation

$$\frac{du}{dt} + TF[C^{-1}(u)] + G[C^{-1}(u)] = B, \qquad t \geqq 0 \quad [u(0) = u_0] \quad (9)$$

II. STATEMENT OF RESULTS, AND EXAMPLES

We need the following definitions.

Definition 1: A real matrix M of arbitrary order n is *strongly column-sum dominant* if and only if for all $j = 1, 2, \ldots, n$

$$m_{jj} - \sum_{i \neq j} | m_{ij} | > 0.$$

An important property of T is that it is strongly column-sum dominant.

Definition 2: We shall say that a real matrix M of order $2p$ is an element of $\mathfrak{D}$ if and only if there exists a diagonal matrix diag $(d_1, d_2, \cdots, d_{2p})$ with each $d_i > 0$ such that

$$\alpha_f^{(k)} < \frac{d_{2k-1}}{d_{2k}} < \frac{1}{\alpha_r^{(k)}}$$

for $k = 1, 2, \cdots, p$, and diag $(d_1, d_2, \cdots, d_{2p})$ M is strongly column-sum dominant.

Our main result* concerning equation (9) is:

---

* Proofs of all results in this section are given in Section III.

Theorem 1: *If* $G \in \mathfrak{D}$, *and* $u_a(\cdot)$ *and* $u_b(\cdot)$ *satisfy*

$$\frac{du_a}{dt} + TF[C^{-1}(u_a)] + G[C^{-1}(u_a)] = B_a(t), \qquad t \geqq 0 \tag{10}$$

$$\frac{du_b}{dt} + TF[C^{-1}(u_b)] + G[C^{-1}(u_b)] = B_b(t), \qquad t \geqq 0 \tag{11}$$

with $B_a \in \mathfrak{B}$ *and* $B_b \in \mathfrak{B}$, *and if* $[B_a(t) - B_b(t)] \to \theta$ *(the zero vector of* E^{2p}*) as* $t \to \infty$, *then* $[u_a(t) - u_b(t)] \to \theta$ *as* $t \to \infty$.

An interesting corollary of Theorem 1 is

Corollary 1: *Referring to equation* (9), *if* $G \in \mathfrak{D}$, *and if there exists a constant vector* B_∞ *such that* $[B(t) - B_\infty] \to \theta$ *as* $t \to \infty$, *then there exists a constant vector* u_∞ *such that* $[u(t) - u_\infty] \to \theta$ *as* $t \to \infty$, *and* u_∞ *is independent of the initial condition* u_0. *In particular, if* $B_\infty = \theta$, *then* $u_\infty = \theta$.

It is interesting to observe that $G \in \mathfrak{D}$ whenever the base leads of all transistors are connected together and there is a resistor between the emitter and base, and between the collector and base, of every transistor, for then G is strongly column-sum dominant. Also it is easy to give examples of conductance matrices which are not strongly column-sum dominant, and which belong to $\mathfrak{D}$. For instance, for the network of Fig. 3.

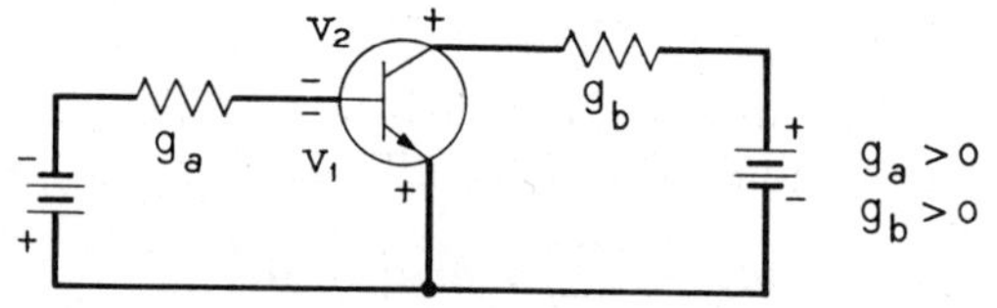

Fig. 3 — Single-transistor network.

$$G = \begin{bmatrix} g_a + g_b & -g_b \\ -g_b & g_b \end{bmatrix}$$

and diag $(d_1, d_2)G$ is strongly column-sum dominant for $d_2 = 1$ and some d_1 such that

$$\alpha_f < d_1 < \frac{1}{\alpha_r}.^*$$

* More generally, G of order $2p$ with positive diagonal elements belongs to $\mathfrak{D}$ whenever it is possible to obtain a strongly column-sum dominant matrix from G by adding an arbitrarily small positive quantity to a single diagonal element.

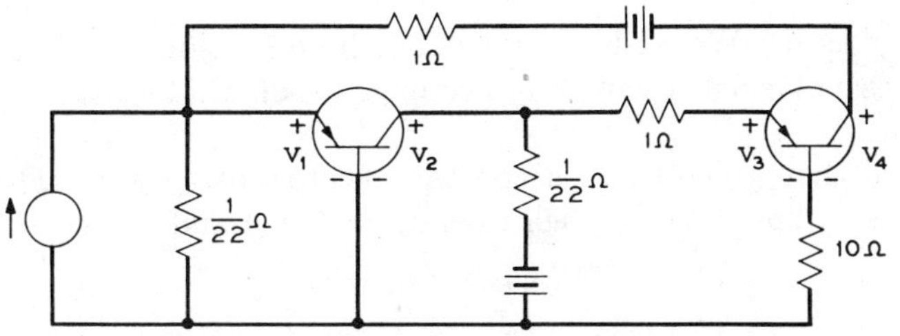

Fig. 4 — A two-transistor circuit.

As another example, consider the circuit of Fig. 4, for which

$$G = \frac{1}{21}\begin{bmatrix} 473 & -10 & 10 & -11 \\ -10 & 473 & -11 & 10 \\ 10 & -11 & 11 & -10 \\ -11 & 10 & -10 & 11 \end{bmatrix}.$$

Since diag (1, 1, 22, 22)G is strongly column-sum dominant, $G \in \mathfrak{D}$.

Finally, for the network shown in Fig. 5,

$$G = \frac{1}{21}\begin{bmatrix} 11 & -10 & 10 & -11 \\ -10 & 11 & -11 & 10 \\ 10 & -11 & 11 & -10 \\ -11 & 10 & -10 & 11 \end{bmatrix}.$$

In this case, G is obviously singular and hence does not belong to $\mathfrak{D}$. Suppose that the source current of Fig. 5 $i_0(t)$ is a constant and that the transistor functions $f_1(\cdot)$, $f_2(\cdot)$, $f_3(\cdot)$, and $f_4(\cdot)$ are all bounded from below by the constant b (this is certainly an assumption consistent with our earlier assumptions and with the character of transistor models

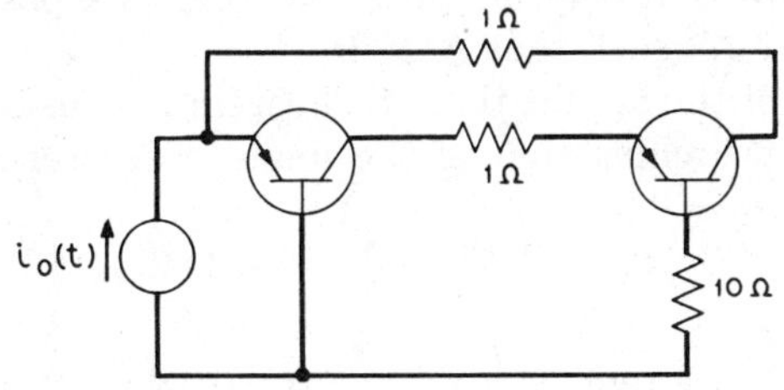

Fig. 5 — Transistor circuit for which the dc equations may have no solution.

ordinarily used.) We wish to show that here for sufficiently small i_0, there does not exist a constant vector u_∞ such that $[u(t) - u_\infty] \to \theta$ as $t \to \infty$.

Suppose that $u(t) \to u_\infty$, a constant vector, as $t \to \infty$. Then there would exist a $2p$-vector v_∞ such that $u_\infty = C(v_\infty)$ and

$$TF(v_\infty) + Gv_\infty = B$$

with $B = (i_0, 0, 0, \cdots, 0)^{tr}$. Let η denote the $2p$-row-vector $(1, 1, 1, \cdots, 1)$. Then

$$\eta TF(v_\infty) + \eta Gv_\infty = \eta B.$$

But $\eta Gv_\infty = 0$, and hence

$$i_0 = \sum_{k=1}^{p} [1 - \alpha_f^{(k)}] f_{2k-1}(v_{\infty 2k-1}) + \sum_{k=1}^{p} [1 - \alpha_r^{(k)}] f_{2k}(v_{\infty 2k})$$

which does not possess a solution v_∞ if

$$i_0 < b \sum_{k=1}^{p} [1 - \alpha_f^{(k)}] + [1 - \alpha_r^{(k)}].$$

2.1 *Estimation of the Rate of Decay of Transients*

Theorem 2: If the hypotheses of Corollary 1 are satisfied with $B(t) = B_\infty$ *for* $t \geqq 0$, *then*

$$\sum_{j=1}^{2p} d_j \mid u_j(t) - u_{\infty j} \mid \leqq \exp(-\underline{K}t) \sum_{j=1}^{2p} d_j \mid u_j(0) - u_{\infty j} \mid, \qquad t \geqq 0$$

for every set of positive constants $d_1, d_2, \cdots, d_{2p}$ *such that*

$$0 < \underline{K} \triangleq \min_j \min \left\{ \frac{1}{\tau_j} (1 - \tilde{d}_j d_j^{-1} \alpha_j), \frac{1}{c_j} (g_{jj} - \sum_{i \neq j} d_i d_j^{-1} \mid g_{ij} \mid) \right\}$$

in which $-\alpha_j$ *is the nonzero off-diagonal term in the jth column of* T, *and* $\tilde{d}_j = d_{j+1}$ *for* j *odd and* $\tilde{d}_j = d_{j-1}$ *for* j *even.*

It is easy to show that $G \varepsilon \mathfrak{D}$ implies that there are positive constants d_j, $j = 1, 2, \cdots, 2p$, such that $\underline{K} > 0$.

As an example of the application of Theorem 2, consider the problem of estimating the switching time of the single-transistor inverter circuit of Fig. 6 in which $\alpha_f = 0.968$, $c_e = 2 \times 10^{-12}$ *fd*, $\tau_e = 1.7 \times 10^{-10}$ second, $\alpha_r = 0.583$, $c_c = 1.7 \times 10^{-12}$ *fd*, and $\tau_c = 2.62 \times 10^{-8}$ second. Here (in mhos)

$$G = \begin{bmatrix} 1.1886 \times 10^{-3} & -1.01215 \times 10^{-3} \\ -1.01215 \times 10^{-3} & 1.01215 \times 10^{-3} \end{bmatrix}$$

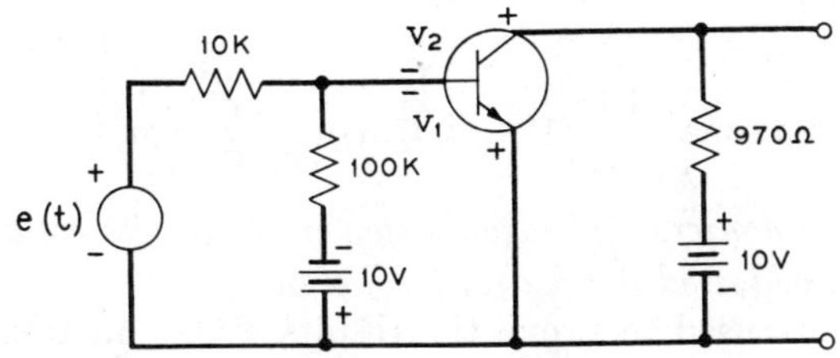

Fig. 6 — Practical logical-inverter circuit.

which takes into account a bulk base resistance of 280 ohms and a bulk collector resistance of 18 ohms. The circuit is initially at steady state with $e(t) = 0.3$ volt for $t < 0$. For $t \geqq 0$, $e(t) = 10$ volts, and as $t \to \infty$, $u(t) \to u_\infty$, some constant vector. With $d_2 = 1$, the number $\underline{K}$ is the smallest of the four quantities: $0.58(1 - 0.968d_1^{-1}) \times 10^{10}$, $0.5(1.1886 - 1.01215d_1^{-1}) \times 10^9$, $0.3815(1 - 0.583\text{d}_1) \times 10^8$, and $0.58(1.01215)(1 - d_1) \times 10^9$.

It is clear that d_1 must satisfy $0.968 < d_1 < 1$ in order that $\underline{K} > 0$. Then optimal choice of d_1 (that is, the choice that yields the largest value of $\underline{K}$) is approximately 0.9709. For $d_1 = 0.9709$, $\underline{K} = 1.66 \times 10^7$. Let the "charge switching time" t_s denote the smallest value of t such that $\sum_{i=1}^{2} | u_i(t) - u_{\infty i} |$ is less than or equal to two percent of $\sum_{i=1}^{2} | u_i(0) - u_{\infty i} |$ for all $t \geqq t_s$. Then our upper bound on t_s is approximately $4 \times (1.66)^{-1} \times 10^{-7} \approx 241$ nanoseconds. The actual value of t_s, as determined by numerically integrating the system of two nonlinear differential equations is approximately 57 nanoseconds. Thus, for this circuit, Theorem 2 provides a very easily evaluated and useful upper bound on t_s.

Finally, we state a result which provides an often rather conservative but easily evaluated *lower bound* on the rate of decay of transients.

Theorem 3: With B a constant real 2p-vector, let

$$\frac{du}{dt} + TF[C^{-1}(u)] + GC^{-1}(u) = B, \qquad t \geqq 0.$$

If there exists a constant 2p-vector u_∞ such that $[u(t) - u_\infty] \to \theta$ as $t \to \infty$, then for any choice of positive constants d_j, $j = 1, 2, \cdots, 2p$:

$$\sum_{j=1}^{2p} d_j \mid u_j(t) - u_{\infty j} \mid \geqq \exp(-\bar{K}t) \sum_{j=1}^{2p} d_j \mid u_j(0) - u_{\infty j} \mid, \qquad t \geqq 0$$

in which

$$\bar{K} = \max_{j} \max \left\{ \frac{1}{\tau_j} (1 + \tilde{d}_j d_j^{-1} \alpha_j), \frac{1}{c_j} \sum_{i=1}^{2p} d_i d_j^{-1} \mid g_{ij} \mid \right\}$$

where $-\alpha_j$ *is the nonzero off-diagonal element in the jth column of* T, *and* $\tilde{d}_j = d_{j+1}$ *for j odd, and* $\tilde{d}_j = d_{j-1}$ *for j even.*

The arguments used to prove the results stated in this section can be modified in a straightforward manner to prove far more general results concerning networks that contain diodes, capacitors, and inductors, in addition to the elements of the structure of Fig. 1. Some of these more general results are described in Section IV.

III. PROOFS

3.1 *Proof of Theorem 1*

We first show that

$$F[C^{-1}(u_a)] - F[C^{-1}(u_b)] = D_1(u_a - u_b),\ t \geqq 0 \tag{12}$$

and

$$C^{-1}(u_a) - C^{-1}(u_b) = D_2(u_a - u_b),\ t \geqq 0 \tag{13}$$

with D_1 and D_2 diagonal matrices dependent on t and possessing some special properties.

For $j = 1, 2, \cdots, 2p$, let $g_j(u_{aj}) = [C^{-1}(u_a)]_j$ and $g_j(u_{bj}) = [C^{-1}(u_b)]_j$. Then, using equation (2),

$$u_{aj} - u_{bj} = c_j[g_j(u_{aj}) - g_j(u_{bj})] + \tau_j\{f_j[g_j(u_{aj})] - f_j[g_j(u_{bj})]\}.$$

Thus if $u_{aj} \neq u_{bj}$,

$$\frac{f_j[g_j(u_{aj})] - f_j[g_j(u_{bj})]}{u_{aj} - u_{bj}} = \frac{r_j(u_{aj}, u_{bj})}{c_j + \tau_j r_j(u_{aj}, u_{bj})},$$

in which (for $u_{aj} \neq u_{bj}$)

$$r_j(u_{aj}, u_{bj}) = \frac{f_j[g_j(u_{aj})] - f_j[g_j(u_{bj})]}{g_j(u_{aj}) - g_j(u_{bj})}.$$

In a similar manner we find that for all $u_{aj} \neq u_{bj}$:

$$\frac{g_j(u_{aj}) - g_j(u_{bj})}{u_{aj} - u_{bj}} = \frac{1}{c_j + \tau_j r(u_{aj}, u_{bj})}.$$

Now, let us define for $j = 1, 2, \cdots, 2p$

$$r_j(u_{aj}, u_{bj}) = f_j'[g_j(u_{aj})]$$

when $u_{aj} = u_{bj}$. Then since u_{aj} and u_{bj} are continuous on $[0, \infty)$, it follows (see Appendix A) that $r_j(u_{aj}, u_{bj})$ is continuous on $[0, \infty)$. Since $r_j(u_{aj}, u_{bj})$ is nonnegative, it is clear that both

$$\frac{r_j(u_{aj}, u_{bj})}{c_j + \tau_j r_j(u_{aj}, u_{bj})}$$

and

$$\frac{1}{c_j + \tau_j r_j(u_{aj}, u_{bj})}$$

are continuous on $[0, \infty)$. Moreover equations (12) and (13) are satisfied with

$$D_1 = \operatorname{diag}\left\{\frac{r_j(u_{aj}, u_{bj})}{c_j + \tau_j r_j(u_{aj}, u_{bj})}\right\} \tag{14}$$

$$D_2 = \operatorname{diag}\left\{\frac{1}{c_j + \tau_j r_j(u_{aj}, u_{bj})}\right\}. \tag{15}$$

At this point we have

$$\frac{d}{dt}(u_a - u_b) + (TD_1 + GD_2)(u_a - u_b) = B_a - B_b, \qquad t \geqq 0 \tag{16}$$

with $TD_1 + GD_2$ continuous on $[0, \infty)$.

We need the following lemma.

Lemma 1: Let $M(\cdot)$ be a continuous real $n \times n$ matrix-valued function of t defined on $[0, \infty)$ such that there exist positive constants ϵ and $c_1, c_2, \cdots, c_n$, with the property that for $j = 1, 2, \cdots, n$ and all $t \geqq 0$*

$$m_{jj} - \sum_{i \neq j} c_i c_j^{-1} \mid m_{ij} \mid \geqq \epsilon.$$

Let x be a differentiable real n-vector-valued function on $[0, \infty)$ such that

$$\frac{dx}{dt} + Mx = 0, \qquad t \geqq 0.$$

Then there exists a constant k such that for $i = 1, 2, \cdots, n$, and all $t \geqq 0$

$$\mid x_i(t) \mid \leqq k \exp(-\epsilon t).$$

Moreover, k depends only on the c_i and the initial values $x_i(0)$.

---

* In Ref. 4, Rosenbrock states a similar result, but does not give a rigorous proof. He considers the case in which $c_j = 1$ for $j = 1, 2, \cdots, n$.

Proof of Lemma 1: Let the functional s be defined in terms of an arbitrary continuously differentiable scalar function $\varphi(\cdot)$ by

$$\begin{aligned} s(\varphi)(t) &= 1 \quad \text{if} \quad \varphi(t) > 0 \quad \text{or} \quad \text{if} \quad \varphi(t) = 0 \quad \text{and} \quad \varphi'(t) > 0 \\ &= -1 \quad \text{if} \quad \varphi(t) < 0 \quad \text{or} \quad \text{if} \quad \varphi(t) = 0 \quad \text{and} \quad \varphi'(t) < 0 \\ &= 0 \quad \text{if} \quad \varphi(t) = 0 \quad \text{and} \quad \varphi'(t) = 0. \end{aligned}$$

Then for $t \geqq 0$,

$$\begin{aligned} \sum_i c_i s(x_i)(t) x_i'(t) &= -\sum_i c_i s(x_i)(t) \sum_j m_{ij} x_j \\ &= -\sum_j x_j \sum_i c_i s(x_i)(t) m_{ij} \\ &= -\sum_j x_j c_j s(x_j)(t) m_{jj} - \sum_j x_j \sum_{i \neq j} c_i s(x_i)(t) m_{ij} \\ &\leqq -\sum_j c_j m_{jj} \mid x_j \mid + \sum_j \mid x_j \mid \sum_{i \neq j} c_i \mid m_{ij} \mid \\ &\leqq -\epsilon \sum_j \mid c_j x_j \mid . \end{aligned}$$

But $\sum_i c_i s(x_i)(t) x_i'$ is equal to $\frac{d^+}{dt} \sum_i \mid c_i x_i \mid$, the right-hand derivative of $\sum_i \mid c_i x_i \mid$ [see Appendix B; the derivative of $\mid x_i \mid$ need not exist at points t at which $x_i(t) = 0$]. Therefore

$$\frac{d^+}{dt} \sum_j \mid c_j x_j \mid \leqq -\epsilon \sum_j \mid c_j x_j \mid, \qquad t \geqq 0$$

from which it follows that

$$\sum_j \mid c_j x_j(t) \mid \leqq \exp(-\epsilon t) \sum_j \mid c_j x_j(0) \mid, \qquad t \geqq 0. \ \square$$

If $M(\cdot)$ satisfies the conditions of Lemma 1, then it is easy to show that the unique continuously differentiable $n \times n$ matrix-valued function X defined on $[0, \infty)$ which satisfies

$$\frac{dX}{dt} + MX = 0, \qquad t \geqq 0 \quad [X(0) = I]$$

possesses the property that (for any norm $\|\cdot\|$ on E^n) there exists a constant K_1 such that

$$\| X(t) X(\tau)^{-1} \| \leqq K_1 \exp[-\epsilon(t - \tau)]$$

for all $t \geqq \tau$.

Returning now to equation (16), assume that $[TD_1 + GD_2]$ satisfies the conditions on $M(\cdot)$ of Lemma 1. Then with Y the solution of

$$\frac{dY}{dt} + [TD_1 + GD_2]Y = 0, \qquad t \geqq 0 \quad [Y(0) = I]$$

we have

$$u_a(t) - u_b(t) = Y(t)[u_a(0) - u_b(0)] + \int_0^t Y(t)Y(\tau)^{-1}[B_a(\tau) - B_b(\tau)]\, d\tau, \qquad t \geqq 0.$$

Therefore, for $t \geqq 0$

$$\begin{aligned} \| u_a(t) - u_b(t) \| &\leqq \| Y(t)[u_a(0) - u_b(0)] \| \\ &\quad + \int_0^t \| Y(t)Y(\tau)^{-1} \| \cdot \| B_a(\tau) - B_b(\tau) \|\, d\tau \\ &\leqq \| Y(t)[u_a(0) - u_b(0)] \| \\ &\quad + K_2 \int_0^t \exp[-\epsilon(t - \tau)]\, \| B_a(\tau) - B_b(\tau) \|\, d\tau \end{aligned}$$

for some positive constant K_2. Since $\| B_a(\tau) - B_b(\tau) \| \rightarrow 0$ as $\tau \rightarrow \infty$, it follows that $\| u_a(t) - u_b(t) \| \rightarrow 0$ as $t \rightarrow \infty$.

It remains only to prove that $[TD_1 + GD_2]$ meets the conditions imposed on $M(\cdot)$ of Lemma 1. Since $G \in \mathfrak{D}$, there exists a diagonal matrix diag $(d_1, d_2, \cdots, d_{2p})$ with $d_j > 0$ for $j = 1, 2, \cdots, 2p$ and

$$\alpha_f^{(k)} < \frac{d_{2k-1}}{d_{2k}} < \frac{1}{\alpha_r^{(k)}}$$

for $k = 1, 2, \cdots, p$ such that both

$$\text{diag}\,(d_1, d_2, \cdots, d_{2p})G$$

and

$$\text{diag}\,(d_1, d_2, \cdots, d_{2p})T$$

are strongly column-sum dominant. Thus for $j = 1, 2, \cdots, 2p$

$$t_{jj} - \sum_{i \neq j} d_i d_j^{-1} \mid t_{ij} \mid > 0$$

$$g_{jj} - \sum_{i \neq j} d_i d_j^{-1} \mid g_{ij} \mid > 0.$$

Let $W = TD_1 + GD_2$. Then, for $j = 1, 2, \cdots, 2p$,

$$w_{jj} = t_{jj} \frac{r_j}{c_j + \tau_j r_j} + g_{jj} \frac{1}{c_j + \tau_j r_j}$$

and

$$\sum_{i \neq j} d_i d_j^{-1} \mid w_{ij} \mid = \sum_{i \neq j} d_i d_j^{-1} \left| t_{ij} \frac{r_j}{c_j + \tau_j r_j} + g_{ij} \frac{1}{c_j + \tau_j r_j} \right| .$$

Therefore

$$w_{jj} - \sum_{i \neq j} d_i d_j^{-1} \mid w_{ij} \mid \geqq \frac{r_j}{c_j + \tau_j r_j} (t_{jj} - \sum_{i \neq j} d_i d_j^{-1} \mid t_{ij} \mid) + \frac{1}{c_j + \tau_j r_j} (g_{jj} - \sum_{i \neq j} d_i d_j^{-1} \mid g_{ij} \mid). \quad (17)$$

Since $r_j \geqq 0$, the right side of equation (17) is bounded from below by some positive constant ϵ uniformly in t and j. □

3.2 *Proof of Corollary 1*

By Corollary 3 of Ref. 5 there exists a unique $v \varepsilon E^{2p}$ such that

$$TF(v) + Gv = B_\infty \quad (18)$$

whenever G is such that all principal minors of $T^{-1}G$ are positive. In Reference 5 it is proved that $T^{-1}G$ will have this property if $T^{-1}G$ can be written as $A^{-1}B$ with both A and B stongly column-sum dominant.

Let $H = \text{diag}(d_1, d_2, \cdots, d_{2p})G$ be strongly column-sum dominant with all $d_i > 0$ and

$$\alpha_f^{(k)} < \frac{d_{2k-1}}{d_{2k}} < \frac{1}{\alpha_r^{(k)}}$$

for $k = 1, 2, \cdots, p$. Then $U \triangleq \text{diag}(d_1, d_2, \cdots, d_{2p})T$ is strongly column-sum dominant, and $T^{-1}G = U^{-1}H$, which proves that equation (18) possesses a unique solution v.

With v the solution of equation (18), let $u_\infty = C(v)$. Clearly if $B_\infty = \theta$, then $u_\infty = \theta$. Let u_b satisfy

$$\frac{du_b}{dt} + TF[C^{-1}(u_b)] + G[C^{-1}(u_b)] = B_\infty, \qquad t \geqq 0$$

with $u_b(0) = u_\infty$. Of course, $u_b(t) = u_\infty$ for all $t \geqq 0$. By Theorem 1, $[u(t) - u_\infty] \rightarrow \theta$ as $t \rightarrow \infty$, independent of u_0.

3.3 *Proof of Theorem 2*

Following the proofs of Theorem 1 and Corollary 1,

$$\frac{d}{dt}(u - u_\infty) + (TD_1 + GD_2)(u - u_\infty) = 0, \qquad t \geqq 0$$

in which

$$D_1 = \operatorname{diag}\left\{\frac{r_j(u_j\,,\,u_{\infty j})}{c_j + \tau_j r_j(u_j\,,\,u_{\infty j})}\right\}$$

and

$$D_2 = \operatorname{diag}\left\{\frac{1}{c_j + \tau_j r_j(u_j\,,\,u_{\infty j})}\right\}.$$

Therefore

$$\frac{d^+}{dt}\sum_i d_i \mid u_i(t) - u_{\infty i} \mid \leqq -\underline{K} \sum_i d_i \mid u_i(0) - u_{\infty i} \mid, \qquad t \geqq 0$$

in which

$$\underline{K} = \min_i \min \left\{\frac{1}{\tau_j}(t_{jj} - \sum_{i \neq j} d_i d_j^{-1} \mid t_{ij} \mid), \frac{1}{c_j}(g_{jj} - \sum_{i \neq j} d_i d_j^{-1} \mid g_{ij} \mid)\right\}.$$

But for $j = 1, 2, \cdots, 2p$

$$t_{jj} - \sum_{i \neq j} d_i d_j^{-1} \mid t_{ij} \mid = 1 - \tilde{d}_j d_j^{-1} \alpha_j \,. \;\square$$

3.4 *Proof of Theorem 3*

Since $TF[C^{-1}(z)] + GC^{-1}(z)$ depends continuously on $z \in E^{2p}$, u_∞ satisfies (see Ref. 6)

$$TF[C^{-1}(u_\infty)] + GC^{-1}(u_\infty) = B.$$

Therefore, following the proofs of Theorem 1 and Corollary 1,

$$\frac{d}{dt}(u - u_\infty) + (TD_1 + GD_2)(u - u_\infty) = 0, \qquad t \geqq 0$$

in which

$$D_1 = \operatorname{diag}\left\{\frac{r_j(u_j\,,\,u_{\infty j})}{c_j + \tau_j r_j(u_j\,,\,u_{\infty j})}\right\}$$

and

$$D_2 = \operatorname{diag}\left\{\frac{1}{c_j + \tau_j r_j(u_j\,,\,u_{\infty j})}\right\}.$$

For any $z \in E^{2p}$, let $\| z \|$ denote $\sum_i d_i |z_i|$. Then, for $t \geqq 0$

$$\left\| \frac{d}{dt}(u - u_\infty) \right\| = \| (TD_1 + GD_2)(u - u_\infty) \|$$

$$\leqq \max_i \left\{ (1 + \tilde{d}_i d_i^{-1} \alpha_i) \frac{r_i(u_i, u_{\infty i})}{c_i + \tau_i r_i(u_i, u_{\infty i})} + \sum_{i=1}^{2p} d_i d_j^{-1} \mid g_{ij} \mid \frac{1}{c_j + \tau_j r_j(u_j, u_{\infty j})} \right\} \| u(t) - u_\infty \|.$$

But, since $r_i(u_i, u_{\infty i}) \geqq 0$,

$$\bar{K} \geqq \max_i \left\{ (1 + \tilde{d}_i d_i^{-1} \alpha_i) \frac{r_i(u_i, u_{\infty i})}{c_i + \tau_i r_i(u_i, u_{\infty i})} + \sum_{i=1}^{2p} d_i d_j^{-1} \mid g_{ij} \mid \frac{1}{c_j + \tau_j r_j(u_j, u_{\infty j})} \right\}.$$

Thus

$$\left\| \frac{d}{dt}(u - u_\infty) \right\| \leqq \bar{K} \| u - u_\infty \|, \qquad t \geqq 0. \tag{19}$$

Clearly,

$$\left\| \frac{d}{dt}(u - u_\infty) \right\| = \lim_{\epsilon \to 0+} \frac{1}{\epsilon} \| u(t + \epsilon) - u_\infty - u(t) + u_\infty \|, \qquad t \geqq 0$$

Also, for $t \geqq 0$, the limit

$$\lim_{\epsilon \to 0+} \frac{1}{\epsilon} [\| u(t) - u_\infty \| - \| u(t + \epsilon) - u_\infty \|]$$

exists and is equal to $-\frac{d^+}{dt} \| u - u_\infty \|$ in which as before $\frac{d^+}{dt}$ denotes the right-hand derivative (see Appendix B). But, since for any $\epsilon > 0$ and $t \geqq 0$,

$$\| u(t) - u_\infty \| - \| u(t + \epsilon) - u_\infty \| \leqq \| u(t + \epsilon) - u_\infty - u(t) + u_\infty \|,$$

we have

$$-\frac{d^+}{dt} \| u - u_\infty \| \leqq \left\| \frac{d}{dt}(u - u_\infty) \right\|, \qquad t \geqq 0. \tag{20}$$

Therefore, using equations (19) and (20),

$$\frac{d^+}{dt} \| u - u_\infty \| \geqq -\bar{K} \| u - u_\infty \|, \qquad t \geqq 0$$

and, for $t \geqq 0$,

$$|| u - u_\infty || \geqq \exp(-\bar{K}t) \, || u(0) - u_\infty ||. \, \square$$

IV. A SIGNIFICANT EXTENSION

We can easily extend our results to cover an interesting class of networks containing diodes, capacitors (not necessarily linear), and (not necessarily linear) inductors, in addition to the elements of the Fig. 1 network.

Let each diode be represented by a model of the type shown in Fig. 7 in which

$$i_d = \frac{d}{dt}[c_d v_d + \tau_d f_d(v_d)] + f_d(v_d),$$

with c_d and τ_d positive constants. Assume that $f_d(\cdot)$ satisfies the conditions placed on $f_e(\cdot)$ and $f_c(\cdot)$ of the transistor model. Let there be q diodes and let v_{2p+k} and i_{2p+k} ($k = 1, 2, \cdots, q$) be the voltage and current associated with the kth diode.

Suppose that the kth capacitor (we assume that there are r capacitors) is governed by

$$\frac{d}{dt}[c_{2p+q+k}(v_{2p+q+k})] = i_{2p+q+k}$$

for $k = 1, 2, \cdots, r$, where $c_{2p+q+k}(\cdot)$ is a strictly-monotone-increasing continuously-differentiable mapping of E^1 *onto* itself such that $c_{2p+q+k}(0) = 0$ and the slope of $c_{2p+q+k}(\cdot)$ is uniformly bounded from above and from below by positive constants.

Finally, let there be s inductors which introduce constraints

$$\frac{d}{dt}[l_{2p+q+r+k}(i_{2p+q+r+k})] = v_{2p+q+r+k}$$

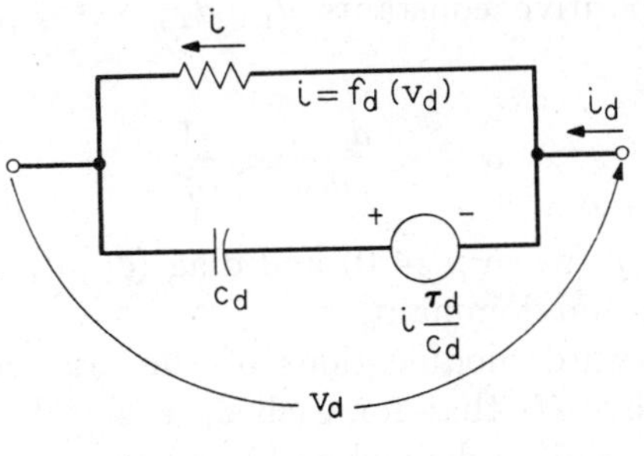

Fig. 7 — Diode model.

for $k = 1, 2, \cdots, s$, in which each $l_{2p+q+r+k}(\cdot)$ is a function of the same type as the $c_{2p+q+k}(\cdot)$.

Assume that the linear resistive portion of the network introduces the constraint

$$\tilde{\imath} = -H\tilde{v} + B, \qquad B \in \mathcal{B}$$

in which $\tilde{\imath} = (i_1, i_2, \cdots, i_{2+p+q+r}, v_{2p+q+r+1}, \cdots, v_{2p+q+r+s})^{tr}$, $\tilde{v} = (v_1, v_2, \cdots, v_{2p+q+r}, i_{2p+q+r+1}, \cdots, i_{2p+q+r+s})^{tr}$, and H is a constant hybrid-parameter matrix of order $(2p + q + r + s)$. Then

$$\frac{d}{dt}[\tilde{C}(\tilde{v})] + \tilde{T}\tilde{F}(\tilde{v}) + H\tilde{v} = B$$

where

$$\begin{aligned}
[\tilde{C}(\tilde{v})]_j &= [C(v)]_j, & j &= 1, 2, \cdots, 2p \\
&= c_j v_j + \tau_j f_j(v_j), & j &= 2p+1, 2p+2, \cdots, 2p+q \\
&= c_j(v_j), & j &= 2p+q+1, \cdots, 2p+q+r \\
&= l_j(i_j), & j &= 2p+q+r+1, \cdots, 2p+q+r+s;
\end{aligned}$$

$\tilde{T}$ is the direct sum of matrices $T \oplus I_q \oplus 0_{r+s}$, in which I_q is the identity matrix of order q and 0_{r+s} is the zero matrix of order $(r + s)$, and

$$\begin{aligned}
[\tilde{F}(\tilde{v})]_j &= [F(v)]_j, & j &= 1, 2, \cdots, 2p \\
&= f_j(v_j), & j &= 2p+1, \cdots, 2p+q.
\end{aligned}$$

Under our assumptions $\tilde{C}(\cdot)^{-1}$ exists and, with $\tilde{u} = \tilde{C}(\tilde{v})$,

$$\frac{d\tilde{u}}{dt} + \tilde{T}\tilde{F}[\tilde{C}^{-1}(\tilde{u})] + H\tilde{C}^{-1}(\tilde{u}) = B. \tag{21}$$

Let $\tilde{\mathfrak{D}}$ denote the set of all real matrices M of order $(2p + q + r + s)$ such that there exist positive constants $d_1, d_2, \cdots, d_{2p+q+r+s}$ with the property that

$$\alpha_f^{(k)} < \frac{d_{2k-1}}{d_{2k}} < \frac{1}{\alpha_r^{(k)}}$$

for $k = 1, 2, \cdots, p$ (when $p \neq 0$) and diag $(d_1, d_2, \cdots, d_{2p+q+r+s})M$ is strongly column-sum dominant.

With straightforward modifications of the arguments already presented, we can prove (i) that for each $\tilde{u}_0 \in E^{(2p+q+r+s)}$ equation (21) possesses a unique solution defined on $[0, \infty)$ such that $\tilde{u}(0) = \tilde{u}_0$, and (*ii*) the analogs of Theorems 1, 2, and 3 and Corollary 1. To be more specific, the analogs of Theorem 1, Corollary 1, and Theorem 2 are:

Theorem 1′: If $H \in \tilde{\mathfrak{D}}$, and $\tilde{u}_a$ and $\tilde{u}_b$ are solutions of equation (21) *with $B = B_a$ and $B = B_b$, respectively, for $t \geqq 0$, and if $[B_a(t) - B_b(t)] \to \theta$ [the zero vector of $E^{(2p+q+r+s)}$] as $t \to \infty$ with $B_a \in \mathfrak{B}$ and $B_b \in \mathfrak{B}$, then $[\tilde{u}_a(t) - \tilde{u}_b(t)] \to \theta$ as $t \to \infty$.*

Corollary 1′: Referring to equation (21), *if $H \in \tilde{\mathfrak{D}}$, and if there exists a constant vector B_∞ such that $[B(t) - B_\infty] \to \theta$ as $t \to \infty$, then there exists a constant vector $\tilde{u}_\infty$ such that $[\tilde{u}(t) - \tilde{u}_\infty] \to \theta$ as $t \to \infty$, and $\tilde{u}_\infty$ is independent of the initial condition $\tilde{u}_0$. In particular, if $B_\infty = \theta$, then $\tilde{u}_\infty = \theta$.*

Theorem 2′: If the hypotheses of Corollary 1′ are satisfied with $B(t) = B_\infty$ for $t \geqq 0$, then with $j_0 = (2p + q + r + s)$, we have

$$\sum_{j=1}^{j_0} d_j \mid \tilde{u}_j(t) - u_{\infty j} \mid \leqq \exp(-\tilde{\underline{K}}t) \sum_{j=1}^{j_0} d_j \mid \tilde{u}_j(0) - \tilde{u}_{\infty j} \mid, \qquad t \geqq 0$$

for every set of positive constants $d_1, d_2, \cdots, d_{2p+q+r+s}$ such that $0 < \tilde{\underline{K}} = \min\{\tilde{\underline{K}}_1, \tilde{\underline{K}}_2, \tilde{\underline{K}}_3\}$ where

$$\tilde{\underline{K}}_1 = \min_{1 \leqq j \leqq 2p} \min \left\{ \frac{1}{\tau_j}(1 - \tilde{d}_j d_j^{-1} \alpha_j), \frac{1}{c_j}\Big(g_{jj} - \sum_{i \neq j} d_i d_j^{-1} \mid g_{ij} \mid\Big) \right\}$$

$$\tilde{\underline{K}}_2 = \min_{2p+1 \leqq j \leqq 2p+q} \min \left\{ \frac{1}{\tau_j}, \frac{1}{c_j}\Big(g_{jj} - \sum_{i \neq j} d_i d_j^{-1} \mid g_{ij} \mid\Big) \right\}$$

$$\tilde{\underline{K}}_3 = \min_{2p+q+1 \leqq j \leqq 2p+q+r+s} \left\{ \frac{1}{s_j}\Big(g_{jj} - \sum_{i \neq j} d_i d_j^{-1} \mid g_{ij} \mid\Big) \right\}$$

in which $s_j = \sup c_j'(\cdot)$ for $j = 2p + q + 1, \cdots, 2p + q + r$; $s_j = \sup l_j'(\cdot)$ for $j = 2p + q + r + 1, \cdots, 2p + q + r + s$; $-\alpha_j$ is the nonzero off-diagonal term in the jth column of T; and $\tilde{d}_j = d_{j+1}$ for j odd and $\tilde{d}_j = d_{j-1}$ for j even. Moreover there exists one such set of constants $\{d_j\}$.

V. FINAL COMMENTS

The results presented here are quite encouraging in that they are concerned with the equations of reasonably realistic nonlinear network models, and provide some understanding of a precise nature in an area where there is a great need for many results of similar type.

VI. ACKNOWLEDGMENT

The writer is indebted to H. Shichman for supplying the data and computer results for the practical circuit of Fig. 6.

APPENDIX A

Proof that $r_i(u_{ai}, u_{bi})$ is continuous.

It is clear that $r_i(u_{ai}, u_{bi})$ is continuous at each point t such that $u_{ai}(t) \neq u_{bi}(t)$. Suppose now that t is such that $u_{ai}(t) = u_{bi}(t)$, and let $\epsilon > 0$ be given. Since u_{ai}, u_{bi}, g and f_i' are continuous, there exists $\delta_1 > 0$ such that

$$| f_i' \{g_i[u_{ai}(t+\eta)]\} - f_i' \{g_i[u_{ai}(t)]\} | \leqq \epsilon$$

for all $| \eta | \leqq \delta_1$. Then for $| \eta | \leqq \delta_1$ either $u_{ai}(t+\eta) = u_{bi}(t+\eta)$ in which case

$$| r_i[u_{ai}(t+\eta), u_{bi}(t+\eta)] - r_i[u_{ai}(t), u_{bi}(t)] | \leqq \epsilon$$

or $u_{bi}(t+\eta) \neq u_{bi}(t+\eta)$ and (using the mean-value theorem)

$$\begin{aligned} r_i[u_{ai}(t+\eta), u_{bi}(t+\eta)] &= \frac{f_i\{g_i[u_{ai}(t+\eta)]\} - f_i\{g_i[u_{bi}(t+\eta)]\}}{g_i[u_{ai}(t+\eta)] - g_i[u_{bi}(t+\eta)]} \\ &= f_i'(\xi) \end{aligned}$$

in which

$$\begin{aligned} | \xi - g_i[u_{ai}(t)] | \leqq \max \{ & | g_i[u_{ai}(t+\eta)] \\ & - g_i[u_{ai}(t)] |, | g_i[u_{bi}(t+\eta)] - g_i[u_{ai}(t)] | \}. \end{aligned}$$

In the latter case, there exists $\delta_2 > 0$ such that $| f'(\xi) - f' \{g_i[u_{ai}(t)]\} | \leqq \epsilon$ for all $| \eta | \leqq \delta_2$. Thus for all $| \eta | \leqq \min \{\delta_1, \delta_2\}$, we have

$$| r_i[u_{ai}(t+\eta), u_{bi}(t+\eta)] - r[u_{ai}(t), u_{bi}(t)] | \leqq \epsilon.$$

APPENDIX B

Proof that the Right-Hand Derivative of $| x_i |$ exists and is equal to $s(x_i)(t)x_i'$.

If t is a point such that $x_i(t) \neq 0$, then it is clear that

$$\frac{d^+}{dt} | x_i | = s(x_i)(t)x_i'(t).$$

At t such that $x_i(t) = 0$ and $x_i'(t) \neq 0$,

$$s(x_i)(t)x_i' = \lim_{\epsilon \to 0} s(x_i)(t) \frac{x_i(t+\epsilon)}{\epsilon} = \frac{d^+}{dt} | x_i |.$$

Finally if $x_i(t) = 0$ and $x_i'(t) = 0$, then

$$0 = \lim_{\substack{\epsilon \to 0+ \\ t \leq \xi \leq t+\epsilon}} | x_i'(\xi) | = \lim_{\epsilon \to 0+} \frac{| x_i(t+\epsilon) |}{\epsilon} = \frac{d^+}{dt} | x_i |,$$

since x_i is continuously differentiable.

REFERENCES

1. Gummel, H. K., "A Charge-Control Transistor Model for Network Analysis Programs," Proc. IEEE, *56,* No. 4 (April 1968), p. 751.
2. Koehler, D., "The Charge-Control Concept in the Form of Equivalent Circuits, Representing a Link Between the Classic Large Signal Diode and Transistor Models," B.S.T.J., *46,* No. 3 (March 1967), pp. 523–576.
3. Nemytskii, V. V. and Stepanov, V. V., *Qualitative Theory of Differential Equations,* Princeton, New Jersey: Princeton University Press, 1960, pp. 9–13.
4. Rosenbrock, H. H., "A Method of Investigating Stability." Proc. 2nd International Federation Automatic Control Congress, Basel, Switzerland, 1963, pp. 590–594.
5. Sandberg, I. W. and Willson, A. N., "Some Theorems on Properties of DC Equations of Nonlinear Networks," B.S.T.J., this issue, pp. 1–34.
6. Bellman, R., *Stability Theory of Differential Equations,* New York: McGraw-Hill, 1953, p. 77.

Theorems on the Computation of the Transient Response of Nonlinear Networks Containing Transistors and Diodes

By I. W. SANDBERG

(Manuscript received June 15, 1970)

We consider in detail the nonlinear equations encountered at each time step when certain implicit numerical-integration algorithms are used. In terms of only the properties of the Jacobian matrix of the pertinent set of differential equations, we present necessary and sufficient conditions for the existence and uniqueness of the solution of the nonlinear equations for all continuous forcing functions and any given step size. Since engineers often think about dynamic nonlinear transistor network problems in terms of the eigenvalues of the relevant Jacobian matrix, the results described are of immediate conceptual value. In particular, it is possible to carry out the algorithms whenever the conditions presented are satisfied.

Several other types of results are also presented. For example, for a special but significant and useful numerical-integration formula, theorems are proved concerning properties of the computed sequence such as the extent to which the sequence is relatively immune to small local errors introduced at each step as a result of the fact that it is ordinarily not possible to compute the solution of a certain equation exactly.

All of the results are concerned with network models that are often used in computer simulations. In fact, we heavily exploit some special properties possessed by the nonlinear functions associated with such models.

I. INTRODUCTION

The set P_0 of all real square matrices each with all principal minors nonnegative plays a key role in the study[1-3] of nonlinear equations of the form $F(x) + Ax = B$, and more generally[4] of equations of the form $CF(x) + Ax = B$, in which $F(\cdot)$ is a "diagonal monotone-nondecreasing mapping" of real Euclidean n-space E^n into itself, A and C are real

$n \times n$ matrices and B is an element of E^n. Such equations arise in the dc analysis of transistor networks, the computation of the transient response of transistor networks, and the numerical solution of certain nonlinear partial-differential equations.

In Ref. 3 a nonuniqueness theorem is proved which focuses attention on a simple special property of transistor-type nonlinearities. It shows that for any transistor-type exponential $F(\cdot)$ the equation $F(x) + Ax = B$ has at least two solutions x for some $B \in E^n$ whenever $A \notin P_0$. The theorem shows that some earlier conditions[1,2] for the existence of a unique solution cannot be improved by taking into account more information concerning the nonlinearities, and therefore makes more clear that the set of matrices P_0 plays a basic role in the theory of nonlinear transistor networks. Ref. 3 also contains material concerned with the convergence of algorithms for computing the solution of $F(x) + Ax = B$ as well as of more general equations, and some related problems concerning the numerical integration of the ordinary differential equations which govern the transient response of nonlinear transistor networks are considered briefly.

The primary purpose of this paper is to present the results of a continuation of the numerical integration study initiated in Ref. 3. Here we further exploit the special property of transistor-type exponential nonlinearities used in Ref. 3.

We consider in detail the nonlinear equations encountered at each time step when certain implicit numerical-integration algorithms are used, and, in terms of only the properties of the Jacobian matrix of the pertinent set of differential equations, we present necessary and sufficient conditions for the existence and uniqueness of the solution of the nonlinear equations for all continuous forcing functions and any given step size. Since engineers often think about dynamic nonlinear transistor network problems in terms of the location of the eigenvalues of the relevant Jacobian matrix, the results described in Section 2.2 are of immediate conceptual value. In particular, these results are of a very different character than those that appear in the literature, and whenever the conditions presented are satisfied, it is possible to carry out the algorithms. Under the assumption that the conditions are satisfied, we also show that there are convergent algorithms for solving the nonlinear equations, and that the Jacobian matrix of the nonlinear equations is essentially always at least weakly well-conditioned in a significant sense.

A part of Section 2.3 reports on a general result concerning conditions under which it is possible to invert nonlinear mappings in E^n. More explicitly, we show that a proposition proved by G. H. Meyer enables us to give a short proof of a new theorem which is a considerably stronger result than that described and used in Ref. 11.

We also present a set of results concerning properties of an important class of transistor-diode networks for which certain implicit numerical-integration algorithms can be carried out for all values of the step size, and, for a special but significant and useful numerical-integration formula, theorems are proved concerning some properties of the computed sequence such as the extent to which the sequence is relatively immune to small local errors introduced at each step as a result of the fact that it is ordinarily not possible to compute the solution of a certain equation exactly.

Finally, in addition to other results, we present new theorems concerning the existence of solutions of the nonlinear dc equation under very realistic assumptions from the viewpoint of models often used in computer simulations.†

Section II contains a detailed discussion of the results and their significance.

II. TRANSIENT RESPONSE OF TRANSISTOR-DIODE NETWORKS AND IMPLICIT NUMERICAL-INTEGRATION FORMULAS

2.1 *Introduction*

We shall consider explicitly only networks containing transistors, diodes, and resistors. However, the material to be presented can be extended to take into account other types of elements as well. In addition, we shall focus attention on the use of linear multipoint integration formulas of closed (i.e., of implicit) type, since such formulas are of considerable use in connection with the typically "stiff systems" of differential equations encountered.

A very large class of networks containing resistors, transistors, and diodes modeled in a standard manner is governed by the equation[5,‡]

$$\frac{du}{dt} + TF[C^{-1}(u)] + GC^{-1}(u) = B(t), \qquad t \geqq 0 \tag{1}$$

† Results concerning the dc equation are directly relevant to the problem of computing the transient response to the extent that in order to numerically integrate the differential equations it is ordinarily necessary to first solve a dc problem to determine the initial conditions.

‡ As a practical matter, the models of transistors and diodes employed here are often used in computer simulations. Of course in some cases it is necessary to use more complicated models.

with $G = \hat{G}(I + R\hat{G})^{-1}$ and where, assuming that there are q diodes and p transistors,

(i) $T = T_1 \oplus T_2 \oplus \cdots \oplus T_p \oplus I_q$, the direct sum of the identity matrix of order q and p 2×2 matrices T_k in which

$$T_k = \begin{bmatrix} 1 & -\alpha_r^{(k)} \\ -\alpha_f^{(k)} & 1 \end{bmatrix}$$

with $0 < \alpha_r^{(k)} < 1$ and $0 < \alpha_f^{(k)} < 1$ for $k = 1, 2, \cdots, p$.

(ii) $R = R_1 \oplus R_2 \oplus \cdots \oplus R_p \oplus R_0$, the direct sum of a diagonal matrix $R_0 = \text{diag}\,(r_1, r_2, \cdots, r_q)$ with $r_k \geqq 0$ for $k = 1, 2, \cdots, q$ and p 2×2 matrices R_k in which for all $k = 1, 2, \cdots, p$

$$R_k = \begin{bmatrix} r_e^{(k)} + r_b^{(k)} & r_b^{(k)} \\ r_b^{(k)} & r_c^{(k)} + r_b^{(k)} \end{bmatrix}$$

with $r_e^{(k)} \geqq 0$, $r_b^{(k)} \geqq 0$, and $r_c^{(k)} \geqq 0$. (The matrix R takes into account the presence of bulk resistance in series with the diodes and the emitter, base, and collector leads of the transistors.)

(iii) $\hat{G}$ is the short-circuit conductance matrix associated with the resistors of the network. (It does not take into account the bulk resistances of the semiconductor devices.)

(iv) $F(\cdot)$ is a mapping of $E^{(2p+q)}$ into $E^{(2p+q)}$ defined by the condition that

$$F(x) = [f_1(x_1), f_2(x_2), \cdots, f_{2p+q}(x_{2p+q})]^{tr}$$

for all $x \,\varepsilon\, E^{(2p+q)}$ with each $f_i(\cdot)$ a continuously-differentiable mapping of E^1 into E^1 such that $f_i'(\alpha) > 0$ for all $\alpha \,\varepsilon\, E^1$.

(v) $C^{-1}(\cdot)$ is the inverse of the mapping $C(\cdot)$, of $E^{(2p+q)}$ into itself, defined by

$$C(x) = cx + \tau F(x)$$

for all $x \,\varepsilon\, E^{(2p+q)}$ with $c = \text{diag}\,(c_1, c_2, \cdots, c_{(2p+q)})$, $\tau = \text{diag}\,(\tau_1, \tau_2, \cdots, \tau_{(2p+q)})$, and with each τ_i and each c_i a positive constant.

(vi) $B(t)$ is a $(2p + q)$-vector which takes into account the voltage and current generators present in the network, and

(vii) u is related to v the vector of ideal-junction voltages of the semiconductor devices (v does not take in account the voltage drops across the bulk resistors) through $C(v) = u$ for all $v \,\varepsilon\, E^{(2p+q)}$

Equation (1) is equivalent to†

† In Ref. 5 it is shown if $B(\cdot)$ is a continuous mapping of $[0, \infty)$ into $E^{(2p+q)}$, then for any initial condition $u^{(0)} \,\varepsilon\, E^{(2p+q)}$ there exists a unique continuous $(2p + q)$-vector-valued function $u(\cdot)$ such that $u(0) = u^{(0)}$ and (1) is satisfied for all $t > 0$.

$$\dot{u} + f(u, t) = \theta_{(2p+q)}, \qquad t \geqq 0 \tag{2}$$

in which of course

$$f(u, t) = TF[C^{-1}(u)] + GC^{-1}(u) - B(t) \tag{3}$$

and $\theta_{(2p+q)}$ is the zero vector of order $(2p + q)$.

It is well known that certain specializations of the general multipoint formula[6,7]

$$y_{n+1} = \sum_{k=0}^{r} a_k y_{n-k} + h \sum_{k=-1}^{r} b_k \tilde{y}_{n-k} \tag{4}$$

in which

$$\tilde{y}_{n-k} = -f(y_{n-k}, (n - k)h) \tag{5}$$

can be used as a basis for computing the solution of equation (2). Here h, a positive number, is the step size, the a_k and the b_k are real numbers, and of course y_n is the approximation to $u(nh)$ for $n \geqq 1$.

In the literature dealing with formulas of the type (4) in connection with systems of equations of the type (2), information concerning the location of the eigenvalues of the Jacobian matrix J_u of $f(u, t)$ with respect to u plays an important role in determining whether or not a given formula will be (in some suitable sense) stable. In particular, an assumption often made is that all of the eigenvalues of J_u lie in the strict right-half plane for all $t \geqq 0$ and all u. For $f(u, t)$ given by equation (3), we have

$$J_u = T \operatorname{diag} \left\{ \frac{f_j'[g_j(u_j)]}{c_j + \tau_j f_j'[g_j(u_j)]} \right\} + G \operatorname{diag} \left\{ \frac{1}{c_j + \tau_j f_j'[g_j(u_j)]} \right\} \tag{6}$$

in which for $j = 1, 2, \cdots, (2p + q)$ $g_j(u_j)$ is the jth component of $C^{-1}(u)$. Thus here J_u is a matrix of the form

$$TD_1 + GD_2 \tag{7}$$

where D_1 and D_2 are diagonal matrices with positive diagonal elements. A simple result concerning (7), Theorem 4 of Ref. 3, asserts that if there exists a diagonal matrix D with positive diagonal elements such that†

(*i*) DT is strongly column-sum dominant, and
(*ii*) DG is weakly column-sum dominant,

then for all diagonal matrices D_1 and D_2 with positive diagonal elements,

† The terms "strongly-column sum dominant" and "weakly column-sum dominant" are reasonably standard. However, they are defined in Section III.

all eigenvalues of (7) lie in the strict right-half plane. This condition on T and G is often satisfied.†

The subclass of numerical integration formulas (4) defined by the condition that $b_{-1} > 0$ are of considerable use[8-10] in applications involving the typically "stiff systems" of differential equations encountered in the analysis of nonlinear transistor networks. With $b_{-1} > 0$, y_{n+1} is defined *implicitly* through

$$y_{n+1} + hb_{-1}f(y_{n+1}, (n+1)h) = \sum_{k=0}^{r} a_k y_{n-k} + h \sum_{k=0}^{r} b_k \tilde{y}_{n-k}$$

in which the right side depends on y_{n-k} only for $k \in \{0, 1, 2, \cdots, r\}$, and for $f(u, t)$ given by equation (3), we have

$$y_{n+1} + hb_{-1}\{TF[C^{-1}(y_{n+1})] + GC^{-1}(y_{n+1})\} = q_n \tag{8}$$

in which

$$q_n = \sum_{k=0}^{r} a_k y_{n-k} + h \sum_{k=0}^{r} b_k \tilde{y}_{n-k} + hb_{-1}B[(n+1)h].$$

Obviously, the numerical integration formula (8) makes sense only if there exists for each n a $y_{n+1} \in E^{(2p+q)}$ such that (8) is satisfied.

2.2 *The Jacobian Matrix J_u and Necessary and Sufficient Conditions for the Existence of a Unique Solution y_{n+1} of (8) for All $q_n \in E^{(2p+q)}$*

Here we shall make the additional assumption that the functions $f_i(\cdot)$ are such that the mapping $F(\cdot)$ belongs to the set $\mathfrak{F}_0^{(2p+q)}$ defined in Section 3.1. This assumption is satisfied whenever the $f_i(\cdot)$ are the usual Ebers–Moll exponential-type nonlinearities. That is, $\mathfrak{F}_0^{(2p+q)}$ contains all of the mappings $F(\cdot)$ such that for each j

$$f_j(x_j) = a_j[\exp(b_j x_j) - 1] \quad \text{or} \quad f_j(x_j) = a_j[1 - \exp(-b_j x_j)]$$

for all $x_j \in E^1$ with a_j and b_j positive constants.

Our first result, Theorem 1 of Section III, is a rather strong result concerning the relation between properties of the Jacobian matrix J_u and properties of equation (8). Let Ξ denote the set of all real numbers σ such that $\det(\sigma I + J_u) = 0$ for some $u \in E^{(2p+q)}$. In other words, let Ξ denote the set of all real numbers σ such that $-\sigma$ is an eigenvalue of J_u at some point u. According to Theorem 1, equation (8) possesses a unique solution y_{n+1} for each $q_n \in E^{(2p+q)}$ (and hence each $B[(n+1)h] \in E^{(2p+q)}$) if and only if $(hb_{-1})^{-1} \notin \Xi$, and also if $(hb_{-1})^{-1} \in \Xi$ then equation (8) possesses at least two solutions for some $q_n \in E^{(2p+q)}$ (and hence for

† See Ref. 5 for examples.

some $B[(n+1)h] \in E^{(2p+q)}$). Therefore, in particular, equation (8) possesses a unique solution for all $q_n \in E^{(2p+q)}$ and all $h \in (0, \bar{h}]$, in which $\bar{h}$ is an arbitrary positive constant, if and only if the intersection of the interval $[(\bar{h}b_{-1})^{-1}, \infty)$ and Ξ is the null set, and equation (8) possesses a unique solution for all $q_n \in E^{(2p+q)}$ and all $h > 0$ if and only if Ξ contains no points of the interval $(0, \infty)$. Finally, as a somewhat peripheral matter, according to Theorem 1, the dc equation $TF(v) + Gv = B$ has at most one solution v for each $B \in E^{(2p+q)}$ if and only if $0 \notin \Xi$.

The statements made in the preceding paragraph are surprising to the extent that on the one hand they are rather definitive and on the other hand they involve only the location of the real eigenvalues of J_u.† Since engineers often find it helpful to think about nonlinear systems in terms of the location of the eigenvalues of a pertinent Jacobian matrix, it is also of interest to note here that equation (8) can possess more than one solution y_{n+1} for some q_n and some $h > 0$ only if the transistor-diode network is locally exponentially unstable at some operating point, that is, only if at some operating point u, $-J_u$ has a real positive eigenvalue.

2.3 *Existence of Convergent Algorithms for Computing the Solution of* (8)

Throughout this section we assume that the $f_i(\cdot)$ are such that the additional condition that $F(\cdot) \in \mathfrak{F}_0^{(2p+q)}$ is satisfied.

Whenever $(hb_{-1})^{-1}$ is not contained in the set Ξ of Section 2.2, equation (8), which we shall write as $Q(y_{n+1}) = q_n$, possesses a unique solution y_{n+1} for any $q_n \in E^{(2p+q)}$. We show here that when $(hb_{-1})^{-1} \notin \Xi$ and each $f_i(\cdot)$ is twice continuously differentiable on E^1,‡ there exist steepest descent as well as Newton-type algorithms each of which generates a sequence in $E^{(2p+q)}$ which converges to y_{n+1}.

Assume that $(hb_{-1}) \notin \Xi$. The Jacobian matrix $(I + hb_{-1}J_{y_{n+1}})$ of $Q(\cdot)$ satisfies

$$\det (I + hb_{-1}J_{y_{n+1}}) \neq 0 \quad \text{for all} \quad y_{n+1} \in E^{(2p+q)} \tag{9}$$

Hence $Q(\cdot)$ is a local homeomorphism on $E^{(2p+q)}$ and since there exists a unique $y_{n+1} \in E^{(2p+q)}$ such that $Q(y_{n+1}) = q_n$ for each $q_n \in E^{(2p+q)}$, $Q(\cdot)$

---

† Indeed, while we can write (8) as $Q(y_{n+1}) = q_n$ with $Q(\cdot)$ a continuously-differentiable mapping of $E^{(2p+q)}$ into itself with Jacobian matrix $(I + hb_{-1}J_{y_{n+1}})$ recall that for $R(\cdot)$ a general continuously-differentiable mapping of E^n into itself with Jacobian matrix J, $\det J \neq 0$ throughout E^n does not imply that (and is not implied by the statement that) for each $x \in E^n$ there exists a unique $y \in E^n$ such that $R(y) = x$, even for $n = 1$.

‡ This differentiability condition is obviously satisfied if the $f_j(\cdot)$ are the usual exponential functions.

is a homeomorphism of $E^{(2p+q)}$ onto itself. Thus, with $\|\cdot\|$ any norm on $E^{(2p+q)}$,

$$\| Q(y) \| \to \infty \quad \text{as} \quad \| y \| \to \infty.^{\dagger}$$

Let $R(\cdot)$ be defined by the condition that $R(y) = Q(y) - q_n$ for all $y \in E^{(2p+q)}$. Then $R(\cdot)$ satisfies $\| R(y) \| \to \infty$ as $\| y \| \to \infty$ and the determinant of the Jacobian matrix of $R(\cdot)$ does not vanish throughout $E^{(2p+q)}$. Therefore, assuming that $R(\cdot)$ is twice continuously differentiable on $E^{(2p+q)}$, it follows (see the Appendix) that the solution y_{n+1} of $R(y_{n+1}) = \theta_{(2p+q)}$ can be computed by using certain steepest descent or Newton-type algorithms.

2.4 *The Jacobian Matrix $(I + hb_{-1}J_{y_{n+1}})$, and Inversion of Nonlinear Operators on E^n and Jacobian Matrices*

As in Section 2.3, let the additional condition that $F(\cdot) \in \mathfrak{F}_0^{(2p+q)}$ be satisfied and let $Q(\cdot)$ be the mapping of $E^{(2p+q)}$ into itself with the property that equation (8) can be written as $Q(y_{n+1}) = q_n$. According to Theorem 2 of Section III the Jacobian matrix $(I + hb_{-1}J_{y_{n+1}})$ possesses the property that there exists a constant $\epsilon > 0$ such that

$$\det (I + hb_{-1}J_{y_{n+1}}) \geqq \epsilon \quad \text{for all} \quad y_{n+1} \in E^{(2p+q)} \tag{10}$$

if and only if the matrix

$$[(hb_{-1})^{-1}\tau + T]^{-1}[(hb_{-1})^{-1}c + G],$$

which we shall call S, belongs to the set P of all real square matrices each with all principal minors positive. Thus when $S \in P$ the matrix $(I + hb_{-1}J_{y_{n+1}})$ is well conditioned in at least the weak sense of (10). This fact is of some interest for two reasons. First, certain standard algorithms require that the matrix $(I + hb_{-1}J_{y_{n+1}})$ be inverted along a sequence of points $\{y_{n+1}^{(k)}\}$ in order to compute the solution y_{n+1} of equation (8), and, secondly, Theorem 3 of Section III shows that if $\det [(hb_{-1})^{-1}I + J_u] \neq 0$ for all $u \in E^{(2p+q)}$ and all $(hb_{-1})^{-1} \in \mathcal{G}'$ in which $\mathcal{G}'$ denotes either $(0, \infty)$ or any interval contained in $(0, \infty)$, then $S \in P$ for all but at most a finite number of points $(hb_{-1})^{-1}$ contained in $\mathcal{G}'$. Therefore, referring to the material of Section 2.2, if $Q(y_{n+1}) = q_n$ possesses a unique solution y_{n+1} for all $q_n \in E^{(2p+q)}$ and all $(hb_{-1})^{-1} \in \mathcal{G}'$, then $(I + hb_{-1}J_{y_{n+1}})$ is at least weakly well conditioned at all but at most a finite number of points contained in $\mathcal{G}'$.

† Since $Q(\cdot)$ is a homeomorphism of $E^{(2p+q)}$ onto itself, $Q(\cdot)^{-1}$ exists and is continuous. Therefore, the image of any closed ball in $E^{(2p+q)}$ under $Q(\cdot)^{-1}$ is contained in some closed ball in $E^{(2p+q)}$, and hence $\| Q(y) \| \to \infty$ as $\|y\| \to \infty$.

Since the elements of $(I + hb_{-1}J_{y_{n+1}})$ are bounded on $y_{n+1} \varepsilon E^{(2p+q)}$, it follows from a theorem described by M. Vehovec[11] that for each $q_n \varepsilon E^{(2p+q)}$ there exists a unique $y_{n+1} \varepsilon E^{(2p+q)}$ such that $Q(y_{n+1}) = q_n$ if $S \varepsilon P$. More explicitly, the theorem described† by Vohovec asserts that if $R(\cdot)$ is a continuously-differentiable mapping of E^n into E^n with $J(R)_q$ the Jacobian matrix of $R(\cdot)$ at an arbitrary point $q \varepsilon E^n$, if the elements of $J(R)_q$ are bounded on E^n, and if there exists a positive constant ϵ such that $\det J(R)_q \geqq \epsilon$ for all $q \varepsilon E^n$, then $R(\cdot)$ is a homeomorphism. Thus, using the theorem of Ref. 11 and Theorems 2 and 3 of Section III, we are able to show that if $\det [(hb_{-1})^{-1}I + J_u] \neq 0$ for all $u \varepsilon E^{(2p+q)}$ and all $(hb_{-1})^{-1} \varepsilon \mathcal{G}'$, then for all but at most a finite number of points $(hb_{-1})^{-1} \varepsilon \mathcal{G}'$, (8) possesses a unique solution y_{n+1} for each $q_n \varepsilon E^{(2p+q)}$. Although this result is obviously much weaker than the existence proposition presented in Section 2.2, it shows that the theorem of Ref. 11 can be exploited to provide some insight in connection with the specific problem considered here.

The theorem of Ref. 11 is of interest primarily because the key hypothesis concerns only the determinant $\det J(R)_q$ (as opposed to the condition of Palais‡ that $\| R(q) \| \to \infty$ as $\| q \| \to \infty$). Theorem 4 of Section III is a general result which is considerably stronger than the theorem of Ref. 11. It shows that the condition of the theorem of Ref. 11 that there exist a positive constant ϵ such that $\det J(R)_q \geqq \epsilon$ for all q can be replaced with the condition that there exist real constants $a > 0$ and $b \geqq 0$ such that

$$\det J(R)_q \geqq \frac{1}{a + b \| q \|} \quad \text{for all} \quad q \varepsilon E^n.$$

2.5 *A Class of Networks for Which* (8) *Possesses a Unique Solution for All Values of the Step Size*

There is an interesting class of transistor-diode-resistor networks with the property that for each network in the class, equation (8) possesses a unique solution for all $h > 0$ (i.e., for all $h > 0$, all $q_n \varepsilon E^{(2p+q)}$, and all diagonal matrices c and τ with positive diagonal elements). In order to define and discuss that class, consider the dc equation $TF(v) + Gv = B$ in which v is the $(2p + q)$-vector of semiconductor ideal-junction voltages and $B \varepsilon E^{(2p+q)}$. If $p > 0$ and the matrix R of Section 2.1 is the zero matrix, v_1 is the emitter-to-base voltage of transistor one, v_2 is the collector-to-base voltage of transistor one, and so forth. By port

† According to Vehovec, the theorem was recently proved by I. Vidar, and the proof is expected to appear in the journal Glasnik Matematicki.

‡ See Ref. 12 and the appendix of Ref. 13. Here $\|\cdot\|$ denotes any norm on E^n.

j of the transistor-diode-resistor network we mean the terminal pair between which the voltage v_j appears. Again we shall make the assumption that $F(\cdot) \in \mathfrak{F}_0^{(2p+q)}$.

In Ref. 3 it is proved that $TF(v) + Gv = B$ possesses at most one solution v for each $B \in E^{(2p+q)}$ if and only if $T^{-1}G \in P_0$. It is also proved in Ref. 3 that equation (8) possesses a unique solution y_{n+1} for each $q_n \in E^{(2p+q)}$ and each $h > 0$ if $M^{-1}G \in P_0$ for all $M \in \mathfrak{I}(T)$ in which here $\mathfrak{I}(T)$ denotes the set of all real matrices having the same form as T and with the "α's" of M not larger than those of T.† In other words, it was also proved in Ref. 3 that equation (8) possesses a unique solution y_{n+1} for each $q_n \in E^{(2p+q)}$ and each $h > 0$ if the dc equation possesses at most one solution for each $B \in E^{(2p+q)}$ for "the original set of α's as well as for an arbitrary set of not-larger α's." Before proceeding, and for the sake of completeness, we mention here that the same result can be obtained by way of the approach of Section 2.2; a direct corollary of Theorem 5 of Section III, Corollary 1, shows that if $M^{-1}G \in P_0$ for all $M \in \mathfrak{I}(T)$, then $\det(\sigma I + J_u) \neq 0$ for all real $\sigma \geqq 0$ and all $u \in E^{(2p+q)}$.

Theorem 5 of Section III provides considerable information concerning the nature of the class of networks for which $M^{-1}G \in P_0$ for all $M \in \mathfrak{I}(T)$. In particular, the theorem shows that $M^{-1}G \in P_0$ for all $M \in \mathfrak{I}(T)$ if and only if $M^{-1}G \in P_0$ for all $M \in \mathfrak{I}_0(T)$ in which $\mathfrak{I}_0(T)$ is the set of all 2^{2p} real square matrices M having the same form as T and with each "α" of M either zero or the corresponding "α" of T.‡ The theorem also shows that "$M^{-1}G \in P_0$ for all $M \in \mathfrak{I}(T)$" is equivalent to each of six other statements involving T and G. For example, according to Theorem 5, we have $M^{-1}G \in P_0$ for all $M \in \mathfrak{I}(T)$ if and only if either $T^{-1}(G + D) \in P_0$ for all diagonal matrices D with positive diagonal elements, which has an obvious network interpretation in terms of the addition of resistors to the network characterized by G, or $T^{-1}G \in P_0$ and $(T_w)^{-1}G_w \in P_0$ for all pairs of matrices T_w and G_w obtained from T and G, respectively, by deleting an arbitrary set w of rows, and the same set of columns, of both T and G.

When the matrix R of Section 2.1 is the zero matrix, the last condition on T and G of the preceding paragraph also has a simple network interpretation: Given T and G, we have $T^{-1}G \in P_0$, and any network obtained from the network characterized by T and G by short-circuiting an arbitrary set w of at most all but one of the $(2p + q)$ semiconductor junctions possesses the following property. With respect to the voltage vector v_w associated with the junctions not short-circuited, and with

† See Definition 4 of Section III for a precise definition of $\mathfrak{I}(T)$.
‡ See Definition 5 of Section III for a precise definition of $\mathfrak{I}_0(T)$.

the components of v_w taken in the same order as those of v, the "new T and G" matrices† T_w and G_w satisfy $(T_w)^{-1}G_w \in P_0$. As reasonable as this condition or any of the other seven equivalent conditions of Theorem 5 might seem, and even though, as Theorem 6 of Section III shows, $T^{-1}G \in P_0$ implies that $(T_w)^{-1}G_w \in P_0$ whenever w has the property that if the port number associated with one junction of a given transistor is contained in w, then the port number associated with the other junction of that transistor is also contained in w, it is the case that there are transistor-diode-resistor networks for which $T^{-1}G \in P_0$ and $M^{-1}G \notin P_0$ for some $M \in \mathfrak{J}(T)$. In fact, Ref. 14 presents an example in which $p = 3$, $q = 0$, $T^{-1}G \in P_0$, and $T^{-1}(G + D) \notin P_0$ for some diagonal matrix D with positive diagonal elements. However, the class of networks for which $T^{-1}G \in P_0$ implies that $M^{-1}G \in P_0$ for all $M \in \mathfrak{J}(T)$ is clearly quite large; it obviously includes all networks in which $p = 0$, it includes all networks in which the base terminals of all transistors are connected to a common point, and as Theorem 7 of Section III shows, the class includes all networks in which $T^{-1}G \in P_0$ and $p = 1$ or $p = 2$.††

2.6 *Results Concerning the Numerical-Integration Formula* $y_{n+1} = y_n + h\tilde{y}_{n+1}$

The general multipoint formula (4) reduces to the well-known implicit numerical-integration formula $y_{n+1} = y_n + h\tilde{y}_{n+1}$ when $a_0 = b_{-1} = 1$, $b_0 = 0$, and $a_k = b_k = 0$ for $k = 1, 2, \cdots, r$. For that important special case, and with $\tilde{y}_{n+1}$ given by equations (3) and (5), $\{y_{n+1}\}$ is defined implicitly through

$$y_{n+1} + h\{TF[C^{-1}(y_{n+1})] + GC^{-1}(y_{n+1})\} = y_n + hB_n \tag{11}$$

for all $n \geqq 0$, in which $B_n = B[(n + 1)h]$. Here we describe some detailed results concerning the relation between the sequences $\{y_{n+1}\}$ and $\{B_n\}$. We assume throughout this section that G is such that there exists a diagonal matrix D with positive diagonal elements with the property that both DT and DG are strongly column-sum dominant. This condition, which is often satisfied,§ guarantees that there exists a unique solution‡ y_{n+1} of equation (11) for each $(y_n + hB_n) \in E^{(2p+q)}$.

† It is a simple matter to show that the "new T and G" matrices are T_w and G_w.
†† It is proved in Ref. 14 that if $q = 0$ and if $p = 1$ or $p = 2$, then $T^{-1}G \in P_0$ implies that $T^{-1}(G + D) \in P_0$ for all diagonal matrices with positive diagonal elements. Thus, by the equivalence of statements (*i*) and (*v*) of Theorem 5 of Section III, it follows at once that if $T^{-1}G \in P_0$ then $M^{-1}G \in P_0$ for all $M \in \mathfrak{J}(T)$ if $q = 0$ and $p = 1$ or $p = 2$. The proof of essentially the same end result given here is of a very different nature and is quite short.
§ See Ref. 5 for examples.
‡ A result mentioned in Section 2.1 implies that if DT and DG are both strongly column-sum dominant, then det $[(h)^{-1}I + J_u] \neq 0$ for all $u \in E^{(2p+q)}$ and all $h > 0$.

Let $\|\cdot\|_1$ be defined by the condition that $\| v \|_1 = \sum_{i=1}^{(2p+q)} | v_i |$ for all $v \in E^{(2p+q)}$. According to Theorem 8 of Section III, there exists a positive constant δ depending only on the c_i, the τ_i, T, G, and D such that

$$\| Dy_n \|_1 \leqq (1 + \delta h)^{-n} \| Dy_0 \|_1 + h \sum_{k=1}^{n} (1 + \delta h)^{-k} \| DB_{(n-k)} \|_1$$

for all $n \geqq 1$. Therefore, it follows that *for all* $h > 0$, the sequence $y_1, y_2, \cdots$ is bounded whenever the sequence $B_1, B_2, \cdots$ is bounded, and $y_1, y_2, \cdots$ approaches $\theta_{(2p+q)}$ the zero vector of $E^{(2p+q)}$ whenever $B_1, B_2, \cdots$ approaches $\theta_{(2p+q)}$.

Typically at each step an iterative algorithm is employed to compute the solution y_{n+1} of equation (11). Since it is ordinarily not possible to compute y_{n+1} with infinite precision, it is important to consider the effects of the errors which are introduced. While, ideally, we would like to determine the sequence $\{y_{n+1}\}$ defined by equation (11) and some initial-condition vector y_0, suppose that we determine instead a sequence $\{\hat{y}_{n+1}\}$ such that, with ϵ an arbitrary positive constant, $\| D(\hat{y}_n - y_n^*) \|_1 \leqq \epsilon$ for all $n \geqq 1$ and

$$y_{n+1}^* + h\{TF[C^{-1}(y_{n+1}^*)] + GC^{-1}(y_{n+1}^*)\} = \hat{y}_n + hB_n \tag{12}$$

for all $n \geqq 0$. That is, suppose that at each step the *local* error $\| D(\hat{y}_n - y_n^*) \|_1$ in solving for "y_{n+1}" is at most ϵ. Then, according to Theorem 8, and with δ the positive constant referred to above,

$$\| D(y_n - \hat{y}_n) \|_1 \leqq (1 + \delta h)^{-n} \| D(y_0 - \hat{y}_0) \|_1 + \epsilon \sum_{k=0}^{n} (1 + \delta h)^{-k} \quad \text{for all} \quad n \geqq 1$$

in which $\hat{y}_0$ is the approximation to y_0. Therefore, given an arbitrarily small positive constant ρ, *for any* $h > 0$ it is possible to choose $\hat{y}_0$ and $\epsilon > 0$ such that the accumulated-error vector $(y_n - \hat{y}_n)$ satisfies $\| y_n - \hat{y}_n \|_1 \leqq \rho$ *for all* $n \geqq 1$.

Finally, Theorem 9 of Section III provides us with a conceptually interesting uniform bound on the norm of the difference between corresponding elements of the sequences $\{y_n\}$ and $\{u_n\}$ in which $u_n = u(nh)$ for all $n \geqq 0$ and $u(\cdot)$ satisfies the differential equation (1). According to Theorem 9, there exist positive constants δ and ρ, both independent of h, such that

$$\| D(u_n - y_n) \|_1 \leqq (1 + \delta h)^{-n} \| D(u_0 - y_0) \|_1 + \rho h$$

for all $n \geqq 1$, assuming that the elements of $B(\cdot)$ and $(d/dt)B(\cdot)$ are bounded and continuous on $[0, \infty)$. In particular, if $y_0 = u_0$ we see that there exists a positive constant ρ', independent of h, such that $\| u_n - y_n \|_1 \leqq \rho' h$ for all $n \geqq 1$, provided only that the assumptions of this section are satisfied and that $B(\cdot)$ and $(d/dt)B(\cdot)$ are bounded and continuous on $[0, \infty)$.

2.7 *Conditions Which Imply That* $T^{-1}\hat{G}(I + R\hat{G})^{-1} \varepsilon P_0$

In this section and in Section 2.8 we present some results concerning properties of the dc equation $TF(v) + Gv = B$. These results are directly relevant to the problem of computing the transient response of transistor-diode networks to the extent that in order to numerically integrate the differential equation (1) it is ordinarily necessary to first solve a dc problem to determine the initial conditions.

As indicated in Section 2.1, $G = \hat{G}(I + R\hat{G})^{-1}$ in which R takes into account the bulk resistances associated with the semiconductor devices. Here we present some material concerning conditions which imply that $T^{-1}\hat{G}(I + R\hat{G})^{-1}$ belongs to P_0.

Let $p > 0$. Theorem 10 of Section III asserts that $T^{-1}\hat{G}(I + R\hat{G})^{-1} \varepsilon P_0$ whenever $T^{-1}\hat{G} \varepsilon P_0$ and R satisfies

$$\alpha_r^{(k)}(1 - \alpha_r^{(k)})^{-1} r_e^{(k)} = r_b^{(k)}$$
$$\alpha_f^{(k)}(1 - \alpha_f^{(k)})^{-1} r_c^{(k)} = r_b^{(k)}$$

for $k = 1, 2, \cdots, p$. This rather special result shows that if $F(\cdot)$ satisfies the additional condition that $F(\cdot)$ belongs to the set $\mathfrak{F}_0^{(2p+q)}$ defined in Section 3.1, and if the network associated with T and $\hat{G}$ possesses the property that there is at most one solution v of the dc equation $TF(v) + \hat{G}v = B$ for each $B \varepsilon E^{(2p+q)}$, then it is always possible to add certain resistors of positive value in series with each transistor lead such that the dc equation of the resulting network possesses at most one solution.

Theorem 11 of Section III directs attention to the fact that there is a nontrivial class of transistor networks for which $T^{-1}\hat{G}(I + R\hat{G})^{-1} \varepsilon P_0$ for all R. According to Theorem 11, if $p > 0$ and $\hat{G}$ is such that $T^{-1}\hat{G} \varepsilon P_0$ for all "α's" (i.e., for all $\alpha_r^{(k)}$ and $\alpha_f^{(k)}$ belonging to (0, 1)), then for any particular set of "α's" $T^{-1}\hat{G}(I + R\hat{G})^{-1} \varepsilon P_0$ for all R.†

Given T, an interesting characterization of the class of short-circuit-conductance matrices $\hat{G}$ such that $M^{-1}\hat{G} \varepsilon P_0$ for all $M \varepsilon \mathfrak{I}(T)$ is provided by Theorem 12 of Section III.‡ According to Theorem 12, $M^{-1}\hat{G} \varepsilon P_0$ for all $M \varepsilon \mathfrak{I}(T)$ if and only if $T^{-1}\hat{G}(I + R\hat{G})^{-1} \varepsilon P_0$ for all R satisfying certain inequality-type conditions. In particular, if the base-lead

† A similar result is proved in Ref. 2 under the assumption that $\hat{G}$ is not singular.
‡ The set $\mathfrak{I}(T)$ is described in Section 2.5.

resistance of each transistor is taken to be zero, then $M^{-1}\hat{G} \in P_0$ for all $M \in \mathfrak{I}(T)$ implies that $T^{-1}\hat{G}(I + R\hat{G})^{-1} \in P_0$ *for all* nonnegative values of each emitter-lead resistor and each collector-lead resistor.

2.8 *Ebers-Moll Models and the Existence of a Solution of $TF(v) + Gv = B$*

In Section III, a set $\mathfrak{F}_3$ of mappings $F(\cdot)$ is defined such that each element of $\mathfrak{F}_3$ possesses certain important properties possessed by an arbitrary $F(\cdot)$ of the type that arises when an Ebers-Moll exponential-nonlinear-function model is used for each transistor and diode. In contrast with the set of all $F(\cdot)$ such that each $f_i(\cdot)$ is a strictly-monotone-increasing mapping of E^1 *onto* E^1, an arbitrary element $F(\cdot)$ of $\mathfrak{F}_3$ possesses the properties that for each j, $f_j(\cdot)$ is bounded on either $[0, \infty)$ or $(-\infty, 0]$, and the two nonlinear functions associated with the same transistor are *both* bounded on either $[0, \infty)$ or $(-\infty, 0]$. The set $\mathfrak{F}_3$ is contained in $\mathfrak{F}_0^{(2p+q)}$ and contains every Ebers-Moll exponential-nonlinear-function-type $F(\cdot)$.

The first part of Theorem 13 of Section III asserts that the equation $TF(v) + Gv = B$ possesses a unique solution v for each $F(\cdot) \in \mathfrak{F}_3$ and each $B \in E^{(2p+q)}$ if and only if $T^{-1}G \in P_0$ *and* $\det G \neq 0$. It is the "only if" part of this proposition which is the new result presented here. The proof exploits some special properties of transformerless resistor networks; it shows that if $T^{-1}G \in P_0$ but $\det G = 0$, then there are functions $t(\cdot)$ and $d(\cdot)$, both functions taking on only the values 1 or -1, such that there is no solution v of $TF(v) + Gv = B$ for some $B \in E^{(2p+q)}$ *for any* set of Ebers-Moll-modeled transistors and diodes with the property that for all k transistor k is a pnp device (as opposed to a npn device) if and only if $t(k) = 1$, and for all j diode j is a p-n junction if and only if $d(j) = 1$.†

The discussion of the preceding paragraph concerning the proof of Theorem 13 shows that it is not possible to make stronger assertions concerning the existence of a unique solution of $TF(v) + Gv = B$ for all $B \in E^{(2p+q)}$ for Ebers-Moll-modeled transistors and diodes unless we take into account more information about the nature of the semiconductor junctions. A good deal of progress in this direction has recently been made, and we state here without proof the following complete result dealing with diode-resistor networks.

Theorem 14:‡ *Let $p = 0$ and $q > 0$. Let $F(\cdot) \in \mathfrak{F}_3$ (see Definition* 12 *of*

† In contrast, the proof of the "only if" part of Theorem 3 of Ref. 1 shows that if $A \notin P_0$ then there is a mapping $F(\cdot)$ with each $f_j(\cdot)$ a *linear* function such that $F(x) + Ax = B$ does not possess a unique solution for all $B \in E^n$.

‡ The proof of Theorem 14 will be presented in a subsequent paper.

Section 3.31), *and for* $j = 1, 2, \cdots, q$ *let* s_j *equal either* 1 *or* -1 *depending on whether* $f_j(\cdot)$ *is bounded on* $[0, \infty)$ *or* $(-\infty, 0]$, *respectively. Then, with* A *any real symmetric nonnegative-definite matrix of order* q, *there exists a unique solution* v *of* $F(v) + Av = B$ *for all* $B \in E^q$ *if and only if there is no real* q*-vector* η *such that* $\eta \neq \theta_q$, $A\eta = \theta_q$, *and* $\eta \in S$, *in which*

$$S = \{y\colon y \in E^q \text{ and } y_j s_j \geqq 0 \text{ for } j = 1, 2, \cdots, q\}.^{\ddagger}$$

III. THEOREMS AND PROOFS

3.1 *Notation and Definitions*

Throughout Section III,

(*i*) unless stated otherwise, p and q denote nonnegative integers such that $(p + q) > 0$, and n denotes an arbitrary positive integer;

(*ii*) the set of all real n-vectors is denoted by E^n, θ is the zero element of E^n, and if $v \in E^n$ and j is an integer such that $1 \leqq j \leqq n$, then v_j denotes the jth component of v;

(*iii*) $\| v \| = (\sum_{j=1}^n v_j^2)^{1/2}$ and $\| v \|_1 = \sum_{j=1}^n | v_j |$ for all $v \in E^n$; for any real $n \times n$ matrix M, $\| M \|$ denotes sup $\{m\colon \| Mx \| \leqq m \| x \|, x \in E^n\}$;

(*iv*) the transpose of an arbitrary (not necessarily square) matrix M is denoted by M^{tr};

(*v*) I_n denotes the identity matrix of order n, and I denotes the identity matrix of order determined by the context in which the symbol is used; if $Q_1, Q_2, \cdots, Q_n$ are square matrices, then $Q_1 \oplus Q_2 \oplus \cdots \oplus Q_n$ denotes the direct sum of $Q_1, Q_2, \cdots, Q_n$, in the order indicated;

(*vi*) if D is a real diagonal matrix, then $D > 0 (D \geqq 0)$ means that the diagonal elements of D are positive (nonnegative); and

(*vii*) we say that a real $n \times n$ matrix M is strongly (weakly) column-sum dominant if and only if for $j = 1, 2, \cdots, n$

$$m_{jj} > (\geqq) \sum_{i \neq j} | m_{ij} |.$$

Definition 1: The set of all real square matrices M such that every principal minor of M is nonnegative (positive) is denoted by $P_0(P)$.

Definition 2: Let $\mathfrak{F}_0^{(2p+q)}$ denote that collection of mappings of $E^{(2p+q)}$ into itself defined by: $F(\cdot) \in \mathfrak{F}_0^{(2p+q)}$ if and only if there exist for $j =$

‡ In the network case, $A = G$, and it is often possible to determine by inspection whether or not there exists an $\eta \neq \theta_q$ such that $G\eta = \theta_q$ and $\eta \in S$.

1, 2, $\cdots$, $(2p + q)$ continuous functions $f_j(\cdot)$ mapping E^1 into E^1 such that for each $x \in E^{(2p+q)}$, $F(x) = [f_1(x_1), f_2(x_2), \cdots , f_{(2p+q)}(x_{(2p+q)})]^{tr}$, and

$$(i) \quad \inf_{\alpha \in (-\infty, \infty)} [f_j(\alpha + \beta) - f_j(\alpha)] = 0,$$

$$(ii) \quad \sup_{\alpha \in (-\infty, \infty)} [f_j(\alpha + \beta) - f_j(\alpha)] = +\infty$$

for all $\beta > 0$ and all $j = 1, 2, \cdots , (2p + q)$.

Definition 3: Let $\mathfrak{J}$ denote the set of all real matrices M such that $M = M_1 \oplus M_2 \oplus \cdots \oplus M_p \oplus I_q$ with

$$M_k = \begin{bmatrix} 1 & -\alpha_r^{(k)} \\ -\alpha_f^{(k)} & 1 \end{bmatrix},$$

$0 \leqq \alpha_r^{(k)} < 1$, and $0 \leqq \alpha_f^{(k)} < 1$ for all $k = 1, 2, \cdots , p$. As suggested, if $q = 0$, then $M = M_1 \oplus M_2 \oplus \cdots \oplus M_p$, while if $p = 0$, then $M = I_q$.

Assumption 1: Throughout Section III, G denotes a real nonnegative-definite matrix of order $(2p + q)$.

A tool that we shall use often is:

Lemma 1: A real square matrix M is an element of P_0 if and only if det $(D + M) \neq 0$ *for all real diagonal matrices $D > 0$.*

Lemma 1 is proved in Ref. 2.

3.2 *Theorem 1: Let $F(\cdot) \in \mathfrak{F}_0^{(2p+q)}$ with each $f_j(\cdot)$ continuously differentiable on $(-\infty, \infty)$ and $f_j'(\alpha) > 0$ for all $\alpha \in (-\infty, \infty)$. Let $T \in \mathfrak{J}$, let $C(\cdot)$ [that is, $c + \tau F(\cdot)$], G, and J_u be as defined in Section 2.1, and let σ be a real nonnegative constant. Then*

$$\sigma y + TF[C^{-1}(y)] + GC^{-1}(y) = r \tag{13}$$

possesses at most one solution y for each $r \in E^{(2p+q)}$ if and only if

$$\det (\sigma I + J_u) \neq 0 \quad \text{for all} \quad u \in E^{(2p+q)}, \tag{14}$$

and if $\sigma > 0$ and condition (14) is satisfied then for each $r \in E^{(2p+q)}$ there exists a solution y of (13).

3.3 *Proof of Theorem 1*

We have

$$\det(\sigma I + J_u)$$
$$= \det(\sigma I + TF'[g(u)]\{c + \tau F'[g(u)]\}^{-1} + G\{c + \tau F'[g(u)]\}^{-1})$$
$$= \det\{c + \tau F'[g(u)]\}^{-1}\cdot\det\{\sigma c + \sigma\tau F'[g(u)] + TF'[g(u)] + G\},$$

in which $g(\cdot)$ is the mapping of $E^{(2p+q)}$ onto itself defined by $g(u) = C^{-1}(u)$ for all $u \in E^{(2p+q)}$, and $F'[g(u)] = \text{diag}\{f_i'[g_i(u_i)]\}$. Since $\det\{c + \tau F'[g(u)]\} > 0$ for all u, $\det(\sigma I + J_u) \neq 0$ for all u if and only if

$$\det\{(\sigma\tau + T)F'[g(u)] + (\sigma c + G)\} \neq 0 \quad \text{for all } u.$$

For each j $g_j(\cdot)$ maps E^1 onto E^1, and since $F(\cdot) \in \mathfrak{F}_0^{(2p+q)}$ with each $f_i(\cdot)$ continuously differentiable on $(-\infty, \infty)$ and $f_i'(\alpha) > 0$ for all $\alpha \in (-\infty, \infty)$, the image of E^1 under the mapping $f_j'[g_j(\cdot)]$ is $(0, \infty)$† for all j. Thus, by Lemma 1 (since $\det(\sigma\tau + T) \neq 0$)$(\sigma\tau + T)^{-1}(\sigma c + G) \in P_0$ if and only if

$$\det(\sigma I + J_u) \neq 0 \quad \text{for all} \quad u. \tag{15}$$

The equation

$$\sigma y + TF[C^{-1}(y)] + GC^{-1}(y) = r$$

possesses a solution y if and only if $x = C^{-1}(y)$ satisfies

$$\sigma C(x) + TF(x) + Gx = r,$$

that is, if and only if

$$(\sigma\tau + T)F(x) + (\sigma c + G)x = r. \tag{16}$$

But equation (16) possesses at most one solution for each $r \in E^{(2p+q)}$ if and only if $(\sigma\tau + T)^{-1}(\sigma c + G) \in P_0$ (see pp. 105–107 of Ref. 3) and hence if and only if condition (15) is met.

Suppose now that $\sigma > 0$. Since G is nonnegative definite, $\det(\sigma c + G) \neq 0$. If condition (15) is satisfied then $(\sigma\tau + T)^{-1}(\sigma c + G) \in P_0$ and hence for each $r \in E^{(2p+q)}$, equation (16) possesses a solution x (see p. 99 of Ref. 3).□

3.4 *Theorem 2: Let $T \in \mathfrak{J}$, and let $F(\cdot) \in \mathfrak{F}_0^{(2p+q)}$ with each $f_i(\cdot)$ continuously differentiable on $(-\infty, \infty)$ and $f_i'(\alpha) > 0$ for all $\alpha \in (-\infty, \infty)$. Then for each $\sigma \geqq 0$ there exists a positive constant ϵ such that $\det(\sigma I + J_u) \geqq \epsilon$ for all $u \in E^{(2p+q)}$ if and only if $(\sigma\tau + T)^{-1}(\sigma c + G) \in P$.*

---

† For any $\beta > 0$ and any $\alpha \in (-\infty, \infty)$, $f_j(\alpha + \beta) - f_j(\alpha) = \beta f_j'(\delta)$ for some $\delta \in [\alpha, \alpha + \beta]$.

3.5 *Proof of Theorem* 2

We have

$$\det(\sigma I + J_u)$$
$$= \det(\sigma I + TF'[g(u)]\{c + \tau F'[g(u)]\}^{-1} + G\{c + \tau F'[g(u)]\}^{-1})$$
$$= \det\{c + \tau F'[g(u)]\}^{-1}\cdot\det\{(\sigma\tau + T)F'[g(u)] + (\sigma c + G)\}$$
$$= \det(\sigma\tau + T)\frac{\det(F'[g(u)] + A)}{\prod_{j=1}^{(2p+q)}(c_j + \tau_j f_j'[g_j(u_j)])} \tag{17}$$

in which $A = (\sigma\tau + T)^{-1}(\sigma c + G)$.

For each sequence $e_1, e_2, \cdots, e_{(2p+q)}$ with each e_j either zero or unity and $e_1, e_2, \cdots, e_{(2p+q)}$ not the sequence $1, 1, \cdots, 1$: let $m_{e_1,e_2,\ldots,e_{(2p+q)}}$ denote the determinant obtained from A by deleting rows $\rho_1, \rho_2, \cdots, \rho_l$ and columns $\rho_1, \rho_2, \cdots, \rho_l$ in which $\{\rho_1, \rho_2, \cdots, \rho_l\} = \{j: e_j = 1\}$. Thus for each sequence $e_1, e_2, \cdots, e_{(2p+q)}$ other than the sequence $1, 1, \cdots, 1$ $m_{e_1,e_2,\cdots,e_{(2p+q)}}$ is a principal minor of A. Let $m_{1,1,\cdots,1} = 1$, and let $d_j = f_j'[g_j(u_j)]$ for all j. Then by a standard expression[15] for the determinant of the sum of two matrices

$$\det(F'[g(u)] + A) = \sum{}' d_1^{e_1}d_2^{e_2}\cdots d_{(2p+q)}^{e_{(2p+q)}}m_{e_1,e_2,\cdots,e_{(2p+q)}}$$

in which $\sum'$ denotes a summation over all $2^{(2p+q)}$ sequences $e_1, e_2, \cdots, e_{(2p+q)}$ and $d_j^0 = 1$ for all j. It is clear that

$$\prod_{j=1}^{(2p+q)}(c_j + \tau_j f'[g_j(u_j)]) = \sum{}' d_1^{e_1}d_2^{e_2}\cdots d_{(2p+q)}^{e_{(2p+q)}}c_{e_1,e_2,\cdots,e_{(2p+q)}}$$

in which each $c_{e_1,e_2,\cdots,e_{(2p+q)}}$ is a positive constant. Thus with $\eta = \det(\sigma\tau + T)$,

$$\eta^{-1}\det(\sigma I + J_u) = \frac{\sum' d_1^{e_1}d_2^{e_2}\cdots d_{(2p+q)}^{e_{(2p+q)}}m_{e_1,e_2,\cdots,e_{(2p+q)}}}{\sum' d_1^{e_1}d_2^{e_2}\cdots d_{(2p+q)}^{e_{(2p+q)}}c_{e_1,e_2,\cdots,e_{(2p+q)}}}. \tag{18}$$

Suppose that all principal minors of A are positive. Then there is a positive constant δ such that

$$m_{e_1,e_2,\cdots,e_{(2p+q)}} \geqq \delta c_{e_1,e_2,\cdots,e_{(2p+q)}}$$

for all $e_1, e_2, \cdots, e_{(2p+q)}$ and hence (since $d_j > 0$ for all j) $\det(\sigma I + J_u) \geqq \eta\delta$ for all $u \in E^{(2p+q)}$.

As in the proof of Theorem 1, the range of each $d_j = f_j'[g_j(u_j)]$ is $(0, \infty)$, and for any positive constants $p_1, p_2, \cdots, p_{(2p+q)}$ there exists a $u \in E^{(2p+q)}$ such that $d_j = p_j$ for all j. If $A \notin P$ then at least one principal minor of A is not positive. If $A \notin P_0$, then $\det(F'[g(u) + A]) = 0$ for some u. Therefore to complete the proof it is sufficient to show that if $A \in P_0$ but $A \notin P$ then there is no constant $\epsilon > 0$ such that $\det(\sigma I + J_u) \geq \epsilon$ for all u.

With $A \in P_0$ and $A \notin P$, for at least one sequence $e'_1, e'_2, \cdots, e'_{(2p+q)}$

$$m_{e'_1, e'_2, \cdots, e'_{(2p+q)}} = 0.$$

If $\det A = m_{0,0,\cdots,0} = 0$ we have

$$\inf_{u \in E^{(2p+q)}} \det(\sigma I + J_u) = 0$$

since $\det(\sigma I + J_u) \to 0$ as $d_j \to 0$ for all j. Suppose now that $\det A > 0$ and that $m_{e'_1, e'_2, \cdots, e'_{(2p+q)}} = 0$ for some sequence $e'_1, e'_2, \cdots, e'_{(2p+q)}$. Then with $d_j = d$ for all j for which $e'_j = 1$ and $d_j = d^{-1}$ for all j for which $e'_j = 0$, we have [see equation (18)] $\det(\sigma I + J_u) \to 0$ as $d \to \infty$. □

3.6 *Theorem 3: Let* $T \in \mathfrak{J}$, *let* $F(\cdot) \in \mathfrak{F}_0^{(2p+q)}$ *with each* $f_i(\cdot)$ *continuously differentiable on* $(-\infty, \infty)$ *and* $f'_i(\alpha) > 0$ *for all* $\alpha \in (-\infty, \infty)$, *and let* $\mathcal{G}$ *denote* $[0, \infty)$ *or an interval contained in* $[0, \infty)$. *Then for all but at most a finite number of points* σ *contained in* $\mathcal{G}$, *there is a real constant* $\epsilon_\sigma > 0$ *such that* $\det(\sigma I + J_u) \geqq \epsilon_\sigma$ *for all* $u \in E^{(2p+q)}$ *if and only if* $\det(\sigma I + J_u) \neq 0$ *for all* $\sigma \in \mathcal{G}$ *and all* $u \in E^{(2p+q)}$.

3.7 *Proof of Theorem* 3

As in the proof of Theorem 1, $(\sigma\tau + T)^{-1}(\sigma c + G) \in P_0$ for all $\sigma \in \mathcal{G}$ if and only if $\det(\sigma I + J_u) \neq 0$ for all $\sigma \in \mathcal{G}$ and all u. We shall also use the fact that since $\det(\sigma\tau + T) > 0$ for all $\sigma \geqq 0$, each principal minor of $(\sigma\tau + T)^{-1}(\sigma c + G)$ is a finite-valued rational function of σ for all $\sigma \geqq 0$.

(if) If $\det(\sigma I + J_u) \neq 0$ for all u and all $\sigma \in \mathcal{G}$, then $(\sigma\tau + T)^{-1}(\sigma c + G) \in P_0$ for all $\sigma \in \mathcal{G}$. It is clear that $(\sigma\tau + T)^{-1}(\sigma c + G) \in P$ for all sufficiently large $\sigma > 0$. Thus each principal minor of $(\sigma\tau + T)^{-1}(\sigma c + G)$ is nonnegative for all $\sigma \in \mathcal{G}$ and is positive for all sufficiently large $\sigma > 0$. They are therefore positive for all but at most a finite number of values of $\sigma \in \mathcal{G}$. Thus, by Theorem 2, if $\det(\sigma I + J_u) \neq 0$ for all $\sigma \in \mathcal{G}$ and all u there exist for all but at most a finite number of points $\sigma \in \mathcal{G}$ a positive constant ϵ_σ such that $\det(\sigma I + J_u) \geqq \epsilon_\sigma$ for all u.

(only if) If $\det(\sigma I + J_u) = 0$ for some $\sigma \in \mathcal{G}$ and some u, then, for that σ, $(\sigma\tau + T)^{-1}(\sigma c + G) \notin P_0$. That is, for that σ at least one principal minor of $(\sigma\tau + T)^{-1}(\sigma c + G)$ is negative. This means that $(\sigma\tau + T)^{-1}(\sigma c + G) \notin P_0$ for all σ contained in some interval $\mathcal{G}' \subset \mathcal{G}$, and by Theorem 2, for all $\sigma \in \mathcal{G}'$ there is no $\epsilon_\sigma > 0$ such that $\det(I + J_u) \geqq \epsilon_\sigma$ for all u. □

3.8 *Theorem 4: Let $R(\cdot)$ be a continuously differentiable mapping of E^n into E^n, and let $J(R)_q$ denote the Jacobian matrix of $R(\cdot)$ at an arbitrary point $q \in E^n$. If the elements of $J(R)_q$ are bounded on E^n, and if there exist real constants $a > 0$ and $b \geqq 0$ such that* $\det J(R)_q \geqq (a + b \, \| q \|)^{-1}$ *for all $q \in E^n$, then $R(\cdot)$ is a homeomorphism of E^n onto E^n.*

3.9 *Proof of Theorem 4*

If Ref. 16 Meyer proves† that $R(\cdot)$ is a homeomorphism of E^n onto E^n if $J(R)_q^{-1}$ exists for all $q \in E^n$ and there exist real constants $\alpha > 0$ and $\beta \geq 0$ such that $\| J(R)_q^{-1} \| \leq \alpha + \beta \| q \|$ for all $q \in E^n$.

With q an arbitrary element of E^n, let $\lambda_1, \lambda_2, \cdots, \lambda_n$ denote the eigenvalues of $J(R)_q^{\mathrm{tr}} J(R)_q$, and let $\lambda_1 = \min_j \{\lambda_j\}$. Then $\lambda_1 \lambda_2 \cdots \lambda_n = [\det J(R)_q]^2 \geqq (a + b \, \| q \|)^{-2}$, and since the elements of $J(R)_q$ are bounded on E^n, there is a constant $\lambda > 0$ such that $\lambda_j \leqq \lambda$ for all j and all $q \in E^n$. Thus

$$(\lambda_1)^{1/2} \geqq \lambda^{-(1/2)(n-1)} (a + b \| q \|)^{-1} \tag{19}$$

for all q. For any $x \in E^n$ and any $q \in E^n$, $x^{\mathrm{tr}} J(R)_q^{\mathrm{tr}} J(R)_q x \geqq \lambda_1 x^{\mathrm{tr}} x$; that is,

$$\| J(R)_q x \| \geqq (\lambda_1)^{1/2} \| x \| \geqq \lambda^{-(1/2)(n-1)} (a + b \| q \|)^{-1} \| x \|.$$

With $x = J(R)_q^{-1} y$ in which y is an arbitrary element of E^n, we have

$$\| J(R)_q^{-1} y \| \leqq \lambda^{(1/2)(n-1)} (a + b \| q \|) \, \| y \|,$$

which shows that our hypothesis concerning $\det J(R)_q$ ensures that Meyer's condition on $\| J(R)_q^{-1} \|$ is satisfied. □

3.10 *Some Further Definitions*

Definition 4: For each $T \in \mathfrak{I}$, let $\mathfrak{I}(T)$ denote the set of all matrices M such that $M = M_1 \oplus M_2 \oplus \cdots \oplus M_p \oplus I_q$ with

$$M_k = \begin{bmatrix} 1 & -\delta_r^{(k)} \\ -\delta_f^{(k)} & 1 \end{bmatrix}$$

and

$$0 < \delta_r^{(k)} \leqq \alpha_r^{(k)} \quad \text{if} \quad \alpha_r^{(k)} > 0 \quad \text{and} \quad \delta_r^{(k)} = 0 \quad \text{if} \quad \alpha_r^{(k)} = 0,$$

$$0 < \delta_f^{(k)} \leqq \alpha_f^{(k)} \quad \text{if} \quad \alpha_f^{(k)} > 0 \quad \text{and} \quad \delta_f^{(k)} = 0 \quad \text{if} \quad \alpha_f^{(k)} = 0,$$

for all $k = 1, 2, \cdots, p$. As suggested, if $q = 0$, then $M = M_1 \oplus M_2 \oplus \cdots \oplus M_p$, while if $p = 0$, then $M = I_q$.

† Meyer's result is a generalization of a well-known result of Hadamard.[17] Hadamard proved that $R(\cdot)$ is a homeomorphism if $J(R)_q^{-1}$ exists for all $q \in E^n$ and satisfies $\| J(R)_q^{-1} \| \leqq \alpha$ for all $q \in E^n$ for some positive constant α.[17]

Definition 5: For each $T \varepsilon \mathfrak{J}$, let $\mathfrak{J}_0(T)$ denote the set of all 2^{2p} matrices M such that $M = M_1 \oplus M_2 \oplus \cdots \oplus M_p \oplus I_q$ with

$$M_k = \begin{bmatrix} 1 & -\delta_r^{(k)} \\ -\delta_f^{(k)} & 1 \end{bmatrix}$$

and

$$\delta_r^{(k)} = \alpha_r^{(k)} \quad \text{or} \quad \delta_r^{(k)} = 0,$$

$$\delta_f^{(k)} = \alpha_f^{(k)} \quad \text{or} \quad \delta_f^{(k)} = 0,$$

for all $k = 1, 2, \cdots, p$. As suggested, if $q = 0$, then $M = M_1 \oplus M_2 \oplus \cdots \oplus M_p$, while if $p = 0$, then $M = I_q$.

Definition 6: Let $Q_{(2p+q)}$ denote the family of all $2^{(2p+q)} - 1$ sets $w = \{i_1, i_2, \cdots, i_r\}$, including the null set, such that $r < (2p + q)$ and $w \subset \{1, 2, \cdots, (2p + q)\}$.

Definition 7: For M an arbitrary square matrix of order $(2p + q)$, and for each $w \varepsilon Q_{(2p+q)}$, let M_w denote the principal submatrix obtained from M by deleting rows $i_1, i_2, \cdots, i_r$ and columns $i_1, i_2, \cdots, i_r$. (If w is the null set, then $M_w = M$.)

Definition 8: For each $j \varepsilon \{1, 2, \cdots, (2p + q)\}$, let U_j denote the $(2p + q)$-column-vector with unity in the jth position and zeros in all other positions.

Definition 9: For each $T \varepsilon \mathfrak{J}$ and each $w \varepsilon Q_{(2p+q)}$, let T^w denote the matrix obtained from T by replacing the jth column of T with U_j for all $j \varepsilon w$.

3.11 *Theorem 5: Let $T \varepsilon \mathfrak{J}$. Then the following statements are equivalent.*

(*i*) $M^{-1}G \varepsilon P_0$ *for all* $M \varepsilon \mathfrak{J}(T)$.
(*ii*) $(D_a + T)^{-1}(D_b + G) \varepsilon P_0$ *for all diagonal* $D_a \geqq 0$ *and all diagonal* $D_b \geqq 0$.
(*iii*) $T^{-1}(G + D) \varepsilon P_0$ *for all diagonal* $D \geqq 0$.
(*iv*) $(D_a + T)^{-1}(D_b + G) \varepsilon P_0$ *for all diagonal* $D_a > 0$ *and all diagonal* $D_b > 0$.
(*v*) $T^{-1}(G + D) \varepsilon P_0$ *for all diagonal* $D > 0$.
(*vi*) $(T_w)^{-1}G_w \varepsilon P_0$ *for all* $w \varepsilon Q_{(2p+q)}$.
(*vii*) $[(T^w)^{-1}G]_w \varepsilon P_0$ *for all* $w \varepsilon Q_{(2p+q)}$.
(*viii*) $M^{-1}G \varepsilon P_0$ *for all* $M \varepsilon \mathfrak{J}_0(T)$.

3.12 *Proof of Theorem 5*

[(*i*) and (*ii*) are equivalent]

By Lemma 1, $(D_a + T)^{-1}(D_b + G) \varepsilon P_0$ if and only if $\det [(D_a + T)^{-1} (D_b + G) + D] \neq 0$ for all diagonal $D > 0$. Thus $(D_a + T)^{-1}(D_b + G) \varepsilon P_0$ for all $D_a \geqq 0$ and all $D_b \geqq 0$ if and only if

$$\det [(D_b D^{-1} + D_a + T)D + G] \neq 0$$

for all $D_a \geqq 0$, all $D_b \geqq 0$, and all $D > 0$, and hence if and only if

$$\det [(\Lambda + T)D + G] \neq 0$$

for all diagonal $\Lambda \geqq 0$ and $D > 0$. Let $T_\Lambda = (\Lambda + T)(I + \Lambda)^{-1}$. Then $(D_a + T)^{-1}(D_b + G) \varepsilon P_0$ for all $D_a \geqq 0$ and all $D_b \geqq 0$ if and only if

$$\det [T_\Lambda(I + \Lambda)D + G] \neq 0$$

for all $\Lambda \geqq 0$ and all $D > 0$, and hence if and only if $\det (T_\Lambda \hat{D} + G) \neq 0$ for all diagonal $\hat{D} > 0$ and all $\Lambda \geqq 0$. By Lemma 1, this means that $T_\Lambda^{-1} G \varepsilon P_0$ for all $\Lambda \geqq 0$ if and only if $(D_a + T)^{-1}(D_b + G) \varepsilon P_0$ for all $D_a \geqq 0$ and all $D_b \geqq 0$. We observe that $T_\Lambda = (T_\Lambda)_1 \oplus (T_\Lambda)_2 \oplus \cdots \oplus (T_\Lambda)_p \oplus I_q$ in which, with $\Lambda = \text{diag}\ (\lambda_1, \lambda_2, \cdots, \lambda_{(2p+q)})$,

$$(T_\Lambda)_k = \begin{bmatrix} 1 & \dfrac{-\alpha_r^{(k)}}{1 + \lambda_{2k}} \\ \dfrac{-\alpha_f^{(k)}}{1 + \lambda_{2k-1}} & 1 \end{bmatrix}$$

for $k = 1, 2, \cdots, p$. Thus for each $\Lambda \geqq 0$, $T_\Lambda \varepsilon \mathfrak{I}(T)$; and if M is an arbitrary element of $\mathfrak{I}(T)$, there is a $\Lambda \geqq 0$ such that $M = T_\Lambda$. Therefore $(D_a + T)^{-1}(D_b + G) \varepsilon P_0$ for all $D_a \geqq 0$ and all $D_b \geqq 0$ if and only if $M^{-1}G \varepsilon P_0$ for all $M \varepsilon \mathfrak{I}(T)$.

[(*i*) and (*iii*) are equivalent]

Repeat the proof of "(*i*) is equivalent to (*ii*)" with each statement that $D_a \geqq 0$ replaced with $D_a = \text{diag}\ (0, 0, \cdots, 0)$.

[(*ii*) and (*iv*) are equivalent and (*iii*) and (*v*) are equivalent]

Suppose that (*ii*) and (*iv*) are not equivalent. Then $(D_a + T)^{-1}(D_b + G) \varepsilon P_0$ for all $D_a > 0$ and all $D_b > 0$, and for some $D_a^* \geqq 0$ and some $D_b^* \geqq 0$, with $D_a^* \not> 0$ or $D_b^* \not> 0$ or $D_a^* \not> 0$ and $D_b^* \not> 0$, $(D_a^* + T)^{-1}(D_b^* + G) \not\varepsilon P_0$. Thus some principal minor of $(D_a^* + T)^{-1}(D_b^* + G)$, and hence of $(D_a^* + T)^{-1}(D_b^* + G) \det (D_a^* + T)$, is negative. Let

$m(D_a^*, D_b^*)$ be some negative principal minor of $(D_a^* + T)^{-1}(D_b^* + G) \det (D_a^* + T)$, and let $m(D_a^* + \epsilon I, D_b^* + \epsilon I)$ be the corresponding principal minor of $(D_a^* + \epsilon I + T)^{-1}(D_b^* + \epsilon I + G) \det (D_a^* + \epsilon I + T)$ for all real $\epsilon \geqq 0$. Thus $m(D_a^* + \epsilon I, D_b^* + \epsilon I)$ is a polynomial $p(\epsilon)$ in ϵ for $\epsilon \geqq 0$, and $p(\epsilon) \geqq 0$ for all $\epsilon > 0$. Therefore $p(0) \geqq 0$, which contradicts $m(D_a^*, D_b^*) < 0$.

A proof that (*iii*) and (*v*) are equivalent can be obtained by modifying the previous paragraph in an obvious manner.

[(*vi*) is equivalent to (*v*)]

By Lemma 1, $T^{-1}(G + D) \varepsilon P_0$ for all diagonal $D > 0$ if and only if $\det [T^{-1}(G + D) + D^*] \neq 0$ for all diagonal $D^* > 0$ and $D > 0$, and hence if and only if $\det (G + TD^* + D) \neq 0$ for all $D^* > 0$ and all $D > 0$. Therefore, by Lemma 1, $T^{-1}(G + D) \varepsilon P_0$ for all $D > 0$ if and only if $(G + TD^*) \varepsilon P_0$ for all $D^* > 0$, that is, if and only if $\det [G_w + (TD^*)_w] \geqq 0$ for all $w \varepsilon Q_{(2p+q)}$ and all $D^* > 0$. Since $(TD^*)_w = T_w D_w^*$, we see that $T^{-1}(G + D) \varepsilon P_0$ for all $D > 0$ if and only if

$$\det [(T_w)^{-1}G_w + D_w^*] \geqq 0 \quad \text{for all} \quad w \varepsilon Q_{(2p+q)} \quad \text{and all} \quad D^* > 0. \tag{20}$$

But, by Lemma 2 (which follows) condition (20) is equivalent to the condition that $\det [(T_w)^{-1}G_w + D_w^*] > 0$, and hence that $\det [(T_w)^{-1}G_w + D_w^*] \neq 0$, for all $w \varepsilon Q_{(2p+q)}$ and all $D^* > 0$. Thus by Lemma 1, $T^{-1}(G + D) \varepsilon P_0$ for all $D > 0$ if and only if $(T_w)^{-1}G_w \varepsilon P_0$ for all $w \varepsilon Q_{(2p+q)}$.

Lemma 2: If A is a real square matrix of order n such that $\det (D + A) = 0$ *for some diagonal* $D > 0$, *then* $\det (D + A) < 0$ *for some diagonal* $D > 0$.

Proof: Using the notation of the proof of Theorem 2,

$$\det (D + A) = \sum{}' d_1^{e_1} d_2^{e_2} \cdots d_n^{e_n} m_{e_1, e_2, \cdots, e_n} \tag{21}$$

for all $D > 0$. Since $m_{1,1,\cdots,1} = 1$, if $\det (D + A) = 0$ for some $D > 0$, then for at least one sequence $e_1', e_2', \cdots, e_n'$ we have $m_{e'_1, e'_2, \cdots, e'_n} < 0$. If $m_{0,0,\cdots,0} = \det A < 0$, then there exists a positive constant σ_1 such that $\det (D + A) < 0$ whenever $0 < d_j < \sigma_1$ for all j. If $\det A \geqq 0$, then, with $d_j = d$ for all j such that $e_j' = 1$ and $d_j = d^{-1}$ for all j such that $e_j' = 0$, there exists a positive constant σ_2 such that $\det (D + A) < 0$ for all $d > \sigma_2$ [see (21)]. □

[(*vi*) and (*vii*) are equivalent]

We shall prove that

$$[(T^w)^{-1}G]_w = (T_w)^{-1}G_w \quad \text{for all} \quad w \varepsilon Q_{(2p+q)}. \tag{22}$$

Obviously the equality of (22) is satisfied if w is the null set.

It is convenient to introduce the following notation. Let u denote the 1×1 matrix containing the entry 1. Let φ denote what might be called the empty matrix, a matrix with no rows or columns; by this we mean that φ is to be interpreted in the following manner: $\varphi \oplus \varphi = \varphi$, $I_s = \varphi$ when $s = 0$, $\varphi^{-1} = \varphi$, and if M_1 and M_2 are any two (ordinary) matrices, then $\varphi \oplus M_1 = M_1$, $M_1 \oplus \varphi = M_1$, and $M_1 \oplus \varphi \oplus M_2 = M_1 \oplus M_2$.

Let $w \,\varepsilon\, Q_{(2p+q)}$ and let w not be the null set. The matrix T can be written as the direct sum $T_1 \oplus T_2 \oplus \cdots \oplus T_p \oplus I_q$. In terms of u and φ, $T_w = t_1 \oplus t_2 \oplus \cdots \oplus t_p \oplus I_s$, in which $s = q - \bar{q}$ where $\bar{q}$ is the number of elements contained in the intersection of the sets w and $\{2p + 1, 2p + 2, \cdots, 2p + q\}$, and for $k = 1, 2, \cdots, p$: $t_k = T_k$ if both $(2k - 1)$ and $2k$ are not elements of w, $t_k = \varphi$ if both $(2k - 1)$ and $2k$ are elements of w, and $t_k = u$ if either $(2k - 1) \,\varepsilon\, w$ and $2k \notin w$ or $(2k - 1) \notin w$ and $2k \,\varepsilon\, w$. Thus $(T_w)^{-1} = t_1^{-1} \oplus t_2^{-1} \oplus \cdots \oplus t_p^{-1} \oplus I_s$. But $(T^w)^{-1} = \hat{T}_1^{-1} \oplus \hat{T}_2^{-1} \oplus \cdots \oplus \hat{T}_p^{-1} \oplus I_q$, in which for $k = 1, 2, \cdots, p$: $\hat{T}_k = T_k$ if both $(2k - 1)$ and $2k$ are not elements of w,

$$\hat{T}_k^{-1} = \begin{bmatrix} 1 & 0 \\ 0 & 1 \end{bmatrix}$$

if both $(2k - 1)$ and $2k$ are contained in w,

$$\hat{T}_k^{-1} = \begin{bmatrix} 1 & \alpha_r^{(k)} \\ 0 & 1 \end{bmatrix}$$

if $(2k - 1) \,\varepsilon\, w$ and $2k \notin w$, and

$$\hat{T}_k^{-1} = \begin{bmatrix} 1 & 0 \\ \alpha_f^{(k)} & 1 \end{bmatrix}$$

if $(2k - 1) \notin w$ and $2k \,\varepsilon\, w$. Thus we see that $[(T^w)^{-1}]_w = (T_w)^{-1}$. Let ${}_{(w)}(T^w)^{-1}$ denote the $(2p + q - r) \times (2p + q)$ matrix obtained from $(T^w)^{-1}$ by deleting rows $i_1, i_2, \cdots, i_r$. But all elements of columns $i_1, i_2, \cdots, i_r$ of ${}_{(w)}(T^w)^{-1}$ are zeros, and hence, with $G_{(w)}$ the matrix obtained from G by deleting columns $i_1, i_2, \cdots, i_r$,

$$\begin{aligned} [(T^w)^{-1}G]_w &= {}_{(w)}(T^w)^{-1}G_{(w)} \\ &= [(T^w)^{-1}]_w G_w = (T_w)^{-1}G_w. \end{aligned}$$

[(*viii*) and (*i*) are equivalent]

If $M^{-1}G \,\varepsilon\, P_0$ for all $M \,\varepsilon\, \mathfrak{I}_0(T)$, then $[(T^w)^{-1}G]_w \,\varepsilon\, P_0$ for all $w \,\varepsilon\, Q_{(2p+q)}$. Thus, statement (*viii*) implies statement (*vii*). Since we have proved that (*vii*) is equivalent to (*i*), it suffices to prove that (*i*) implies (*viii*).

Suppose that $M^{-1}G \varepsilon P_0$ for all $M \varepsilon \mathfrak{I}(T)$. Let $\hat{M}$ be an arbitrary element of $\mathfrak{I}_0(T)$. Then $[\hat{M} + \delta(T - \hat{M})] \varepsilon \mathfrak{I}(T)$ for all $\delta \varepsilon (0, 1]$, and therefore $[\hat{M} + \delta(T - \hat{M})]^{-1}G \varepsilon P_0$ for all $\delta \varepsilon (0, 1]$. At this point a continuity-type argument similar to that used in the proof of [(*ii*) and (*iv*) are equivalent] shows that $\hat{M}^{-1}G \varepsilon P_0$. □

3.13 *Corollary 1 (Corollary to Theorem 5):*

If $T \varepsilon \mathfrak{I}$ *and* $M^{-1}G \varepsilon P_0$ *for all* $M \varepsilon \mathfrak{I}(T)$, *then* $\det(\sigma I + J_u) \neq 0$ *for all* $\sigma \geqq 0$ *and all* $u \varepsilon E^{(2p+q)}$ *provided that for all* j $f_j(\cdot)$ *is continuously differentiable on* $(-\infty, \infty)$ *and* $f_j'(\alpha) > 0$ *for all* $\alpha \varepsilon (-\infty, \infty)$.

3.14 *Proof of Corollary 1.*

If $T \varepsilon \mathfrak{I}$ and $M^{-1}G \varepsilon P_0$ for all $M \varepsilon \mathfrak{I}(T)$, then, by the equivalence of (*i*) and (*ii*) of Theorem 5, $(\sigma\tau + T)^{-1}(\sigma c + G) \varepsilon P_0$ for all $\sigma \geqq 0$. The first portion of the proof of Theorem 1 shows that if $(\sigma\tau + T)^{-1}(\sigma c + G) \varepsilon P_0$ for all $\sigma \geqq 0$ and if for all j $f_j(\cdot)$ is continuously differentiable on $(-\infty, \infty)$ and $f_j'(\alpha) > 0$ for all $\alpha \varepsilon (-\infty, \infty)$, then $\det(\sigma I + J_u) \neq 0$ for all $\sigma \geqq 0$ and all $u \varepsilon E^{(2p+q)}$.

3.15 *Definition 10:* For $p > 0$ let $Q'_{(2p+q)}$ denote the subset of $Q_{(2p+q)}$ containing all sets w belonging to $Q_{(2p+q)}$ such that w is not the null set and $2k \varepsilon w$ if and only if $(2k - 1) \varepsilon w$ for $k = 1, 2, \cdots, p$. For $p = 0$, let $Q'_{(2p+q)}$ denote the family of all sets contained in $Q_{(2p+q)}$ with the exception of the null set.

3.16 *Theorem 6: If* $T \varepsilon \mathfrak{I}$ *and* $T^{-1}G \varepsilon P_0$, *then* $(T_w)^{-1}G_w \varepsilon P_0$ *for all* $w \varepsilon Q'_{(2p+q)}$.

3.17 *Proof of Theorem 6*

Let $T \varepsilon \mathfrak{I}$, and let $T^{-1}G \varepsilon P_0$. By Lemma 1, $\det(TD + G) \neq 0$ (and hence $\det(TD + G) > 0$) for all diagonal $D > 0$. Let $w = \{i_1, i_2, \cdots, i_r\} \varepsilon Q'_{(2p+q)}$, and let $d_{i_k} = d$ for $k = 1, 2, \cdots, r$.

It may be the case that $(TD + G)$ is a block matrix of the form

$$\begin{bmatrix} (TD + G)_w & H_{12} \\ H_{21} & (d\hat{T} + H_{22}) \end{bmatrix} \tag{23}$$

in which $\hat{T}$ is a direct sum of all 2×2 and 1×1 block matrices on the diagonal of T which do not appear in T_w, and H_{12}, H_{21}, and H_{22} are independent of D. Clearly $\det \hat{T} > 0$. If $(TD + G)$ is not of the form (23), then by a sequence of interchanges of rows and corresponding columns of $(TD + G)$ we obtain a matrix of that form.

Thus, for some $\hat{T}$ of the form indicated above and for the corresponding constant matrices H_{12}, H_{21}, and H_{22} whose elements are elements of G,

$$\det (TD + G) = \det \begin{bmatrix} (TD+G)_w & H_{12} \\ H_{12} & (d\hat{T} + H_{22}) \end{bmatrix}$$

for all $d_j > 0$ for $j \not\varepsilon w$. For all sufficiently large $d > 0$, $\det (d\hat{T} + H_{22}) > 0$, and then

$$0 < \det (TD + G) = \det (d\hat{T} + H_{22}) \cdot \det [(TD + G)_w - H_{12}(d\hat{T} + H_{22})^{-1} H_{21}]$$

for all $d_j > 0$ for $j \not\varepsilon w$. Since $H_{12}(d\hat{T} + H_{22})^{-1}H_{21}$ approaches the zero matrix of order $(2p + q - r)$ as $d \rightarrow \infty$, we must have $\det (TD + G)_w \geqq 0$ for all $d_j > 0$ for $j \not\varepsilon w$. Therefore, since $(TD)_w = T_w D_w$, we must have $\det (T_w D_w + G_w) \geqq 0$ for all $D_w > 0$. But this means (see Lemma 2) that $\det (T_w D_w + G_w) \neq 0$ for all $D_w > 0$. Thus, by Lemma 1, $(T_w)^{-1} G_w \varepsilon P_0$. □

3.18 *Theorem 7: If* $T \varepsilon \mathfrak{I}$ *with* $p = 1$ *or* $p = 2$, *and if* $T^{-1}G \varepsilon P_0$ *with* G *the short-circuit conductance matrix of a transformerless positive-element resistance network, then* $(T_w)^{-1}G_w \varepsilon P_0$ *for all* $w \varepsilon Q_{(2p+q)}$.

3.19 *Proof of Theorem 7*

Suppose that $T^{-1}G \varepsilon P_0$ with $p = 2$. Theorem 6 asserts that $(T_w)^{-1}G_w \varepsilon P_0$ for all $w \varepsilon Q'_{(2p+q)}$. But, aside from the null set, the sets $w = \{i_1, i_2, \cdots, i_r\}$ that are contained in $Q_{(2p+q)}$ but not in $Q'_{(2p+q)}$ possess the property that $T_w = T_1 \oplus I_{(2+q-r)}$, or $T_w = u \oplus T_2 \oplus I_{(1+q-r)}$ where u is the 1×1 matrix containing the element 1, or $T_w = I_{(4+q-r)}$.

If $T_w = I_{(4+q-r)}$, then obviously $(T_w)^{-1}G_w \varepsilon P_0$. If $T_w = T_1 \oplus I_{(2+q-r)}$, then for any $D_w = \operatorname{diag} [D_2 \oplus D_{(2+q-r)}]$ with $D_2 > 0$ and $D_{(2+q-r)} > 0$ diagonal matrices of order 2 and $(2 + q - r)$ respectively,

$$\det (T_w D_w + G_w) = \begin{bmatrix} T_1 D_2 + G_{11} & G_{12} \\ G_{21} & D_{(2+q-r)} + G_{22} \end{bmatrix} \tag{24}$$

in which G_{11}, G_{12}, G_{21}, and G_{22} are the appropriate block matrices of G_w. Since $\det [D_{(2+q-r)} + G_{22}] > 0$, we have

$$\det (T_w D_w + G_w) = \det [D_{(2+q-r)} + G_{22}] \cdot \det \{T_1 D_2 + G_{11} - G_{12}[D_{(2+q-r)} + G_{22}]^{-1} G_{21}\}.$$

But $G_{11} - G_{12}[D_{(2+q-r)} + G_{22}]^{-1}G_{21}$ is the short-circuit conductance matrix of a transformerless common-ground 2-port network; it is of the form

$$\begin{bmatrix} g_{11} & -g_{12} \\ -g_{12} & g_{22} \end{bmatrix}$$

with $g_{11} \geqq 0$, $g_{22} \geqq 0$, $g_{12} \geqq 0$, $g_{11} \geqq g_{12}$, and $g_{22} \geqq g_{12}$. Therefore[1]

$$\det \{T_1D_2 + G_{11} - G_{12}[D_{(2+q-r)} + G_{22}]^{-1}G_{21}\} > 0$$

for all $D_2 > 0$ and all $D_{(2+q-r)} > 0$, $\det (T_wD_w + G_w) \neq 0$ for all $D_w > 0$, and hence, by Lemma 1, $(T_w)^{-1}G_w \in P_0$. Finally, the case in which $T_w = u \oplus T_2 \oplus I_{(1+q-r)}$ can be treated in a manner similar to that used to show that $(T_w)^{-1}G_w \in P_0$ when $T_w = T_1 \oplus I_{(2+q-r)}$, since, with w such that $T_w = u \oplus T_2 \oplus I_{(1+q-r)}$, and with D an arbitrary diagonal matrix of order $(4 + q - r)$, a sequence of interchanges of rows and corresponding columns of $(T_wD + G_w)$ can be performed to obtain a matrix of the type that appears on the right side of equation (24). Therefore $(T_w)^{-1}G_w \in P_0$ for all $w \in Q_{(2p+q)}$.

When $p = 1$, aside from the null set, the sets $w = \{i_1, i_2, \cdots, i_r\}$ that are contained in $Q_{(2p+q)}$ but not in $Q'_{(2p+q)}$ possess the property that $T_w = I_{(2+q-r)}$ and obviously when $T_w = I_{(2+q-r)}$, $(T_w)^{-1}G_w \in P_0$. □

3.20 *Theorem 8: Let $T \in \mathfrak{J}$ and let G possess the property that for some diagonal matrix $D > 0$, both DT and DG are strongly-column-sum dominant. For each $j = 1, 2, \cdots, (2p + q)$ let $f_j(\cdot)$ be a continuous monotone-nondecreasing mapping of E^1 into itself such that $f_j(0) = 0$, let $h \in (0, \infty)$, and, with $F(\cdot)$ and $C(\cdot)$ defined relative to the $f_j(\cdot)$ as in Section 2.1, suppose that the sequences $\{y_n\}$ and $\{w_n\}$ in $E^{(2p+q)}$ satisfy*

$$y_{n+1} + h\{TF[C^{-1}(y_{n+1})] + GC^{-1}(y_{n+1})\} = y_n + w_n$$

for all $n \geqq 0$. Then there exists a positive constant δ depending only on the c_j, the τ_j, T, G, and D such that

$$(i) \qquad \| Dy_n \|_1 \leqq (1 + \delta h)^{-n} \| Dy_0 \|_1 + \sum_{k=1}^{n} (1 + \delta h)^{-k} \| Dw_{(n-k)} \|_1$$

for all $n \geqq 1$, and

$$(ii) \qquad \| D(y_n - \tilde{y}_n) \|_1 \leqq (1 + \delta h)^{-n} \| D(y_0 - \tilde{y}_0) \|_1 + \epsilon \sum_{k=0}^{n} (1 + \delta h)^{-k}$$

for all $n \geqq 1$, in which $\{\tilde{y}_n\}$ is any sequence in $E^{(2p+q)}$ with the property that $\| D(\tilde{y}_n - y_n^*) \|_1 \leqq \epsilon$ for all $n \geqq 1$ with ϵ a positive constant and the *sequence* $\{y_n^*\}$ *such that*

$$y_{n+1}^* + h\{TF[C^{-1}(y_{n+1}^*)] + GC^{-1}(y_{n+1}^*)\} = \tilde{y}_n + w_n$$

for all $n \geqq 0$.

3.21 *Proof of Theorem 8*

We shall first prove part (*ii*). With D such that DT and DG are strongly-column-sum dominant, we have for all $n \geqq 0$

$$Dy_{n+1} + h\{DTF[C^{-1}(y_{n+1})] + DGC^{-1}(y_{n+1})\} = Dy_n + Dw_n$$

and

$$Dy_{n+1}^* + h\{DTF[C^{-1}(y_{n+1}^*)] + DGC^{-1}(y_{n+1}^*)\} = Dy_n^* + D(\tilde{y}_n - y_n^*) + Dw_n$$

in which we shall take y_0^* to be $\tilde{y}_0$. As in the proof of Theorem 2 of Ref. 3, we write

$$F[C^{-1}(y_{n+1})] - F[C^{-1}(y_{n+1}^*)] = \operatorname{diag}\left(\frac{r(n)_j}{c_j + \tau_j r(n)_j}\right)(y_{n+1} - y_{n+1}^*) \quad (25)$$

and

$$C^{-1}(y_{n+1}) - C^{-1}(y_{n+1}^*) = \operatorname{diag}\left(\frac{1}{c_j + \tau_j r(n)_j}\right)(y_{n+1} - y_{n+1}^*) \quad (26)$$

in which $r(n)_j$ depends on the jth components of y_{n+1} and y_{n+1}^*, and $r(n)_j \geqq 0$ for all $n \geqq 0$ and all j.

Thus, with $Q = DTD^{-1}$ and $R = DGD^{-1}$,

$$\left\{I + hQ \operatorname{diag}\left(\frac{r(n)_j}{c_j + \tau_j r(n)_j}\right) + hR \operatorname{diag}\left(\frac{1}{c_j + \tau_j r(n)_j}\right)\right\} D(y_{n+1} - y_{n+1}^*)$$
$$= D(y_n - y_n^*) - D(\tilde{y}_n - y_n^*)$$

for all $n \geqq 0$. At this point we shall use the proposition that if M is any real matrix of order $(2p + q)$ with the property that there exists a positive constant η such that $m_{jj} - \sum_{i \neq j} |m_{ij}| \geqq \eta$ for all j, then $\| Mx \|_1 \geqq \eta \| x \|_1$ for all $x \in E^{(2p+q)}$. Now let

$$M = \left\{I + hQ \operatorname{diag}\left(\frac{r(n)_j}{c_j + \tau_j r(n)_j}\right) + hR \operatorname{diag}\left(\frac{1}{c_j + \tau_j r(n)_j}\right)\right\}$$

for arbitrary $n \geqq 0$. Then for arbitrary j

$$m_{jj} - \sum_{i \neq j} |m_{ij}| = 1 + hq_{jj}\left(\frac{r(n)_j}{c_j + \tau_j r(n)_j}\right) + hr_{jj}\left(\frac{1}{c_j + \tau_j r(n)_j}\right)$$
$$- h \sum_{i \neq j} \left| q_{ij} \frac{r(n)_j}{c_j + \tau_j r(n)_j} + r_{ij} \frac{1}{c_j + \tau_j r(n)_j} \right|$$

$$\geqq 1 + h\left(q_{ii} - \sum_{i \neq j} |q_{ij}|\right) \frac{r(n)_j}{c_j + \tau_j r(n)_j}$$

$$+ h\left(r_{jj} - \sum_{i \neq j} |r_{ij}|\right) \frac{1}{c_j + \tau_j r(n)_j}$$

$$\geqq 1 + \delta h,$$

in which

$$\delta = \min \left\{ \min_i c_i^{-1}\left(r_{ii} - \sum_{i \neq j} | r_{ij} |\right), \quad \min_i \tau_i^{-1}\left(q_{ii} - \sum_{i \neq j} | q_{ij} |\right)\right\}.$$

Therefore

$$\| D(y_{n+1} - y_{n+1}^*) \|_1$$
$$\leqq (1 + \delta h)^{-1} \| D(y_n - y_n^*) - D(\tilde{y}_n - y_n^*) \|_1$$
$$\leqq (1 + \delta h)^{-1} \| D(y_n - y_n^*) \|_1 + (1 + \delta h)^{-1} \| D(\tilde{y}_n - y_n^*) \|_1$$
$$\leqq (1 + \delta h)^{-1} \| D(y_n - y_n^*) \|_1 + \epsilon(1 + \delta h)^{-1}$$

for all $n \geqq 0$, and hence

$$\| D(y_n - y_n^*) \|_1 \leqq (1 + \delta h)^{-n} \| D(y_0 - y_0^*) \|_1 + \epsilon \sum_{k=1}^{n} (1 + \delta h)^{-k}$$

for all $n \geqq 1$. Finally, since $\| D(y_n - \tilde{y}_n) \|_1 \leqq \| D(y_n - y_n^*) \|_1 + \| D(y_n^* - \tilde{y}_n) \|_1 \leqq \| D(y_n - y_n^*) \|_1 + \epsilon$, and since $y_0^* = \tilde{y}_0$,

$$\| D(y_n - \tilde{y}_n) \|_1 \leqq (1 + \delta h)^{-n} \| D(y_0 - \tilde{y}_0) \|_1 + \varepsilon \sum_{k=0}^{n} (1 + \delta h)^{-k}$$

for all $n \geqq 1$, which completes the proof of part (*ii*) of the theorem.

The proof of part (*i*) is similar to that of part (*ii*). Using

$$Dy_{n+1} + h\{DTF[C^{-1}(y_{n+1})] + DGC^{-1}(y_{n+1})\} = Dy_n + Dw_n$$

for all $n \geqq 0$, and equations (25) and (26) with $y_{n+1}^* = \theta$ for all n, we find that

$$\| Dy_{n+1} \|_1 \leqq (1 + h\,\delta)^{-1} \| Dy_n \|_1 + (1 + h\,\delta)^{-1} \| Dw_n \|_1$$

for all $n \geqq 0$. Therefore

$$\| Dy_n \|_1 \leqq (1 + h\delta)^{-n} \| Dy_0 \|_1 + \sum_{k=1}^{n} (1 + h\delta)^{-k} \| Dw_{(n-k)} \|_1$$

for all $n \geqq 1$. □

3.22 *Theorem 9:* *Let $T \in \mathfrak{I}$ and let G possess the property that for some diagonal matrix $D > 0$, both DT and DG are strongly-column-sum dominant. Let $B(\cdot)$ denote a real continuously-differentiable $(2p + q)$-vector-valued function of t for $t \in [0, \infty)$ such that both $B(\cdot)$ and $(d/dt)B(\cdot)$ are bounded on $[0, \infty)$. With $F(\cdot)$ such that each $f_i(0) = 0$, and with $C(\cdot)$ defined relative to $F(\cdot)$ as in Section 2.1, let $u(\cdot)$ satisfy*

$$\frac{du}{dt} + TF[C^{-1}(u)] + GC^{-1}(u) = B(t), \qquad t \geqq 0$$

and, with h an arbitrary positive constant, let u_n denote $u(nh)$ for all $n \geqq 0$. Let $\{y_n\}$ be a sequence in $E^{(2p+q)}$ such that

$$y_{n+1} + h\{TF[C^{-1}(y_{n+1})] + GC^{-1}(y_{n+1})\} = y_n + hB[(n + 1)h], \quad n \geqq 0.$$

Then there exist positive constants δ and ρ, both independent of h, such that

$$\| D(u_n - y_n) \|_1 \leqq (1 + \delta h)^{-n} \| D(u_0 - y_0) \|_1 + \rho h$$

for all $n \geqq 1$.

3.23 *Proof of Theorem 9*

The sequence $\{u_n\}$ satisfies

$$u_{n+1} + h\{TF[C^{-1}(u_{n+1})] + GC^{-1}(u_{n+1})\}$$
$$= u_n + B[(n + 1)h] + \xi_n, \qquad n \geqq 0$$

in which ξ_n is often referred to as "the local-truncation error at step n." We shall first bound ξ_n.

Since $B(\cdot)$ is bounded on $[0, \infty)$, and since for some $D > 0$, both DT and DG are strongly-column-sum dominant, a direct modification of the proof of Theorem 1 of Ref. 5 shows that $u(\cdot)$ is bounded on $[0, \infty)$; and hence since

$$\frac{d^2u}{dt^2} = J_u\{TF[C^{-1}(u)] + GC^{-1}(u)\} - J_uB(t) + \frac{d}{dt}B(t), \qquad t \geqq 0 \tag{27}$$

with $(d/dt)B(\cdot)$ and the elements of the Jacobian matrix J_u bounded, it is clear that (d^2u/dt^2) is bounded on $[0, \infty)$. By the usual Taylor-series-type argument we can show that for arbitrary $n \geqq 0$, $\xi_n = \frac{1}{2}h^2U_n$ in which for each j the jth component of U_n is the jth component of (d^2u/dt^2) evaluated at some point contained in the interval $[nh, (n + 1)h]$. Thus there exists a positive constant ρ_1 such that

$$\| D\xi_n \|_1 \leqq \tfrac{1}{2}h^2\rho_1 \quad \text{for all} \quad n \geqq 0. \tag{28}$$

Therefore, using (28) and the equations

$$u_{n+1} + h\{TF[C^{-1}(u_{n+1})] + GC^{-1}(u_{n+1})\} = u_n + B[(n+1)h] + \xi_n, \qquad n \geqq 0$$

$$y_{n+1} + h\{TF[C^{-1}(y_{n+1})] + GC^{-1}(y_{n+1})\} = y_n + B[(n+1)h], \qquad n \geqq 0$$

by an argument similar to that used in the proof of part (*ii*) of Theorem 8, and with δ as defined there, we find that

$$\| D(u_{n+1} - y_{n+1}) \|_1 \leqq (1 + \delta h)^{-1} \| D(u_n - y_n) \|_1 + (1 + \delta h)^{-1}\tfrac{1}{2}h^2\rho_1$$

for all $n \geqq 0$, and hence that

$$\| D(u_n - y_n) \|_1 \leqq (1 + \delta h)^{-n} \| D(u_0 - y_0) \|_1 + \tfrac{1}{2}h^2\rho_1 \sum_{k=1}^{n} (1 + \delta h)^{-k}$$

$$\leqq (1 + \delta h)^{-n} \| D(u_0 - y_0) \|_1 + \tfrac{1}{2}h^2\rho_1 \sum_{k=1}^{\infty} (1 + \delta h)^{-k}$$

$$\leqq (1 + \delta h)^{-n} \| D(u_0 - y_0) \|_1 + \tfrac{1}{2}h\delta^{-1}\rho_1$$

for all $n \geqq 1$. □

3.24 *Definition 11:* Let $R = R_1 \oplus R_2 \oplus \cdots \oplus R_p \oplus R_0$ in which $R_0 = \text{diag}\,(r_1, r_2, \cdots, r_q)$ with $r_j \geqq 0$ for $j = 1, 2, \cdots, q$ and

$$R_k = \begin{bmatrix} r_e^{(k)} + r_b^{(k)} & r_b^{(k)} \\ r_b^{(k)} & r_c^{(k)} + r_b^{(k)} \end{bmatrix}$$

with $r_e^{(k)} \geqq 0$, $r_b^{(k)} \geqq 0$, and $r_c^{(k)} \geqq 0$ for all $k = 1, 2, \cdots, p$. As suggested, if $q = 0$, then $R = R_1 \oplus R_2 \oplus \cdots \oplus R_p$, while if $p = 0$, then $R = R_0$.

3.25 *Theorem 10: Let $T \in \mathfrak{I}$. If $p > 0$ and if R satisfies*

$$\alpha_r^{(k)}(1 - \alpha_r^{(k)})^{-1}r_e^{(k)} = r_b^{(k)}$$

$$\alpha_f^{(k)}(1 - \alpha_f^{(k)})^{-1}r_c^{(k)} = r_b^{(k)}$$

for $k = 1, 2, \cdots, p$, then $T^{-1}G(I + RG)^{-1} \in P_0$ whenever $T^{-1}G \in P_0$.

3.26 *Proof of Theorem 10*

By Lemma 1, $T^{-1}G(I + RG)^{-1} \in P_0$ if and only if

$$\det\,[T^{-1}G(I + RG)^{-1} + D^*] \neq 0 \tag{29}$$

for all diagonal $D^* > 0$. But (29) is satisfied if and only if

$$\det\,(T^{-1}G + D^*RG + D^*) \neq 0.$$

Here, since

$$\alpha_r^{(k)}(1 - \alpha_r^{(k)})^{-1}r_e^{(k)} = r_b^{(k)}$$
$$\alpha_f^{(k)}(1 - \alpha_f^{(k)})^{-1}r_c^{(k)} = r_b^{(k)}$$

for $k = 1, 2, \cdots, p$ we have $R = DT^{-1}$ for some diagonal matrix $D \geqq 0$. Thus (29) is satisfied if and only if

$$\det [(I + DD^*)T^{-1}G + D^*] \neq 0.$$

When $T^{-1}G \in P_0$ we have

$$\det (T^{-1}G + \tilde{D}) \neq 0$$

for all diagonal $\tilde{D} > 0$. Thus (29) is satisfied for all $D^* > 0$ whenever $T^{-1}G \in P_0$. □

3.27 *Theorem 11: If $M^{-1}G \in P_0$ for all $M \in \mathfrak{I}$, then for any $T \in \mathfrak{I}$, $T^{-1}G(I + RG)^{-1} \in P_0$ for all R.*

3.28 *Proof of Theorem 11*

Let $T \in \mathfrak{I}$. As in the proof of Theorem 10, $T^{-1}G(I + RG)^{-1} \in P_0$ if and only if

$$\det [(T^{-1} + D^*R)G + D^*] \neq 0$$

for all diagonal $D^* > 0$. It is a simple matter to verify that for each $D^* > 0$ and each R there exists an $\tilde{M} \in \mathfrak{I}$ and a diagonal matrix $D > 0$ such that $(T^{-1} + D^*R) = D\tilde{M}^{-1}$. Since $M^{-1}G \in P_0$ for all $M \in \mathfrak{I}$, we have (by Lemma 1)

$$\det (D\tilde{M}^{-1}G + D^*) \neq 0$$

for all $D^* > 0$. □

3.29 *Theorem 12: Let $T \in \mathfrak{I}$ with $p > 0$ and $q \geqq 0$. Then $M^{-1}G \in P_0$ for all $M \in \mathfrak{I}(T)$ if and only if $T^{-1}G(I + RG)^{-1} \in P_0$ for all R such that*

$$\alpha_r^{(k)}(1 - \alpha_r^{(k)})^{-1}r_e^{(k)} \geqq r_b^{(k)}$$
$$\alpha_f^{(k)}(1 - \alpha_f^{(k)})^{-1}r_c^{(k)} \geqq r_b^{(k)}$$

for $k = 1, 2, \cdots, p$ and $r_j \geqq 0$ for all j such that $1 \leqq j \leqq q$.

3.30 *Proof of Theorem 12*

As in the proof of Theorem 10, $T^{-1}G(I + RG)^{-1} \in P_0$ if and only if

$$\det (T^{-1}G + D^*RG + D^*) \neq 0 \qquad (30)$$

for all diagonal $D^* > 0$. The inequalities $r_j \geqq 0$ for all j such that $1 \leqq j \leqq q$ and

$$\alpha_r^{(k)}(1 - \alpha_r^{(k)})^{-1} r_e^{(k)} \geqq r_b^{(k)}$$

$$\alpha_f^{(k)}(1 - \alpha_f^{(k)})^{-1} r_c^{(k)} \geqq r_b^{(k)}$$

for $k = 1, 2, \cdots, p$ are equivalent to the condition that $R = D_1 T^{-1} + D_2$ for some diagonal matrix $D_2 \geqq 0$ and some diagonal matrix $D_1 \in S$, in which S is the set of all diagonal matrices $D \geqq 0$ such that DT^{-1} is symmetric. Hence $T^{-1}G(I + RG)^{-1} \in P_0$ for all such R if and only if

$$\det \{[(I + D_1 D^*)T^{-1} + D^* D_2]G + D^*\} \neq 0 \tag{31}$$

for all diagonal $D^* > 0$, $D_2 \geqq 0$, and $D_1 \in S$.

Let $\Lambda = \text{diag}\,(\lambda_1, \lambda_2, \cdots, \lambda_{(2p+q)})$ be such that

$$D_2 = D^{*-1}\,\Delta^{-1}\Lambda(I + D_1 D^*)$$

in which

$$\Delta = \text{diag}\,(\delta_1, \delta_1, \delta_2, \delta_2, \cdots, \delta_p, \delta_p) \oplus I_q$$

if $q > 0$, $\Delta = \text{diag}\,(\delta_1, \delta_1, \delta_2, \delta_2, \cdots, \delta_p, \delta_p)$ if $q = 0$, and

$$\delta_k = 1 - \alpha_f^{(k)}\alpha_r^{(k)} \quad \text{for} \quad k = 1, 2, \cdots, p.$$

The left side of (31) is

$$\det [(I + D_1 D^*)(T^{-1} + \Delta^{-1}\Lambda)G + D^*]$$

which can be written as

$$\det [(I + D_1 D^*)\,\Delta^{-1}(I + \Lambda)\,\Delta_\Lambda T_\Lambda^{-1} G + D^*] \tag{32}$$

with

$$T_\Lambda^{-1} = \Delta_\Lambda^{-1}\,\Delta(I + \Lambda)^{-1}(T^{-1} + \Delta^{-1}\Lambda)$$

and

$$\Delta_\Lambda = \text{diag}\,(\delta_1', \delta_1', \delta_2', \delta_2', \cdots, \delta_p', \delta_p') \oplus I_q$$

if $q > 0$ and $\Delta_\Lambda = \text{diag}\,(\delta_1', \delta_1', \delta_2', \delta_2', \cdots, \delta_p', \delta_p')$ if $q = 0$, in which for $k = 1, 2, \cdots, p$

$$\delta_k' = 1 - \alpha_f^{(k)}\alpha_r^{(k)}(1 + \lambda_{(2k-1)})^{-1}(1 + \lambda_{2k})^{-1}.$$

But (32) vanishes if and only if $\det (T_\Lambda^{-1}G + \tilde{D})$ vanishes, in which $\tilde{D} = \Delta_\Lambda^{-1}(I + \Lambda)^{-1}\,\Delta(I + D_1 D^*)^{-1} D^*$. We observe that $\tilde{D}$ is a positive diagonal matrix and that given any diagonal $\tilde{D}' > 0$ and given any

$\Lambda \geqq 0$ we can choose $D^* > 0$ and $D_1 \in S$ so that $\tilde{D} = \tilde{D}'$. Thus $T^{-1}G(I + RG)^{-1} \in P_0$ for all $R = (D_1 T^{-1} + D_2)$ with $D_1 \in S$ and $D_2 \geqq 0$ if and only if

$$\det (T_\Lambda^{-1} G + \tilde{D}) \neq 0$$

for all $\Lambda \geqq 0$ and $\tilde{D} > 0$, that is, if and only if $T_\Lambda^{-1} G \in P_0$ for all $\Lambda \geqq 0$ (see Lemma 1 of Section 3.1). But

$$T_\Lambda = T_1 \oplus T_2 \oplus \cdots \oplus T_p \oplus I_q \quad \text{if} \quad q > 0$$

and

$$T_\Lambda = T_1 \oplus T_2 \oplus \cdots \oplus T_p \quad \text{if} \quad q = 0$$

with

$$T_k = \begin{bmatrix} 1 & \dfrac{-\alpha_r^{(k)}}{1 + \lambda_{2k-1}} \\ \dfrac{-\alpha_f^{(k)}}{1 + \lambda_{2k}} & 1 \end{bmatrix}$$

for all $k = 1, 2, \cdots, p$. Therefore $T^{-1}G(I + RG)^{-1} \in P_0$ for all $R = (D_1 T^{-1} + D_2)$ with $D_2 \geqq 0$ and $D_1 \in S$ if and only if $M^{-1}G \in P_0$ for all $M \in \mathfrak{J}(T)$. □

3.31 *Definition 12:* Let $\mathfrak{F}_3$ denote the set of all $F(\cdot)$ such that

(*i*) $F(\cdot) \in \mathfrak{F}_0^{(2p+q)}$, and
(*ii*) for each $j = 1, 2, \cdots, (2p + q)$ there exists a real constant β_j such that $f_j(\cdot)$ is a strictly-monotone-increasing mapping of E^1 onto either (β_j, ∞) or $(-\infty, \beta_j)$, and
(*iii*) whenever $p > 0$, $f_{(2k-1)}(\cdot)$ and $f_{2k}(\cdot)$ are both bounded on either $[0, \infty)$ or $(-\infty, 0]$ for $k = 1, 2, \cdots, p$.

3.32 *Theorem 13: Let $T \in \mathfrak{J}$, and, referring to the network of Fig. 1 in which it is assumed that R (see Section 2.1) is the zero matrix, let G denote the short-circuit conductance matrix of the linear portion of the network. (The linear portion is assumed to contain only sources and linear resistors of nonnegative resistance.) Then the equation $F(x) + T^{-1}Gx = B$ possesses a unique solution x for each $F(\cdot) \in \mathfrak{F}_3$ and each $B \in E^{(2p+q)}$ if and only if $T^{-1}G \in P_0$ and* $\det G \neq 0$. *If $T^{-1}G \in P_0$ and* $\det G = 0$, *then there exists a real $(2p + q)$-vector η such that (i) $\eta \neq \theta$, and for some $F(\cdot) \in \mathfrak{F}_3$ all of the components of $F(\alpha\eta)$ are bounded on $\alpha \in [0, \infty)$, and (ii) for any $F(\cdot) \in \mathfrak{F}_3$ with the property that all of the components of $F(\alpha\eta)$ are bounded on $\alpha \in [0, \infty)$ the equation $F(x) + T^{-1}Gx = B$ does not possess a solution for some $B \in E^{(2p+q)}$.*

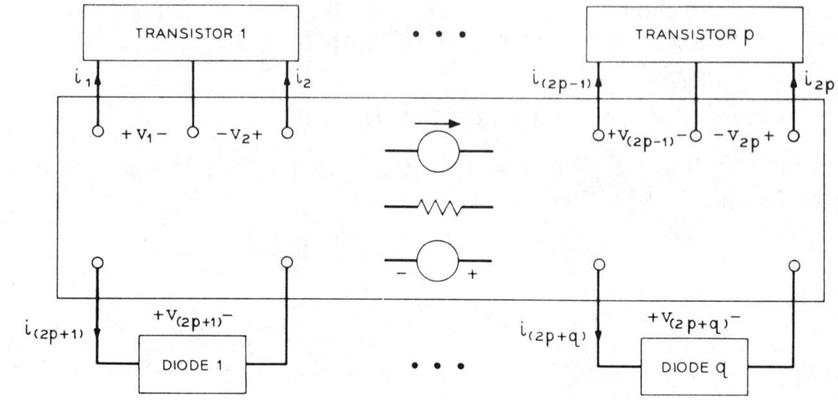

Fig. 1—General network containing transistors, diodes, resistors, and sources.

3.33 *Proof of Theorem 13*

(if) If $T^{-1}G \varepsilon P_0$ with det $T^{-1}G \neq 0$, and if $F(\cdot) \varepsilon \mathfrak{F}_3$, then, since each $f_i(\cdot)$ is a strictly-monotone-increasing mapping of E^1 onto (β_i, ∞) or $(-\infty, \beta_i)$ for some real constant β_i, by Theorem 4 of Ref. 2, the equation $F(x) + T^{-1}Gx = B$ possesses a unique solution x for each $B \varepsilon E^{(2p+q)}$.

(only if) Assume that $T^{-1}G \not\varepsilon P_0$. Then since $\mathfrak{F}_3$ is contained in $\mathfrak{F}_0^{(2p+q)}$, by Theorem 1 of Ref. 3, for each $F(\cdot) \varepsilon \mathfrak{F}_3$ there exists a $B \varepsilon E^{(2p+q)}$ such that there are at least two solutions x of $F(x) + T^{-1}Gx = B$.

Assume now that $T^{-1}G \varepsilon P_0$ and that det $G = 0$. We shall use the proposition that if $R(\cdot)$ is any continuous mapping of $E^{(2p+q)}$ into itself, then $R(\cdot)$ is a homeomorphism of $E^{(2p+q)}$ onto itself if and only if $R(\cdot)$ is a local homeomorphism on $E^{(2p+q)}$ and $\| R(x) \| \to \infty$ as $\| x \| \to \infty$.†

Let $R(\cdot)$ be defined by the condition that $R(x) = F(x) + T^{-1}Gx$ for all $x \varepsilon E^{(2p+q)}$. For any $F(\cdot) \varepsilon \mathfrak{F}_3$ the operator $R(\cdot)$ is a local homeomorphism on $E^{(2p+q)}$, since with $F(\cdot)$ such that each $f_i(\cdot)$ is a strictly-monotone-increasing mapping of E^1 *onto* E^1 the mapping $[F(\cdot) + T^{-1}G]$ is a homeomorphism of $E^{(2p+q)}$ onto itself.[1] In addition, for any $F(\cdot) \varepsilon \mathfrak{F}_3$ and any $B \varepsilon E^{(2p+q)}$, there is at most one $x \varepsilon E^{)2p+q)}$ such that $R(x) = B$.[1]

Let us suppose that for each $B \varepsilon E^{(2p+q)}$ and each $F(\cdot) \varepsilon \mathfrak{F}_3$ there exists a solution x of $R(x) = B$. Then for all $F(\cdot) \varepsilon \mathfrak{F}_3$, $R(\cdot)$ is a homeomorphism of $E^{(2p+q)}$ onto itself, and hence for all $F(\cdot) \varepsilon \mathfrak{F}_3$ $\| R(x) \| \to \infty$ as $\| x \| \to \infty$. But, by Lemma 3 (which appears below) $E^{(2p+q)}$ contains a vector η such that $\eta \neq \theta$, $\eta_j \varepsilon \{0, +1, -1\}$ for all j, and $G\eta = \theta$; and if

† See Ref. 12 and the appendix of Ref. 13.

$p > 0$, η satisfies $\eta_{(2k-1)}\eta_{2k} \geqq 0$ for all $k = 1, 2, \cdots, p$. Let $\mathfrak{F}_3(\eta)$ denote the subset of $\mathfrak{F}_3$ containing all elements $F(\cdot)$ with the property that $f_j(\alpha\eta_j)$ is bounded on $\alpha \in [0, \infty)$ for all $j = 1, 2, \cdots, (2p + q)$. Since $\eta_{(2k-1)}\eta_{2k} \geqq 0$ for all $k = 1, 2, \cdots, p$ when $p > 0$, it is clear that $\mathfrak{F}_3(\eta)$ is not empty. However, for any $F(\cdot) \in \mathfrak{F}_3(\eta)$ we have $\| R(\alpha\eta) \| = \| F(\alpha\eta) \|$ with $\| F(\alpha\eta) \|$ bounded on $\alpha \in [0, \infty)$, which contradicts the assumption that there exists a solution x of $R(x) = B$ for each $F(\cdot) \in \mathfrak{F}_3$ and each $B \in E^{(2p+q)}$.

Lemma 3: Let G be the short-circuit conductance matrix of the linear portion of the network of Fig. 1. *If* $\det G = 0$, *then there exists a vector* $\eta \in E^{(2p+q)}$ *such that* $G\eta = \theta$, $\eta \neq \theta$, *and* $\eta_j \in \{0, +1, -1\}$ *for all* $j = 1, 2, \cdots, (2p + q)$; *and if* $p > 0$ η *also satisfies* $\eta_{(2k-1)}\eta_{2k} \geqq 0$ *for* $k = 1, 2, \cdots, p$.

Proof of Lemma 3:

Let N denote the $(2p + q)$-port resistor network obtained from the network of Fig. 1 by removing all transistors and diodes and by setting the value of each source to zero. The short-circuit conductance matrix G possesses the property that if $v \in E^{(2p+q)}$ denotes the vector of port voltages of N and $i \in E^{(2p+q)}$ denotes the corresponding vector of port currents (with polarities as indicated in Fig. 1), then $i = -Gv$.

Let $\det G = 0$. Then the open-circuit resistance matrix of N does not exist. Therefore there exists a port ℓ of N such that there is no path through resistors of N that connects the two terminals of port ℓ when all other ports are open-circuited. Let a one-volt source be placed at port ℓ so that $v_\ell = 1$. Then when all ports j of N with $j \neq \ell$ are open-circuited, $i_\ell = 0$ and there is zero current in every resistor of N. Let S denote a set of port numbers of N with the following properties. The number ℓ is not contained in S and when all ports j with $j \in S$ are short-circuited and all ports j with $j \notin S \cup \{\ell\}$ are open-circuited then zero current flows through the one-volt source; when any port $j_1 \notin S \cup \{\ell\}$ and all ports j with $j \in S$ are short-circuited and all ports j with $j \notin S \cup \{\ell, j_1\}$ are open-circuited then nonzero current flows through the one-volt source. It is clear that such a set S exists (with the understanding that S might be the null set). In general S contains r port numbers where $0 \leqq r \leqq (2p + q - 1)$.

If $r = (2p + q - 1)$, then with $v_\ell = 1$ and with all remaining components of v equal to zero, we have $Gv = \theta$. Obviously in this case we can take the vector η of the statement of Lemma 3 to be v.

If $r \neq (2p + q - 1)$, then, with $v_\ell = 1$, with $v_j = 0$ for all $j \in S$, and with all ports $j \notin S \cup \{\ell\}$ open-circuited, there exists for each $j \notin S \cup \{\ell\}$ some path through the one-volt source and the resistors of N that connects the two terminals of port j. Therefore when $r \neq (2p + q - 1)$, when all ports $j \notin S \cup \{\ell\}$ are open circuited, when $v_\ell = 1$, and when $v_j = 0$ for all $j \in S$, the open-circuit voltage v_j at each port j with $j \notin S \cup \{\ell\}$ is well defined and nonzero. Since no current flows in any resistor of N when $v_\ell = 1$, $v_j = 0$ for all $j \in S$, and all ports $j \notin S \cup \{\ell\}$ are open-circuited, it follows that $v_j \in \{-1, +1\}$ for all $j \notin S$. With $v_\ell = 1$, with $v_j = 0$ for all $j \in S$, and with v_j the corresponding open-circuit voltage for each $j \notin S \cup \{\ell\}$, we have $Gv = \theta$. When $p > 0$, the vector v also satisfies the condition that $v_{(2k-1)}v_{2k} \geqq 0$ for all $k = 1, 2, \cdots, p$ since if $v_{(2k-1)}v_{2k}$ were negative for some k, then for that k $v_{(2k-1)} = 1$ and $v_{2k} = -1$ or $v_{(2k-1)} = -1$ and $v_{2k} = 1$; in either case $| v_{(2k-1)} - v_{2k} | = 2$ which contradicts the proposition that a network of nonnegative resistors can have no voltage gain. □

APPENDIX*

A theorem due to R. S. Palais† asserts that if $R(\cdot)$ is a continuously-differentiable mapping of E^n into itself with values $R(q)$ for $q \in E^n$, then $R(\cdot)$ is a diffeomorphism‡ of E^n onto itself if and only if

(*i*) $\det J_q \neq 0$ for all $q \in E^n$, in which J_q is the Jacobian matrix of $R(\cdot)$ with respect to q, and
(*ii*) $\| R(q) \| \to \infty$ as $\| q \| \to \infty$.

If $R(\cdot)$ is any twice-continuously-differentiable mapping of E^n into itself such that conditions (*i*) and (*ii*) of Palais' theorem are satisfied, then E^n contains a unique element x such that $R(x) = \theta$ in which θ is the zero element of E^n, and there are steepest decent as well as Newton-type algorithms each of which generates a sequence in E^n that converges to x. To show this, let [18] $f(y) = \| R(y) \|^2$ for all $y \in E^n$ in which $\|\cdot\|$ denotes the usual Euclidean norm (i.e., the square-root of the sum of squares). Since condition (*i*) of Palais' theorem is satisfied, the gradient ∇f of $f(\cdot)$ satisfies $(\nabla f)(y) \neq \theta$ unless $f(y) = 0$,§ and since condition (*ii*) of Palais' theorem is satisfied, the set $S = \{y \in E^n : f(y) \leqq f(x^{(0)})\}$ is bounded for any $x^{(0)} \in E^n$. Therefore we may appeal to, for example, the theorem of page 43 of Ref. 18 according to which for any $x^{(0)} \in E^n$, for any member of a certain class of mappings $\varphi(\cdot)$ of S

* The material of this appendix together with some misprints appears in Ref. 3.
† See Ref. 12 and the appendix of Ref. 13.
‡ A diffeomorphism of E_n onto itself is a continuously differentiable mapping of E_n into E_n which possesses a continuously differentiable inverse.
§ Here we have used the fact that $(\nabla f)(y) = 2J_y{}^{tr}R(y)$ for all $y \in E_n$.[18]

into E^n, and for suitably chosen constants $\gamma_0, \gamma_1, \cdots$, the sequence $x^{(0)}, x^{(1)}, \cdots$ defined by

$$x^{(k+1)} = x^{(k)} + \gamma_k \varphi(x^{(k)}) \quad \text{for all} \quad k \geqq 0$$

belongs to S and is such that $\| R(x^{(k)}) \| \rightarrow 0$ as $k \rightarrow \infty$. However, since $R^{-1}(\cdot)$ exists and is continuous, it follows from

$$x^{(k)} = R^{-1}[R(x^{(k)})] \quad \text{for all} \quad k \geqq 0$$

and the fact that $R(x^{(k)}) \rightarrow \theta$ as $k \rightarrow \infty$, that $\lim_{k\to\infty} x^{(k)}$ exists and

$$\lim_{k\to\infty} x^{(k)} = R^{-1}(\theta),$$

which means that $\lim_{k\to\infty} x^{(k)}$ is the unique solution x of $R(y) = \theta$.

REFERENCES

1. Sandberg, I. W., and Willson, A. N., Jr., "Some Theorems on Properties of DC Equations of Nonlinear Networks," B.S.T.J., *48*, No. 1 (January 1969), pp. 1–34.
2. Sandberg, I. W., and Willson, A. N., Jr., "Some Network-Theoretic Properties of Nonlinear DC Transistor Networks," B.S.T.J., *48*, No. 5 (May-June 1969), pp. 1293–1312.
3. Sandberg, I. W., "Theorems on the Analysis of Nonlinear Transistor Networks," B.S.T.J., *49*, No. 1 (January 1970), pp. 95–114.
4. Willson, A. N., Jr., "New Theorems on the Equations of Nonlinear DC Transistor Networks," B.S.T.J., this issue, pp. 1713–1738.
5. Sandberg, I. W., "Some Theorems on the Dynamic Response of Nonlinear Transistor Networks," B.S.T.J., *48*, No. 1 (January 1969), pp. 35–54.
6. Hamming, R. W., *Numerical Methods for Scientists and Engineers*, New York: McGraw-Hill Book Co., (1962).
7. Ralston, A. A., *A First Course in Numerical Analysis*, New York: McGraw-Hill Book Co., (1965).
8. Hachtel, G. D., and Rohrer, R. A., "Techniques for the Optimal Design and Synthesis of Switching Circuits," Proc. of the IEEE, *55*, No. 11 (November 1967), pp. 1864–1876.
9. Sandberg, I. W., and Shichman, H., "Numerical Integration of Systems of Stiff Nonlinear Differential Equations," B.S.T.J., *47*, No. 4 (April 1968), pp. 511–527.
10. Calahan, D. A., "Efficient Numerical Analysis of Non-Linear Circuits," Proc. Sixth Ann. Allerton Conf. on Circuit and System Theory, University of Illinois, 1968, pp. 321–331.
11. Vehovec, M., "Simple Criterion for the Global Regularity of Vector-Valued Functions," Elec. Letters, *5*, No. 26 (December 1969), pp. 680–681.
12. Palais, R. S., "Natural Operations on Differential Forms," Trans. Amer. Math. Soc., *92*, No. 1 (1959), pp. 125–141.
13. Holzmann, C. A., and Liu, R., "On the Dynamical Equations of Nonlinear Networks with n-coupled elements," Proc. Third Ann. Allerton Conf. on Circuit and System Theory, University of Illinois, 1965, pp. 536–545.
14. Mitra, D., Sandberg, I. W., and Gopinath, B., "A Note on a Curious Property of the Equations of Nonlinear Networks Containing Transistors," to be published.
15. Muir, T., *A Treatise on the Theory of Determinants*, New York: Dover Publications, Inc., (1960), pp. 31–33.
16. Meyer, G. H., "On Solving Nonlinear Equations with a One-Parameter Operator Imbedding," Tech. Rep. 67–50, Comp. Science Center, University of Maryland, College Park, 1967.
17. Hadamard, J., "Sur Les Transformations Ponctuelles," Bull. Soc. Math. France, 48, (1920), pp. 13–27.
18. Goldstein, A. A., *Constructive Real Analysis*, New York: Harper and Row (1967), pp. 41–45.

On the Equations of Nonlinear Networks

THOMAS E. STERN, MEMBER, IEEE

Abstract—**A systematic method of selecting state variables and formulating the normal form differential equations of general classes of nonlinear networks is presented. It is shown that the key problem in the formulation of the differential equations is the calculation of certain functions which appear in these equations. Iterative procedures for the calculation of these functions are given for certain classes of networks, and it is shown that the conditions for convergence of these procedures are closely related to certain physical properties of the network. These procedures are readily adaptable to machine computation and can be made an integral part of any numerical method used for the solution of the differential equations of the network.**

I. Introduction

A BASIC STEP in the resolution of problems in network theory is the characterization of the network in terms of a set of state variables $\mathbf{x}$[1] and the formulation of a set of differential equations in the normal form $\dot{\mathbf{x}} = \mathbf{f}(\mathbf{x}, t)$ that define the behavior of these state variables. The problem was first considered for linear networks by Bashkow [1] and more recently, in more generality, by Bryant [2], [3]. Certain types of nonlinear networks have also been considered by Brayton and Moser [4], Liu and Auth [5], and Desoer and Katzenelson [6]. The purpose of this paper is to perform the same task for some broad classes of nonlinear networks, with special emphasis on the problem of computing the function $\mathbf{f}$ appearing in the normal form equations. Computational questions are of prime importance here since, even when it can be shown that a given network can be characterized in terms of normal form equations, the function $\mathbf{f}$ often cannot be expressed in closed form. Among the questions we shall attempt to answer are: 1) is it possible to characterize a given network in terms of differential equations in normal form? 2) if it is possible, do these equations possess unique solutions?

Sections II and III of this paper deal with the formulation of the normal form equations for fairly general classes of networks and with questions of existence and uniqueness of solutions. In Section IV, certain special (but nonetheless broad) classes of networks are considered in more detail, the primary concern being methods of computing the function $\mathbf{f}$.

II. Selection of State Variables

We assume that the network is defined in terms of a directed nonseparable graph containing inductive, capacitive, and resistive branches. The branch variable relationships are given by equations of the form

$$\mathbf{f}_L(\mathbf{i}_L, \boldsymbol{\lambda}_L, t) = \mathbf{0}, \quad \mathbf{f}_C(\mathbf{e}_C, \mathbf{q}_C, t) = \mathbf{0}, \quad \mathbf{f}_R(\mathbf{i}_R, \mathbf{e}_R, t) = \mathbf{0} \quad (1)$$

where $\mathbf{i}_L$, $\boldsymbol{\lambda}_L$, and $\mathbf{f}_L$ are l-vectors representing, respectively, the inductor currents, flux-linkages, and the mutual constraints among them. The capacitor voltages $\mathbf{e}_C$ and charges $\mathbf{q}_C$, and the resistor currents $\mathbf{i}_R$ and voltages $\mathbf{e}_R$, are related in a similar fashion. Note that a "network element" may consist of several branches. For example, a triode might be represented by two coupled resistive branches. This type of representation admits very general types of elements. Resistive elements, for example, would include fixed and time-varying current and voltage sources, linear and nonlinear controlled sources, such as vacuum tubes and transistors, ideal transformers, negative resistance and conductance elements, etc.

We shall choose as state variables, linear combinations of capacitor charges $\mathbf{q}$ and linear combinations of inductor flux-linkages $\boldsymbol{\lambda}$. These variables must be chosen in such a way as to constitute a "complete set."

Definition 1: A set of $\mathbf{x}$ of network variables will be called *complete*, if to every value of $\mathbf{x}$, there correspond unique values of all branch variables.

To define the state variables, we begin by choosing an appropriate tree T in the manner of Bryant [2]. Based on T, we divide the branches of the network into six disjoint subsets S_α, S_β, S_γ, S_δ, S_ϵ, S_ζ, containing, respectively, the capacitive chords, the resistive chords, the inductive chords, the capacitive tree-branches, the resistive tree-branches, and the inductive tree-branches of T. We number the branches consecutively, starting with the members of S_α and indicate the variables associated with the branches in a given subset by vectors with the appropriate subscripts. Thus, the currents, voltages, and flux-linkages of the members of S_γ are denoted by $\mathbf{i}_\gamma$, $\mathbf{e}_\gamma$, $\boldsymbol{\lambda}_\gamma$, respectively.

Bryant has shown that his method of tree construction always leads to a fundamental loop matrix B of the form

$$B = [I, F] = \left[\begin{array}{ccc|ccc} I_{\alpha\alpha} & 0 & 0 & F_{\alpha\delta} & 0 & 0 \\ 0 & I_{\beta\beta} & 0 & F_{\beta\delta} & F_{\beta\epsilon} & 0 \\ 0 & 0 & I_{\gamma\gamma} & F_{\gamma\delta} & F_{\gamma\epsilon} & F_{\gamma\zeta} \end{array}\right] \quad (2)$$

with corresponding fundamental cutset matrix $Q = [-F^t, I]$. The dimensions of the submatrices of B are determined by the order of their associated subsets. For example, if S_α and S_δ contain k and r elements, respectively, then $F_{\alpha\delta}$ is $k \times r$. Letting $\mathbf{e} = (\mathbf{e}_\alpha, \mathbf{e}_\beta, \mathbf{e}_\gamma, \mathbf{e}_\delta, \mathbf{e}_\epsilon, \mathbf{e}_\zeta)$ and $\mathbf{i} = (\mathbf{i}_\alpha, \mathbf{i}_\beta, \mathbf{i}_\gamma, \mathbf{i}_\delta, \mathbf{i}_\epsilon, \mathbf{i}_\zeta)$, Kirchhoff's voltage and current laws are

$$B\mathbf{e} = \mathbf{0}, \qquad Q\mathbf{i} = \mathbf{0}. \quad (3)$$

Manuscript received September 29, 1964; revised September 20, 1965. This work was partially supported by the National Science Foundation under grants NSF GP-533, 2789, and 14514, and by the U. S. Navy under contract NONR 42599(04).

The author is with the Department of Electrical Engineering, Columbia University, New York, N. Y. He is currently a Fulbright Research Scholar at the Faculté des Sciences, l'Université de Paris, France.

[1] Boldface variables are to be interpreted as column vectors.

Reprinted from *IEEE Transactions on Circuit Theory*, vol. CT-13, pp. 74–81, March 1966.

We now define as our state variables, $\mathbf{q}$ the cutset charges and $\boldsymbol{\lambda}$ the loop flux-linkages, where

$$\mathbf{q} = \mathbf{q}_\delta - F^t_{\alpha\delta}\mathbf{q}_\alpha, \qquad \boldsymbol{\lambda} = \boldsymbol{\lambda}_\gamma + F_{\gamma\zeta}\boldsymbol{\lambda}_\zeta. \tag{4}$$

In order to determine whether the set $(\mathbf{q}, \boldsymbol{\lambda})$ is complete, we must examine the following three sets of implicit equations.

Inductor equations

$$\mathbf{f}_L(\mathbf{i}_L, \boldsymbol{\lambda}_L, t) = \mathbf{0}, \quad \mathbf{i}_L = (\mathbf{i}_\gamma, \mathbf{i}_\zeta), \quad \boldsymbol{\lambda}_L = (\boldsymbol{\lambda}_\gamma, \boldsymbol{\lambda}_\zeta) \tag{L}$$
$$\boldsymbol{\lambda} - \boldsymbol{\lambda}_\gamma - F_{\gamma\zeta}\boldsymbol{\lambda}_\zeta = \mathbf{0}, \quad \mathbf{i}_\zeta - F^t_{\gamma\zeta}\mathbf{i}_\gamma = \mathbf{0}.$$

Capacitor equations

$$\mathbf{f}_C(\mathbf{e}_C, \mathbf{q}_C, t) = \mathbf{0}, \quad \mathbf{e}_C = (\mathbf{e}_\alpha, \mathbf{e}_\delta), \quad \mathbf{q}_C = (\mathbf{q}_\alpha, \mathbf{q}_\delta) \tag{C}$$
$$\mathbf{q} - \mathbf{q}_\delta + F^t_{\alpha\delta}\mathbf{q}_\alpha = \mathbf{0}, \qquad \mathbf{e}_\alpha + F_{\alpha\delta}\mathbf{e}_\delta = \mathbf{0}.$$

Resistor equations

$$\mathbf{f}_R(\mathbf{i}_R, \mathbf{e}_R, t) = \mathbf{0}, \quad \mathbf{i}_R = (\mathbf{i}_\beta, \mathbf{i}_\epsilon), \quad \mathbf{e}_R = (\mathbf{e}_\beta, \mathbf{e}_\epsilon) \tag{R}$$
$$\mathbf{i}_\epsilon - F^t_{\beta\epsilon}\mathbf{i}_\beta - F^t_{\gamma\epsilon}\mathbf{i}_\gamma = \mathbf{0}, \quad \mathbf{e}_\beta + F_{\beta\epsilon}\mathbf{e}_\epsilon + F_{\beta\delta}\mathbf{e}_\delta = \mathbf{0}.$$

Since the above equations contain all the branch variables, it can be seen that the state variables $(\mathbf{q}, \boldsymbol{\lambda})$ defined in (4) constitute a complete set if and only if for every value of $(\mathbf{q}, \boldsymbol{\lambda}, t)$ (L), (C), and (R) possess unique solutions for all variables in terms of $(\mathbf{q}, \boldsymbol{\lambda}, t)$. It should be observed that $(\mathbf{q}, \boldsymbol{\lambda})$ will *not* constitute a complete set if either 1) there are some values of $(\mathbf{q}, \boldsymbol{\lambda}, t)$ for which more than one solution of (L), (C), and (R) exists; or 2) there are some values for which no solution exists. The first condition occurs, for example, in a circuit consisting of a current-controlled negative resistance in parallel with a linear capacitor, and the second condition occurs when a perfectly saturating inductor is present in the network.

The above procedure for selecting state variables does not lead to a unique set, since it depends upon the choice of the tree T. However, it can be shown that any set of state variables chosen in this manner is related to any other set by a nonsingular linear transformation, and thus, all such sets are complete if and only if any one of them is complete.

Given only the general form of (L), (C), and (R), it is difficult to come to any useful conclusions regarding existence of unique solutions. We, therefore, defer this question to Section IV, where certain important special cases will be considered. It should be noted, however, that this is the crucial problem in formulating the network equations.

III. Differential Equations: Existence and Uniqueness of Solutions

Assuming that $(\mathbf{q}, \boldsymbol{\lambda})$ constitute a complete set, it is possible to write

$$\mathbf{i}_\gamma = \mathbf{I}_\gamma(\boldsymbol{\lambda}, t), \qquad \mathbf{e}_\delta = \mathbf{E}_\delta(\mathbf{q}, t) \tag{5}$$
$$\mathbf{i}_\beta = \mathbf{I}_\beta[\mathbf{I}_\gamma(\boldsymbol{\lambda}, t), \mathbf{E}_\delta(\mathbf{q}, t), t], \quad \mathbf{e}_\epsilon = \mathbf{E}_\epsilon[\mathbf{I}_\gamma(\boldsymbol{\lambda}, t), \mathbf{E}_\delta(\mathbf{q}, t), t]$$

where (5) represents single-valued functions whose domain is the entire $(\mathbf{q}, \boldsymbol{\lambda}, t)$ space. Utilizing Kirchhoff's laws and (4), these functions can be combined to yield the normal form equations

$$\dot{\mathbf{q}} = F^t_{\beta\delta}\mathbf{I}_\beta[\mathbf{I}_\gamma(\boldsymbol{\lambda}, t), \mathbf{E}_\delta(\mathbf{q}, t), t] + F^t_{\gamma\delta}\mathbf{I}_\gamma(\boldsymbol{\lambda}, t) \tag{E}$$
$$\dot{\boldsymbol{\lambda}} = -F_{\gamma\epsilon}\mathbf{E}_\epsilon[\mathbf{I}_\gamma(\boldsymbol{\lambda}, t), \mathbf{E}_\delta(\mathbf{q}, t), t] - F_{\gamma\delta}\mathbf{E}_\delta(\mathbf{q}, t).$$

The manner in which the normal form equations were constructed is conveniently illustrated in terms of the block diagram in Fig. 1. In the figure, the elements of the various subsets have been grouped into three individual blocks indicated by the appropriate symbol for resistors, capacitors, or inductors. Each block has two sets of ports corresponding to the manner in which its individual branches have been divided into subsets of chords and tree-branches. (The heavy lines in the diagram indicate sets of interconnections of vector variables.) The interconnections among the branches are indicated by other blocks labeled for appropriate submatrices of B. Note that these blocks can be interpreted as banks of ideal transformers with unity turns-ratios. It will be observed that the variables which appear in the normal form equations (E) are only those which appear at the external terminals of the three dashed enclosures in the figure. Those which appear inside the dashed lines are subsidiary variables which are eliminated in the process of solving (L), (C), and (R). Solution of (L), (C), and (R) is therefore equivalent to finding the terminal characteristics of the three dashed blocks shown in the figure.

We may now state

Theorem 1: Given a network N and a complete set of state variables $(\mathbf{q}, \boldsymbol{\lambda})$ as defined above, if the functions $\mathbf{I}_\gamma$, $\mathbf{E}_\delta$, $\mathbf{I}_\beta$, $\mathbf{E}_\epsilon$ have bounded continuous first partial derivatives with respect to $(\mathbf{q}, \boldsymbol{\lambda})$ and are continuous in t for all $(\mathbf{q}, \boldsymbol{\lambda}, t)$, then (E) possesses unique solutions for all t and all initial conditions.

Proof: The conditions of the theorem are more than enough to ensure the existence of a Lipschitz condition over the whole $(\mathbf{q}, \boldsymbol{\lambda}, t)$ space, and hence, the conditions of the standard Cauchy-Lipschitz uniqueness theorem are fulfilled everywhere [7]. The fact that bounded derivatives have been assumed means that solutions may be continued indefinitely forward or backward in time; i.e., solutions with "finite escape time" are ruled out.

Although the conditions of Theorem 1 are more restrictive than necessary, they result in a simple statement of the theorem and do not rule out any situations of significant physical interest which might possibly be included under weaker assumptions.

In order for the above procedure to be of some engineering interest, we must have some method for determining whether the state variables constitute a complete set and some computational scheme for obtaining the functions of (5) which are solutions of (L), (C), and (R). In Section IV, we consider these problems for three important classes of networks: 1) reciprocal networks, 2) networks solvable by contraction mapping techniques, and 3) networks solvable by linear operations.

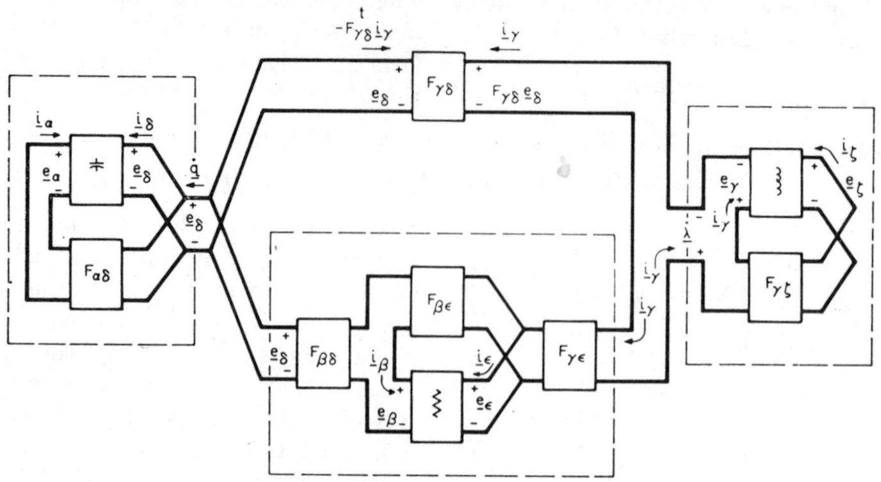

Fig. 1. Block diagram for normal form equations.

IV. Solution of Equations (L), (C), and (R)

In the sequel, we shall make frequent use of certain incremental parameter matrices for network elements. For example, in the case of a multi-port resistive element, if the branch voltages $\mathbf{e}_R$ are expressible explicitly as differentiable functions of $(\mathbf{i}_R, t)$, then the associated incremental resistance matrix is defined as $R(\mathbf{i}_R, t) = (\partial \mathbf{e}_R/\partial \mathbf{i}_R)$.[2] Similarly, if the relations are expressible in the inverse form, an incremental conductance matrix $G(\mathbf{e}_R, t) = (\partial \mathbf{i}_R/\partial \mathbf{e}_R)$ is defined. Finally, if $(\mathbf{i}_\beta, \mathbf{e}_\epsilon)$ are given explicitly as differentiable functions of $(\mathbf{e}_\beta, \mathbf{i}_\epsilon, t)$, then we define a hybrid incremental resistance matrix

$$M(\mathbf{e}_\beta, \mathbf{i}_\epsilon, t) = \frac{\partial(\mathbf{i}_\beta, \mathbf{e}_\epsilon)}{\partial(\mathbf{e}_\beta, \mathbf{i}_\epsilon)} = \begin{bmatrix} M_{\beta\beta} & M_{\beta\epsilon} \\ M_{\epsilon\beta} & M_{\epsilon\epsilon} \end{bmatrix}. \qquad (6)$$

A. Reciprocal Networks

For our purposes reciprocity shall be defined in terms of the symmetry of certain incremental parameter matrices. Thus, we shall say that a resistive element is reciprocal if its associated matrix R, G, or

$$\begin{bmatrix} M_{\beta\beta} & M_{\beta\epsilon} \\ -M_{\epsilon\beta} & -M_{\epsilon\epsilon} \end{bmatrix}$$

is symmetric for all values of its arguments. (It is easily shown that if one of these three matrices is symmetric, then any of the others which exist are also symmetric.) Reciprocity is defined similarly for inductive and capacitive elements. A network will be called reciprocal if all of its elements are reciprocal. (Thus, any network containing only one-port elements is reciprocal.)

In this section, we shall show that when a network is at least "partially reciprocal" a convergent iteration scheme can often be used to solve (L), (C), and (R). To this end, we need the following definition and theorem.

Definition 2: A square matrix $A(\mathbf{x}, t)$ will be called uniformly positive definite (u.p.d.) in $\mathbf{x}$ if it is symmetric; and if for each t, there exists a $\mu > 0$ such that $\mathbf{u}^t A \mathbf{u} \geq \mu \mathbf{u}^t \mathbf{u}$ for all $\mathbf{u}$ and $\mathbf{x}$.

Theorem 2: Given a function $\mathbf{y} = \mathbf{Y}(\mathbf{x}, t)$, continuous in t with continuous bounded partials in $\mathbf{x}$, if $(\partial \mathbf{Y}/\partial \mathbf{x})$ is u.p.d. in $\mathbf{x}$, then

a) there exists a unique (inverse) function $\mathbf{X}(\mathbf{y}, t)$ continuous in t with continuous bounded partials in $\mathbf{y}$ such that

$$\mathbf{x} = \mathbf{X}[\mathbf{Y}(\mathbf{x}, t), t] \quad \text{for all} \quad (\mathbf{x}, t)$$

and

$$\mathbf{y} = \mathbf{Y}[\mathbf{X}(\mathbf{y}, t), t] \quad \text{for all} \quad (\mathbf{y}, t);$$

b) the function $\mathbf{X}$ may be evaluated for all values of $(\mathbf{y}, t)$ by the following convergent iteration formula:

$$\mathbf{x}_{k+1} = \mathbf{x}_k - h[\mathbf{Y}(\mathbf{x}_k, t) - \mathbf{y}] \qquad (7)$$

where h is a positive constant. (See Appendix for proof.)

Now let us observe that (L), (C), and (R) are each of the same basic form. Thus, any results obtained, say, for (R) will apply with minor modifications to (L) and (C). Hence, to avoid unnecessary repetition, all results given in the remainder of the paper will be stated only in terms of (R). (The resistor equations have been chosen as a model, since most of the commonly used nonlinear elements are resistive in nature.) The reader can easily carry over any of the discussion which follows to the case of (L) or (C).

We shall partition S_β into two subsets $S_{\beta'}$ and $S_{\beta''}$, where $S_{\beta'}$ contains all those resistive chords of the tree T associated with fundamental loops containing at least one resistive tree-branch, and $S_{\beta''}$ contains all other resistive chords.[3] Similarly, let $S_{\epsilon'}$ contain all those resistive tree-branches associated with fundamental cutsets containing at least one resistive chord, and let $S_{\epsilon''}$ con-

[2] Given a function of the form $\mathbf{y} = \mathbf{Y}(\mathbf{x}, t)$, we denote the matrix of partial derivatives $[\partial Y_i/\partial x_j]$ by $(\partial \mathbf{Y}/\partial \mathbf{x})$.

[3] This method of partitioning is similar to that used in [6].

tain all other resistive tree-branches. Assume that the branches are ordered so that

$$\mathbf{i}_\beta = \begin{bmatrix} \mathbf{i}_{\beta'} \\ \mathbf{i}_{\beta''} \end{bmatrix}, \quad \mathbf{e}_\beta = \begin{bmatrix} \mathbf{e}_{\beta'} \\ \mathbf{e}_{\beta''} \end{bmatrix}, \quad \mathbf{i}_\epsilon = \begin{bmatrix} \mathbf{i}_{\epsilon'} \\ \mathbf{i}_{\beta''} \end{bmatrix},$$

$$\mathbf{e}_\epsilon = \begin{bmatrix} \mathbf{e}_{\epsilon'} \\ \mathbf{e}_{\epsilon''} \end{bmatrix}, \quad F_{\beta\epsilon} = \left[\begin{array}{c|c} F_{\beta'\epsilon'} & 0 \\ \hline 0 & 0 \end{array}\right].$$

Now for the next result, we assume that the resistive branch relations conform to one of the two following cases:

Case I: $(\mathbf{i}_{\beta'}, \mathbf{i}_{\beta''}, \mathbf{i}_{\epsilon'}, \mathbf{e}_{\epsilon''})$ are given as explicit functions of $(\mathbf{e}_{\beta'}, \mathbf{e}_{\beta''}, \mathbf{e}_{\epsilon'}, \mathbf{i}_{\epsilon''}, t)$,

Case II: $(\mathbf{e}_{\beta'}, \mathbf{i}_{\beta''}, \mathbf{e}_{\epsilon'}, \mathbf{e}_{\epsilon''})$ are given as explicit functions of $(\mathbf{i}_{\beta'}, \mathbf{e}_{\beta''}, \mathbf{i}_{\epsilon'}, \mathbf{i}_{\epsilon''}, t)$,

where all functions are assumed continuous in t with continuous bounded partials in all other arguments. Under the above assumptions, a careful examination of (R) reveals that the double-primed variables are expressed explicitly in terms of $\mathbf{i}_\gamma$, $\mathbf{e}_\delta$, and the primed variables. Theorem 2 may be adapted to the task of solving for the remaining (primed) variables as follows.

Corollary 2R1:

a) Let the resistive branch relations be expressible in the form of Case I above, and let the matrix

$$[-F^t_{\beta'\epsilon'}, I_{\epsilon'\epsilon'}]\left[\frac{\partial(\mathbf{i}_{\beta'}, \mathbf{i}_{\epsilon'})}{\partial(\mathbf{e}_{\beta'}, \mathbf{e}_{\epsilon'})}\right]\begin{bmatrix} -F_{\beta'\epsilon'} \\ I_{\epsilon'\epsilon'} \end{bmatrix} \tag{8}$$

be u.p.d. in $(\mathbf{e}_\beta, \mathbf{e}_{\epsilon'}, \mathbf{i}_{\epsilon''})$; or

b) let the resistive branch relations be expressible in the form of Case II above, and let the matrix

$$[I_{\beta'\beta'}, F_{\beta'\epsilon'}]\left[\frac{\partial(\mathbf{e}_{\beta'}, \mathbf{e}_{\epsilon'})}{\partial(\mathbf{i}_{\beta'}, \mathbf{i}_{\epsilon'})}\right]\begin{bmatrix} I_{\beta'\beta'} \\ F^t_{\beta'\epsilon'} \end{bmatrix} \tag{9}$$

be u.p.d. in $(\mathbf{i}_{\beta'}, \mathbf{e}_{\beta''}, \mathbf{i}_\epsilon)$; then there exist unique functions $\mathbf{I}_\beta(\mathbf{i}_\gamma, \mathbf{e}_\delta, t)$, $\mathbf{E}_\epsilon(\mathbf{i}_\gamma, \mathbf{e}_\delta, t)$, continuous in t with continuous bounded partials in $(\mathbf{i}_\gamma, \mathbf{e}_\delta)$ which satisfy (R). Furthermore, these functions can be obtained by a convergent iteration formula of the form of (7). (See Appendix for proof.)

In the special case where there is no coupling between members of $S_{\beta'}$ and $S_{\epsilon'}$, the matrices (8) and (9) take the simpler forms

$$[F^t_{\beta'\epsilon'}(\partial\mathbf{i}_{\beta'}/\partial\mathbf{e}_{\beta'})F_{\beta'\epsilon'} + (\partial\mathbf{i}_{\epsilon'}/\partial\mathbf{e}_{\epsilon'})]$$

and

$$[(\partial\mathbf{e}_{\beta'}/\partial\mathbf{i}_{\beta'}) + F_{\beta'\epsilon'}(\partial\mathbf{e}_{\epsilon'}/\partial\mathbf{i}_{\epsilon'})F^t_{\beta'\epsilon'}],$$

respectively.

The matrices which appear in Corollary 2R1 are, respectively, the incremental cutset conductance and loop resistance matrices for those portions of the network in which resistive elements interact with each other. The imposition of the u.p.d. condition on these matrices means, roughly speaking, that they must behave locally as if they represent linear passive reciprocal resistive networks. Note that individual elements are *not* required to fulfill a u.p.d. condition and that no u.p.d. restrictions are placed on resistive elements in $S_{\beta''}$ and $S_{\epsilon''}$. Thus, for example, a wide variety of network configurations containing voltage and current-controlled negative resistances would satisfy the conditions of the corollary.

Another useful result, which does not depend upon the aforementioned method of partitioning follows.

Corollary 2R2: Given a set of resistive elements whose branch relations are continuous in t with continuous bounded and u.p.d. incremental resistance or conductance matrix, (R) can be solved uniquely by iteration, no matter how these elements are imbedded in a network. Furthermore, the solutions will have the smoothness properties specified in Corollary 2R1. (See Appendix for proof.)

Note that Corollary 2R2 requires that all resistive elements be reciprocal, while Corollary 2R1 requires only that certain portions of the network "look" reciprocal when viewed from certain ports. The more restrictive conditions stated in Corollary 2R2 were imposed so that conclusions concerning solvability could be drawn without reference to the topology of the network. It should be observed that the above corollaries [and their analogues for (L) and (C)] indicate conditions under which we can compute unique solutions of (L), (C), and (R), possessing the smoothness properties required in Theorem 1. Therefore, the corollaries define a class of networks for which we may construct a set of normal form equations possessing unique solutions.

B. Solution by Contraction Mapping

In this section we shall delineate a class of networks in which, by certain simple manipulations, (L), (C), and (R) can each be converted to the form of contraction mappings. We then make use of the well-known properties of these mappings to prove the existence of unique solutions of (L), (C), and (R) and to demonstrate convergent iteration procedures for obtaining these solutions. We shall use the following definition and theorem.

Definition 3: A square matrix $W(\mathbf{u}, t)$ will be said to be uniformly Hadamard (u.H.) in $\mathbf{u}$ if, for each t, there exists a $\mu > 0$ such that

$$W_{ii} - \sum_{i \neq j} |W_{ij}| \geq \mu \quad \text{for all } \mathbf{u} \text{ and all } i.$$

Note that both the u.p.d. and u.H. properties are closely related to properties of parameter matrices of linear passive networks. (Constant dominant matrices are special cases of u.H. matrices.)

Theorem 3: Consider an implicit equation of the form

$$\mathbf{x} = \mathbf{h}(A\mathbf{x} + \mathbf{y}, t) \tag{10}$$

where the function $\mathbf{h}(\mathbf{u}, t)$ is continuous in t with continuous bounded derivatives in $\mathbf{u}$; $\mathbf{h}$, an n-vector; $\mathbf{u}$ and $\mathbf{y}$, k-vectors; and A, a $k \times n$ matrix. If the matrix $W =$

$[I - (\partial \mathbf{h}/\partial \mathbf{u})A]$ is u.H. in $\mathbf{u}$, then (10) has a unique solution: $\mathbf{x} = \mathbf{g}(\mathbf{y}, t)$ continuous in t with continuous bounded derivatives in $\mathbf{y}$. Furthermore, this solution can be computed using a convergent iteration of the form

$$\mathbf{x}_{k+1} = \mathbf{x}_k - \Delta[\mathbf{h}(A\mathbf{x}_k + \mathbf{y}, t) - \mathbf{x}_k] \tag{11}$$

where Δ is a constant nonsingular diagonal matrix (chosen in such a way as to make (11) a contraction). See Appendix for proof.

Theorem 3 may be adapted to the solution of (L), (C), and (R) in various ways. We shall consider only one example here (for the case of (R)).

Corollary 3R1: Let the resistive branch variables $(\mathbf{i}_\beta, \mathbf{e}_\epsilon)$ be explicitly expressible in terms of $(\mathbf{e}_\beta, \mathbf{i}_\epsilon, t)$ where all functions are assumed continuous in t with continuous bounded derivatives in all other variables. Furthermore, let there be no coupling between members of $S_{\beta'}$ and $S_{\epsilon'}$ (where the branches have been partitioned in the manner of Section IV–A). Define the matrices

$$W_1 = [I_{\beta'\beta'} + M_{\beta'\beta'}F_{\beta'\epsilon'}M_{\epsilon'\epsilon'}F^t_{\beta'\epsilon'}],$$
$$W_2 = [I_{\epsilon'\epsilon'} + M_{\epsilon'\epsilon'}F^t_{\beta'\epsilon'}M_{\beta'\beta'}F_{\beta'\epsilon'}]. \tag{12}$$

If either W_1 or W_2 is u.H. in $(\mathbf{e}_\beta, \mathbf{i}_\epsilon)$, then there exist unique functions $\mathbf{I}_\beta(\mathbf{i}_\gamma, \mathbf{e}_\delta, t)$, $\mathbf{E}_\epsilon(\mathbf{i}_\gamma, \mathbf{e}_\delta, t)$ continuous in t with continuous bounded derivatives in $(\mathbf{i}_\gamma, \mathbf{e}_\delta)$ which satisfy (R). Furthermore, these functions can be obtained by a convergent iteration formula of the form of (11). (See Appendix for proof.)

The matrix W_1 appearing in Corollary 3R1 is directly related to the incremental loop resistance matrix for those loops in which resistive elements interact with each other. Imposition of the u.H. condition on W_1 means that the magnitude of the total incremental self-resistance of each of these loops must be greater than the sum of the magnitudes of the incremental mutual resistances between loops of the same type. An analogous interpretation holds for W_2.

C. Equations Solvable by Linear Operations

In many cases (L), (C), and (R) can be solved by linear operations alone. In such cases, the conditions stated above can be considerably relaxed. Some of these situations are considered below.

Linear Networks: Let the resistive branch relations be expressible in the linear form,

$$\begin{bmatrix} \mathbf{i}_\beta \\ \mathbf{e}_\epsilon \end{bmatrix} = [M(t)]\begin{bmatrix} \mathbf{e}_\beta \\ \mathbf{i}_\epsilon \end{bmatrix} + \begin{bmatrix} \mathbf{J}(t) \\ \mathbf{V}(t) \end{bmatrix}$$

where $\mathbf{J}(t)$ and $\mathbf{V}(t)$ represent independent current and voltage sources. (Effects of dependent sources are included in M.) Then it can be shown by direct substitution that (R) is uniquely solvable, provided that the matrix

$$\left[I - M\begin{bmatrix} 0 & -F_{\beta\epsilon} \\ F^t_{\beta\epsilon} & 0 \end{bmatrix}\right]$$

is nonsingular. Assuming that this is the case and that analogous conditions hold for the inductive and capacitive elements, (E) can be written explicitly in the linear form $\dot{\mathbf{x}} = A(t)\mathbf{x} + \mathbf{b}(t)$.

Complete Networks: Networks in which S_α and S_ζ are empty and $F_{\beta\epsilon} = 0$ are called "complete," (See [4]). In terms of the graph, this means that there are no all-inductor cutsets, no all-capacitor loops, and there are no loops in which resistive branches appear as both chords and tree-branches. In such cases, it will be observed that (L), (C), and (R) can be solved by direct substitution, provided that the branch relations are expressible in the explicit forms

$$\mathbf{i}_L = \mathbf{i}_L(\boldsymbol{\lambda}_L, t) \qquad \mathbf{i}_\beta = \mathbf{i}_\beta(\mathbf{e}_\beta, \mathbf{i}_\epsilon, t)$$
$$\mathbf{e}_C = \mathbf{e}_C(\mathbf{q}_C, t) \qquad \mathbf{e}_\epsilon = \mathbf{e}_\epsilon(\mathbf{e}_\beta, \mathbf{i}_\epsilon, t).$$

Others: One can conceive of many other special cases. We shall merely state one example. Suppose the resistive branch variables $(\mathbf{i}_\beta, \mathbf{e}_\epsilon)$ are explicitly expressible in terms of $(\mathbf{e}_\beta, \mathbf{i}_\epsilon, t)$ and that the primed branch relations can be expressed in the form

$$\mathbf{i}_{\beta'} = \mathbf{f}(\mathbf{e}_\beta, \mathbf{i}_{\epsilon''}, t) + M_{\beta'\epsilon'}\mathbf{i}_{\epsilon'},$$
$$\mathbf{e}_{\epsilon'} = M_{\epsilon'\beta'}\mathbf{e}_{\beta'} + \mathbf{g}(\mathbf{e}_{\beta''}, \mathbf{i}_{\epsilon''}, t). \tag{13}$$

Then it can be shown by direct substitution that (R) is uniquely solvable provided that the matrices $[I - M_{\beta'\epsilon'}F^t_{\beta'\epsilon'}]$, $[I + M_{\epsilon'\beta'}F_{\beta'\epsilon'}]$ are each nonsingular.

V. An Example

The results of Section IV will now be illustrated with the aid of a fairly simple example. For the network of Fig. 2(a), an appropriate tree T has been chosen as indicated in Fig. 2(b). The symbols in parentheses in the figure indicate the subsets to which each branch belongs. Subject to this choice of T, the state variables are

$$\mathbf{q} = (q_5 + q_1, q_6), \qquad \boldsymbol{\lambda} = \lambda_4 + \lambda_8,$$

and (R) takes the form

$$\mathbf{f}_R(\mathbf{i}_R, \mathbf{e}_R, t) = \mathbf{0}, \qquad e_2 + e_5 + e_7 = 0,$$
$$-i_2 + i_4 + i_7 = 0, \qquad e_3 - e_6 = 0$$

where

$$\mathbf{e}_R = (e_2, e_3, e_7)$$

and

$$\mathbf{i}_R = (i_2, i_3, i_7).$$

We shall consider several possible choices of the resistive branches, noting first that Branch 3, being a member of $S_{\beta''}$, can be ignored provided that its branch relation can be solved explicitly for i_3. We shall, henceforth, make that assumption.

1) Assume that Branch 2 is a linear positive fixed resistor with resistance R_2, and Branch 7 is a current-

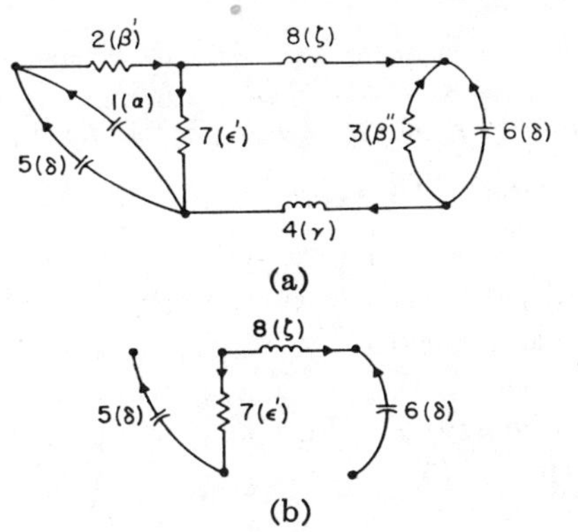

Fig. 2. (a) Network. (b) Tree T.

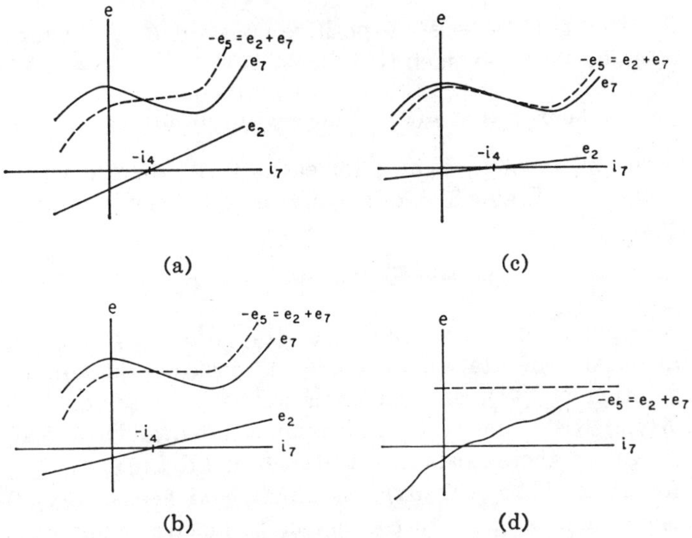

Fig. 3. Composite branch relations.

controlled negative resistance (CCNR) with incremental resistance $R_7(i_7)$. Attempting to apply Corollary 2R1 to this case, the pertinent matrix (9) becomes a single element $[R_2 + R_7(i_7)]$. Solvability of (R), therefore, depends only on the relative magnitudes of R_2 and R_7. The composite function $(e_2 + e_7)$ is plotted as a function of i_7 in Fig. 3(a) for an arbitrary fixed value of i_4. Note that in this case, $[R_2 + R_7(i_7)]$ is positive and bounded away from zero for all i_7. Thus, the conditions of Corollary 2R1 are fulfilled, and (R) possesses the desired solution. It should be clear from the figure that every value of (e_5, i_4) determines a unique value of i_2.

2) In Fig. 3(b), the situation is the same as in 1) above, except that the value of R_2 has been decreased so that it is exactly equal in magnitude to the largest value of the negative slope of $e_7(i_7)$. Hence, the composite characteristic has a point of zero slope, and its inverse has a point of infinite slope. Under such circumstances, the resultant network differential equations will not obey a Lipschitz condition, and hence, multiple solutions may exist. Clearly, the u.p.d. condition of Corollary 2R1 is violated in this case.

3) In Fig. 3(c), R_2 has been decreased still further so that the composite characteristic has a region of negative slope with the result that multiple solutions of (R) occur. Again, the u.p.d. condition is violated.

4) In Fig. 3(d), different resistive elements have been used in Branches 2 and 7; and the composite characteristic has positive slope everywhere, but its slope is not bounded away from zero. As a consequence, there are values of capacitor voltage e_5 which are prohibited. Again the u.p.d. condition does not hold.

5) Let Branch 2 be a voltage-controlled negative resistance, and let Branch 7 be a CCNR. Since Corollaries 2R1 and 2R2 do not apply in this case, we try Corollary 3R1. The pertinent matrix is $W_1(e_2, i_7) = [1 + G_2(e_2)R_7(i_7)]$, where G_2 and R_7 represent the incremental conductance of Branch 2 and incremental resistance of Branch 7, respectively. The u.H. condition requires that $(1 + G_2R_7)$ be positive and bounded away from zero for all (e_2, i_7). This condition is just sufficient to ensure the existence of a unique solution with the desired smoothness properties.

6) Let Branches 2 and 7 constitute two coils of an ideal transformer so that the primed branch relations take the form

$$i_2 = ni_7, \qquad e_7 = -ne_2$$

where n is the turns-ratio. This is a special case of (13); and thus, (R) can be solved uniquely, provided that the pertinent matrices (which, in this case, are each the single element $(1 - n)$) are nonsingular. Note that as $n \to 1$, the matrices become singular, the capacitors 1 and 5 are effectively short-circuited, and thus, the chosen state variables do not form a complete set.

VI. Conclusions

A systematic method of selecting state variables and formulating the normal form equations of nonlinear networks has been presented. For the classes of networks considered, procedures have been indicated for solving the necessary nonlinear implicit equations by iteration. Conditions for convergence of the iterations are closely related to certain physical properties of the network. These procedures are readily adaptable to machine computation and, in fact, can be made an integral part of any numerical method used for solving the differential equations of the network.

Appendix

Proof of Theorem 2: To prove part (*a*), we must show that for each $(\mathbf{y}, t)$, there exists a unique value of $\mathbf{x}$ satisfying

$$\mathbf{G}(\mathbf{x}, t) = \mathbf{Y}(\mathbf{x}, t) - \mathbf{y} = \mathbf{0}. \tag{14}$$

Letting

$$\phi(\mathbf{x}, t) = \int_{\mathbf{x}_l}^{\mathbf{x}} \mathbf{G}^t(\mathbf{u}, t)\, d\mathbf{u} \tag{15}$$

and noting that $\partial\mathbf{G}/\partial\mathbf{x}$ is u.p.d., we find that ϕ is strictly convex in $\mathbf{x}$ (See [9], p. 92ff.) and has the property that for each t, there exists a positive constant β, a constant α, and a vector $\mathbf{b}$, such that

$$\phi(\mathbf{x}, t) + \alpha \geq \beta(\mathbf{x} - \mathbf{b})^t(\mathbf{x} - \mathbf{b}) \quad \text{for all} \quad \mathbf{x} \tag{16}$$

(See [10] p. 83ff.). Thus, for each $(\mathbf{y}, t)$, there exists a point $\mathbf{x} = \mathbf{a}$ at which $\phi(\mathbf{x}, t)$ has a global minimum. At this point,

$$[\partial\phi/\partial\mathbf{x}]^t = \mathbf{Y}(\mathbf{a}, t) - \mathbf{y} = \mathbf{0},$$

showing that $\mathbf{a}$ is a solution of (14). Since ϕ can have, at most, one stationary point, $\mathbf{a}$ must be the unique solution of (14), and hence, a unique inverse function $\mathbf{X}(\mathbf{y}, t)$ exists. Since $(\partial\mathbf{Y}/\partial\mathbf{x})$ is assumed u.p.d. and bounded, for each t, there exists an $\epsilon > 0$ such that det $(\partial\mathbf{Y}/\partial\mathbf{x}) \geq \epsilon$ for all $\mathbf{x}$. Thus, $(\partial\mathbf{Y}/\partial\mathbf{x})^{-1}$ is u.p.d., and hence, $\mathbf{X}(\mathbf{y}, t)$ is continuous in t with continuous bounded partials in $\mathbf{y}$. This completes the proof of part (a).

To prove part (b), we show that for sufficiently small positive h, the sequence $\{\mathbf{x}_k\}$ generated by (7) converges from any initial trial $\mathbf{x}_0$ to the point $\mathbf{x} = \mathbf{a}$ which is the unique solution of (14). (In the proof, we shall simplify notation by omitting the variable t, since it is treated merely as a constant parameter throughout.)

Take $\mathbf{x}_l = \mathbf{a}$ as the lower limit of integration in (15), so that ϕ is everywhere non-negative. Letting $\Delta\phi(\mathbf{x}_k) = \phi(\mathbf{x}_{k+1}) - \phi(\mathbf{x}_k)$, we may use Taylor's formula and (7) to obtain

$$\Delta\phi(\mathbf{x}_k) = -h\mathbf{G}^t(\mathbf{x}_k)\mathbf{G}(\mathbf{x}_k) + \tfrac{1}{2}h^2\mathbf{G}^t(\mathbf{x}_k)[\partial\mathbf{G}/\partial\mathbf{x}]_{\xi_k}\mathbf{G}(\mathbf{x}_k)$$

$$\xi_k = \lambda\mathbf{x}_k + (1 - \lambda)\mathbf{x}_{k+1} \qquad 0 < \lambda < 1.$$

Since $(\partial\mathbf{G}/\partial\mathbf{x})$ is bounded, h can always be chosen sufficiently small (and positive), so that for all $\mathbf{x}_k$

$$\Delta\phi(\mathbf{x}_k) \leq -M\mathbf{G}^t(\mathbf{x}_k)\mathbf{G}(\mathbf{x}_k), \qquad M > 0. \tag{17}$$

Now for any $\mathbf{x}_0$, there exists an R such that

$$\phi(\mathbf{x}) > \phi(\mathbf{x}_0) \quad \text{for all} \quad \mathbf{x} : ||\mathbf{x} - \mathbf{a}|| > R. \tag{18}$$

Given any $\epsilon > 0$, consider the set, $S : \epsilon \leq ||\mathbf{x} - \mathbf{a}|| \leq R$. On S, the function $\mathbf{G}^t(\mathbf{x})\mathbf{G}(\mathbf{x})$ has a minimum $b > 0$, and thus,

$$\Delta\phi(\mathbf{x}_k) \leq -Mb < 0 \quad \text{for all} \quad \mathbf{x}_k \in S.$$

Assuming that $\mathbf{x}_k \in S$ for all k,

$$\phi(\mathbf{x}_k) \leq \phi(\mathbf{x}_0) - kMb \quad \text{for all} \quad k.$$

This implies that $\phi(\mathbf{x}_k)$ will eventually become negative (an impossibility) contradicting our assumption. Thus, for sufficiently large k, $\mathbf{x}_k$ must remain outside of S. But from condition (17) and (18), $||\mathbf{x}_k - \mathbf{a}|| \leq R$ for all k. Thus, there must exist an N such that $k > N$ implies $||\mathbf{x}_k - \mathbf{a}|| < \epsilon$, which completes the proof of convergence.

Proof of Corollary 2R1: We prove only part (a). The proof for part (b) is completely analogous.

Substituting the branch relations for the primed variables into (R) and combining equations, we have

$$F^t_{\gamma\epsilon'}\mathbf{i}_\gamma$$
$$= -F^t_{\beta'\epsilon'}\mathbf{i}_{\beta'}\,(-F_{\beta'\epsilon'}\mathbf{e}_{\epsilon'} - F_{\beta'\delta}\mathbf{e}_\delta, -F_{\beta''\delta}\mathbf{e}_\delta, \mathbf{e}_{\epsilon'}, F^t_{\gamma\epsilon''}\mathbf{i}_\gamma, t)$$
$$+ \mathbf{i}_{\epsilon'}(-F_{\beta'\epsilon'}\mathbf{e}_{\epsilon'} - F_{\beta'\delta}\mathbf{e}_\delta, -F_{\beta''\delta}\mathbf{e}_\delta, \mathbf{e}_{\epsilon'}, F^t_{\gamma\epsilon''}\mathbf{i}_\gamma, t) \tag{19}$$

where the F_{ij} appearing in (19) represent submatrices of those in (2), defined by the partitioning of S_β and S_ϵ. Making the identification $\mathbf{y} = F^t_{\gamma\epsilon'}\mathbf{i}_\gamma$, $\mathbf{x} = \mathbf{e}_{\epsilon'}$, we have a function of the form $\mathbf{y} = \mathbf{Y}(\mathbf{x}, t)$ which must be solved for $\mathbf{x}$ as in Theorem 2. (We consider all variables on the right-hand side of (6) other than $\mathbf{e}_{\epsilon'}$ as fixed parameters.) Noting that

$$(\partial\mathbf{Y}/\partial\mathbf{x}) = [-F^t_{\beta'\epsilon'}, I_{\epsilon'\epsilon'}]\left[\frac{\partial(\mathbf{i}_{\beta'}, \mathbf{i}_{\epsilon'})}{\partial(\mathbf{e}_{\beta'}, \mathbf{e}_{\epsilon'})}\right]\begin{bmatrix} -F_{\beta'\epsilon'} \\ I_{\epsilon'\epsilon'} \end{bmatrix}$$

we find that (19) satisfies the conditions of Theorem 2 and can, therefore, be solved uniquely for $\mathbf{e}_{\epsilon'}$ using a convergent iteration formula of the form of (7). The remaining variables are determined by direct substitution into (R). Thus, two functions $\mathbf{E}_\epsilon$ and $\mathbf{I}_\beta$ can be determined which, by Theorem 2 (and the assumed boundedness, continuity, and u.p.d. conditions), must be continuous in t with continuous bounded partials in $\mathbf{i}_\gamma$ and $\mathbf{e}_\delta$. This completes the proof.

Proof of Corollary 2R2: It is easily shown (using Theorem 2) that the conditions of the corollary imply that the branch relations can be expressed in the forms required by Corollary 2R1, where the matrix $\partial(\mathbf{i}_{\beta'}, \mathbf{i}_{\epsilon'})/\partial(\mathbf{e}_{\beta'}, \mathbf{e}_{\epsilon'})$ is continuous in t, bounded, and u.p.d. in $(\mathbf{e}_\beta, \mathbf{e}_{\epsilon'}, \mathbf{i}_{\epsilon''})$. Thus, if we can show that the matrix (8) is also u.p.d., we can invoke Corollary 2R1 to prove 2R2. Clearly (8) is symmetric; furthermore, for each t, there exists a $\mu > 0$ such that

$$\mathbf{u}^t[-F^t_{\beta'\epsilon'}, I_{\epsilon'\epsilon'}]\left[\frac{\partial(\mathbf{i}_{\beta'}, \mathbf{i}_{\epsilon'})}{\partial(\mathbf{e}_{\beta'}, \mathbf{e}_{\epsilon'})}\right]\begin{bmatrix} -F_{\beta'\epsilon'} \\ I_{\epsilon'\epsilon'} \end{bmatrix}\mathbf{u}$$
$$\geq \mu\mathbf{u}^t[F^t_{\beta'\epsilon'}F_{\beta'\epsilon'} + I_{\epsilon'\epsilon'}]\mathbf{u}$$

for all $(\mathbf{e}_\beta, \mathbf{e}_{\epsilon'}, \mathbf{i}_{\epsilon''}, \mathbf{u})$. Since any matrix of the form $[F^tF + I]$ is positive definite, we have

$$\mathbf{u}^t[F^t_{\beta'\epsilon'}F_{\beta'\epsilon'} + I_{\epsilon'\epsilon'}]\mathbf{u} \geq \lambda\mathbf{u}^t\mathbf{u} \quad \text{for all} \quad \mathbf{u},$$

where λ is positive. The two inequalities above show that (8) is u.p.d., completing the proof.

Proof of Theorem 3: (We shall omit the variable t to simplify the notation.) Define the function

$$\boldsymbol{\psi}(\mathbf{x}) = \mathbf{x} - \Delta[\mathbf{h}(A\mathbf{x} + \mathbf{y}) - \mathbf{x}]$$

where $\Delta = \text{diag}\,(-1/M_1, -1/M_2, \cdots, -1/M_n)$, and for each i, M_i is any upper bound of $w_{ii}(\mathbf{x})$. Then we have

$$\sum_{j=1}^{n} |\partial\psi_i/\partial x_j| = 1 - \frac{w_{ii}}{M_i} + \sum_{j\neq i}\left|\frac{w_{ij}}{M_i}\right| \qquad i = 1, 2, \cdots, n.$$

But from the u.H. condition on W, there exists a $\mu > 0$ such that

$$0 < \frac{\mu}{M_i} < \left[\frac{w_{ii}}{M_i} - \sum_{j\neq i}\left|\frac{w_{ij}}{M_i}\right|\right] \leq 1 \quad \text{for all} \quad \mathbf{x}$$
$$i = 1, 2, \cdots, n.$$

Thus,

$$\sum_{j=1}^{n} |\partial \psi_i / \partial x_j| \leq \left|1 - \frac{\mu}{M_i}\right| < 1 \quad \text{for all} \quad \mathbf{x}$$

$$i = 1, 2, \cdots, n. \qquad (20)$$

Condition (20) implies that $\boldsymbol{\psi}(\mathbf{x})$ is a contraction. (This assertion follows easily from the mean value theorem.) Hence, the iteration formula of (11) converges (for each $\mathbf{y}$ t) to the unique solution of (10).

It is easily shown that the u.H. property on W implies that for each t, there exists an $\epsilon > 0$ such that $\det [W] > \epsilon$ for all $\mathbf{x}$. This fact, together with the assumed smoothness and boundedness of $\mathbf{h}$, is sufficient to assure similar properties for $\mathbf{g}(\mathbf{y}, t)$. This completes the proof.

Proof of Corollary 3R1: We prove only the case involving W_1. (The proof for W_2 is completely analogous.) The primed branch relations may be written in the form

$$\mathbf{i}_{\beta'} = \mathbf{I}_{\beta'}(\mathbf{e}_{\beta'}, \mathbf{e}_{\beta''}, \mathbf{i}_{\epsilon''}, t), \qquad \mathbf{e}_{\epsilon'} = \mathbf{E}_{\epsilon'}(\mathbf{e}_{\beta''}, \mathbf{i}_{\epsilon'}, \mathbf{i}_{\epsilon''}, t),$$

and these relations may be substituted into (R) and combined to give

$$\mathbf{i}_{\beta'} = \mathbf{I}_{\beta'}[-F_{\beta'\epsilon'}\mathbf{E}_{\epsilon'}(-F_{\beta''\delta}\mathbf{e}_\delta, F^t_{\beta'\epsilon'}\mathbf{i}_{\beta'} + F^t_{\gamma\epsilon'}\mathbf{i}_\gamma, F^t_{\gamma\epsilon''}\mathbf{i}_\gamma, t)$$
$$- F_{\beta'\delta}\mathbf{e}_\delta, - F_{\beta''\delta}\mathbf{e}_\delta, F^t_{\gamma\epsilon''}\mathbf{i}_\gamma, t] \qquad (21)$$

which is of the form of (10), where

$$\mathbf{x} \rightarrow \mathbf{i}_{\beta'}, \qquad \mathbf{y} \rightarrow F^t_{\gamma\epsilon'}\mathbf{i}_\gamma,$$

$$A \rightarrow F^t_{\beta'\epsilon'}, \qquad W = [I - (\partial \mathbf{h}/\partial \mathbf{u})A] \rightarrow W_1.$$

By hypothesis, W_1 is u.H.; and thus by Theorem 3, (21) can be solved uniquely for $\mathbf{i}_{\beta'}$ by iteration. The resultant function $\mathbf{i}_{\beta'} = \mathbf{z}(\mathbf{i}_\gamma, \mathbf{e}_\delta, t)$ may then be substituted into (R) to obtain the remaining variables. Thus, we can determine $\mathbf{i}_\beta$ and $\mathbf{e}_\epsilon$ as functions of $(\mathbf{i}_\gamma, \mathbf{e}_\delta, t)$. By Theorem 3 (and the assumed boundedness, continuity and u.H. conditions), these functions must be continuous in t with continuous bounded partials in $(\mathbf{i}_\gamma, \mathbf{e}_\delta)$. This completes the proof.

Acknowledgment

The author wishes to express his thanks to S. J. Oh of Columbia University, New York, N. Y., who collaborated on the proofs of the theorems.

References

[1] T. R. Bashkow, "The *A* matrix, a new network description," *IRE Trans. on Circuit Theory*, vol. CT-4, pp. 117–119, September 1957.
[2] P. R. Bryant, "The order of complexity of electrical networks," *IEEE (London)*, Monograph 335E, June 1959.
[3] ——, "The explicit form of Bashkow's *A* matrix," *IRE Trans. on Circuit Theory (Correspondence)*, vol. CT-9, pp. 303–306, September 1962.
[4] R. K. Brayton and J. K. Moser, "A theory of nonlinear networks (I and II)," *Quart. Appl. Math.*, vol. 22, pp. 1–33, April 1964, and pp. 81–104, July 1964.
[5] R. W. Liu and L. V. Auth, "Qualitative synthesis of nonlinear networks," *1963 Proc. 1st Annual Allerton Conf. on Circuit and System Theory*, pp. 330–343.
[6] C. A. Desoer and J. Katzenelson, "Nonlinear RLC networks," *Bell Sys. Tech. J.*, vol. 44, January 1965.
[7] E. A. Coddington and N. Levinson, *Theory of Ordinary Differential Equations*. New York: McGraw-Hill, 1955.
[8] T. M. Apostol, *Mathematical Analysis*. Reading, Mass.: Addison-Wesley, 1957.
[9] W. H. Fleming, *Functions of Several Variables* (preliminary edition). Reading, Mass.: Addison-Wesley, 1963.
[10] T. L. Saaty and J. Bram, *Nonlinear Mathematics*. New York: McGraw-Hill, 1964.

Normal Form and Stability of a Class of Coupled Nonlinear Networks

P. P. VARAIYA AND R. LIU, MEMBER, IEEE

***Abstract*—This paper is concerned with the qualitative properties of a subclass of the class of nonlinear *R, L, C* networks for which there may be mutual coupling among elements of like type. Sufficient conditions are given which guarantee that such a network may be characterized by a set of nonlinear equations in normal form, which indicates a unique response for any excitation. Finally, a study of the stability of networks of this class indicates that for passive networks their asymptotic stability is in one-to-one correspondence with their "weak observability" at the resistor terminals.**

I. Introduction and Notation

THIS PAPER is concerned with the properties of a class of coupled nonlinear R, L, C networks. We first impose conditions on the network topology and on the network elements which insure a unique response defined by a set of differential equations in the normal form. Reviews of this subject are given in references [2], [4], and [9]. The nature of the conditions' results is similar to that given by Desoer and Katzenelson [1] and by Holzmann and Liu [2]. A novel part of the proof is a constructive method of solving a class of nonlinear algebraic equations. Next we study the stability of this class of networks and relate the question of asymptotic stability to the notion of observability [3]. We show that for passive networks with linear inductors and linear capacitors, the notion of asymptotic stability coincides with that of observability at the resistor terminals. For nonlinear passive networks this equivalence is obtained if we suitably weaken the definition of observability.

As far as possible, the notation used here is that of Kuh [4]. Thus let $\mathfrak{N}$ be a nonseparable connected network, and let $\mathfrak{T}$ be a normal tree of $\mathfrak{N}$.[1] We assume that each tree branch is in parallel with a current source and each link contains a voltage source. The sources are assumed independent. We denote the link element voltages (currents) by v_R, v_S, v_L, (i_R, i_S, i_L); the voltage sources, in the links, by e_R, e_S, e_L; the tree branch element voltages (currents) by $v_G, v_C, v_\Gamma (i_G, i_C, i_\Gamma)$; and the current sources, across the tree branches, by j_G, j_C, and j_Γ. Then the Kirchhoff voltage law is given by (see [4])

$$\begin{aligned} v_S + F_{SC}v_C &= e_S, \\ v_R + F_{RC}v_C + F_{RG}v_G &= e_R, \end{aligned} \tag{1}$$

and

$$v_L + F_{LC}v_C + F_{LG}v_G + F_{L\Gamma}v_\Gamma = e_L;$$

and the Kirchhoff current law is given by

$$\begin{aligned} i_C - F_{SC}^T i_S - F_{RC}^T i_R - F_{LC}^T i_L &= j_C, \\ i_G - F_{RG}^T i_R - F_{LG}^T i_L &= j_G, \end{aligned} \tag{2}$$

and

$$i_\Gamma - F_{L\Gamma}^T i_L = j_\Gamma.$$

II. Normal Form

The following conditions are imposed throughout. Let μ stand for R, L, or C. Then there is a normal tree $\mathfrak{T}$ of $\mathfrak{N}$ such that

*C*1. μ-elements in the links are coupled among themselves. Dually, the μ-elements in the tree branches are coupled among themselves. Examples of coupled-elements are given in references [4] and [9].

*C*2. The link resistive and capacitive elements are voltage-controlled and link inductive elements are flux-controlled, whereas in the tree branches the resistive and inductive elements are current-controlled and capacitive elements are charge-controlled. More explicitly, *C*1 and *C*2 become

$$i_R = \hat{i}_R(v_R), \qquad i_L = \hat{i}_L(\phi_L), \quad \text{and} \quad q_S = \hat{q}_S(v_S) \tag{3}$$[2]

and

$$v_G = \hat{v}_G(i_G) \qquad \phi_\Gamma = \hat{\phi}_\Gamma(i_\Gamma) \quad \text{and} \quad v_C = \hat{v}_C(q_C).$$

Equations (1)–(3) can be conveniently rewritten as

$$\begin{aligned} v_R + F_{RG}\hat{v}_G(i_G) &= e_R^* \triangleq e_R - F_{RC}v_C, \\ -F_{RG}^T\hat{i}_R(v_R) + i_G &= j_G^* \triangleq j_G + F_{LG}^T i_L, \end{aligned} \tag{R}$$

$$\begin{aligned} \phi_L + F_{L\Gamma}\hat{\phi}_\Gamma(i_\Gamma) &\triangleq \phi, \\ -F_{L\Gamma}^T\hat{i}_L(\phi_L) + i_\Gamma &\triangleq j_\Gamma, \end{aligned} \tag{L}$$

$$v_S + F_{SC}\hat{v}_C(q_C) = e_S, \tag{C}$$

Manuscript received January 7, 1966; revised August 8, 1966. The research reported herein was supported in part by the Joint Services Electronics Program (U. S. Army, U. S. Navy, and U. S. Air Force) under Grant AF-AFOSR-139-65, and by the National Aeronautics and Space Admininistration under Grant NSG-354, supplement 2.

P. P. Varaiya is with the Department of Electrical Engineering, University of California, Berkeley, Calif.

R. Liu is with the Department of Electrical Engineering, University of Notre Dame, Notre Dame, Ind. From September, 1965, until June, 1966, he was a visiting Associate Professor in the Department of Electrical Engineering of the University of California.

[1] A normal tree is one that contans the maximum number of capacitances and the minimum number of inductances possible (see references [8] and [9]).

[2] To avoid notational problems we assume that the network is time-invariant, although the sources are time-variable. The same procedure applies for time-varying networks.

Reprinted from *IEEE Transactions on Circuit Theory*, vol. CT-13, pp. 413–418, December 1966.

and

$$-F_{SC}^T\hat{q}_S(v_S) + q_C \triangleq q.$$

We also have the set of differential equations

$$\dot{q} = F_{LC}^T i_L + F_{RC}^T i_R + j_C \tag{D}$$

and

$$\dot{\phi} = -F_{LC}v_C - F_{LG}v_G + e_L.$$

We remark that q is the vector of the fundamental cutset charges and ϕ is the vector of the fundamental loop fluxes. We also notice that the equations R, L, and C are of the following form:

$$x + Af(y) = u \tag{*}$$

and

$$-A^T g(x) + y = v,$$

where $x, u \in R^n$; $y, v \in R^m$; $f : R^m \to R^m$; $g : R^n \to R^n$; and A is a fixed $n \times m$ matrix. We now state some sufficient conditions on f and g such that (*) has a unique solution in x and y for each value of u and v.

Theorem 1.1: If the functions f and g satisfy conditions $H1$ and $H2$ or they satisfy conditions $H1$ and $H3$, then (*) has a unique solution.

$H1$. f and g are continuously differentiable and the Jacobian matrices $F(y) \triangleq (\partial f/\partial y)(y)$ and $G(x) \triangleq (\partial g/\partial x)(x)$ are positive semidefinite[3] for all x and y.

$H2$. Either $F(y)$ is a symmetric positive definite matrix for all y or $G(x)$ is a symmetric positive definite matrix for all x.

$H3$. Either $F(y)$ is diagonal for all y or $G(x)$ is diagonal for all x.

Proof: We wish to determine the solutions (if any) to the set of equations

$$\begin{aligned} x + Af(y) &= u \\ -A^T g(x) + y &= v. \end{aligned} \tag{*}$$

Define

$$\alpha(x, y) \triangleq \tfrac{1}{2}\{\|x + Af(y) - u\|^2 + \|-A^T g(x) + y - v\|^2\}.$$

Then x and y solve (*) if and only if $\alpha(x, y) = 0$. Consider the differential equations

$$\frac{dx}{dt} = -\frac{\partial\alpha}{\partial x} = -\{(x + Af(y) - u) - [A^T G(x)]^T(-A^T g(x) + y - v)\}$$

and

$$\frac{dy}{dt} = -\frac{\partial\alpha}{\alpha y} = -\{[AF(y)]^T(x + Af(y) - u) + (-A^T g(x) + y - v)\}.$$

Along all solutions of the differential equations, we have

$$\frac{d\alpha}{dt} = \frac{\partial\alpha}{\partial x}\frac{dx}{dt} + \frac{\partial\alpha}{\partial y}\frac{dy}{dt} = -\left\|\frac{\partial\alpha}{\partial x}\right\|^2 - \left\|\frac{\partial\alpha}{\partial y}\right\|^2 \le 0.$$

Define

$$z = -A^T g(x) + y - v$$

and

$$w = x + Af(y) - u;$$

then $d\alpha/dt = 0$ if and only if

$$\begin{bmatrix} I & -(A^T G(x))^T \\ (AF(y))^T & I \end{bmatrix}\begin{bmatrix} w \\ z \end{bmatrix} \triangleq \begin{bmatrix} M \end{bmatrix}\begin{bmatrix} w \\ z \end{bmatrix} = 0. \tag{4}$$

Now

$$\det(M) = \det(I + AF(y)A^T G(x)) = \det(I + A^T G(x) AF(y)).$$

In the Appendix we show that either $H2$ or $H3$ implies $\det M \ge 1$. Hence given either $H2$ or $H3$ (4) holds if and only if $w = 0$ and $z = 0$, i.e, $d\alpha/dt = 0$ if and only if x and y solve (*). Also $\det M \ge 1$ implies that $\alpha(x, y) \to \infty$ as $\|x\| + \|y\| \to \infty$. A theorem of Liapunov [5] shows that (*) has at least one solution. We now prove uniqueness. Suppose (x_1, y_1) and (x_2, y_2) solve (*) for some fixed u and v. Then

$$(x_1 - x_2) + A(f(y_1) - f(y_2)) = 0 \tag{5}$$

and

$$-A^T(g(x_1) - g(x_2)) + (y_1 - y_2) = 0.$$

Consider one-dimensional arcs $x(\theta)$ and $y(\theta)$, $0 \le \theta \le 1$ given by

$$x(\theta) = \theta x_1 + (1 - \theta)x_2$$

and

$$y(\theta) = \theta y_1 + (1 - \theta)y_2.$$

Then $dx/d\theta = x_1 - x_2$ and $dy/d\theta = y_1 - y_2$. Furthermore, (5) is equivalent to the following:

$$(x_1 - x_2) + A\int_{\theta=0}^{1} F(y(\theta))(y_1 - y_2)\, d\theta = 0$$

and

$$-A^T\int_{\theta=0}^{1} G(x(\theta))(x_1 - x_2)\, d\theta + (y_1 - y_2) = 0.$$

But $\int_{\theta=0}^{1} F(y(\theta))\, d\theta$ and $\int_{\theta=0}^{1} G(x(\theta))\, d\theta$ have the same properties as $F(y)$ and $G(x)$, respectively, so that (6) holds if and only if $x_1 = x_2$ and $y_1 = y_2$. Q.E.D.

In the following we assume that f and g satisfy the hypotheses of Theorem 1.1.

[3] An $n \times n$ matrix M is positive semidefinite if $\langle x, Mx\rangle \ge 0$ for all x. It is positive definite if $\langle x, Mx\rangle > 0$ for all $x \ne 0$.

Corollary 1.1: The solution of (*) can be obtained as the limit of the solution of a globally asymptotically stable differential equation.

Remark: If we suitably bound the norms of the matrices $F(y)$ and $G(x)$, then the differential equation can be replaced by a difference equation. (See Katzenelson and Seitelman [6].)

Let the unique solutions of (*) be denoted by $x = \tilde{x}(u, v)$ and $y = \tilde{y}(u, v)$, and let $\eta(u, v) \triangleq g(\tilde{x}(u, v))$ and $\xi(u, v) \triangleq f(\tilde{y}(u, v))$. The proofs of the next two corollaries are given in the Appendix.

Corollary 1.2:

a) $(\partial\eta(u, v)/\partial u)$ is positive semidefinite; it is positive definite if $G(x)$ is positive definite for all x. Dually $(\partial\xi(u, v)/\partial v)$ is positive semidefinite; it is positive definite if $F(y)$ is positive definite for all y.

b) If $G(x)$ and $F(y)$ are symmetric matrices, then $(\partial\eta(u, v)/\partial u)$ and $(\partial\xi(u, v)/\partial v)$ are also symmetric.

Corollary 1.3:

a) If $||u|| + ||v|| \to \infty$ then $||x|| + ||y|| \to \infty$.

b) If f and g also satisfy

$$\langle x, g(x)\rangle > 0 \quad \text{for} \quad x \neq 0$$

and

$$\langle y, f(y)\rangle > 0 \quad \text{for} \quad y \neq 0$$

then

$$\langle u, \eta(u, v)\rangle + \langle v, \xi(u, v)\rangle > 0 \quad \text{if} \quad u \neq 0 \quad \text{or} \quad v \neq 0.$$

We now impose the conditions of Theorem 1.1 on the network characteristics.

Theorem 2.1:[4] If each of the equations (L), (C), and (R) satisfies the hypotheses of Theorem 1.1, then the network response is unique and is defined by a differential equation in normal form.

Proof: Theorem 1.1 implies that the equations (R), (L), and (C) can be solved giving,

$$v_G = \tilde{v}_G(e_R^*, j_G^*), \qquad i_R = \tilde{\imath}_R(e_R^*, j_G^*), \tag{7}$$

$$i_L = \tilde{\imath}_L(\phi, j_\Gamma), \quad \text{and} \quad v_C = \tilde{v}_C(e_S, q). \tag{8}$$

Furthermore since $e_R^* \triangleq e_R - F_{RC}v_C$ and $j_G^* \triangleq j_G + F_{LG}^T i_L$ we can obtain [using (8)], v_G and i_R in terms of q, ϕ, e_R, and j_G. Substituting these functions in (D) we obtain the right-hand side of (D) as a function of q, ϕ, and the sources. Further, since all the functions in (7) and (8) are continuously differentiable, the solution of the final form of (D) exists and is unique.[5] Q.E.D.

[4] If the elements are uncoupled this theorem is equivalent to the result of Desoer and Katzenelson [1], except that we require differentiability of the characteristics.

[5] The existence and uniqueness of the solution hold on some nonvanishing interval (t_0, t_1) where t_0 is any initial time. (See reference [11]).

Conditions $C1$ and $C2$ impose important restrictions on the allowable network configurations. Thus the resistive elements can only be coupled among themselves and those which are voltage-controlled must form links, whereas those which are current-controlled must be tree branches. Thus for example, a current-controlled voltage source must have both its input and output branches in the tree. To illustrate Theorem 2.1 we derive the normal form equations for the network of Fig. 1.

The gyrator is considered as a pair of nonlinear, coupled, controlled resistors with characteristic: $v_{G1} = v_1(i_{G1}, i_{G2})$ $v_{G2} = v_2(i_{G1}, i_{G3})$. The third resistor is current-controlled with its characteristic expressed as $v_{G3} = v_3(i_{G3})$. The two inductors are coupled and flux-controlled with characteristics $i_{Lj} = i_j(\phi_{L1}, \phi_{L2})$ $j = 1, 2$. The nonlinear capacitors are uncoupled and charge-controlled, with $v_{Ck} = \bar{v}_k(q_{Ck})$ $k = 1, 2, 3$. The normal tree is clearly unique and consists of all the capacitive and resistive branches. Corresponding to (R), (L), (C) and (D)we have

$$\begin{bmatrix} i_{G1} \\ i_{G2} \\ i_{G3} \end{bmatrix} = j_G^* \triangleq \begin{bmatrix} 1 & 0 \\ 0 & 1 \\ 0 & -1 \end{bmatrix} \begin{bmatrix} i_{L1} \\ i_{L2} \end{bmatrix} \tag{R'}$$

$$\begin{bmatrix} \phi_{L1} \\ \phi_{L2} \end{bmatrix} \triangleq \phi \tag{L'}$$

$$\begin{bmatrix} q_{C1} \\ q_{C2} \\ q_{C3} \end{bmatrix} \triangleq q \tag{C'}$$

and

$$\dot{q} = \begin{bmatrix} 1 & 0 \\ 0 & 1 \\ 1 & 1 \end{bmatrix} \begin{bmatrix} i_{L1} \\ i_{L2} \end{bmatrix}$$

$$\dot{\phi} = -\begin{bmatrix} 1 & 0 & 1 \\ 0 & 1 & 1 \end{bmatrix} \begin{bmatrix} v_{C1} \\ v_{C2} \\ v_{C3} \end{bmatrix} - \begin{bmatrix} 1 & 0 & 0 \\ 0 & 1 & -1 \end{bmatrix} \begin{bmatrix} v_{G1} \\ v_{G2} \\ v_{G3} \end{bmatrix} + \begin{bmatrix} e \\ 0 \\ 0 \end{bmatrix}. \tag{D'}$$

Solving (R'), (L'), and (C'), and substituting in (D') we obtain the normal form equations:

$$\dot{q}_{C1} = i_1(\phi_{L1}, \phi_{L2})$$

$$\dot{q}_{C2} = i_2(\phi_{L1}, \phi_{L2})$$

$$\dot{q}_{C3} = i_1(\phi_{L1}, \phi_{L2}) + i_2(\phi_{L1}, \phi_{L2})$$

$$\dot{\phi}_{L1} = -(\bar{v}_1(q_{C1}) + \bar{v}_3(q_{C3})) - v_1(i_1(\phi_{L1}, \phi_{L2}), i_2(\phi_{L1}, \phi_{L2})) + e$$

$$\dot{\phi}_{L2} = -(\bar{v}_2(q_{C2}) + \bar{v}_3(q_{C3})) - v_2(i_1(\phi_{L1}, \phi_{L2}), i_2(\phi_{L1}, \phi_{L2})) + v_3(-i_2(\phi_{L1}, \phi_{L2})).$$

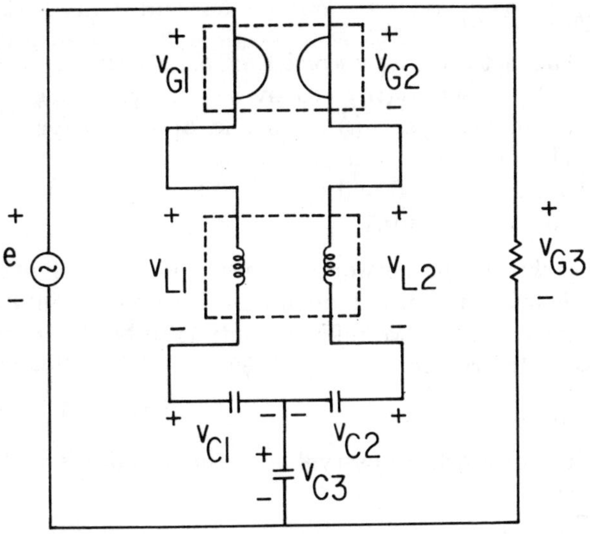

Fig. 1.

III. Stability

A. From now on we assume that all the sources are identically zero. We also suppose that the network satisfies all the hypotheses of Theorem 2.1 and in addition the following conditions:

*H*4. The resistors are passive, i.e.,

$$\langle v_R, \hat{\imath}_R(v_R)\rangle > 0 \quad if \quad v_R \neq 0$$

and

$$\langle i_G, \hat{v}_G(i_G)\rangle > 0 \quad \text{if} \quad i_G \neq 0.$$

*H*5. The inductors and capacitors are passive, i.e.,

$$\langle i_\Gamma, \hat{\phi}_\Gamma(i_\Gamma)\rangle > 0 \quad \text{if} \quad i_\Gamma \neq 0,$$

$$\langle \phi_L, \hat{\imath}_L(\phi_L)\rangle > 0 \quad \text{if} \quad \phi_L \neq 0,$$

$$\langle v_S, \hat{q}_S(v_S)\rangle > 0 \quad \text{if} \quad v_S \neq 0,$$

and

$$\langle q_C, \hat{v}_C(q_C)\rangle > 0 \quad \text{if} \quad q_C \neq 0.$$

Furthermore, it will be assumed that the Jacobian matrices associated with the inductors and capacitors, i.e., the matrices $\partial\phi_\Gamma/\partial i_\Gamma$, $\partial i_L/\partial\phi_L$, $\partial q_S/\partial v_S$, and $\partial v_C/\partial q_C$ are symmetric.

By the well-known theorem on exact differential forms [7] we obtain

Lemma 3.1: The hypothesis *H*5 implies that there are real-valued functions p_Γ, p_L, p_S, and p_C of the variables i_Γ, ϕ_L, v_S and q_C such that,

$$\frac{\partial p_\Gamma}{\partial i_\Gamma} = \phi_\Gamma, \quad \frac{\partial p_L}{\partial \phi_L} = i_L, \quad \frac{\partial p_S}{\partial v_S} = q_S, \quad \text{and} \quad \frac{\partial p_C}{\partial q_C} = v_S.$$

Furthermore, each of the p_i is positive definite, i.e., $p_i \geq 0$ and $p_i(x_i) = 0$ if and only if $x_i = 0$.

We now make assumption

*H*6. The tree branch capacitors and the link inductors are realistic, i.e.,

$$p_C(q_C) + p_L(\phi_L) \to \infty \quad \text{as} \quad \|q_C\| + \|\phi_L\| \to \infty.$$

Remark: *H*5 is equivalent to saying that the inductors and capacitors represent a conservative system. *H*6 is equivalent to saying that as the charges in the tree capacitors or the flux in the link inductors become unbounded, the associated stored energy also becomes unbounded.

B. A built-in Liapunov function: Let q and ϕ be fixed and consider,

$$p(q, \phi) \triangleq \int_0^q \langle v_C, dq'\rangle + \int_0^\phi \langle i_L, d\phi'\rangle. \tag{9}$$

We first remark that assumption *H*5 and Corollary 1.2b imply that $\partial v_C/\partial q$ and $\partial i_L/\partial\phi$ are symmetric matrices so that the integral in (9) is independent of the path of integration. Therefore $p(q, \phi)$ is well defined.

Lemma 3.2:

a) $p(q, \phi) \geq 0$ for all q, ϕ and $p(q, \phi) = 0$ if and only if $q = 0$ and $\phi = 0$.

b) Furthermore, $p(q, \phi) \to \infty$ as $\|q\| + \|\phi\| \to \infty$.

Proof:

a) By Corollary 1.2a, $\partial v_C/\partial q$ and $\partial i_L/\partial\phi$ are positive semidefinite matrices. This condition implies that $p(q, \phi) \geq 0$. By *H*5 and Corollary 1.3b, $p(q, \phi) = 0$ if and only if $q = 0$ and $\phi = 0$.

b) By Corollary 1.3a, $\|q\| + \|\phi\| \to \infty$ implies that $\|q_C\| + \|v_S\| \to \infty$. Now from (9) and (L) and (C) we have

$$p(q, \phi) = \int_0^{q_C} \langle v_C, dq_C'\rangle + \int_0^{q_S} \langle v_S, dq_S'\rangle + \int_0^{\phi_L} \langle i_L, d\phi_L'\rangle + \int_0^{\phi_L} \langle i_\Gamma, d\phi_\Gamma'\rangle$$

and

$$p(q, \phi) = p_C(q_C) + p_L(\phi_L) + \int_0^{q_S} \langle v_S, dq_S'\rangle + \int_0^{\phi_\Gamma} \langle i_\Gamma, d\phi_\Gamma'\rangle. \tag{10}$$

Now,

$$\int_0^{q_S} \langle v_S, dq_S'\rangle + \int_0^{\phi_\Gamma} \langle i_\Gamma, d\phi_\Gamma'\rangle = \langle v_S, q_S\rangle + \langle i_\Gamma, \phi_\Gamma\rangle - p_S(v_S) - p_\Gamma(i_\Gamma) \geq 0$$

by *H*5. Also by *H*6, $p_C(q_C) + p_L(\phi_L) \to \infty$ as $\|q_C\| + \|\phi_L\| \to \infty$. Hence $p(q, \phi) \to \infty$ as $\|q\| + \|\phi\| \to \infty$.

Q.E.D.

Suppose $q(t)$ and $\phi(t)$ is the solution of the normal form starting in the initial condition $q(0)$ and $\phi(0)$. Let $p(t) \triangleq p(q(t), \phi(t))$. Then,

$$\frac{dp}{dt} = \frac{\partial p}{\partial q}\frac{dq}{dt} + \frac{\partial p}{\partial \phi}\frac{d\phi}{dt}$$
$$= -\langle v_R(t), i_R(t)\rangle - \langle v_G(t), i_G(t)\rangle \text{ by (D), (R), (L), (C)}$$
$$\leq 0 \text{ for all } t$$

and

$$\frac{dp}{dt} = 0 \text{ if and only if } v_R = 0 \quad \text{and} \quad i_G = 0.$$

The last two statements follow from $H4$. We therefore have

Theorem 3.1:

a) The zero-input response of a network satisfying the conditions of Theorem 2.1 and $H4$—$H6$ is bounded.

b) The network is globally asymptotically stable[6] if and only if $v_R(t) \equiv 0$ and $i_G(t) \equiv 0$ for all t implies that $q(t) \equiv 0$ and $\phi(t) \equiv 0$ for all t.

Proof:

a) Let $q(0)$ and $\phi(0)$ be the initial state and $p(t) = p(q(t), \phi(t))$. Then since $dp/dt \leq 0$ we have $p(t) \leq p(0)$ for all t. Since $p(q, \phi) \to \infty$ as $||q|| + ||\phi|| \to \infty$ we have $q(t)$ and $\phi(t)$ bounded for all t.

b) Consider the set of solutions of the normal form for which p is a constant, i.e., $dp/dt \equiv 0$ for all t, or equivalently $v_R(t) \equiv 0$ and $i_G(t) \equiv 0$. By (D) and (R) this set is identical to the solutions of the following pair of equations:

$$\dot{q} = F_{LC}^T i_L, \qquad (11)$$
$$\dot{\phi} = -F_{LC} v_C,$$

and

$$F_{RC} v_C(t) \equiv 0 \quad \text{and} \quad F_{LG}^T i_L(t) \equiv 0. \qquad (12)$$

It is well known [5] that all the trajectories of the network state q and ϕ converge to the trajectories which satisfy (11) and (12). Hence the network is globally asymptotically stable if and only if (11) and (12) have the trivial solution $q(t) \equiv 0$ and $\phi(t) \equiv 0$. Q.E.D.

Remark:

1) It is worth noticing that the entire positive limiting set [5] of the network under zero-input is contained in the solutions of equations (11) and (12); hence the positive limiting set depends only on the location of the resistors and not on their characteristics.

2) It is worth mentioning that even if all the hypotheses of Theorem 3.1 *except* $H6$ are satisfied, then the theorem does not hold. The reader is referred to [12] where the authors give an example of a passive, lossy, time-invariant series R, L, C network which under zero-input conditions gives an *unbounded* solution for certain initial conditions.

[6] Same definition as completely stable as given in reference [5].

Definition 3.1:

a) The network is *observable at the resistor terminals* if $v_G(t) = 0$ and $i_R(t) = 0$ over a nonvanishing time interval implies that $q(t) \equiv 0$ and $\phi(t) \equiv 0$ for all t.

b) The network is *weakly observable at the resistor terminals* if $v_G(t) \equiv 0$ and $i_R(t) \equiv 0$ for all t implies that $q(t) \equiv 0$ and $\phi(t) \equiv 0$.

Remark: If the network is linear, the two definitions coincide with the usual definition for observability [3] of linear systems if we take the output to be the voltage across all the resistors.

Corollary 3.2:[7]

a) The network is observable $\Rightarrow$ the network is weakly observable.

b) The network is globally asymptotically stable if and only if it is weakly observable at the resistor terminals.

c) For a network with *linear* inductors and *linear* capacitors but nonlinear resistors the network is asymptotically stable if and only if it is observable at the resistor terminals. Note that in this case asymptotic stability implies global asymptotic stability.

Proof:

a) Follows from the definition;

b) is equivalent to Theorem 3.1b); and

c) is a well-known fact about time-invariant linear differential equations. Q.E.D.

Corollary 3.2c gives a useful stability criterion for networks with linear inductors and capacitors. Suppose that the inductor and capacitor characteristics can be expressed as $i_L = \Gamma\phi_L$, $\phi_\Gamma = Li_\Gamma$, $v_c = Sq_c$, $q_S = Cv_S$ where Γ, L, S and C are positive definite symmetric matrices. Then simple manipulations yield, $\phi = \mathcal{L}i_L$ and $q = \mathcal{C}v_C$ where $\mathcal{L} = [\Gamma^{-1} + F_{L\Gamma}LF_{L\Gamma}^T]$ and $\mathcal{C} = [S^{-1} + F_{SC}^T CF_{SC}]$. Substituting in (11) and (12) yields

$$\mathcal{C}\dot{v}_C = F_{LC}^T i_L, \qquad (13)$$
$$\mathcal{L}\dot{i}_L = -F_{LC} v_C,$$

and

$$F_{RC} v_C = 0 \quad \text{and} \quad F_{LG}^T i_L = 0. \qquad (14)$$

Let

$$A \triangleq \begin{bmatrix} 0 & \mathcal{C}^{-1}F_{LC}^T \\ -\mathcal{L}^{-1}F_{LC} & 0 \end{bmatrix} \quad \text{and} \quad B = \begin{bmatrix} F_{RC} & 0 \\ 0 & F_{LG}^T \end{bmatrix}. \qquad (15)$$

Then (13) and (14) are equivalent to

$$\dot{x} = Ax$$

and

$$Bx = 0.$$

[7] The authors are grateful to Professor C. A. Desoer for pointing out the relation between stability and observability.

By the well-known conditions [3] for observability of a linear time-invariant system we obtain

Corollary 3.3: If the inductors and capacitors are *linear*, then the network is asymptotically stable if and only if the columns of the matrix

$$[B^T, (BA)^T, \cdots, (BA^{n-1})^T]$$

spans R^n where A and B are defined in (15) and A has dimension $n \times n$.

Appendix

In this appendix, we will sketch the proof of the following facts.

1) Either $H2$ or $H3$ implies $\det M \geq 1$.

2) Corollary 1.2.

3) Corollary 1.3.

To prove the first statement it is sufficient to prove the following two assertions.

Let X and Y be two $n \times n$ real constant positive semidefinite (not necessarily symmetric) matrices.

Assertion 1: If X is diagonal then $\det (I + XY) \geq 1$.

Proof: Applying a theorem given in reference 10,[8] we have $\det (I + XY) = 1 + \sum_{i=1}^{n-1} sp_i(XY) + \det XY$ where $sp_i(XY)$ means "sum of all principal minors of XY of order i." It is straightforward to show that $sp_i(XY) \geq 0$ and $\det XY \geq 0$.

Assertion 2: If X is positive definite and symmetric then $\det (I + XY) \geq 1$.

Proof: Let P be a nonsingular matrix such that $P^{-1}X^{-1}P = \Lambda$ where Λ is a positive definite diagonal real matrix. Then

$$\begin{aligned}\det (I + XY) &= \det X \det (X^{-1} + Y) \\ &= \det X \det (\Lambda + P^{-1}YP) \\ &= \det X \det \Lambda \det (I + \Lambda^{-1}P^{-1}YP).\end{aligned}$$

[8] The authors are grateful to Harold R. Hall for pointing out the applicability of this result to our problem.

Now, $\det X \det \Lambda = 1$, Λ^{-1} is a positive definite diagonal matrix and $P^{-1}YP$ is positive semidefinite. Thus, the result follows from Assertion 1.

Proof of Corollary 1.2: Taking partial derivatives of (*) with respect to u, and eliminating $\partial \tilde{y}/\partial u$, we obtain

$$(I + GAFA^T) \frac{\partial \eta}{\partial u} = G.$$

Now

$$(I + FAFA^T) \frac{\partial \eta}{\partial u} (I + GAFA^T)^T = G + FAF^TA^TG^T,$$

which is positive semidefinite if F and G are, and is symmetric if F and G are. Consequently, Corollary 1.2 follows from the fact that $(I + GAFA^T)$ is nonsingular. There is a similar proof for $\partial \xi/\partial v$.

Proof of Corollary 1.3:

a) Follows from the assumption that the functions on the left-hand side of (*) are continuous, and

b) follows from the direct evaluation by use of (*).

References

[1] C. A. Desoer and J. Katzenelson, "Nonlinear RLC networks," *Bell Sys. Tech. J.*, vol. 44, pp. 161–198, January 1965.
[2] C. A. Holzmann and R. Liu, "On the dynamical equations of nonlinear networks with n-coupled elements," *1965 Proc. Third Ann. Allerton Conf.*, pp. 536–545.
[3] L. A. Zadeh and C. A. Desoer, *Linear System Theory.* New York: McGraw-Hill, 1963, pp. 501–504.
[4] E. S. Kuh, "Representation of nonlinear networks," *1965 Proc. NEC*, vol. 21, pp. 702–706.
[5] J. LaSalle and S. Lefschetz, *Stability by Liapunov's Direct Method with Applications.* New York: Academic, 1961, pp. 56–67.
[6] J. Katzenelson and L. H. Seitelman, "An iterative solution of nonlinear resistive networks," *1965 Proc. Third Ann. Allerton Conf.*, pp. 647–657.
[7] W. H. Fleming, *Functions of General Variables.* Reading, Mass: Addison-Wesley, 1965, pp. 67–69 and p. 279.
[8] P. R. Bryant, "The order of complexity of electrical networks," *Proc. IEE* (London), vol. 106C, pp. 174–188, 1959.
[9] E. S. Kuh and R. A. Rohrer, "The state variable approach to network analysis," *Proc. IEEE*, vol. 53, pp. 672–686, July 1965.
[10] A. C. Aitken, *Determinants and Matrices*, 9th ed. Edinburgh: Oliver and Boyd, 1962, p. 88.
[11] E. A. Coddington and N. Levinson, *Theory of Ordinary Differential Equations.* New York: McGraw-Hill, 1955, p. 22.
[12] C. A. Desoer, R. Liu, and L. V. Auth, Jr., "Linearity vs. nonlinearity and asymptotic stability in the large," *IEEE Trans. on Circuit Theory*, vol. CT-12, pp. 117–118, March 1965.

Global Inverse Function Theorem

FELIX F. WU AND CHARLES A. DESOER

Abstract—A simple proof of global inverse function theorem in R^n is given. A global homeomorphic version of the theorem is proved first. A global diffeomorphic version follows by an application of the classical local inverse function theorem.

INTRODUCTION

The problem of determining that a given function from R^n into R^n has an inverse is very useful in applications. In 1959, Palais established the necessary and sufficient condition for a function to be a diffeomorphism of R^n onto itself, which appeared as an episode in a paper [1] dealing with the determination of spaces of intertwining operators on differential forms. This global version of the classical inverse function theorem has been applied widely in nonlinear network theory [2]–[12] and is generally referred to as Palais' Theorem by circuit theorists. Palais originally stated it without proof as a corollary in [1]. We believe that a simple proof of this useful theorem will be helpful to the readers of this TRANSACTIONS.

This correspondence presents a proof that is intuitively appealing and easily understood with a modest background in mathematical analysis [13], [14]. We first prove the necessary and sufficient condition for a global homeomorphism [2]; the case of a global diffeomorphism follows easily by an application of the classical inverse function theorem.

THEOREM

Let f be a map from R^n into R^n, then f is a homeomorphism[1] of R^n onto R^n if and only if f is

1) a local homeomorphism[2]
2) a proper map.[3]

Remark: In applications, when one seeks to calculate operating points, the circuit equations are of the form $f(x)=y$, where y denotes the inputs (bias sources) and x the chosen set of independent variables (node voltages, or loop currents, $\cdots$). For a given input y, any x satisfying $f(x)=y$ is a corresponding operating point. The local homeomorphism condition 1) means that for any sufficiently small change in bias, there is a unique change in operating point which depends continuously on the change in bias. On the other hand, the continuity of f implies that the inverse image of any closed set is closed; therefore, for condition 2), it suffices to check that the inverse image of any bounded set is a bounded set. Physically, this condition means that any set of bias sources which lie in any fixed ball in R^n gives rise to operating points which lie in some fixed ball of finite radius.

Proof

$\Rightarrow$By assumption f is a (global) homeomorphism; hence it is a local homeomorphism. Because f^{-1} is continuous, it maps any compact set into a compact set [13, p. 78], [14, theorem 4.1, p. 207].

$\Leftarrow$We prove this in three steps: 1) f is surjective (onto), 2) f is injective (one-to-one), 3) f^{-1} is continuous. To facilitate the presentation, we denote the domain of f by X and the range by Y; of course, $X=Y=R^n$.

1) *Surjective:* Let Y_1 be the image of f, i.e., $Y_1=f(X)$, or more specifically, $Y_1=\{y\in Y | f^{-1}(y)$ is a nonempty subset of $X\}$. We know that Y is connected. If Y_1 is both open and closed, knowing also the fact that Y_1 is not empty, we can conclude that $Y_1=Y$ [13, p. 59], i.e., f is surjective.

a) Y_1 open: Let $y_1\in Y_1$, then there exists an $x_1\in X$ such that $f(x_1)=y_1$. Now f is a local homeomorphism means that there exist open neighborhoods U of x_1 and V of y_1 such that f is a homeomorphism from U onto V. So $V\subset Y_1$ and Y_1 is thus open.

b) Y_1 closed: Let y be an accumulation point of Y_1. Then there exists a sequence $\{y_i\}_1^\infty$ with $y_i\in Y_1 \forall i$, and $y_i\rightarrow y$. Consider $K=\{y_i\}_1^\infty\cup\{y\}$, which is clearly closed and bounded in Y, hence compact [13, p. 58], [14, theorem 4.5, p. 208]. By assumption, $f^{-1}(K)$ is compact in X. Now pick $x_i\in f^{-1}(y_i)$. $\{x_i\}_1^\infty$ is a sequence in a compact set $f^{-1}(K)$ in a metric space R^n, therefore, $\{x_i\}_1^\infty$ has a convergent subsequence, say $\{x_{i_j}\}_{j=1}^\infty\rightarrow x$ [13, p. 56], [14, theorem 4.4, p. 208]. But $\{f(x_{i_j})\}_{j=1}^\infty$ is a subsequence of $\{y_i\}_1^\infty$, therefore converges to the same limit y. f is a continuous map because it is a local homeomorphism, hence

$$f(x) = f(\lim_{j\rightarrow\infty} x_{i_j}) = \lim_{j\rightarrow\infty} f(x_{i_j}) = y.$$

Therefore $y\in Y_1$. Hence Y_1 is closed [13, p. 47], [14, p. 203].

2) *Injective:* Suppose that f is not injective, hence there exist two distinct points x_1, x_2 such that $f(x_1)=f(x_2)$. Without loss of generality, we can assume $f(x_1)=f(x_2)=0$. Let α: $[0, 1]\rightarrow X$ be defined by $\alpha(t)=(1-t)x_1+tx_2$ and $\beta=f\circ\alpha$. Geometrically, α is the line segment joining x_1 to x_2 and β, its image in Y under f, is a closed curve through 0 (Fig. 1). Let B: $[0, 1]\times[0, 1]\rightarrow Y$ be defined by $B(t, \tau)=(1-\tau)\beta(t)$. Thus for each τ, $B(\cdot, \tau)$ is obtained by shrinking the closed curve β toward the origin. The rough idea of the proof is to shrink the curve β toward the origin, the corresponding curve α will be continuously deformed into some curve joining x_1 to x_2; the contradiction will be reached in the limit when β degenerates into a single point.

a) Construction of the inverse image of B (Fig. 2): Let us define for each t a map $A(t, \cdot)$: $[0, 1]\rightarrow X$ by the following process of piecing together the local inverses of B. First let $A(t, 0)=\alpha(t)$. Since f is a local homeomorphism, there exist homeomorphic neighborhoods of $\alpha(t)$ and $\beta(t)$, U_1 and V_1, respectively. Define $A(t, \cdot)$: $[0, \tau_1]\rightarrow U_1$ to be the local inverse image of $B(t, \tau)$ for $\tau\in[0, \tau_1]$ where τ_1 is so chosen that $B(t, \tau)\in V_1$, for all $\tau\in[0, \tau_1]$. Thus we have $f(A(t, \tau_1))=B(t, \tau_1)$ and we can define $A(t, \cdot)$ on $[\tau_1, \tau_2]$ with $\tau_2>\tau_1$ as the local inverse of $B(t, \cdot)$ around $B(t, \tau_1)$. Repeat the same procedure; at each step, we extend τ from τ_k to τ_{k+1} with $\tau_{k+1}>\tau_k$. We are going to show by contradiction that the domain of $A(t, \cdot)$ can always be extended to include 1. Suppose that the above process fails to do so. Then the increasing sequence $\{\tau_k\}$ is bounded by 1 and has a least upper bound T, so $\{\tau_k\}\rightarrow T\leq 1$. But $B(t, \cdot)$ is continuous

$$\lim_{k\rightarrow\infty} B(t, \tau_k) = B(t, T)$$

and since $\{A(t, \tau_k)\}_{k=1}^\infty$ is a sequence in a compact set

$$f^{-1}(B(t, \tau):\tau \in [0, 1])$$

it has a subsequence converging to a limit $A(t, T)$. Now because f is continuous, $f(A(t, T))=B(t, T)$, hence the domain of $A(t, \cdot)$ is extended to include T; moreover, in the case when $T<1$, it can even be extended beyond T by local homeomorphism. Thus we can define a map A: $[0, 1]\times[0, 1]\rightarrow X$ with the property that $f\circ A=B$, and also $A(0, \tau)=x_1$, $A(1, \tau)=x_2$, $\forall\tau$.

b) Continuity of $A(\cdot, \tau)$: We will show that for each τ, $A(\cdot, \tau)$: $[0, 1]\rightarrow X$ is continuous by open-set arguments [13, p. 70], [14, pp. 201–202]. Let $\mathcal{O}$ be any open set in X and let its inverse image under $A(\cdot, \tau)$ be denoted by $\mathfrak{I}$; equivalently, $\mathfrak{I}$ is the inverse image under $A(\cdot, \tau)$ of the intersection of $\mathcal{O}$ with the image of $A(\cdot, \tau)$. Now for each t, let the homeomorphic neighborhoods of $A(t, \tau)$ and $B(t, \tau)$ be U_t and V_t, respectively. Note that

$$\bigcup_{t\in[0,1]} (U_t \cap \mathcal{O})$$

has $\mathfrak{I}$ as its inverse image under $A(\cdot, \tau)$ and

Manuscript received May 12, 1971; revised July 31, 1971. This research was sponsored by National Science Foundation under Grant GK-10656X.
The authors are with the Department of Electrical Engineering and Computer Sciences and the Electronics Research Laboratory, University of California, Berkeley, Calif. 94720.

[1] A homeomorphism of X onto Y is, by definition, a continuous bijective map f: $X\rightarrow Y$ such that f^{-1} is also continuous.

[2] A map f: $X\rightarrow Y$ is said to be a local homeomorphism if whenever $x\in X$ and $y\in Y$ are such that $f(x)=y$, then there exist open neighborhoods U of x and V of y such that f restricted to U is a homeomorphism of U onto V.

[3] A continuous map is said to be proper if the inverse image of any compact set is compact.

Reprinted from *IEEE Transactions on Circuit Theory*, vol. CT-19, pp. 199–201, March 1972.

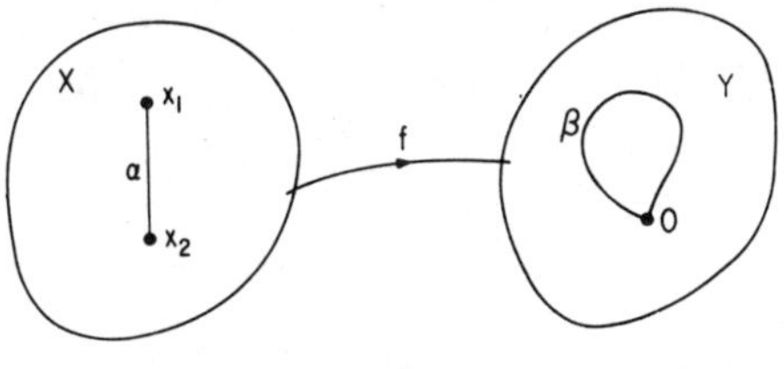

Fig. 1. For proof by contradiction, it is assumed that $x_1 \neq x_2$ and that $f(x_1) = f(x_2) = 0$. The line segment α which joins x_1 to x_2 is mapped by f onto closed curve β.

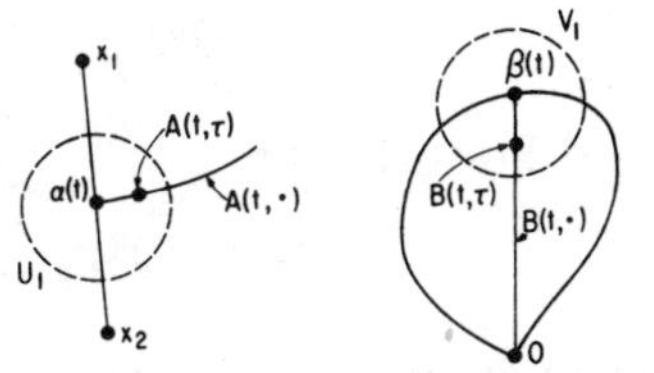

Fig. 2. As τ goes from 0 to 1, $B(t, \tau) = (1-\tau)$, $\beta(t)$ travels in a straight line from $\beta(t)$ to 0. For the same fixed t, the corresponding curve $A(t, \tau)$ is constructed from $B(t, \tau)$ by successive local homeomorphisms.

$$\bigcup_{t \in [0,1]} f(U_t \cap \mathcal{O})$$

also has $\mathfrak{J}$ as its inverse image under $B(\cdot, \tau)$ since $B = f \circ A$. But $f(U_t \cap \mathcal{O})$ are open in Y, so does

$$\bigcup_{t \in [0,1]} f(U_t \cap \mathcal{O}).$$

Because $B(\cdot, \tau)$ is continuous, $\mathfrak{J}$ is open in [0, 1]. This completes our proof that $A(\cdot, \tau)$ is continuous.

Now for $\tau = 1$, $B(t, 1) = 0$, $\forall t$. Geometrically, this is done by shrinking β to the origin. The corresponding $A(t, 1)$ is still a continuous curve joining two distinct points x_1 and x_2. But the inverse image of a single point under a local homeomorphism f can not be a continuous curve. To demonstrate this, suppose it were true: if every neighborhood of x_1 would contain points of $f^{-1}(0)$ other than x_1 itself, then it would be impossible for homeomorphic neighborhoods of x_1 and 0 to exist. Thus we have proved that f is injective.

Remark: Here we have in fact tacitly constructed a covering homotopy A of B [15, theorem 3, p. 59].

3) *Continuity of f^{-1}:* Recall that continuity is a local property [13, p. 68], [14, pp. 201–202]. The fact that f^{-1} exists globally (by 1) and 2)) together with the local homeomorphism assumption asserts that f^{-1} is continuous.

Q.E.D.

Lemma 1

Let f be a continuous map from R^n into R^n, then f is a proper map if and only if

$$\lim_{\|x\| \to \infty} \|f(x)\| = \infty.$$

Proof

$\Rightarrow$By contradiction. Suppose that there is a sequence $\{x_k\}$ with $\|x_k\| \to \infty$, yet $\|f(x_k)\| \leq M < \infty$. Consider the closed and bounded ball $B_M = \{y \mid \|y\| \leq M\}$, because f is proper, $f^{-1}(B_M)$ is compact. However, $\{x_k\}$ is contained in $f^{-1}(B_M)$, but $\|x_k\| \to \infty$ contradicts the compactness of $f^{-1}(B_M)$.

$\Leftarrow$$f$ is continuous implies that for each closed set K, $f^{-1}(K)$ is closed. Suppose K is bounded yet $f^{-1}(K)$ is not, then there exists a sequence $\{x_k\}$ in $f^{-1}(K)$ with $\|x_k\| \to \infty$. Clearly $\{f(x_k)\} \subset K$. But by assumption $\|f(x_k)\| \to \infty$, which contradicts boundedness of K.

Q.E.D.

Lemma 2

Let f be a C^k map $(k \geq 1)$ from R^n into R^n, then f is a local C^k diffeomorphism[4] if and only if det $Df(x) \neq 0$, $\forall x \in R^n$.

[4] A C^k map is, by definition, a map with continuous derivatives up to order k. A C^k diffeomorphism is, by definition, a bijective C^k map such that the inverse is also C^k.

Proof

This is the well-known classical local inverse function theorem [13, p. 211 and example 17, p. 217], [14, p. 167].

Corollary

Let f be a C^k map from R^n into R^n, then f is a C^k diffeomorphism if and only if

1) $\det Df(x) \neq 0, \quad \forall x \in R^n$
2) $\lim_{\|x\| \to \infty} \|f(x)\| = \infty.$

Remark: To see that the condition (2) is needed for global inverse, consider a) $(x, y) \mapsto (e^x, ye^{-x})$, and b) $(x, y) \mapsto (e^x \sin y, e^x \cos y)$. Both have nonsingular Jacobians everywhere but neither satisfies condition 2). Note that a) maps R^2 onto the open right half plane, so it is not surjective; and b) maps each strip $2(k-1)\pi < y \leq 2k\pi$, k = integer, onto $R^2 - \{(0, 0)\}$, so it is not injective.

Proof

It follows from the Theorem, Lemmas 1 and 2, as well as the fact that differentiability is a local property [13, p. 198], [14, p. 142].

Q.E.D.

Acknowledgment

The authors would like to thank Prof. A. Weinstein for his helpful discussion.

Prof. L. O. Chua and one of the reviewers have called to our attention that there is an alternative proof of the theorem via continuation property in [16, pp. 132–137].

References

[1] R. S. Palais, "Natural operations on differential forms," *Trans. Amer. Math. Soc.*, vol. 92, pp. 125–141, 1959.
[2] C. A. Holzmann and R. Liu, "On the dynamical equations of nonlinear networks with n-coupled elements," in *Proc. 3rd Annu. Allerton Conf. Circuit and System Theory*, pp. 536–545, 1965.
[3] T. Ohtsuki and H. Watanabe, "State-variable analysis of *RLC* networks containing nonlinear coupling elements," *IEEE Trans. Circuit Theory*, vol. CT-16, pp. 26–38, Feb. 1969.
[4] I. W. Sandberg and A. N. Willson, Jr., "Some theorems on properties of DC equations of nonlinear networks," *Bell Syst. Tech. J.*, vol. 48, pp. 1–34, Jan. 1969.
[5] I. W. Sandberg, "Theorems on the analysis of nonlinear transistor networks," *Bell Syst. Tech. J.*, vol. 49, pp. 95–114, Jan. 1970.
[6] ——, "Theorems on the computation of the transient responses of nonlinear networks containing transistors and diodes," *Bell Syst. Tech. J.*, vol. 49, pp. 1739–1776, Oct. 1970.
[7] ——, "Necessary and sufficient conditions for the global invertibility of certain nonlinear operators that arise in the analysis of networks," *IEEE Trans. Circuit Theory*, vol. CT-18, pp. 260–263, Mar. 1971.
[8] E. S. Kuh and I. N. Hajj, "Nonlinear circuit theory: Resistive networks," *Proc. IEEE*, vol. 59, pp. 340–355, Mar. 1971.
[9] T. Fujisawa and E. S. Kuh, "Some results on existence and uniqueness of solution of nonlinear networks," *IEEE Trans. Circuit Theory*, vol. CT-18, pp. 501–506, Sept. 1971.
[10] ——, "Piecewise-linear theory of nonlinear networks," to be published in *SIAM J. Appl. Math.*, Jan. 1972.
[11] L. O. Chua and Y. F. Lam, "Foundations of nonlinear network theory," Purdue Univ., Lafayette, Ind., Tech. Rep. TR-EE 70-22, June 1970.
[12] L. O. Chua, "Linear transformation converter and its application to the synthesis of nonlinear networks," *IEEE Trans. Circuit Theory*, vol. CT-17, pp. 584–594, Nov. 1970.
[13] M. Rosenlicht, *Introduction to Analysis*. Glenview, Ill.: Scott, Foresman and Co., 1968.
[14] L. H. Loomis and S. Sternberg, *Advanced Calculus*. Reading, Mass.: Addison-Wesley, 1968.
[15] I. M. Singer and J. A. Thorpe, *Lecture Notes on Elementary Topology and Geometry*. Glenview, Ill.: Scott, Foresman and Co., 1967.
[16] J. M. Ortega and W. C. Rheinboldt, *Iterative Solution of Nonlinear Equations in Several Variables*. New York: Academic Press, 1970.

Global Homeomorphism of Vector-Valued Functions*

LEON O. CHUA

Department of Electrical Engineering and Computer Sciences, University of California, Berkeley, California 94720

AND

YING-FAI LAM

Department of Electrical Engineering, University of Waterloo, Waterloo, Ontario, Canada

Submitted by L. A. Zadeh

1. INTRODUCTION

In 1959, Palais [1] proved that the necessary and sufficient conditions for a function $f : R^n \to R^n$ to be a diffeomorphism of R^n onto itself are

(1) $\det J_f(x) \neq 0$ and

(2) $\lim_{\|x\| \to \infty} \|f(x)\| = \infty$,

where R^n denotes the Euclidean n-space, $J_f(x)$ denotes the Jacobian matrix of f, and $\| x \|^2 = \sum_{i=1}^n x_i^2$. This powerful theorem has played a fundamental role in many recent research works in nonlinear network theory [2–9]. On many occasions, however, Palais' theorem is used only to show that a function possesses a continuous global inverse f^{-1}, and the differentiability of the inverse map is not really essential. Since there exist many globally one-to-one functions which fail to satisfy condition (1) of Palais' theorem, our objective in this paper is to derive weaker global inversion theorems which do not require the Jacobian to be nonzero everywhere. Since we will be concerned exclusively with functions from R^n into R^n, the following classical theorem due to Brouwer shows that the class of globally one-to-one and continuous functions from R^n into R^n is identical to the class of globally homeomorphic functions from R^n into R^n:

THEOREM 1.1 (Invariance of Domain [10]). *If A is open in R^n and $f : A \to R^n$ is one-to-one and continuous, then $f(A)$ is open and f is a homeomorphism.*

* This work was supported by the National Science Foundation, Grant GK 2988 and GK-32236.

Reprinted with permission from *Journal of Mathematical Analysis and Applications*, vol. 39, pp. 600–624, September 1972. Copyright © 1972 by Academic Press, Inc.

In view of this theorem, we will be concerned with theorems on global homeomorphisms in this paper. Unless otherwise stated, all functions are assumed to be from the Euclidean n-space R^n into R^n and are of class C^1.

Our main result in Section 2 consists of a weaker form of Palais' Theorem for global homeomorphic onto functions which allows the Jacobian to vanish on a set of isolated points. In Section 3, we study the properties of a class of global homeomorphic functions which arise frequently in nonlinear network theory; namely, the class of "increasing functions." Several theorems will be presented which guarantee that a vector-valued function is increasing. Unlike the results in Section 2, most of the theorems in this section are valid not only for vector-valued functions from R^n into R^n, but also from an open convex subset $K \subset R^n$ into R^n. The hypothesis in these theorems clearly reveals that the class of increasing functions is a natural generalization of the class of functions which are expressible as the gradient of a strictly convex scalar-potential function [11–12]. The main result in this section consists of a sufficient condition which replaces the requirement that the Jacobian matrix be positive definite by an "almost-positive definite" requirement to be defined in Section 3. Finally, in Section 4 we consider a class of "quasi-increasing" functions which need not be increasing or decreasing, but are nevertheless globally homeomorphic.

2. Sufficient Conditions for Global-Homeomorphic onto Functions

A mapping $f: X \to Y$ is called a "local homeomorphism" if for each $x \in X$, a neighborhood of x is mapped homeomorphically by f onto a neighborhood of $f(x)$. In order to prove our main theorem on global homeomorphism, it is convenient to introduce the notion of a covering map [1, 13]:

Definition 2.1. Let X and Y be a connected and locally connected topological space. If f is a mapping of X onto Y with the property that each $y \in Y$ has a neighborhood V such that each component of $f^{-1}(V)$ is mapped homeomorphically onto V by f, then f is called a "covering map" and (X, f) is called a "covering space" of the space Y. In this case, the cardinal number n of the set $f^{-1}(y)$ is the same for all $y \in Y$. If n is a finite integer, then f is called a finite covering, or more specifically, an n-covering.

It is well known that every homeomorphic onto function $f: X \to Y$ is a covering map and every covering map is a local homeomorphism [13]. However, the converse is not true: A local homeomorphism need not be a covering map and a covering map need not be a homeomorphic onto function. Hence, an n-covering map lies somewhere in between a local homeomorphism and a global homeomorphism. A 1-covering map, of course, must necessarily be a global homeomorphic onto function. The following standard results on covering maps will be needed in the proof of our global inversion theorem and are reproduced here for handy reference:

LEMMA 2.1 [1]. *Let $f : X \to Y$, $X = Y = R^n$, be a local homeomorphism. A necessary and sufficient condition that f be a finite covering is that*

$$\lim_{\|x\| \to \infty} \| f(x) \| = \infty.$$

Furthermore, if f is a finite covering, then the set

$$Y_m = \{ y \in Y : f^{-1}(y) \text{ contains at least } m \text{ distinct points} \}$$

is either empty or equal to Y for each positive integer m.

LEMMA 2.2 [13]. *Let $f : X \to Y$, $X = Y = R^n$, be a covering map. If A is any component of $f^{-1}(Y)$, then A is open and f restricted to A is a 1-covering of Y, i.e., $f : A \to Y$ is bijective.*

We are now ready to state the main result in this section.

THEOREM 2.1. *Let $f : X \to Y$, $X = Y = R^n$, $n \neq 2$, be a C^1 map. Let $S = \{x \in R^n : \det J_f(x) = 0\}$ and $T = \{x \in R^n : x \notin S\}$. Then the following conditions are sufficient for f to be a homeomorphism of R^n onto R^n:*

(1) $\det J_f(x) > 0$ *for all* $x \in T$ *and* S *is at most a set of isolated points.*

(2) $\lim_{\|x\| \to \infty} \| f(x) \| = \infty$.

Proof. Consider first the case $n = 1$. Suppose $f : R^1 \to R^1$ is not one-to-one. Then there exist two distinct points x_1 and x_2 such that $f(x_1) = f(x_2)$. Let $x_1 < x_2$. Condition (1) implies that there exists an interval (a, b) with $x_1 \leqslant a < b \leqslant x_2$ such that $f'(x) > 0$ on (a, b). Since f is C^1, we can write

$$f(x_2) - f(x_1) = \int_{x_1}^{x_2} f'(x)\, dx \geqslant \int_a^b f'(x)\, dx > 0,$$

which is a contradiction. Hence, f is one-to-one. By Theorem 1.1, f is a homeomorphism on R^1. Moreover, in view of Condition (2) and Lemma 2.1, f is a homeomorphism of R^1 onto R^1.

It remains to consider $n \geqslant 3$. Let p be a point in S. There exists an open neighborhood U_p about p in R^n such that $U_p \cap S = \{p\}$: Since $\det J_f(x) > 0$ on U_p except at the isolated point p where $\det J_f(x) = 0$, it follows from a recent result by Church and Hemmingsen [14] that f is a local homeomorphism on U_p, for each $p \in S$. Since $\det J_f(x) > 0$ for each $x \in T$, f is a local homeomorphism on R^n for all $n \geqslant 3$. In view of Lemma 2.1, we know f is a finite covering. Let A be any component of $f^{-1}(Y)$. Then it follows from

Lemma 2.2 that A is open in R^n and f restricted to A is a bijective map onto R^n. By Theorem 1.1, f restricted to A is a homeomorphism of A onto Y. Hence, if we can show that $A = X$, then we would have completed the proof of this theorem.

Suppose A is a proper subset of X. Since A is open in X, A is an open proper subset of X. Let $b \in X$ be a boundary point of A and let M_b be an open connected neighborhood of $f(b)$. Since f is a finite covering map on X, $f^{-1}(M_b)$ has a finite and nonzero number of components. Let N_b be a component of $f^{-1}(M_b)$ that contains the point b. Let $N_b{}^* = A \cap f^{-1}(M_b)$. Since f is continuous, f^{-1} is open. Hence, both N_b and $N_b{}^*$ are open and connected. Also note that f maps both N_b and $N_b{}^*$ topologically onto M_b. Since N_b is an open set that contains b, the set $N_b \cap A$ is not empty. It follows that $N_b \cap N_b{}^*$ is not empty, for otherwise there will be at least one point x_1 in $N_b \cap A$ and a point x_2 in $N_b{}^*$ such that $f(x_1) = f(x_2) \in M_b$ and f restricted to A will not be one-to-one on A, which is a contradiction. Since both N_b and $N_b{}^*$ are connected, we have $N_b = N_b{}^*$. Hence, b is in $N_b{}^*$ and, therefore, is in A. This implies that A cannot be an open proper subset of X. That is, A is closed in X. We have A is both open and closed in X, and A is nonempty. Therefore, we can conclude that $A = X$. This completes the proof of Theorem 2.1.

We remark that Condition 1 of Theorem 2.1 is not necessary for f to be a homeomorphism on R^n. However, the following lemma shows that Condition 2 of Theorem 2.1 is also a necessary condition:

LEMMA 2.3. *Let $f : X \to Y$, $X = Y = R^n$. If f is a homeomorphic function of R^n onto R^n, then*

$$\lim_{\|x\| \to \infty} \|f(x)\| = \infty.$$

Proof. We first note that if f is a homeomorphic onto function, then f^{-1} is also a homeomorphic onto function from Y to X. Let y be an arbitrary point in Y and $V \subset Y$ be an open connected set containing y. Let $U = f^{-1}(V)$. Since f^{-1} is continuous and V is connected, it follows that U is connected. Since f^{-1} is a homeomorphism on Y, f^{-1} is open on Y, hence U is open in X. It is well known that the restriction of a continuous function to an open connected set is continuous. Hence, f and f^{-1} are continuous mappings on U and V, respectively. In order to show that f maps U onto V homeomorphically, we only need to show that f maps U univalently onto V. But this is trivial because (1) f is one-to-one on X, so f is one-to-one on U, (2) in view of our definition for U, corresponding to each point v in V, there is a u in U such that $f(u) = v$, where $u = f^{-1}(v)$. Since y is an arbitrary point in Y, we have shown that f has the property that for each y in Y, there is a neighborhood V about y such that f maps $f^{-1}(V)$ onto V homeomorphically. This implies that f is a covering map. Moreover, f is a 1-covering map since f is a homeomorphism on X. By Lemma 2.1, we have the conclusion of this lemma.

An immediate consequence of Theorem 2.1 is the following global implicit function theorem:

Corollary 2.1. *Let* $f : X \times Z \to Y$, i.e., $f(x, z) = y$ *where* $x \in X = R^n$, $z \in Z = R^m$, $y \in Y = R^n$, $n \geqslant 1$ *and* $n + m \neq 2$. *Suppose* f *satisfies the following two conditions:*

(1) det $\partial f/\partial x \geqslant 0$ *for all* x *and* z *and* det $\partial f/\partial x = 0$ *on at most a set of isolated points in* $X \times Z$.

(2) $\lim_{\|x\|\to\infty} \|f(x, z)\| = \infty$ *for all* z.

Then there exists a unique continuous function g *such that* $x = g(y, z)$ *for all* (y, z) *in* $Y \times Z$.

Proof. The following proof is virtually identical to that given by Kuh and Hajj [7] for their version of the global inversion theorem which is based upon Palais' theorem. Let us define the following vectors:

$$\hat{x} = \begin{bmatrix} x \\ z \end{bmatrix}, \qquad \hat{f}(\hat{x}) = \hat{f}(x, z) = \begin{bmatrix} f(x, z) \\ z \end{bmatrix} = \begin{bmatrix} y \\ z \end{bmatrix} = \hat{y}.$$

Then

$$J_{\hat{f}}(\hat{x}) = \begin{bmatrix} \dfrac{\partial f}{\partial x} & \dfrac{\partial f}{\partial z} \\ & \\ 0 & 1 \end{bmatrix}$$

and det $J_{\hat{f}}(\hat{x}) = \det \partial f/\partial x$. Now conditions (1) and (2) imply the following:

(1)′ det $J_{\hat{f}}(\hat{x}) \geqslant 0$ for all $\hat{x}$ and det $J_{\hat{f}}(\hat{x}) = 0$ on at most a set of isolated points in $X \times Z$.

(2)′ $\lim_{\|\hat{x}\|\to\infty} \|\hat{f}(\hat{x})\| = \infty$.

It follows from Theorem 2.1 that $\hat{f}$ is a homeomorphism of $R^n \times R^m$ onto itself, and hence we have a unique, continuous function $\hat{g} = \hat{f}^{-1}$ such that $\hat{g}(\hat{y}) = \hat{x}$; i.e.,

$$\begin{bmatrix} x \\ z \end{bmatrix} = \hat{x} = \hat{g}(\hat{y}) = \begin{bmatrix} g(y, z) \\ g^*(y, z) \end{bmatrix}$$

for all $\hat{y} = (y, z)$ in $R^n \times R^m$. Consequently, we obtain the first n components as $x = g(y, z)$ for all x and z, and g is a unique, continuous function for all (y, z) in $R^n \times R^m$. This completes the proof.

By requiring the function f in Theorem 2.1 to be of class C^n, $n \geqslant 3$, it is possible to allow $\det J_f(x) = 0$ on a somewhat larger set which we define next [15]:

DEFINITION 2.2. Let U be a nonempty subset of R^n. Then U is said to be of dimension zero if and only if U is a totally disconnected set. The empty set and only the empty set has dimension -1.

THEOREM 2.2. *Let $f : X \rightarrow Y$, $X = Y = R^n$, $n \neq 2$, be a C^n map. Let $S = \{x \in R^n : \det J_f(x) = 0\}$ and $T = \{x \in R^n : x \in S\}$. Then the following conditions are sufficient for f to be a homeomorphism of R^n onto R^n:*

(1) $\det J_f(x) > 0$ *for all $x \in T$ and S is a set of dimension* 0 *or* -1.

(2) $\lim_{\|x\| \rightarrow \infty} \|f(x)\| = \infty$.

Proof. We first show the theorem is true for the case $n = 1$. From Condition (1), S is a set of dimension 0 or -1, hence, for any two distinct points a and b in R^1, say $a < b$, there is some point c, $a < c < b$, such that $f'(c) > 0$. Since f is C^1, we have f' is continuous. Hence, there is a neighborhood N_c about the point c scuh that for any x in N_c, we have $f'(x) > 0$.

Suppose f is not one-to-one. Then there exists at least two distinct points x_1 and x_2 such that $f(x_1) = f(x_2)$. From the preceding paragraph, we know that there is an interval (d_1, d_2), $x_1 \leqslant d_1 < d_2 \leqslant x_2$ such that $f'(x) > 0$ for all x in (d_1, d_2). But

$$f(x_2) - f(x_1) = \int_{x_1}^{x_2} f'(x)\, dx \geqslant \int_{d_1}^{d_2} f'(x)\, dx > 0.$$

This is absurd. Hence, f is one-to-one on X. Theorem 1.1 says that f is a homeomorphism. In view of Condition (2) and Lemma 2.1, f is a homeomorphism of R^1 onto itself. For the case $n \geqslant 3$, it suffices to prove that f is a local homeomorphism on R^n because the remaining proof then follows exactly that of Theorem 2.1. In this case, the fact that f is a local homeomorphism follows from another recent result by Church [16–17], provided that $n \geqslant 3$. This completes the proof of Theorem 2.2.

We remark that the first part of Condition (1) of Theorems 2.1 and 2.2 can obviously be replaced by $\det J_f(x) < 0$. We also remark that the converse of Theorem 2.2 is true for the case $n = 1$, i.e., Conditions (1) and (2) are both necessary and sufficient for the function f in Theorem 2.2 to be a homeomorphism of R^1 onto R^1. For otherwise, S must have dimension > 0 and hence must contain a nonempty open interval (a, b). This would imply

$$f(x) = f(a) + \int_a^x f'(y)\, dy = f(a), \qquad \forall x \in (a, b),$$

which is absurd. For the case $n \geqslant 3$, the converse of Theorem 2.2, and for that matter, Theorem 2.1, is obviously not true since Condition (1) of both theorems is not necessary. There are many homeomorphic onto mappings with their Jacobian vanishing on an $(n-1)$-dimensional set. The following simple example is a case in point:

Let $f : R^3 \to R^3$ be defined by

$$\begin{aligned} y_1 &= f_1(x) = x_1^3, \\ y_2 &= f_2(x) = x_2^3, \\ y_3 &= f_3(x) = x_1 + x_2 + x_3 . \end{aligned}$$

This function f has a global inverse on R^3; namely,

$$x_1 = y_1^{1/3}, \qquad x_2 = y_2^{1/3},$$

$$x_3 = y_3 - y_1^{1/3} - y_2^{1/3}.$$

Hence, $f : R^3 \to R^3$ is a homeomorphic onto mapping. But $\det J_f(x) = 9x_1^2x_2^2$ vanishes on two 2-dimensional hyperplanes: one defined by $x_1 = 0$ and the other defined by $x_2 = 0$. However, the following theorem gives a partial converse to both Theorems 2.1 and 2.2:

THEOREM 2.3. *Let $f : X \to Y$, $X = Y = R^n$ be a C^n map. If f is a homeomorphism of R^n onto itself, then*

(1) *either* $\det J_f(x) \geqslant 0$ *or* $\det J_f(x) \leqslant 0$, *and there does not exist an n-dimensional open set $N \subset X$ such that* $\det J_f(x) = 0 \ \forall x \in N$.

(2) $\lim_{\|x\| \to \infty} \|f(x)\| = \infty$.

Proof. It suffices to prove only Property (1) because Property (2) follows immediately from Lemma 2.3. Since f is a homeomorphism of R^n onto itself, it is one-to-one and open on X. This implies that f is a light and open map [18]. Moreover, since f is C^n, it follows from Corollary 1.7 in Church [17] that either $\det J_f(x) \geqslant 0$ or $\det J_f(x) \leqslant 0$.

Suppose there is an n-dimensional open set $N \subset X$ such that $\det J_f(x) = 0$ for all x in N.

Let $\hat{x}$ be a point in N and let r be small enough so that $B(\hat{x}, r)$, an open ball centered at $\hat{x}$ with radius r, is contained in N. Then $(\hat{x} + rz)$ is in N for all z in $B(0, 1) \subset R^n$. Hence, $\det J_f(\hat{x} + rz) = 0$ whenever $\|z\| < 1$.

Let us define a mapping $\hat{f} : B(0, 1) \subset R^n \to R^n$ by $\hat{f}(z) \equiv f(\hat{x} + rz)$ for z in $B(0, 1)$. $\hat{f}$ is C^n since f is and $\det J_{\hat{f}}(z) \equiv 0$ on $B(0, 1)$. It had been shown in [6] that if an n-dimensional vector-valued function h is a homeomorphism on an n-dimensional open ball B, then $\det J_h(x) \not\equiv 0$ on B. Hence $\hat{f}$ cannot be

Proof. f is one-to-one on X implies f^{-1} exists on $f(X)$. Moreover, f^{-1} is uniformly continuous on $f(X)$ because given any $\epsilon > 0$, there exists a $\delta > 0$ such that $\|f^{-1}(y_2) - f^{-1}(y_1)\| = \| x_2 - x_1 \| < \epsilon$ whenever $\| y_2 - y_1 \| = \|f(x_2) - f(x_1)\| < \delta$, where we have let $y_i = f(x_i)$, $i = 1, 2$.

Let $\{x_k\}$ be a sequence in X such that $\| x_k \| \to \infty$ as $k \to \infty$ and consider the sequence $\{y_k = f(x_k)\}$. Suppose the sequence $\{y_k\}$ is bounded, i.e., $\| y_k \| = \|f(x_k)\| \leqslant B < \infty$ for all k. Then there exists a Cauchy subsequence $\{y_{k_i}\}$ which converges to a point y_0, where $\| y_0 \| \leqslant B$. From a standard result in analysis [19], we know that the image of a Cauchy sequence under a uniformly continuous mapping is Cauchy. Hence, the subsequence $\{x_{k_i} = f^{-1}(y_{k_i})\}$ is also a Cauchy sequence, and $\lim_{k_i \to \infty} \| x_{k_i} \| \neq \infty$, which is a contradiction. Therefore, the sequence $\{y_k = f(x_k)\}$ cannot be bounded. This implies that $\lim_{\|x\| \to \infty} \|f(x)\| = \infty$. Moreover, since f is one-to-one and continuous, it follows again from Theorem 1.1 and Lemma 2.1 that f is a homeomorphism of R^n onto R^n. This completes the proof.

Remark. Let $f : R^n \to R^n$. If f^{-1} exists on $f(R^n)$ and is uniformly continuous, then we have

$$\lim_{\|x\| \to \infty} \|f(x)\| = \infty.$$

3. Properties and Characterizations of Increasing Functions

Our objective in this section is to characterize an important class of homeomorphic vector-valued functions in Euclidean n-space which are frequently encountered in the physical sciences [20, 21]. These functions are usually called "monotone operators" and have been treated extensively in more general spaces by Minty [22] and Browder [23]. Since the term monotone operator or monotone function has been used in the literature under various different definitions [18, 22–26] which, unfortunately, are not consistent with one another, we have adopted the following concise and unambiguous terminologies:

Definition 3.1. Let $f : X \to Y$, $X \subset R^n$, $Y = R^n$. Let the following inner product be denoted by

$$\langle f(x_1) - f(x_2), x_1 - x_2 \rangle \equiv \alpha(x_1, x_2).$$

Then f is said to be:

(a) increasing on X, or simply an *increasing function* if and only if

$$\alpha(x_1, x_2) > 0, \qquad \forall x_1, x_2 \in X \qquad \text{and} \qquad x_1 \neq x_2;$$

(b) nondecreasing on X, or simply a *nondecreasing function* if and only if

$$\alpha(x_1, x_2) \geqslant 0, \qquad \forall x_1, x_2 \in X;$$

(c) decreasing on X, or simply a *decreasing function* if and only if

$$\alpha(x_1, x_2) < 0, \qquad \forall x_1, x_2 \in X \qquad \text{and} \qquad x_1 \neq x_2;$$

(d) nonincreasing on X, or simply a *nonincreasing function* if and only if

$$\alpha(x_1, x_2) \leqslant 0, \qquad \forall x_1, x_2 \in X.$$

Since each property in this section concerning an increasing (nondecreasing) function has an obvious corresponding property concerning a decreasing (nonincreasing) function, all results will be stated only in terms of increasing or nondecreasing functions.

LEMMA 3.1. *Let $f : U \to R^n$, where U is an open convex subset of R^n.*

(a) *If f is increasing on U, then f is one-to-one on U.*

(b) *If f is continuous and increasing on U, then f is a homeomorphism on U and its inverse function $f^{-1} : f(U) \to U$ is also increasing on $f(U)$.*

Proof. To prove (a), suppose the contrary. Then there exists two distinct points x and y in U such that $f(x) = f(y)$. If we let

$$g(\lambda) \equiv (y - x)^t f(x + \lambda(y - x)),$$

then $g(1) = g(0)$. But f is increasing on U implies that $g(1) - g(0) > 0$, hence the contradiction. To prove (b), we use Theorem 1.1 to assert that f is a homeomorphism on U. To show that f^{-1} is increasing on $f(U)$, let y_1 and y_2 be two arbitrary but distinct points in $f(U)$. Let $y_1 = f(x_1)$ and $y_2 = f(x_2)$. Then

$$\langle f^{-1}(y_1) - f^{-1}(y_2), y_1 - y_2 \rangle = \langle x_1 - x_2, f(x_1) - f(x_2) \rangle > 0.$$

This completes the proof of Lemma 3.1.

DEFINITION 3.2. Let $f : X \to Y$, $X \subset R^n$, $Y = R^n$. Then f is said to be a state function on X, or simply a state function, if, and only if, the Jacobian matrix $J_f(x)$ is symmetric for all $x \in X$.

DEFINITION 3.3. Let $f : U \to R^1$, where U is an open convex subset of R^n. Then f is said to be

(a) strictly convex on U, or simply a *strictly convex function* if, and only if,

$$f(\lambda x + (1 - \lambda) y) < \lambda f(x) + (1 - \lambda) f(y)$$
$$\forall x, y \neq x \in U \quad \text{and} \quad 0 < \lambda < 1.$$

(b) convex on U, or simply a *convex function* if and only if the inequality sign in (a) is replaced by "$\leqslant$".

The following relationship between increasing and strictly convex functions had been proved in [6] and is reproduced here for handy reference:

THEOREM 3.1. *Let U be an open convex subset of R^n and $\varphi : U \to R^1$ be a C^1 map. Let $f = \nabla\varphi$ on U. Then f is increasing on U if and only if φ is strictly convex on U.*

It is well known that a scalar function $\varphi : U \to R^1$ defined on an open convex set is strictly convex if the Hessian matrix $H_\varphi(x)$ of φ [which is also the Jacobian matrix $J_f(x)$ of $f = \nabla\varphi$] is positive definite [26]. Our next objective is to weaken this hypothesis by allowing the Hessian to vanish on a set of isolated points.

LEMMA 3.2. *Let U be an open convex subset of R^1 and let $\varphi : U \to R^1$ be a C^2 scalar function. Let $S = \{x \in U : \varphi''(x) = 0\}$ and $T = \{x \in U : x \notin S\}$. Then φ is strictly convex on U if and only if $\varphi''(x) > 0$ for all x in T and S is at most a set of isolated points.*

Proof. Suppose $\varphi''(x) > 0$ for all x in T and S is a set of isolated points. Then for any two distinct points u and v in U with $u > v$, we have

$$\varphi'(u) - \varphi'(v) = \int_v^u \varphi''(x)\, dx > 0. \tag{3.1}$$

Let $w = (1 - \lambda) p + \lambda q$ where $\lambda \in (0, 1)$, p and q are any two distinct points in U. It follows from (3.1) that

$$\varphi(w) - \varphi(p) = \int_p^w \varphi'(x)\, dx < \varphi'(w)(w - p), \tag{3.2}$$

$$\varphi(q) - q(w) = \int_w^q \varphi'(x)\, dx > \varphi'(w)(q - w). \tag{3.3}$$

Multiplying (3.2) by $(1 - \lambda)$ and (3.3) by λ and adding, we obtain

$$\varphi((1 - \lambda) p + \lambda q) < (1 - \lambda)\varphi(p) + \lambda\varphi(q) \tag{3.4}$$

for all $p \neq q$ in U. Hence φ is strictly convex on U.

To prove the converse, let φ be strictly convex on U and suppose there exists a nonempty interval $(a, b) \subset U$ such that $\varphi''(x) \leqslant 0$ for all x in (a, b). If we let $w = (1 - \lambda) p + \lambda q$ where $\lambda \in (0, 1)$, p and q are any two distinct points in (a, b), then

$$\varphi(w) - \varphi(p) = \int_p^w \varphi'(x)\, dx \geqslant \varphi'(w)\,(w - p), \tag{3.5}$$

$$\varphi(q) - \varphi(w) = \int_w^q \varphi'(x)\, dx \leqslant \varphi'(w)\,(q - w). \tag{3.6}$$

Multiplying (3.5) by $(1 - \lambda)$ and (3.6) by λ and adding, we obtain

$$\varphi((1 - \lambda)\, p + \lambda q) \geqslant (1 - \lambda)\, \varphi(p) + \lambda \varphi(q). \tag{3.7}$$

This implies φ is not strictly convex on U, which is a contradiction. Hence, there cannot exist a nonempty open interval in U on which $\varphi'' \leqslant 0$. Therefore, $\varphi''(x) \leqslant 0$ can occur at most on a set of isolated points. Suppose there is a point C such that $\varphi''(C) < 0$. Since φ'' is continuous, there is an interval N about c such that $\varphi''(x) < 0$ for all x in N. This is a contradiction. Hence, $\varphi''(x) \geqslant 0$ for all x and $\varphi''(x) = 0$ can occur on at most a set of isolated points.

THEOREM 3.2. *Let U be an open convex subset of R^n and let $\varphi : U \to R^1$ be a C^2 scalar function. Then φ is strictly convex on U if and only if the quadratic form*

$$Q(\lambda, x, z) \equiv (z - x)^t\, H_\varphi(x + \lambda(z - x))\,(z - x) \tag{3.8}$$

is positive for all x and z in U, and for all $\lambda \in (0, 1)$ except for at most a set of isolated points on which $Q(\lambda, x, z) = 0$.

Proof. The strict convexity of φ on U is equivalent to the strict convexity of φ on each straight line segment in U. This is equivalent to the strict convexity of the function

$$g(\lambda) = \varphi(x + \lambda(z - x)), \qquad \lambda \in (0, 1),$$

where x and z are any two distinct points on U. By Lemma 3.2, $g(\lambda)$ is strictly convex on (0, 1) if, and only if, $g''(\lambda) > 0$ except for at most a set of isolated points on which $g''(\lambda) = 0$. But

$$g''(\lambda) = (z - x)^t\, H_\varphi(x + \lambda(z - x))\,(z - x) = Q(\lambda, x, z).$$

Hence, g is strictly convex if, and only if $Q(\lambda, x, z)$ is positive for all $\lambda \in (0, 1)$ except for at most a set of isolated points on which $Q(\lambda, x, z) = 0$. This completes the proof of Theorem 3.2.

The result of Theorem 3.2 motivates the following definitions:

DEFINITION 3.4. An $n \times n$ real matrix A which need not be symmetric is said to be positive definite (positive semidefinite) if, and only if, $y^t A y > 0$ ($y^t A y \geqslant 0$) for all $n \times 1$ real vectors $y \neq 0$.

DEFINITION 3.5. Let U be a subset of R^n. An $n \times n$ matrix $A(x)$ is said to be "almost positive definite" on U if and only if, $A(x)$ is positive definite for all x in U except for at most a set of isolated points on which $A(x)$ is positive semidefinite.

COROLLARY 3.1. *Let U be an open convex subset of R^n and let $\varphi : U \to R^1$ be a C^2 map. If the Hessian matrix H_φ of φ is almost positive definite on U, then φ is strictly convex on U.*

Proof. Let x and z be any two distinct points in U, then by hypothesis, we have for any $y \neq 0$ in R^n,

$$y^t H_\varphi(x + \lambda(z - x))\, y \geqslant 0, \qquad \lambda \in (0, 1), \tag{3.9}$$

where the equality sign may hold only on a set of isolated points in (0, 1) that λ takes on. Now if we let $y = z - x$ in (3.9), and use the notation in (3.8), we obtain $Q(\lambda, x, z) > 0$ for all x and $z \neq x$ in U and for all $\lambda \in (0, 1)$, except possibly for a set of isolated points on which $Q(\lambda, x, z) = 0$. Hence, by Theorem 3.2, φ is strictly convex on U. This completes the proof of Corollary 3.1.

It is well known that if f is a C^1 state function defined on an open rectangular subset $U \subset R^n$, then there exists a C^2 scalar potential function $\varphi : U \to R^1$ such that $f = \nabla\varphi$ [26, 27], where an open rectangular subset of R^n is defined to be the set $\{x : a_i < x_i < b_i\,,\ i = 1, 2, \ldots, n\}$. The following theorem is a utilization of state functions:

THEOREM 3.3. *Let U be an open rectangular subset of R^n and let $f : U \to R^n$ be a C^1 state function. Then f is increasing on U if and only if the quadratic form*

$$P(\lambda, u, v) \equiv (v - u)^t\, J_f(u + \lambda(v - u))\, (v - u) \tag{3.10}$$

is positive for all u and $v \neq u$ in U, and for all $\lambda \in (0, 1)$ except for at most a set of isolated points on which $P(\lambda, u, v) = 0$.

Proof. Since f is a C^1 state function, there exists a C^2 potential function $\varphi : U \to R^1$ such that $f = \nabla\varphi$ on U. The conclusion follows immediately from Theorems 3.1 and 3.2. This completes the proof of Theorem 3.3.

COROLLARY 3.2. *Let U be an open rectangular subset of R^n and let $f : U \to R^n$ be a C^1 state function. Then f is increasing on U if $J_f(x)$ is almost positive definite on U.*

Proof. Since f is a C^1 state function, there exists a C^2 potential function $\varphi : U \to R^1$ such that $f = \nabla\varphi$ on U [26–27]. By Corollary 3.1, φ is strictly convex on U and hence f is increasing on U by Theorem 3.1. This completes the proof of Corollary 3.2.

An example of a state function which satisfies the hypothesis of Corollary 3.2 but whose Jacobian matrix is not positive definite is given by

$$f(x) = \begin{bmatrix} f_1(x) \\ f_2(x) \end{bmatrix} = \begin{bmatrix} x_1 + \frac{1}{3}x_1^3 + x_2 \\ x_1 + x_2 + \frac{1}{3}x_2^3 \end{bmatrix}.$$

A simple calculation shows that $\det J_f(x) = 0$ at $x = 0$ and $J_f(x)$ is almost positive definite on R^2. Moreover, f is increasing because

$$\begin{aligned} \langle f(u) - f(v), u - v\rangle = (u_1 - v_1 + u_2 - v_2)^2 \\ + \tfrac{1}{3}(u_1 - v_1)^2 (u_1^2 + u_1v_1 + v_1^2) \\ + \tfrac{1}{3}(u_2 - v_2)^2 (u_2^2 + u_2v_2 + v_2^2) > 0 \end{aligned}$$

whenever $u \neq v$.

However, the converse to Corollary 3.2 is not true and consequently, the converse to Corollary 3.1 is also not true. The following trivial example will bear this out:

Let $\varphi : R^2 \to R^1$ and $f : R^2 \to R^2$ be defined by

$$\varphi(x) = \tfrac{1}{4}x_1^4 + \tfrac{1}{2}x_2^2 + \tfrac{1}{4}x_2^4,$$
$$f_1(x) = x_1^3, \qquad f_2(x) = x_2 + x_2^3.$$

It is obvious that $f = \nabla\varphi$. Trivially, one can show that f is increasing on R^2, hence φ is strictly convex on R^2. But, a straightforward calculation will show that $J_f(x) = H_\varphi(x)$ is positive definite for all x in R^2 except on the line $x_1 = 0$ and on this line, $J_f(x) = H_\varphi(x)$ is positive semidefinite.

If f is not a state function, it will not be possible to generalize the preceding characterization of an increasing function f in terms of the strict convexity of some scalar function φ because only state functions can be expressed as the gradient of a potential function. However, the following theorem shows that a natural generalization of the preceding characterization to nonstate functions can be achieved through their Jacobian matrices.

THEOREM 3.4. *Let U be an open convex subset of R^n and let $f : U \to R^n$ be a C^1 function. If $J_f(x)$ is almost positive definite on U, then f is increasing on U.*

Proof. Let u and v be any two distinct points in U. Then the points on the straight line $v + \lambda(u - v)$ are in U for $\lambda \in [0, 1]$. Let

$$g(\lambda) = (u - v)^t f(v + \lambda(u - v)). \tag{3.11}$$

Since g is a C^1 function on $[0, 1]$ and $J_f(x)$ is almost positive definite on U, we have

$$g(1) - g(0) = \int_0^1 \frac{dg(\lambda)}{d\lambda}\, d\lambda = \int_0^1 (u - v)^t\, J_f(v + \lambda(u - v))\,(u - v)\, d\lambda > 0. \tag{3.12}$$

But

$$g(1) - g(0) = (u - v)^t f(u) - (u - v)^t f(v) = (u - v)^t\,(f(u) - f(v)). \tag{3.13}$$

It follows, therefore, from (3.12) and (3.13) that f is increasing on U. This completes the proof of Theorem 3.4.

An example of a nonstate function which satisfies the hypothesis of Theorem 3.4 is given by

$$f(x) = \begin{bmatrix} f_1(x) \\ f_2(x) \end{bmatrix} = \begin{bmatrix} x_1 + \frac{1}{3}x_1^3 + 2x_2 \\ x_2 + \frac{1}{3}x_2^3 \end{bmatrix}.$$

Let the nonsymmetric Jacobian matrix $J_f(x)$ be resolved into a symmetric part $A(x)$ and a skew-symmetric part $B(x)$

$$\begin{aligned} J_f(x) &= \begin{bmatrix} 1 + x_1^2 & 2 \\ 0 & 1 + x_2^2 \end{bmatrix} = \begin{bmatrix} 1 + x_1^2 & 1 \\ 1 & 1 + x_2^2 \end{bmatrix} + \begin{bmatrix} 0 & 1 \\ -1 & 0 \end{bmatrix} \\ &\equiv A(x) + B(x). \end{aligned}$$

Clearly, $A(x)$ is almost positive definite, and hence so is $J_f(x)$. To verify that f is indeed increasing, we found

$$\begin{aligned} \langle f(u) - f(v), u - v \rangle &= (u_1 - v_1 + u_2 - v_2)^2 + \tfrac{1}{3}\,[(u_1 - v_1)^2\,(u_1^2 + u_1 v_1 + v_1^2) \\ &\quad + (u_2 - v_2)^2\,(u_2^2 + u_2 v_2 + v_2^2)] > 0 \qquad \text{for all } u \neq v. \end{aligned}$$

Although the converse of Theorem 3.4 is not true, the following theorem shows that not much improvement is possible:

THEOREM 3.5. *Let U be an open convex subset of R^n and let $f : U \to R^n$ be a C^1 function. If $J_f(x)$ is positive semidefinite for all x in U, then f is nondecreasing on U.*

Proof. The proof of Theorem 3.5 follows *mutatis mutandis* from the proof of Theorem 3.4 and is, therefore, omitted.

LEMMA 3.3. *Let U be an open convex subset of R^n and $f: U \to R^n$ be a C^1 map. If f is nondecreasing on U, then $J_f(x)$ is positive semidefinite on U.*

Proof. Suppose there exists a point x_0 in U and a pair of distinct vectors x_1 and x_2 in R^n such that the quadratic form

$$Q(x_0, x_1, x_2) \equiv (x_2 - x_1)^t J_f(x_0)(x_2 - x_1)$$

is negative. Then there exists a neighborhood about x_0 which contains an open ball $B(x_0, r) \subset U$ such that for every z in $B(x_0, r)$, we have $Q(z, x_1, x_2) < 0$. Since $B(x_0, r)$ is an n-dimensional ball, for each y in R^n, there exists a vector $\bar{y}$ in $B(x_0, r)$ and $\alpha > 0$ such that $y = \alpha(\bar{y} - x_0)$. Hence, we have two vectors $\bar{x}_1$ and $\bar{x}_2$ in $B(x_0, r)$ such that $x_1 = a_1(\bar{x}_1 - \bar{x}_0)$ and $x_2 = a_2(\bar{x}_2 - \bar{x}_0)$, where $a_1 > 0$, $a_2 > 0$. Let $a = \max\{a_1, a_2\}$. Then we can find $\hat{x}_1$ and $\hat{x}_2$ in $B(x_0, r)$ such that $x_1 = a(\hat{x}_1 - x_0)$ and $x_2 = a(\hat{x}_2 - x_0)$. Thus

$$x_2 - x_1 = a(\hat{x}_2 - \hat{x}_1). \tag{3.14}$$

Since $\hat{x}_1$ and $\hat{x}_2$ are in $B(x_0, r)$, the points $[\hat{x}_1 + \lambda(\hat{x}_2 - \hat{x}_1)]$ with λ in $[0, 1]$ are in $B(x_0, r)$. Hence, for λ in $[0, 1]$, we have

$$Q(\hat{x}_1 + \lambda(\hat{x}_2 - \hat{x}_1), \hat{x}_1, \hat{x}_2) = \frac{1}{a^2} Q(\hat{x}_1 + \lambda(\hat{x}_2 - \hat{x}_1), x_1, x_2) < 0. \tag{3.15}$$

Consider the following scalar function:

$$g(\lambda) \equiv (\hat{x}_2 - \hat{x}_1)^t f(\hat{x}_1 + \lambda(\hat{x}_2 - \hat{x}_1)), \qquad \lambda \in [0, 1].$$

Then, $g(1) - g(0) = (\hat{x}_2 - \hat{x}_1)^t [f(\hat{x}_2) - f(\hat{x}_1)] \geqslant 0$ because f is nondecreasing on U. But $g(1) - g(0) = g'(\lambda^*)$ for some λ^* in $(0, 1)$. Hence, we have

$$\begin{aligned} g(1) - g(0) = g'(\lambda^*) &= (\hat{x}_2 - \hat{x}_1)^t J_f(\hat{x}_1 + \lambda^*(\hat{x}_2 - \hat{x}_1))(\hat{x}_2 - \hat{x}_1) \\ &= Q(\hat{x}_1 + \lambda^*(\hat{x}_2 - \hat{x}_1), \hat{x}_1, \hat{x}_2) < 0, \end{aligned} \tag{3.16}$$

which is a contradiction. Hence, there does not exist a point x_0 in U and a pair of vectors x_1 and x_2 in R^n such that $Q(x_0, x_1, x_2) < 0$. Therefore, $Q(x_0, x_1, x_2) \geqslant 0$ for all x_0 in U and x_1, x_2 in R^n. This implies that $J_f(x)$ is a positive semidefinite matrix for all x in U.

The preceding results can be summarized as follows:

Let U be an open convex subset of R^n and let $f: U \to R^n$ be a C^1 function.

Consider the following properties:

(A) $J_f(x)$ is positive definite on U.

(B) $J_f(x)$ is almost positive definite on U.

(C) f is increasing on U.

(D) f is nondecreasing on U.

(E) $J_f(x)$ is positive semidefinite on U.

Then we have the following results:

$$(A) \Rightarrow (B) \Rightarrow (C) \Rightarrow (D) \Leftrightarrow (E).$$

4. Properties and Characterization of Quasi-Increasing Functions

Our objective in this section is to study the properties of a class of homeomorphic functions, called "quasi-increasing functions," which includes the class of increasing functions as a proper subset. Our main motivation is to derive weaker global inversion theorems which allow the Jacobian of an appropriately transformed homeomorphic function to vanish on hyperplanes, rather than on a set of isolated points.

Definition 4.1. An $n \times n$ matrix A is said to be a class-E matrix if, and only if, each row and each column of A have one and only one nonzero element which is either 1 or -1.

Notice that premultiplication (postmultiplication) of a matrix J by a class-E matrix A is equivalent to a permutation of the rows (columns) of J with certain rows (columns) of J multiplied by -1. In this respect, class-E matrices act like elementary matrices [28]. The following two important properties of class-E matrices can be shown trivially:

1. If E_a and E_b are class-E matrices, then E_bE_a is a class-E matrix. Conversely, if E_c is a class-E matrix, there are two class-E matrices, E_a and E_b such that $E_bE_a = E_c$.
2. Every class-E matrix is an orthogonal matrix.

Definition 4.2. A function $f: R^n \to R^n$ is said to be quasi-increasing if, and only if, there exist two class-E matrices E_a and E_b such that the transformed function $\hat{f}: R^n \to R^n$ defined by $\hat{f}(x) = E_a f \circ (E_b x)$ is an increasing function.

Lemma 4.1. *Every quasi-increasing function f is a homeomorphic function.*

Proof. By definition, the transformed function $\hat{f}(x) = E_a f \circ (E_b x)$ is increasing and hence by Lemma 3.1(b), is homeomorphic in R^n. Since class-E

matrices are nonsingular, the function $f \circ (E_b x) = E_a^{-1} f(x)$ is also homeomorphic in R^n. If we let $z = E_b x$, then $x = E_b^{-1} z$ is well defined and hence $f(z)$ is homeomorphic in R^n. This completes the proof.

The following lemma is an important observation for the development of this section:

LEMMA 4.2. *Let E_a and E_b be two class-E matrices of order n and $f : U \subset R^n \to R^n$ be a C^1 map on U. Then $E_a J_f(x) E_b$ is positive definite (semi-definite) at a point $x \in U$ if, and only if, $E_c J_f(x)$ is positive definite (semi-definite) at x, where $E_c = E_b E_a$.*

Proof. Since the proofs of the positive definite part and the positive semi-definite part are similar, it suffices to prove the positive definite part only.

Since E_b is nonsingular, so is $E_b{}^t$. Hence, for every vector $z \neq 0$ in R^n, there is a vector $s \neq 0$ in R^n such that $z = E_b{}^t s$. So, we have the following:

$$Q(z) \equiv z^t E_a J_f(x) E_b z = s^t E_b E_a J_f(x) E_b E_b{}^t s = s^t E_c J_f(x) s \equiv P(s).$$

This implies that $Q(z) > 0$ for all $z \neq 0$ in R^n if, and only if, $P(s) > 0$ for all $s \neq 0$ in R^n. Hence, our conclusion follows.

In view of Lemma 4.2, we can conclude that if $f : U \subset R^n \to R^n$ and is a C^1 map on U, then $E_a J_f(x) E_b$ is almost positive definite on U if and only if $E_c J_f(x)$ is almost positive definite on U, where $E_c = E_b E_a$, and E_a, E_b are class-E matrices.

THEOREM 4.1. *Let $f : X \to Y$, $X = Y = R^n$, be a C^1 map. Suppose there exists a class-E matrix E_c such that $E_c J_f(x)$ is almost positive definite on R^n, then f is quasi-increasing.*

Proof. Let E_a and E_b be two class-E matrices such that $E_c = E_b E_a$. For each x in X and $y = f(x)$, let $x = E_b u$ and $w = E_a y$. Then

$$w = E_a f \circ (E_b u) \equiv \hat{f}(u)$$

and $\hat{f} : U \to W$, where $U = E_b^{-1} X = R^n$ and $W = E_a Y = R^n$, is a C^1 map. Moreover,

$$J_{\hat{f}}(u) = E_a J_f \circ (E_b u) E_b = E_a J_f(x) E_b. \tag{4.1}$$

It follows from the hypothesis and Lemma 4.2 that $J_{\hat{f}}(u)$ is almost positive definite on R^n. By Theorem 3.4, $\hat{f}$ is an increasing function on R^n. Hence, f is quasi-increasing on R^n and the theorem is proved.

An example of a function $f : R^n \to R^n$ which is neither increasing nor decreasing but which satisfies the hypothesis of Theorem 4.1 is given by

$$f(x) = \begin{bmatrix} f_1(x) \\ f_2(x) \end{bmatrix} = \begin{bmatrix} x_1 + x_2 + x_2{}^3 \\ -x_1 - x_1{}^3 - x_2 \end{bmatrix}.$$

f is neither increasing or decreasing on R^2 because

$$\langle f(a) - f(b), a - b\rangle = -1 \quad \text{when} \quad a_1 = b_1, \quad a_2 = 1, \quad b_2 = 0$$

and

$$\langle f(a) - f(b), a - b\rangle = 1 \quad \text{when} \quad a_2 = b_2, \quad a_1 = 1, \quad b_1 = 0.$$

Let

$$E_c = \begin{bmatrix} 0 & -1 \\ 1 & 0 \end{bmatrix}, \quad E_a = \begin{bmatrix} 0 & -1 \\ -1 & 0 \end{bmatrix} \quad \text{and} \quad E_b = \begin{bmatrix} 1 & 0 \\ 0 & -1 \end{bmatrix}.$$

Then

$$E_c = E_b E_a \quad \text{and} \quad E_c J_f(x) = \begin{bmatrix} 1 + 3x_1^2 & 1 \\ 1 & 1 + 3x_2^2 \end{bmatrix} = A(x).$$

A simple computation will show that $A(x)$ is positive definite except at the point $x_1 = x_2 = 0$ and at this point, $A(x)$ is positive semidefinite. Hence, $A(x)$ is almost positive definite on R^2 and so f is quasi-increasing. Indeed, the function

$$\hat{f}(u) = E_a J_f(E_b u) = \begin{bmatrix} u_1 + u_1^3 - u_2 \\ - u_1 + u_2 + u_2^3 \end{bmatrix}$$

is increasing because

$$\begin{aligned} &\langle \hat{f}(a) - \hat{f}(b), a - b\rangle \\ &\quad = [(a_1 - b_1) - (a_2 - b_2)]^2 + (a_1 - b_1)^2 (a_1 + a_1 b_1 + b_1^2) \\ &\qquad + (a_2 - b_2)^2 (a_2 + a_2 b_2 + b_2^2) > 0 \end{aligned}$$

whenever $a \neq b$.

Our next theorem shows that a function $f : R^n \to R^n$ can be quasi-increasing even if the Jacobian of $\hat{f}$ is allowed to vanish on $(n - 1)$-dimensional hyperplanes.

THEOREM 4.2. *Let $f : X \to Y$, $X = Y = R^n$, be a C^1 map. Suppose there exists a class-E matrix E_c such that $E_c J_f(x)$ is positive definite for all x in R^n except for at most a finite number of hyperplanes of dimension less than or equal to $(n - 1)$ defined as*

$$\sum_{j=1}^{n} p_{ij} x_j = q_j \qquad i = 1, 2,..., m,$$

where $x = (x_1, x_2, ..., x_n)$; p_{ij} and q_j, $j = 1, 2,..., n$, $i = 1, 2,..., m$, are constants. Let

$$S_i = \left\{ x : \sum_{j=1}^{n} p_{ij} x_j = q_j \right\}, \qquad i = 1, 2,..., m.$$

If for any two distinct vectors α *and* β *in each* S_i, $i = 1, 2, \ldots, m$, *we have*

$$\langle E_c[f(\alpha) - f(\beta)], \alpha - \beta \rangle > 0, \tag{4.2}$$

then f is quasi-increasing.

Proof. Let E_a and E_b be two class-E matrices such that $E_c = E_b E_a$. For each x in X and $y = f(x)$, let $x = E_b u$ and $w = E_b y$. Then

$$U = E_b^{-1} X = R^n, \qquad W = E_a Y = R^n$$

and

$$w = E_a y = E_a f(x) = E_a f \circ (E_b u) \equiv \hat{f}(u).$$

That is, $\hat{f}: U \to W$, $U = W = R^n$, is a C^1 map. Let r and s be any two distinct points in R^n. Let

$$g(\lambda) = (r - s)^t \hat{f}(s + \lambda(r - s)), \qquad \lambda \in [0, 1].$$

Then g is a C^1 map on [0, 1]. Hence, we have

$$\begin{aligned} g(1) - g(0) &= \int_0^1 g'(\lambda)\, d\lambda = \int_0^1 (r - s)^t J_{\hat{f}}(s + \lambda(r - s))\,(r - s)\, d\lambda \\ &= \int_0^1 (r - s)^t E_a J_f \circ [E_b(s + \lambda(r - s))]\, E_b(r - s)\, d\lambda \\ &= \int_0^1 (r - s)^t E_b{}^t E_b E_a J_f \circ [E_b(s + \lambda(r - s))]\, E_b(r - s)\, d\lambda \\ &= \int_0^1 (x_r - x_s)^t E_c J_f(x_s + \lambda(x_r - x_s))\,(x_r - x_s)\, d\lambda, \end{aligned} \tag{4.3}$$

where $x_r = F_b r$ and $x_s = E_b s$.

Since $\Gamma \equiv \{x : x = x_s + \lambda(x_r - x_s), \lambda \in [0, 1]\}$ is a straight line connecting the points x_r and $x_s \neq x_r$, only the following three cases can occur:

(a) Γ does not intersect any S_k, $k = 1, 2, \ldots, m$;

(b) Γ intersects some S_k, $k = 1, 2, \ldots, m$;

(c) Γ is contained in some S_k, say $S_{l_1}, S_{l_2}, \ldots, S_{l_p}$, where

$$1 \leqslant l_1, l_2, \ldots, l_p \leqslant m,$$

i.e.,

$$\Gamma \subset S_{l_1} \cap S_{l_2} \cdots \cap S_{l_p}.$$

Notice that a straight line can intersect each hyperplane at most at one point unless it is contained in that hyperplane. If case (a) or (b) occurs, then, except for a finite number of λ in (0, 1) for which the straight line

$$\Gamma = \{x : x = x_s + \lambda(x_r - x_s),\ \lambda \in [0, 1]\}$$

intersects with some hyperplane S_k, the matrix $E_c J_f(x_s + \lambda(x_r - x_s))$ is positive definite. Hence, (4.3) would imply that

$$g(1) - g(0) > 0,$$

that is,

$$(r - s)^t \hat{f}(r) - (r - s)^t \hat{f}(s) = \langle \hat{f}(r) - \hat{f}(s), r - s \rangle > 0. \tag{4.4}$$

If case (c) occurs, then the straight line Γ connecting the points x_r and x_s is contained in $S_{l_1} \cap S_{l_2} \cap \cdots \cap S_{l_p}$. By assumption,

$$\langle E_c[f(x_r) - f(x_s)],\ x_r - x_s \rangle > 0.$$

This implies that

$$\begin{aligned}\langle E_b E_a[f \circ (E_b r) - f \circ (E_b s)],\ E_b r - E_b s \rangle &= \langle E_a f \circ (E_b r) - E_a f \circ (E_b s), r - s \rangle \\ &= \langle \hat{f}(r) - \hat{f}(s), r - s \rangle > 0.\end{aligned}$$

Hence, for any two distinct points r and s in R^n, we have

$$\langle \hat{f}(r) - \hat{f}(s), r - s \rangle > 0.$$

That is, $\hat{f}$ is increasing on R^n and so f is quasi-increasing. This completes the proof of Theorem 4.2.

To illustrate the utility of Theorem 4.2, we will present next an example which applies Theorem 4.2 to show that a function which is neither increasing nor decreasing can nevertheless be quasi-increasing even though it fails to satisfy the hypothesis of Theorem 4.1.

EXAMPLE 4.1. Let $f : R^2 \to R^2$ where

$$f(x) = \begin{bmatrix} x_1 + x_2 + x_2^3 \\ -x_1 - x_2 \end{bmatrix} \equiv \begin{bmatrix} f_1(x_1, x_2) \\ f_2(x_1, x_2) \end{bmatrix}.$$

f is neither increasing nor decreasing because $\langle f(u) - f(v), u - v \rangle = -1$ when $u = (0, 1)$ and $v = (0, 0)$ and $\langle f(u) - f(v), u - v \rangle = 1$ when $u = (1, 1)$ and $v = (2, 1)$. Moreover, the Jacobian of f vanishes on the line $x_2 = 0$. However, if we choose

$$E_c = \begin{bmatrix} 0 & -1 \\ 1 & 0 \end{bmatrix}, \quad E_a = \begin{bmatrix} 0 & -1 \\ -1 & 0 \end{bmatrix}, \quad \text{and} \quad E_b = \begin{bmatrix} 1 & 0 \\ 0 & -1 \end{bmatrix},$$

then it is easily verified that $E_c = E_b E_a$ and $E_c J_f(x)$ is positive definite except on the line $x_2 = 0$, and on this line, $\langle E_c[f(u) - f(v)], (u - v)\rangle > 0$. Hence, by Theorem 4.2, f is quasi-increasing and homeomorphic on R^2. Indeed, the inverse function is given by

$$f^{-1} = \begin{bmatrix} y_2 - (y_1 + y_2)^{1/3} \\ (y_1 + y_2)^{1/3} \end{bmatrix}.$$

By choosing $E_c = E_a = E_b = I$, the identity matrix, Theorem 4.2 becomes a sufficient condition for f to be an increasing function while allowing the Jacobian of $J_f(x)$ to vanish on a larger set than a set of isolated points. The next example illustrates this point.

EXAMPLE 4.2. Let $f : R^2 \to R^2$ where

$$f(x) = \begin{bmatrix} \frac{1}{3}x_1^3 + x_2 \\ -x_1 + x_2 + \frac{1}{3}x_2^3 \end{bmatrix}.$$

It is easily verified that $J_f(x)$ is positive definite except on the line $x_1 = 0$, and on this line, $\langle f(u) - f(v), u - v\rangle > 0$. Hence, by Theorem 4.2, f is an increasing function.

5. Concluding Remarks

Several new theorems have been presented for a function $f : R^n \to R^n$ to be homeomorphic on R^n. Unlike the Palais theorem [1], these global inversion theorems share a common feature in that the Jacobian of f is allowed to vanish on at least a set of isolated points. This feature is also preserved in the global implicit function theorems which follow quite naturally from the inversion theorems. A version of partial converse to the global inverse (implicit) function theorem is also presented.

The class of increasing functions is seen to be a natural generalization of the class of state functions associated with strict convex potential functions. A further generalization leads to the definition of a quasi-increasing function which behaves in many respects like an increasing function.

Acknowledgments

The authors would like to thank Professor Jan Mycielski from the University of Colorado and Professor Edwin Duda from the University of Miami for their assistance in the proof of Theorem 2.1.

References

1. R. S. Palais, Natural operations on differential forms, *Trans. Amer. Math. Soc.* **92** (1959), 125–141.
2. C. A. Holtzmann and R. W. Liu, On the dynamical equations of nonlinear network with *n*-coupled elements, Proc. of the 3rd Annual Allerton Conf. on Circuit and System Theory, pp. 533–545, Univ. of Illinois, Urbana, IL, 1965.
3. T. Ohtsuki and H. Watanabe, State variable analysis of RLC networks containing nonlinear coupling elements, *IEEE Trans. Circuit Theory* **16** (1969), 26–38.
4. I. W. Sandberg and A. N. Wilson, Jr., Some theorems on properties of DC equations of nonlinear networks, *Bell System Tech. J.* **48** (1969), 1–34.
5. I. W. Sandberg, Theorems on the analysis of nonlinear transistor networks, *Bell System Tech. J.* **49** (1970), 95–114.
6. L. O. Chua and Y.-F. Lam, "Foundations of nonlinear network theory, Part I," Technical Report No. TR-EE 70-22, Purdue University, Lafayette, IN, June, 1970.
7. L. O. Chua, The linear transformation converter and its application to the synthesis of nonlinear networks, *IEEE Trans. Circuit Theory* **17** (1970), 584–594.
8. E. S. Kuh and I. N. Hajj, Nonlinear Circuit Theory resistive networks, *Proc. IEEE*, Vol. 59, March (1971), 340-355.
9. T. Fujisawa and E. S. Kuh, Some results on existence and uniqueness of solution of nonlinear networks, *IEEE Trans. Circuit Theory*, **18** (1971), 501–506.
10. L. Bers, "Topology," Courant Institute of Mathematical Science, New York University, New York, 1956.
11. R. K. Brayton and J. K. Moser, A theory of nonlinear networks—I, II, *Quart. Appl. Math.* **22** (1964), 1–33; 81–104.
12. L. O. Chua, "Stationary Principles and Potential Functions for Nonlinear Networks, Parts I and II," Technical Report No. TR-EE 70-35, Purdue University, Lafayette, IN, September, 1970.
13. C. Chevalley, "Theory of Lie Groups," Princeton University Press, Princeton, NJ, 1946.
14. P. T. Church and E. Hemmingsen, Light open maps on *n*-manifolds, *Duke Math. J.* **27** (1960), 527–536.
15. W. Hurewicz and H. Wallman, "Dimension Theory," Princeton University Press, Princeton, NJ, 1948.
16. P. T. Church, Differentiable open maps, *Bull. Amer. Math. Soc.* **68** (1962), 468–469.
17. P. T. Church, Differentiable open maps on *n*-manifolds, *Trans. Amer. Math. Soc.* **109** (1963), 87–100.
18. G. T. Whyburn, "Topological Analysis," Princeton University Press, Princeton, NJ, 1964.
19. W. Rudin, "Principles of Mathematical Analysis," McGraw-Hill, New York, 1964.
20. G. J. Minty, Monotone networks, *Proc. Roy. Soc. Ser. A* **257** (1960), 194–212.
21. C. L. Dolph, Recent developments in some non-self-adjoint problems of mathematical physics, *Bull. Amer. Math. Soc.* **67** (1961), 1–69.
22. G. J. Minty, Monotone (non-linear) operators in Hilbert space, *Duke Math. J.* **29** (1962), 341–346.
23. F. E. Browder, The solvability of nonlinear functional equations, *Duke Math. J.* **30** (1963), 557–566.
24. D. Gale and H. Nikaido, The jacobian matrix and global univalence of mappings, *Math. Ann.* **159** (1965), 81–93.

25. P. T. Church, Differentiable monotone maps on manifolds, *Trans. Amer. Math. Soc.* **128** (1967), 185–205.
26. M. S. Burger and M. S. Burger, "Perspectives in Nonlinearity, An Introduction to Nonlinear Analysis," Benjamin, New York, 1968.
27. T. N. Apostol, "Mathematical Analysis, A Modern Approach to Advanced Calculus," Addison-Wesley, Reading, MA, 1957.
28. L. Mirsky, "An Introduction to Linear Algebra," Oxford University Press, Oxford, 1955.

Some Results on Existence and Uniqueness of Solutions of Nonlinear Networks

TOSHIO FUJISAWA AND ERNEST S. KUH, FELLOW, IEEE

Abstract—This paper deals with nonlinear networks which can be characterized by the equation $f(x) = y$, where $f(\cdot)$ maps the real Euclidean n-space R^n into itself and is assumed to be continuously differentiable. x is a point in R^n and represents a set of chosen network variables, and y is an arbitrary point in R^n and represents the input to the network. The authors derive sufficient conditions for the existence of a unique solution of the equation for all $y \in R^n$ in terms of the Jacobian matrix $\partial f/\partial x$. It is shown that if a set of cofactors of the Jacobian matrix satisfies a "ratio condition," the network has a unique solution. The class of matrices under consideration is a generalization of the class P recently introduced by Fiedler and Pták, and it includes the familiar uniformly positive-definite matrix as a special case.

I. Introduction

It has been known that the analysis of general nonlinear RLC networks depends on the analysis of three one-element-kind subnetworks [1]–[5]. Furthermore, in analyzing the one-element-kind networks, it is important to know the conditions under which the network has a unique solution. Various useful sufficient conditions for existence and uniqueness of solutions have been found by many workers in terms of the element characteristics [1], [4]–[10]. However, by and large, these conditions are restricted to networks of special classes. In this paper nonlinear resistive networks of the most general form will be considered; these include passive as well as active elements, nonlinear resistors as well as coupled elements.

It is well known that most nonlinear resistive networks can be characterized by the equation

$$f(x) = y \tag{1}$$

where $f(\cdot)$ maps the real Euclidean n-space R^n into itself. x is a point in R^n and represents a set of chosen network variables; y is an arbitrary point in R^n and represents an arbitrary set of inputs to the network. Necessary and sufficient conditions for the existence of a unique solution of (1) for all $y \in R^n$ have been investigated by Palais [11], [12]. More specifically, the Palais theorem states that the necessary and sufficient conditions for the mapping $f: R^n \rightarrow R^n$ to be a C^1-diffeomorphism of R^n onto itself are 1) $f(x)$ be continuously differentiable; 2) $\det J \neq 0$ for all x in R^n, where $J = \partial f/\partial x$ is the Jacobian matrix; and 3) $\lim \|f(x)\| = \infty$ as $\|x\| \rightarrow \infty$, where $\|\cdot\|$ is the Euclidean norm. Note that C^1-diffeomorphism implies that f^{-1} is also continuously differentiable.

In the one-dimensional case, the Palais theorem has the following well-known interpretation: the function maps R^1 onto itself and it must have nonzero slope everywhere. The familiar quasi-linear resistor has precisely such characteristics. Duffin [6] in his well-known paper proved that a nonlinear network which is formed by an arbitrary interconnection of quasi-linear resistors has a unique solution for all possible inputs.

Two special cases of (1) have been considered recently. The first one is a sufficient condition due to Ohtsuki and Watanabe [4] and states that if the Jacobian matrix J is uniformly positive definite, there exists a unique solution for all $y \in R^n$. The second deals with a subclass of the equation in (1), which is of the form

$$f(x) = F(x) + Ax = y \tag{2}$$

where $F \in \mathfrak{F}^n$ and $\mathfrak{F}^n$ is defined as a family of all functions F such that for $i = 1, 2, \cdots, n$, the ith component of $F(\cdot)$ is a strictly monotone increasing function of x_i, which maps R^1 onto itself. Sandberg and Willson have shown [9] that the necessary and sufficient condition on the $n \times n$ constant matrix A for (2) to have a unique solution for each $F(\cdot) \in \mathfrak{F}^n$ and each $y \in R^n$ is that A be of class P_0. In their subsequent work more general results have been obtained [13], [14].

Class P_0 and P matrices were introduced by Fiedler and Pták [15], when they considered generalizations of positive definiteness and monotonicity. A constant square matrix is said to be of class P_0 if its determinant and all principal minors are nonnegative. Similarly, a constant square matrix is said to be of class P if its determinant and all principal minors are positive.

In this paper we derive two theorems which state sufficient conditions for f to be a C^1-diffeomorphism. In this case the inverse mapping f^{-1} is continuously differentiable, and hence the dependence of solution x on input y is smooth. The conditions are stated in terms of the Jacobian matrix $J(x) = \partial f/\partial x$. The first is a direct generalization of the matrix of class P, and the second represents a further generalization of the first. Various remarks and examples are given to compare the results with existing ones and to illustrate further implications of the theorems.

Manuscript received June 27, 1970; revised March 3, 1971. This paper was presented at the 1970 IEEE International Symposium on Circuit Theory, Atlanta, Ga., December 14–16, 1970. Research was sponsored by the National Science Foundation under Grants GK-10656X/GF-357.

T. Fujisawa is with the Department of Control Engineering, Osaka University, Toyonaka, Osaka, Japan. He is currently a Visiting Professor of Electrical Engineering and Computer Sciences at the University of California, Berkeley, Calif. 94720.

E. S. Kuh is with the Department of Electrical Engineering and Computer Sciences and the Electronics Research Laboratory, University of California, Berkeley, Calif. 94720.

Reprinted from *IEEE Transactions on Circuit Theory*, vol. CT-18, pp. 501-506, September 1971.

II. Main Theorem on the Existence of a Unique Solution

Theorem 1

Let f be a continuously differentiable mapping of R^n into itself, and let J_k be the matrix consisting of the first k rows and the first k columns of the Jacobian matrix J for $k=1, \cdots, n$. Then, for any $y \in R^n$ there exists one and only one solution of (1) if there exists a positive constant $\epsilon > 0$ such that

$$|\det J_1| \geqq \epsilon, \quad \left|\frac{\det J_2}{\det J_1}\right| \geqq \epsilon, \cdots, \quad \left|\frac{\det J_n}{\det J_{n-1}}\right| \geqq \epsilon \qquad (3)$$

uniformly in all $\mathbf{x} \in R^n$.

The above condition will be referred to as the "ratio condition" for convenience. Before presenting the proof of the theorem, we wish to point out two important facts.

The first fact is that $\det J = \det J_n \neq 0$ implies that for any interior point y in the range of f there exists a neighborhood of y where the continuously differentiable inverse of f is defined. Thus, if f^{-1} is uniquely determined globally, the local inverses are restrictions of this global inverse; hence the global inverse f^{-1} is continuously differentiable. Therefore, the fact that f maps R^n onto itself as a one-to-one correspondance implies that f is a C^1-diffeomorphism.

Second, the ratio condition implies that the following set of principal minors of the Jacobian matrix $J(\mathbf{x})$ are nonzero for all $\mathbf{x}$:

$$\det J_k(\mathbf{x}) \neq 0, \qquad k = 1, 2, \cdots, n. \qquad (4)$$

Note that if the above condition holds, it does not necessarily follow that (1) has a solution. For example, let

$$y_1 = f_1(x_1, x_2) = x_1 \cosh x_2$$

$$y_2 = f_2(x_1, x_2) = \int_0^{x_2} \frac{dz}{\cosh z}.$$

Then

$$J(\mathbf{x}) = \begin{bmatrix} \cosh x_2 & x_1 \sinh x_2 \\ 0 & \dfrac{1}{\cosh x_2} \end{bmatrix}.$$

It is seen that $\det J_1 = \cosh x_2$ and $\det J_2 = 1$; they are nonzero for all $\mathbf{x}$. However, $\det J_2/\det J_1 = 1/\cosh x_2$, which does not satisfy the ratio condition. It is easy to check that the equations $y_1 = f_1(x_1, x_2)$ and $y_2 = f_2(x_1, x_2)$ above do not always have a solution for all y_1 and y_2 because the range of f_2 is bounded.

It is of further interest to note that the condition in (4) does not guarantee the uniqueness of solution.[1] The following example given in [16] illustrates this fact. Let

$$y_1 = f_1(x_1, x_2) = e^{2x_1} - x_2^2 + 3$$

$$y_2 = f_2(x_1, x_2) = 4e^{2x_1}x_2 - x_2^3.$$

[1] This is pointed out by Dr. Alan Willson.

Then $\det J_1 = 2e^{2x_1}$ and $\det J_2 = 8e^{4x_1} + 10e^{2x_1}x_2^2$. Even though the condition in (4) is satisfied, the ratio condition is not. Two solutions exist for the equations at $x_1 = 0$, $x_2 = \pm 2$ for $y_1 = y_2 = 0$.

Other implications of the theorem will be examined in Section III after we give a proof of the main theorem.

Proof: It is easy to see that the statement of the theorem holds for $n=1$, since the condition $|\det J_1| = |df_1/dx_1| \geqq \epsilon > 0$ implies that f_1 is a strictly monotone increasing or decreasing mapping of R^1 onto itself. The theorem is proven by induction.

Under the assumption that the statement is valid for $n = k-1$, the case of $n=k$ is considered. In this case

$$y_i = f_i(x_1, \cdots, x_{k-1}, x_k), \qquad i = 1, \cdots, k \qquad (5)$$

or in vector notation

$$\mathbf{y} = \mathbf{f}(\mathbf{x}). \qquad (6)$$

For the purpose of notational simplicity the following convention is introduced:

$$\mathbf{x}_{-k} = [x_1, \cdots, x_{k-1}]^t$$

$$\mathbf{y}_{-k} = [y_1, \cdots, y_{k-1}]^t$$

$$\mathbf{f}_{-k} = [f_1, \cdots, f_{k-1}]^t \qquad (7)$$

where the superscript t denotes the transpose. The first $k-1$ equations of (6) may be expressed as

$$\mathbf{y}_{-k} = \mathbf{f}_{-k}(\mathbf{x}_{-k}, x_k). \qquad (8)$$

If the value of x_k is kept fixed, the mapping $\mathbf{f}_{-k}$ of R^{k-1} into itself is continuously differentiable. Furthermore, the mapping $\mathbf{f}_{-k}$ with Jacobian matrix $J_{k-1}(\mathbf{x})$ satisfies the ratio condition for $n=k-1$, and therefore $\mathbf{f}_{-k}$ is a C^1-diffeomorphism of R^{k-1} onto itself due to the assumption of mathematical induction. Thus $\mathbf{x}_{-k}$ can be represented as a function of $\mathbf{y}_{-k}$ and x_k as follows:

$$\mathbf{x}_{-k} = \overset{-1}{\mathbf{f}}_{-k}(\mathbf{y}_{-k}, x_k). \qquad (9)$$

Substituting this relation into the kth equation of (5), y_k is represented as a function of $\mathbf{y}_{-k}$ and x_k. The dependence of y_k on x_k can be determined in the following way provided that the value of $\mathbf{y}_{-k}$ is kept fixed [17]. The differentiation of (6) yields

$$d\mathbf{y} = J_k d\mathbf{x}$$

or equivalently

$$\begin{bmatrix} d\mathbf{y}_{-k} \\ dy_k \end{bmatrix} = J_k \begin{bmatrix} d\mathbf{x}_{-k} \\ dx_k \end{bmatrix}. \qquad (10)$$

Since $d\mathbf{y}_{-k} = 0$ and $\det J_k \neq 0$, Cramer's rule can be used to obtain

$$\frac{dy_k}{dx_k} = \frac{\det J_k}{\det J_{k-1}}. \qquad (11)$$

The condition of nonzero minors in (4) implies that if the value of y_{-k} is kept fixed, y_k is a strictly monotone increasing or decreasing function of x_k for $-\infty<x_k<\infty$. The ratio condition (3) implies in addition that the range of y_k covers the whole real line $-\infty<y_k<\infty$.

Let $y^*=[y_1^*, \cdots, y_{k-1}^*, y_k^*]^t$ be an arbitrary given point of R^k. For any value $x_k=s$ there is one and only one point

$$x(s) = [x_1(s), \cdots, x_{k-1}(s), s]^t \tag{12}$$

such that

$$[y_1^*, \cdots, y_{k-1}^*, y_k]^t = f(x(s)). \tag{13}$$

It is now clear from the property of the dependence of y_k on $x_k=s$, discussed in the preceding paragraph, that there exists one and only one value of $s=s^*$ for which the following relation holds:

$$y^* = f(x(s^*)). \tag{14}$$

Thus the point $x(s^*)$ is the unique solution to the equation $y^*=f(x)$. This completes the proof of Theorem 1.

III. Remarks and Special Cases

Remark 1

Ohtsuki and Watanabe have shown [4] that uniform positive definiteness implies C^1-diffeomorphism of f. It is not difficult to show that uniform positive definiteness implies the ratio condition.

Consider the mapping of (1), where the Jacobian matrix $J(x)$ is uniformly positive definite. Then there exists a positive constant $\epsilon>0$ such that

$$z_k^t J_k(x) z_k \geqq \epsilon \|z_k\|^2 \tag{15}$$

for any $x\in R^n$ and $z_k\in R^k$, where $k=1, 2, \cdots, n$. If $k=1$, (15) becomes

$$J_1(x) \geqq \epsilon \tag{16}$$

which is the first inequality in (3). To obtain the remaining conditions, we introduce, for convenience, the following notations. We denote the first k equations of $f(x)=y$ by $\hat{f}(x)=\hat{y}$, and the first k components of x by $\hat{x}$. The following statement is true for $k=2, \cdots, n$: condition (15) implies that $\hat{f}$ with Jacobian matrix J_k is a C^1-diffeomorphism of R^k onto itself provided that the values of $x_{k+1}, \cdots, x_n$ are kept fixed. Furthermore condition (15) implies that the inverse mapping $\hat{f}^{-1}$ satisfies global Lipschitz condition with a Lipschitz constant $1/\epsilon$, i.e.,

$$\|d\hat{x}\| \leqq \frac{1}{\epsilon}\|d\hat{y}\|. \tag{17}$$

In deriving (11), we set all inputs fixed with the exception of y_k; thus $\|d\hat{y}\| = |dy_k|$. From (11) we obtain

$$\|d\hat{y}\| = \frac{\det J_k}{\det J_{k-1}} |dx_k| \leqq \frac{\det J_k}{\det J_{k-1}} \|d\hat{x}\|. \tag{18}$$

Note that the ratio, $\det J_k/\det J_{k-1}$, is always positive because the Jacobian matrix is uniformly positive definite. Combining the inequalities in (17) and (18), we obtain the desired ratio condition for $k=2, \cdots, n$:

$$\frac{\det J_k}{\det J_{k-1}} \geqq \epsilon > 0. \tag{19}$$

Thus we have demonstrated that uniform positive definiteness of the Jacobian matrix implies the ratio condition.

Remark 2

The class P matrix of Fiedler and Pták is closely related to the condition of a set of nonzero minors in (4). If the Jacobian matrix is of class P for all x, obviously condition (4) is satisfied. However, condition (4) allows $\det J_k$ to be either positive or negative. For example, let

$$y_1 = f_1(x_1, x_2) = x_1 + x_1^3$$
$$y_2 = f_2(x_1, x_2) = -x_2.$$

The Jacobian matrix is not of class P, yet it satisfies the ratio condition; therefore, the equation has a unique solution for all y. In addition, it is important to note that a matrix with a set of positive minors ($\det J_k$, $k=1, 2, \cdots, n$) need not be of class P. For example, in

$$J = \begin{bmatrix} 1 & 5 & 0 \\ 0 & 1 & 5 \\ 1 & 5 & 1 \end{bmatrix}$$

one principal minor is negative. All these point out that the ratio condition represents an important extension of the class P matrix so long as one deals with the question of uniqueness of solution of nonlinear equations. Finally, it is interesting to point out that if the Jacobian matrix is of class P for all x, then the solution of the equation $f(x)=y$ is unique if it exists [16].

Remark 3

It will be illustrated later that the following subclass of (1) is very useful in the study of a broad class of nonlinear networks:

$$f(x) = Dx + g(x) = y \tag{20}$$

where D is a diagonal constant matrix with positive elements and g is continuously differentiable. The Jacobian matrix is of the form

$$J(x) = D + M(x) \tag{21}$$

where $M(x) \triangleq \partial g/\partial x$.

Lemma

Let D be a positive diagonal constant matrix and $M(x)$ be of class P_0 for all $x\in R^n$; then the matrix $J(x)=D+M(x)$ satisfies the ratio condition.

Proof: Let $d_1, d_2, \cdots, d_n$ be diagonal elements of D, which are all positive. Let J_k and M_k denote the matrices consisting of the first k rows and the first k columns of J and M, respectively. Let $M_k(-i_1, \cdots, -i_p)$ denote the matrix obtained from M_k by deleting the i_1th, $\cdots$, i_pth rows and columns, where $1 \leqq i_1 < \cdots < i_p \leqq k$. The following relation is easily obtained:

$$\det J_1(x) = d_1 + \det M_1(x) \geqq d_1$$

which gives the first of the ratio condition in (3). For $k>1$

$$\det J_k(x) = \det M_k(x) + \sum_{i_1} d_{i_1} \det M_k(-i_1) + \sum_{i_1<i_2} d_{i_1}d_{i_2} \det M_k(-i_1, -i_2) + \cdots + d_1d_2 \cdots d_{k-1}d_k.$$

Separating the right-hand side into two groups, terms including d_k and others, we obtain

$$\det J_k(x) = d_k \det J_{k-1}(x) + (\text{nonnegative terms}).$$

Hence

$$\det J_k(x)/\det J_{k-1}(x) \geqq d_k$$

uniformly in all x for $k=2, \cdots, n$. By putting $\epsilon = \min(d_1, \cdots, d_n)$, we see that the ratio condition in (3) is satisfied. This completes the proof of the lemma.

From the work of Fielder and Pták [15] we know that if $M(x) \in P_0$ for all x, $J(x) = D + M(x)$ belongs to P for all x. The above lemma shows that, in addition, $J(x)$ satisfies the ratio condition. This lemma leads immediately to the following corollary of Theorem 1.

Corollary 1

If $\partial g/\partial x$ belongs to P_0 for all $x \in R^n$, then (20) has a unique solution for any $y \in R^n$.

Sandberg and Willson [9], [10], [13], [14] also considered the equation of the type

$$BF(x) + Cx = y \tag{22}$$

where B and C are constant square matrices.

For the present let us assume that the ith component of F is a nondecreasing continuously differentiable function of x_i, which maps R^1 *into* itself. Therefore F does not necessarily belong to class $\mathfrak{F}^n$ defined earlier. In this case, the Jacobian matrix of F may vanish at some point, and hence for the purpose of assuring C^1-diffeomorphism the additional assumption $\det C \neq 0$ is needed. Under this assumption (22) can be rewritten as follows:

$$C^{-1}BF(x) + x = C^{-1}y. \tag{23}$$

Then the Jacobian matrix of the left-hand side is

$$1 + C^{-1}BD(x) \tag{24}$$

where $D(x)$ is a nonnegative diagonal matrix for all x. Therefore, $C^{-1}B \in P_0$ assures that $C^{-1}BD(x) \in P_0$ for all x. This demonstrates that $C^{-1}B \in P_0$ is sufficient to guarantee the existence of a unique solution of (22).

Sandberg and Willson made further assumptions on F and were able to show that the condition $C^{-1}B \in P_0$ was necessary as well. A more comprehensive study of the equation in (22) is given in a recent paper of Willson [14]. In the general case of (20), the condition $M(x) \in P_0$ for all x is sufficient but not necessary as the following example illustrates. Let $g(x) = -\frac{1}{2}x$; then $M = -\frac{1}{2}1 \notin P_0$, but the Jacobian matrix $J(x) = 1 + M = \frac{1}{2}1$ satisfies the ratio condition.

Remark 4

Consider another generalization of the form of (20). Let

$$f(x) = F(x) + g(x) = y \tag{25}$$

where $F \in \mathfrak{F}^n$ and, in addition, each component of F is continuously differentiable and has a positive slope everywhere. Let $g(x)$ be continuously differentiable as before. Clearly, F itself defines a C^1-diffeomorphism of R^n onto itself. By putting $z = F(x)$ or equivalently $x = F^{-1}(z)$, we may write (25) in the following form:

$$z + g(F^{-1}(z)) = y. \tag{26}$$

The Jacobian matrix of the left-hand side is

$$1 + M(x)D(z) \tag{27}$$

where $D(z)$ is the Jacobian matrix of F^{-1} and is a positive diagonal matrix which depends on z. $M(x)$ is the Jacobian matrix of $g(x)$ as before. It is easily seen that $M(x) \in P_0$ for all x implies that $M(x)D(z) \in P_0$ for all z. Therefore, from Corollary 1 we obtain the following.

Corollary 2

If $\partial g/\partial x$ belongs to class P_0 for all $x \in R^n$, and if the ith component of $F(\cdot)$ is a continuously differentiable function of x_i which maps R^1 onto itself and has a positive slope for $i = 1, 2, \cdots, n$, then (25) has a unique solution for any $y \in R^n$.

Sandberg and Willson considered the special case where $g(x)$ is a linear function Ax as in (2) [9, theorem 6] and obtained the conclusion that $A \in P_0$ guarantees the existence of a unique solution.

IV. A Network Example

Consider a fairly general class of nonlinear resistive network which has the following properties. 1) There exists a tree such that all tree-branch elements are current controlled and all link elements are voltage controlled. 2) There exist no couplings between the tree-branch elements and the link elements. The network has a unique solution for all possible inputs if a certain matrix which characterizes the network belongs to class P_0. The result can be considered a generalization of early works of Desoer and Katzenelson [1], Varaiya and Liu [3], Ohtsuki and Watanabe [4], and Stern [5].

Let v_l, v_t, i_l, and i_t denote the link voltages, tree-branch voltages, link currents, and tree-branch currents, respectively. Then the branch characteristics are given by

$$v_t = \hat{v}_t(i_t) \tag{28}$$

and

$$i_l = \hat{i}_l(v_l). \tag{29}$$

Note that couplings among the tree-branch elements themselves and couplings among the link elements are allowed. We assume that the functions $\hat{v}_t(\cdot)$ and $\hat{i}_l(\cdot)$ are continuously differentiable. Let $R_t(i_t)=\partial\hat{v}_t/\partial i_t$ and $G_l(v_l)=\partial i_l/\partial v_l$ be the Jacobian matrices for the tree-branch elements and the link elements, respectively. The conditions for uniqueness of solution will be stated in terms of the two Jacobian matrices and the network topology.

Let us denote the fundamental loop matrix and the fundamental cut-set matrices by $B=[1, E]$ and $Q=[-E^t, 1]$, respectively. Then, Kirchhoff's voltage and current laws are expressed as

$$v_l + Ev_t = e \tag{30}$$

and

$$-E^t i_l + i_t = j \tag{31}$$

where e and j are the fundamental loop voltage sources and the fundamental cut-set current sources, respectively. Combining (28)–(31), we obtain

$$i_t - E^t\hat{i}_l(e - E\hat{v}_t(i_t)) = j. \tag{32}$$

With i_t as the unknown variable, the Jacobian matrix of the left-hand side is

$$J(i_t) = 1 + E^t G_l(e - E\hat{v}_t(i_t))ER_t(i_t). \tag{33}$$

According to Corollary 1, (32) has a unique solution for all j if

$$E^t G_l(v_l)ER_t(i_t) \in P_0 \tag{34}$$

for all v_l and i_t.

Further conclusions can be obtained if we restrict the element characteristics, for example, to be uncoupled. In this case G_l and R_t are diagonal matrices. Then, if G_l and R_t are nonnegative diagonal, which implies that both the link and tree-branch elements are monotone nondecreasing, (34) is satisfied. This is what Desoer and Katzenelson have found. On the other hand, if only $R_t(i_t)$ is diagonal and nonnegative for all i_t, then the condition

$$E^t G_l(v_l)E \in P_0 \tag{35}$$

for all v_l guarantees the uniqueness of solution for all input. This further extends the case considered by Varaiya and Liu [3], where G_l is positive semidefinite and R_t is nonnegative diagonal.

V. Generalization

Theorem 1 can be further generalized. By applying interchange of rows and of columns to the Jacobian matrix, the following theorem can be easily proved.

Theorem 2

Let f be a continuously differentiable mapping of R^n into itself. Then, for any $y \in R^n$ there exists one and only one solution of (1) if there exist two permutations $(i_1, i_2, \cdots, i_n)$ and $(j_1, j_2, \cdots, j_n)$ of $(1, 2, \cdots, n)$ and there exists a positive constant $\epsilon > 0$ for which

$$\left|\det J\begin{pmatrix} i_1 \\ j_1 \end{pmatrix}\right| \geqq \epsilon, \quad \left|\frac{\det J\begin{pmatrix} i_1, i_2 \\ j_1, j_2 \end{pmatrix}}{\det J\begin{pmatrix} i_1 \\ j_1 \end{pmatrix}}\right| \geqq \epsilon, \cdots,$$

$$\left|\frac{\det J\begin{pmatrix} i_1, i_2, \cdots, i_n \\ j_1, j_2, \cdots, j_n \end{pmatrix}}{\det J\begin{pmatrix} i_1, i_2, \cdots, i_{n-1} \\ j_1, j_2, \cdots, j_{n-1} \end{pmatrix}}\right| \geqq \epsilon \tag{36}$$

uniformly in all $x \in R^n$, where

$$J\begin{pmatrix} i_1, \cdots, i_k \\ j_1, \cdots, j_k \end{pmatrix}$$

is composed of the i_1th, $\cdots$, i_kth rows and j_1th, $\cdots$, j_kth columns of the Jacobian matrix J, in these orders.

This theorem allows the use of a set of nonprincipal cofactors in the ratio condition and thus enlarges applicability.

Remark 5

Consider the special case of (1):

$$f(x) = Ax + w = y \tag{37}$$

where A is a constant matrix and w is a constant vector. The ratio condition in (36) of Theorem 2 is equivalent to the condition $\det A \neq 0$. This can be seen as follows. For the constant matrix case, $\det J = \det A \neq 0$ implies that there exists at least one cofactor,

$$\det J\begin{pmatrix} i_1, \cdots, i_{n-1} \\ j_1, \cdots, j_{n-1} \end{pmatrix}$$

which is nonzero, for otherwise $\det J$ would be zero. Thus, the last condition in (36)

$$\left|\det J\begin{pmatrix} i_1, \cdots, i_n \\ j_1, \cdots, j_n \end{pmatrix} \Big/ \det J\begin{pmatrix} i_1, \cdots, i_{n-1} \\ j_1, \cdots, j_{n-1} \end{pmatrix}\right| \geqq \epsilon > 0$$

is automatically satisfied. The same argument can be applied to the second to the last condition in (36), etc. Therefore, the statement in Theorem 2 represents the necessary and sufficient condition for the existence of a unique solution of the linear mapping in (37) for all $y \in R^n$.

VI. A Network Interpretation

The ratio conditions in Theorems 1 and 2 have a network interpretation. Consider an n-port resistive network with port current and voltage variables defined as follows: $y_1=i_1$, $y_2=i_2$, $\cdots$, $y_n=i_n$; $x_1=v_1$, $x_2=v_2$, $\cdots$, $x_n=v_n$. The ratio condition in Theorem 1 can then be interpreted in terms of n specific driving-point characteristics:

$$\det J_1 = \left.\frac{\Delta i_1}{\Delta v_1}\right|_{v_2,\cdots,v_n=\text{const}}$$

$$\frac{\det J_2}{\det J_1} = \left.\frac{\Delta i_2}{\Delta v_2}\right|_{\substack{i_1=\text{const}\\ v_3,\cdots,v_n=\text{const}}}$$

$$\cdots\cdots\cdots\cdots$$

$$\frac{\det J_n}{\det J_{n-1}} = \left.\frac{\Delta i_n}{\Delta v_n}\right|_{i_1,\cdots,i_{n-1}=\text{const}}.$$

Similarly the ratio condition in Theorem 2 can be interpreted in terms of n transfer and/or driving-point characteristics:

$$\det J\begin{pmatrix} i_1 \\ j_1 \end{pmatrix} = \left.\frac{\Delta i_{i_1}}{\Delta v_{j_1}}\right|_{v_{j_k}=\text{const}, \quad k\neq 1}$$

$$\frac{\det J\begin{pmatrix} i_1, i_2 \\ j_1, j_2 \end{pmatrix}}{\det J\begin{pmatrix} i_1 \\ j_1 \end{pmatrix}} = \left.\frac{\Delta i_{i_2}}{\Delta v_{j_2}}\right|_{\substack{i_{i_1}=\text{const}\\ v_{j_k}=\text{const}, \quad k\neq 1,2}}$$

$$\cdots\cdots\cdots\cdots$$

$$\frac{\det J\begin{pmatrix} i_1, \cdots, i_n \\ j_1, \cdots, j_n \end{pmatrix}}{\det J\begin{pmatrix} i_1, \cdots, i_{n-1} \\ j_1, \cdots, j_{n-1} \end{pmatrix}} = \left.\frac{\Delta i_{i_n}}{\Delta v_{j_n}}\right|_{i_{i_k}=\text{const}, \quad k\neq n}.$$

Acknowledgment

The authors wish to thank Dr. A. N. Willson, Jr., for his critical review of the paper and his suggestions.

References

[1] C. A. Desoer and J. Katzenelson, "Nonlinear RLC networks," *Bell Syst. Tech. J.*, vol. 44, Jan. 1965, pp. 161–198.

[2] E. S. Kuh, "Representation of nonlinear networks," in *1965 Proc. Nat. Electronics Conf.*, vol. 21, pp. 702–706.

[3] P. P. Varaiya and R. W. Liu, "Normal form and stability of a class of coupled nonlinear networks," *IEEE Trans. Circuit Theory*, vol. CT-13, Dec. 1966, pp. 413–418.

[4] T. Ohtsuki and H. Watanabe, "State-variable analysis of RLC networks containing nonlinear coupling elements," *IEEE Trans. Circuit Theory*, vol. CT-16, Feb. 1969, pp. 26–38.

[5] T. E. Stern, *Theory of Nonlinear Networks and Systems*. Reading, Mass.: Addison-Wesley, 1965.

[6] R. J. Duffin, "Nonlinear networks (I, II, and IIb)," *Bull. Amer. Math. Soc.*, vols. 52–54, 1946–1948, pp. 836–838, 963–971, and 119–127.

[7] L. O. Chua and R. A. Rohrer, "On the dynamic equations of a class of nonlinear RLC networks," *IEEE Trans. Circuit Theory*, vol. CT-12, Dec. 1965, pp. 475–489.

[8] A. N. Willson, Jr., "On the solutions of equations for nonlinear resistive networks," *Bell Syst. Tech. J.*, vol. 47, Oct. 1968, pp. 1755–1773.

[9] I. W. Sandberg and A. N. Willson, Jr., "Some theorems on properties of dc equations of nonlinear networks," *Bell Syst. Tech. J.*, vol. 48, Jan. 1969, pp. 1–34.

[10] ——, "Some network-theoretic properties of nonlinear dc transistor networks," *Bell Syst. Tech. J.*, vol. 48, May–June 1969, pp. 1293–1311.

[11] R. S. Palais, "Natural operations on differential forms," *Trans. Amer. Math. Soc.*, vol. 92, July 1959, pp. 125–141.

[12] C. A. Holzmann and R. W. Liu, "On the dynamical equations of nonlinear networks with n-coupled elements," in *1965 Proc. 3rd Ann. Allerton Conf. Circuit and System Theory*, 1965, pp. 536–545.

[13] I. W. Sandberg, "Theorems on the analysis of nonlinear transistor networks," *Bell Syst. Tech. J.*, vol. 49, Jan. 1970, pp. 95–114.

[14] A. N. Willson, Jr., "New theorems on the equations of nonlinear dc transistor networks," *Bell Syst. Tech. J.*, vol. 49, Oct. 1970, pp. 1713–1738.

[15] M. Fielder and V. Pták, "Some generalizations of positive definiteness and monotonicity," *Numer. Math.*, vol. 9, 1966, pp. 163–172.

[16] H. Nikaido, *Convex Structures and Economic Theory*. New York: Academic Press, 1968, pp. 359–360.

[17] E. S. Kuh and I. Hajj, "Nonlinear circuit theory: Resistive networks," *Proc. IEEE*, vol. 59, Mar. 1971, pp. 340–355.

Author Index

Editor's Biography

Alan N. Willson, Jr. (S'66–M'67–SM'73) was born in Baltimore, Md., on October 16, 1939. He received the B.E.E. degree from the Georgia Institute of Technology, Atlanta, in 1961, and the M.S. and Ph.D. degrees from Syracuse University, Syracuse, N.Y., in 1965 and 1967, respectively.

From 1961 to 1964 he was with the IBM Corporation. He was an Instructor in Electrical Engineering at Syracuse University from 1965 to 1967. From 1967 to 1973 he was a member of the Technical Staff at Bell Laboratories, Murray Hill, N.J. He is presently an Associate Professor of Engineering and Applied Science at the University of California, Los Angeles. He has been engaged in research concerning the stability of distributed circuits, properties of nonlinear networks, theory of active circuits, and digital filters.

Dr. Willson is a member of Eta Kappa Nu, Sigma Xi, Tau Beta Pi, the American Association for the Advancement of Science, and the Society for Industrial and Applied Mathematics. He is currently an Associate Editor of the IEEE TRANSACTIONS ON CIRCUITS AND SYSTEMS.